Instructor's Manual:
PREPARATOR'S GUIDE

for

Biology in the Laboratory
Third Edition

Doris R. Helms
Carl W. Helms

Clemson University

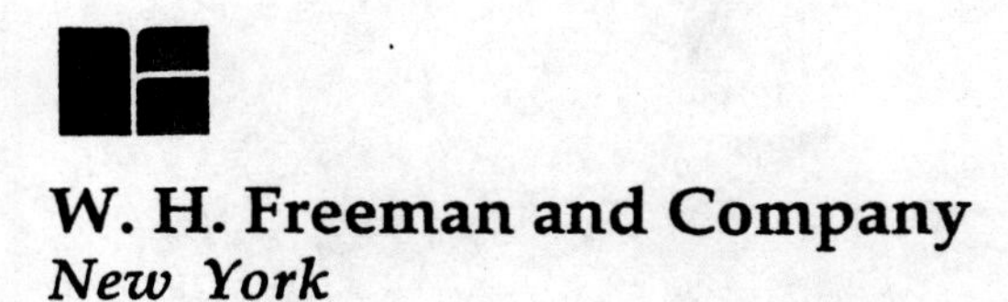

W. H. Freeman and Company
New York

SUPPLEMENTS EDITOR: *Patrick Shriner*
ASSOCIATE EDITOR: *Debra Siegel*
ADMINISTRATIVE ASSISTANT: *Ceserina Pugliese*
COVER DESIGN: *Vicki Tomaselli*
PRODUCTION COORDINATOR: *Paul W. Rohloff*

ISBN: 0-7167-3237-8

Printed in the United States of America

First printing 1998

TABLE OF CONTENTS

FOREWORD

The laboratory exercises presented in *Biology in the Laboratory, 3rd ed.*, were chosen for their educational value and with concern for their ease of preparation. Some laboratories may seem to be quite sophisticated, but with the help of the *Preparator's Guide* you should have no difficulty in purchasing or preparing all the necessary materials. Before attempting any of the laboratory exercises, be sure to read the PREPARATION SUGGESTIONS section, which is intended to provide you with the easiest and most convenient laboratory setups.

Each three-hour laboratory is divided into several exercises, most of which require less than sixty minutes to complete. Some, however, require overnight incubations, and suggestions are made for completing these laboratories in both small and large class settings. See pages xv-xvi for suggestions on selecting single exercises or combining several exercises from different laboratories to tailor *Biology in the Laboratory, 3rd ed.*, to your class schedule and available equipment. For ease of adaptation, the *Preparator's Guide* provides separate listings of the materials and solutions necessary for each exercise.

Biology in the Laboratory, 3rd ed., is complemented by BioBytes software found on the CD-ROM that accompanies each manual. Programs include four simulations, *Alien, Cycle, Dueling Alleles,* and *Seedling*. *Reaction Time* measures voluntary reaction time and performs statistical analysis on the results and is an integral part of several exercises. Directions for installing the software and opening programs are included before we begin our coverage of each laboratory topic. An Instructor's Guide, Student Information, and a Worksheet are included for each of the interactive simulations. The Student Information and Worksheet are also found on the CD-ROM (and computer hard disk following installation). The version included with this manual includes several corrections and one figure omitted from the first printing of the CD-ROM and may be duplicated for distribution if desired. Updated versions are also found at http://www.whfreeman.com/biolab/helms.

For each laboratory topic, the *Preparator's Guide* contains the following information:

I. FOREWORD

A brief summary of all the laboratory activities.

II. TIME REQUIREMENTS

A list of the exercises and an estimate of the time required for each.

III. STUDENT MATERIALS AND EQUIPMENT

This section is designed to help you plan. Only materials used by the students are listed. The numbers or amounts necessary for a class of 24 are given in the last column. These represent a multiple of the Per Student, Per Pair, or Per Group (4 students) column for each item. Pay particular attention to the amounts given for solutions and whether these are needed by each student, pair, group, or class. Materials requiring additional preparation are followed by a bold number in parentheses (**N**). This number corresponds to a more detailed description of the material as presented in section IV.

IV. PREPARATION OF MATERIALS AND SOLUTIONS

All directions for mixing solutions and preparing specific materials are presented in this section. The amounts given are sufficient for a class of 24 students, depending on the containers used to distribute solutions to students. Materials and equipment needed by preparators in preparing for specific experiments are also referenced in this section. For instance, in Laboratory 9, Mitosis, "stained onion root tips," listed under Student Materials and Equipment in Exercise C, is given the reference number **(2)**. Under **(2)** in Section IV, materials, solutions, and the procedure for preparing stained onion root tips are given. Directions for setup and specific cautions are also presented. All directions for mixing solutions are given in g or mg amounts where molar or normal solutions are used. This will make preparation easier and will make it possible for students to help.

V. PREPARATION SUGGESTIONS

This section is arranged by exercise. The short cuts, pit-falls, and additional details found here should help you have a successful laboratory experience with your students.

VI. ORDERING INFORMATION

Only materials that are not usually found in a teaching laboratory are listed in this section. Carolina Biological Supply Company (Carolina Biological), Fisher Scientific (Fisher), Sigma (Sigma), and Ward's Natural History Establishment, Inc. (Ward's) are the main suppliers listed (you may wish to refer to a more complete list of potential suppliers on pages ix-x). Most suppliers now have home pages on the World Wide Web and several have their complete catalog available on-line. It may be worth while to use one of the Web search engines to locate potential suppliers (a search for "sea urchins," for example, will lead you to several of the suppliers we have listed plus an assemblage of useful information)

Equipment and supplies are listed in the order in which they appear within the laboratory exercises. To determine what to order for the academic year, first select the exercises you wish to do. Then check Sections III and IV to identify the necessary materials and equipment. For those materials not already on hand, check Section VI for specific ordering information. You may need to adjust sizes and amounts to your needs—these lists should, however, give you a starting reference.

We assume that you will have a (binocular) compound microscope (some with oil-immersion objectives) as well as a dissecting microscope for every student. Appropriate balances and pH meters are also expected in certain laboratories. Some more specialized equipment, including spectrophotometers, blenders, and dissecting pans and supplies are listed where they are required.

Activities that require prior planning or ordering live materials are listed on pages xi-xiv. This section is designed to assist with your advanced planning.

We hope that *Biology in the Laboratory, 3rd ed.* provides you and your students with many exciting and enjoyable experiences. Please let us know if you have any difficulties with individual exercises. We will be happy to assist if possible. If you have suggestions, a particularly good set of student data or preparation short-cuts or alternatives, please share them with us and others by sending it to us by e-mail so we can include it on our developing Web Homepage, http://www/whfreeman.com/biolab/helms. Check this site for additional information and corrections to *Biology in the Laboratory, 3rd ed.* and these supporting *Instructor's Manuals.*

Doris R. Helms, Professor, Biology Program, 330 Long Hall
864-656-2418
BIOL110@clemson.edu

Carl W. Helms, Professor, Biological Sciences, 132 Long Hall
864-656-4224
cwhelms@clemson.edu

Biology Program, 330 Long Hall, Clemson University, Clemson, SC 29634

December 1997

SUPPLIERS

The following companies and organizations are potential sources for the equipment and supplies listed in the *Preparator's Guide.* Contact information is followed by a URL for access via the World Wide Web (if available). Many suppliers offer contacts to sales and technical support via the Web. E-mail addresses are also given where available. If you need to locate additional vendors, searching for the company name on the Web may yield useful contact information.

Note that several of the suppliers offer complete on-line catalog information (Fisher, VWR) or are in the process of developing this capability (Sigma). Some also permit ordering via the Web. These capabilities will expand rapidly and should be very helpful to you as you locate supplies for your laboratory offerings.

American Type Culture Collection, c/o Sales and Marketing Department, 12301 Parklawn Dr., Rockville, MD 20852
800-638-6507 (Voice), 301-231-5816 (FAX)
http://www.atcc.org
tech@atcc.org

Carolina Biological Supply Company, 2700 York Road, Burlington, NC 27215
800-334-5551 (Voice), 800-222-7112 (FAX)
http://www.carolina.com
carolina@carolina.com

Difco Laboratories, PO Box 331058, Detroit, MI 48232-7058
800-521-0851 (Voice), 313-462-8517 (FAX)
http://difco.com
dorney@difco.com

Edmund Scientific Company, Industrial Optics Division, 101 East Gloucester Pike, Barrington, NJ 08007-01380
609-573-6852 (Voice), 609-573-6840 (FAX)
http://www.edsci.com/
education@edsci.com

Edvotek, Inc., PO Box 1232, West Bethesda, MD 20827-1232
800-338-6835 (Voice), 301-340-0582 (FAX)
http://www.edvotek.com
edvotek@aol.com

Fisher Scientific, 711 Forbes Ave., Pittsburgh, PA 15219-4785
800-766-7000 (Voice), 800-926-1166 (FAX)
http://www.fisher1.com

Fluker Farms, 1333 Plantation Road, Port Allen, LA 70767
504-343-7035 (Voice), 504-336-0671 (FAX)
http://www.flukerfarms.com/

Fotodyne Corporation, 950 Walnut Ridge Dr., Hartland, WI 530029
800-362-3686 (Voice), 414-369-7013 (FAX)
http://www.fotodyne.com/

Gulf Specimen Marine Laboratories, Inc., PO Box 237, Panacea, FL 32346
850-984-5297 (Voice), 850-984-5233 (FAX)
http://www.adsul.com/gulfspecimen
gspecimen@adsul.com

LaMotte Company, PO Box 329, Chestertown, MD 21620
800-344-3100 (Voice), 410-778-6394 (FAX)
http://www.lamotte.com/
mkt@lamotte.com

Life Technologies, 3175 Staley Rd., Grand Island, NY 14072 (supplier of Gibco products)
800-828-6686 (Voice)
http://www.lifetech.com/

Modern Biology, Inc., 111 North 500 West, West Lafayette, IN 27906
800-733-6544 (Voice), 317-449-4577 (FAX)
http://www.modernbio.com/

Pacific Bio-Marine Laboratories, Inc., PO Box 1348, Venice, CA 90294-1348
310-677-1056 (Voice), 310-677-1027 (FAX)
no URL or e-mail

Sigma, PO Box 14508, St. Louis, MO 63178
800-325-3010 (Voice), 800-325-5052 (FAX)
http://www.sigma.sial.com

Pacific Bio-Marine Laboratories, Inc., Box 536, Venice, CA 90291
310-677-1056 (Voice)

VWR Scientific Products, 1210 Goshen Parkway, West Chester, PA 19380
800-932-5000 (Voice), 610-429-9340 (FAX)
http://www.vwrsp.com

Ward's Natural Science Establishment, Inc., PO Box 92912, Rochester, NY 14692-9012
800-962-2660 (Voice), 800-635-8439 (FAX)
http://www.wardsci.com
customer service@wardsci.com

LABORATORIES REQUIRING PRIOR PLANNING OR LIVE MATERIAL

Laboratory 1—Observations and Measurements: The Microscope	
Exercise C	*Elodea, Paramecium* (order live)
Extending Your Investigation	*Elodea, Paramecium* (order live)
Laboratory 6—Prokaryotic Cells	
Exercise B, Part 3	*Escherischia coli,* nutrient agar slant
	Staphylococcus aureus, nutrient agar slant
Exercise B, Part 4	*Oscillatoria, Nostoc, Cylindrospermum* (order)live
Exercise C, Parts 1, 2	*Escherischia coli,* nutrient agar slant
	Serrratia marcescens, nutrient agar slant
	order to arrive at least 1 week before the laboratory to have enough time to prepare broth cultures
Laboratory 7—Eukaryotic Cells	
Exercise A	onion, potato, banana, carrot
Extending Your Investigation	*Elodea* (order live)
Exercise C	*Stentor, Volvox* (order live)
Laboratory 8—Osmosis and Diffusion	
Exercise C	*Chara, Elodea, Spirogyra* (order live)
Laboratory 9—Mitosis	
Exercise C	onion root tips (start 3-4 days prior to the laboratory to allow ample time for fining and staining)
Laboratory 10—Enzymes	
Exercise B	spices for cheesemaking (have students bring spices—announce this during the prior laboratory period)
Laboratory 11—Energetics, Fermentation, and Respiration	
Exercise A, Parts 1, 2	yeast solution (prepare 12 hours before laboratory)
Exercise A, Part 4	yeast solution (prepare 24 hours before laboratory)
Laboratory 12–Photosynthesis	
Exercise A, Parts 1,3	fresh spinach
Exercise B. Part 3	wheat seedlings (plant 2 flats 4-5 days before the laboratory)
Exercise C	variegated *Coleus* (place a plant in the dark for two days (prior to laboratory). Alternatively, cover parts of leaves of a green *Coleus* one week prior to the laboratory.

<u>Laboratory 13</u>—Meiosis: Independent Assortment and Segregation

Exercise D, Parts 1, 2	*Sordaria fimicola* (order wild type and tan). Agar plates should be inoculated 5 days prior to the laboratory or crosses must be ordered 3 weeks in advance.

<u>Laboratory 14</u>—Genes and Chromosomes: Chromosome Mapping

Exercise A	*Sordaria* plates, tan × wild crosses (must be set up 7 days prior to the laboratory, either by students during Laboratory 13 or by the preparator). Crosses must be ordered 3 weeks in advance.
Exercise B	*Drosophila mojavensis.* If flies are ordered, allow 9 days for larvae to develop to a suitable size for chromosome squashes.
Exercise C	*Escherischia coli* donor Hfr and *E. coli* recipient must be available 3-4 days prior to the laboratory in order to start broth cultures
Exercise D	restriction enzyme kit (order 2-3 weeks ahead for delivery 2-3 days so you have time to familiarize yourself with the instructions prior to the laboratory). Store DNA fragments in the refrigerator

<u>Laboratory 15</u>—Human Genetic Traits

Exercise F	forensics kit (order 2-3 weeks ahead of time for delivery 2-3 days prior to the laboratory). Store restriction fragments in refrigerator.

<u>Laboratory 16</u>—DNA Isolation

Exercise B	testis tissue (obtained from a local veterinarian or spay/neuter clinic several weeks (or months) prior to the laboratory to ensure availability.
Exercise C	bacteria paste (prepared ahead). It will take at least a week (or more for large groups of students) to collect enough bacteria unless large batches are used.

<u>Laboratory 18</u>—Molecular Genetics: Recombinant DNA

Exercises B, C	kits (order ahead to ensure availability). If a kit is not used, order live *E. coli* cells ahead so that streak plates can be made at least 3 days prior to the laboratory if the instructor is to prepare competent cells, or 2 days prior to the laboratory if students are to prepare competent cells.

Laboratory 19—Genetic Control of Development and Immune Defenses

Exercise A	sea urchins, live (order)
Exercises B – F	frogs, live (order if desired as a substitute for preserved embryos)
Exercise G	chicken eggs (incubate three groups, one starting 72 hours, another starting 96 hours, and a third starting 5-6 days prior to laboratory—optional)
Exercise H	serum, antiserum, and bovine serum albumin (order one week in advance—store frozen)

Laboratory 21—The Genetic Basis of Evolution II—Diversity

Exercise B	cow, goat, sheep, and horse serum (order 3 weeks prior to laboratory—store in refrigerator)

Laboratory 22—Diversity—Kingdoms Eubacteria, Archaebacteria, and Protista

Exercise A	nutrient agar plates (distribute for students' bacteria experiments during the previous week's laboratory)
Exercise B	*E. coli* and S. *aureus* (order live cultures). Prepare antibiotic sensitivity plates 24-48 hours prior to laboratory.
Extending Your Investigation	
Exercise C	soybean plants (inoculate with *Rhizobium* 8 weeks prior to laboratory)
Exercise D	*Vibrio fischeri* (order live cultures—cultures must be started 14-48 hours prior to the laboratory)
Exercise E	*Nostic, Cylindrospermum, Oscillatoria, Anabaena, Gloeocapsa* (order live)
Exercise F	*Paramecium* (order live)
Exercise G	termites (order live or obtain from wooded area)
Exercise H	*Physarum* (order live—start slime mold cultures 2 days prior to the laboratory)
Exercise I	water mold (start cultures at least 4-5 days prior to the laboratory)
Exercise J	*Euglena, Ectocarpus, Polysiphonia, Corallopsis* (order live)
Exercise K	*Chlamydomonas, Spirogyra, Gonium, Volvox, Zygnema, Stigeoclonium, Ulva, Ulothrix* (order live)
Exercise L	plankton sample (during the previous week's laboratory, assign a student to bring in a plankton sample)

Laboratory 23—Diversity-Fungi and the Nontracheophytes

Exercise A	Distribute bread (in plastic bags) to students during the previous week's laboratory.
Extending Your Investigation	bread mold (Exercise A)
Exercise B	*Sordaria* (order live—start *Sordaria* cultures 7-8 days prior to the laboratory)
Exercise D	*Alloymces* (order live)
Exercises F and G	fungus and lichen collection (assign during the previous week's laboratory)
Exercise H	*Marchantia* and hornworts (order live or collect—if ordering *Marchantia* archegonia and antheridia from a biological supply company, order at least 4 weeks prior to the laboratory)

Laboratory 24—Diversity—The Tracheophytes (Vascular Land Plants)

Exercise A	*Psilotum, Lycopodium, Selaginella, Equisetum,* fern leaves (order live or obtain from a greenhouse)
Exercise B	cycads (order live male and female plants if available or obtain from a greenhouse)
Exercise C	fresh flowers (make arrangements for a variety of fresh flowers for dissection)
	Impatiens (order for pollen germination experiment)

Laboratory 25—Diversity—Porifera, Cnidaria, and Wormlike Invertebrates

Exercise A	*Grantia* (order live)
Exercise B	*Hydra* (order live)
Exercise C, Part 1	planaria (order live)
Exercise C, Part 3	*Tubatrix* (order live)
Exercise D, Part 4	*Lumbricus* (order live)

Laboratory 26—Diversity—Mollusks, Arthropods, and Echinoderms

Exercise A, Part 3	pulmonate snails (order live)
Exercise B, Part 3 (optional)	*Limulus,* hermit or fiddler crabs, spiders (order live or collect them if available)
Extending Your Investigation	clams and mussels (obtain from local grocery store)

Laboratory 28—Plant Anatomy—Roots, Stems, and Leaves

Exercise B	herbaceous plants (ask students to collect weeds or house plants)

Laboratory 29—Angiosperm Development—Fruit, Seeds, Meristems, and Secondary Growth

Exercise A	fruits (check with local grocery story one week prior to laboratory to ensure a variety of fruits for dissection)
Exercise B, Parts 1, 2	*Phaseolus* (lima bean) and *Zea* (corn) seeds (obtain live—soak overnight)
Exercise D	*Phaseolus* (lima bean), *Zea* (corn), and Little Marvel pea seeds (obtain live—germinate 6 days prior to laboratory). You will also need older plants—plant seeds 8-12 days prior to the laboratory.
Exercise E	*Coleus* (obtain live) radish seeds (order live—germinate 2 days prior to the laboratory)
Extending Your Investigation	pea seeds (order live—germinate)
Exercise G (optional)	tobacco seeds (order live—plant 4-6 weeks prior to the laboratory)

Laboratory 30—Water Movement and Mineral Nutrition in Plants

Exercise A	*Zebrina* (obtain live)
Exercise B	barley seeds (plant one week prior to the laboratory)
Exercise C, Part 1	*Coleus* (small live plant—order or obtain locally)
Exercise C, Part 2	*Photinia* (red tip) or privet branches (live, cut)
Exercise E	sunflower seeds (If the mineral nutrition experiment is to be done as a demonstration, prepare, in appropriate solutions, one set of sunflower seeds 9 weeks prior to the laboratory and one set 6 weeks prior to the laboratory. In both cases, and to provide plants for student experiments, plant sunflower seeds 3 weeks prior to placing them in nutrient solutions.

Laboratory 31—Plant Responses to Stimuli

Exercise A, Part 1	sunflower seeds (plant 4 weeks prior to the laboratory—5 weeks if the experiment is started during the previous laboratory)
Exercise A, Part 2	*Coleus* (live plants, 2 per student group)
Exercise A, Part 3	dwarf peas (plant 12 days prior to the laboratory—19 days if the experiment is started during the previous laboratory)
Exercise C, Part 1A	gizzard plants (treat 2-3 days prior to the laboratory)
Exercise D, Parts 1-3	lettuce seed (order)

Laboratory 32—Animal Tissues

Exercise A	frogs (order live)

Laboratories 33-36—The Anatomy of Representative Vertebrates

All	shark, frog, turtle, rat (order preserved) order or gather appropriate demonstration material

Laboratory 37—The Basics of Animal Form: Skin, Bones, and Muscles

Exercise C	chicken bones (demineralize 2 weeks prior to laboratory)
Extending Your Investigation	chicken wings (fresh—obtain from local grocery store)
Exercise E	rabbit psoas muscle preparation (order in advance and store in refrigerator)

Laboratory 38—The Physiology of Circulation

Exercise B	sheep heart (fresh, from local abattoir or order preserved)
Exercise D	goldfish, live (obtain locally)
Exercise G	*Daphnia magna* (order live)

Laboratory 39—Gas Exchange and Respiratory Systems

Exercise B	cow blood (make arrangements to obtain fresh blood from a local abattoir)

Laboratory 40—The Digestive, Excretory, and Reproductive Systems

Exercise C	beef or sheep kidney (make arrangements to obtain fresh blood from a local abattoir)

Laboratory 42—Behavior

Exercise A	isopods (pill bugs) (collect or order live)
Exercise B	*Drosophila* (order live)
Exercise C	mealworms (order live)

Laboratory 43—Communities and Ecosystems

Exercise A	*Paramecium caudatum, P. aurelia* (order and set up live cultures one, two, and three weeks prior to the laboratory).
Exercise B	Students should collect leaf litter before coming to the laboratory. Alternatively, collect leaf litter one week prior to the laboratory and separate organisms for observation in the laboratory.

Laboratory 45—Productivity in an Aquatic Ecosystem

Exercise B	*Chlorella* (order live and set up dilution cultures one week ahead of laboratory). Set up cultures for use in the laboratory 2 to 3 days prior to the laboratory.

SUGGESTED LABORATORY OUTLINE FOR GENERAL BIOLOGY

FIRST SEMESTER

Week 1 Laboratory I **Science—A Process**

Week 2 Laboratory 1 **Observations and Measurements: The Microscope**

Week 3 Laboratory 2 **Observations and Measurements: Measuring Techniques**
- Exercise A The Metric System (take home)
- Exercise B Measuring Mass
- Exercise C Measuring Volume
  - Part 1 The Pipette
  - Part 2 The Graduated Cylinder

Laboratory 3 **pH and Buffers**
- Exercise B Using Indicators to Measure pH
  - Part 1 Making a pH Indicator
  - Part 2 Measuring pH with Cabbage Indicator
- Exercise F Buffers

Week 4 Laboratory 4 **Using the Spectrophotometer**
- Exercise A How the Spectrophotometer Works
  - Part 1 Principles of Spectrophotometry
  - Part 2 Using the Spectrophotometer
  - Part 3 Determining Transmittance and Absorbance

Laboratory 5 **Organic Molecules**
- Exercise A Testing for Carbohydrates
  - Part 1 Benedict's Test for Reducing Sugars
  - Part 2 Lugol's Test for Starch
- Exercise C Testing for Proteins and Amino Acids
  - Part 1 Testing for Protein with the Biuret Reagent

Week 5 Laboratory 6 **Prokaryotic Cells**
- Exercise B Examining Bacterial Cells
  - Part 1 Observing Bacteria Using the Light Microscope
  - Part 2 Observing Bacteria Using the Transmission Electron Microscope (TEM)
  - Part 4 Examining Cyanobacteria
- Exercise C Working with Bacteria
  - Part 1 Techniques for Transferring Cultures

Laboratory 7 **Eukaryotic Cells**
- Exercise A Examining Plant Cells
- Exercise B Examining Animal Cells
  - Part 1 Studying Animal Cells Using Light Microscopy

Week 6	Laboratory 8	**Osmosis and Diffusion**
	Exercise C	Diffusion Across a Selectively Permeable Membrane
	Exercise D	A Look at Osmosis
	Part 1	Measuring Osmotic Potential
	Part 2	Measuring Pressure Potential: The Osmometer
Week 7	Laboratory 9	**Mitosis**
Week 8	Laboratory 10	**Enzymes**
Week 9	Laboratory 11	**Energetics, Fermentation, and Respiration**
Week 10	Laboratory 12	**Photosynthesis**
Week 11	Laboratory 13	**Meiosis: Independent Assortment and Segregation**
	Laboratory 14	**Genes and Chromosomes: Chromosome Mapping**
	Exercise A	Mapping the Genes of *Sordaria fimicola*
Week 12	Laboratory 14	**Genes and Chromosomes: Chromosome Mapping**
	Exercise D	Restriction Endonucleases: Mapping Bacteriophage Lambda
	Laboratory 15	**Human Genetic Traits**
	Exercise A	Human Cytogenetics (take home)
	Exercise C	Mendelian Inheritance in Humans
	Part 1	Monohybrid Crosses in Humans
	Exercise F	Forensic Science: DNA Fingerprinting (combine with Laboratory 14, Exercise D)
Week 13	Laboratory 16	**DNA Isolation**
	Exercise B	DNA Isolation Procedure
	Part 1 or 2	DNA Isolation (Chloroform or No Chloroform)
	Laboratory 17	**DNA—The Genetic Material: Replication, Transcription, and Translation** (assign for independent study in resource center)
	Laboratory 18	**Molecular Genetics: Recombinant DNA**
	Exercise A	Bacterial Transformation: Constructing Recombinant Plasmids (take home)
	Exercise B	Rapid Colony Transformation with pAMP: Ampicillin Resistance
Week 14	Laboratory 19	**Genetic Control of Development and Immune Defenses**
Week 15	Laboratory 20	**The Genetic Basis of Evolution I—Populations**
	Exercise B	Estimating Allelic Frequency from a Single Population Sample
	Exercise C	The Founder Effect as an Example of Genetic Drift
	Exercise D	The Role of Gene Flow in Similarity Between Two Populations

SECOND SEMESTER

Week 1	Laboratory 22	**Diversity—Kingdoms Eubacteria, Archaebacteria, and Protista**
	Exercise A	Morphology of Bacteria
	Exercise E	Diversity and Structure of Cyanobacteria
	Exercise F	Identifying Protozoans
	Laboratory 23	**Diversity—Fungi and the Nontracheophytes**
	Exercise A	Division Zygomycota
	Exercise C	Division Basidiomycota
	Exercise G	Diversity Among the Lichens
	Exercise H	Nontracheotphytes
Week 2	Laboratory 24	**Diversity—The Tracheophytes (Vascular Land Plants)**
Week 3	Laboratory 25	**Diversity—Porifera, Cnidaria, and Wormlike Invertebrates**
Week 4	Laboratory 26	**Diversity—Mollusks, Arthropods, and Echinoderms**
Week 5	Laboratory 27	**Diversity—Phylum Chordata**
Week 6	Laboratory 28	**Plant Anatomy—Roots, Stems, and Leaves**
Week 7	Laboratory 29	**Angiosperm Development—Fruit, Seeds, Meristems, and Secondary Growth**
Week 8	Laboratory 30	**Water Movement and Mineral Nutrition in Plants**
	Laboratory 31	**Plant Responses to Stimuli**
Week 9	Laboratory 33	**Introduction to the Study of Anatomy and the External Anatomy and Integument of Representative Vertebrates**
	alternate	
	Laboratory 32	**Animal Tissues**
	or	
	Laboratory 37	**The Basics of Animal Form: Skin, Bones, and Muscles**
Week 10	Laboratory 34	**The Anatomy of Representative Vertebrates: Behavioral Systems**
	alternate	
	Laboratory 41	**Control—The Nervous System**
Week 11	Laboratory 35	**The Anatomy of Representative Vertebrates: Digestive and Respiratory Systems**
	alternate	
	Laboratory 38	**Physiology of Circulation**
	or	
	Laboratory 39	**Gas Exchange and Respiration**
Week 12	Laboratory 36	**The Anatomy of Representative Vertebrates: Circulatory and Urogenital Systems**
	alternate	
	Laboratory 40	**The Digestive, Excretory, and Reproductive Systems**

Week 13	Laboratory 42 alternate Laboratory 44	**Behavior** **Predator-Prey Relations**
Week 14	Laboratory 43	**Communities and Ecosystems**
Week 15	Laboratory 45	**Productivity in an Aquatic Ecosystem**

Instructor's Manual:
PREPARATOR'S GUIDE

The BioBytes CD-ROM: Installation and Operation

CONTENTS OF THE CD-ROM

The BioBytes CD-ROM has the following components:

1. **Five interactive programs:** *Alien, Dueling Alleles, Cycle, Seedling,* and *Reaction Time.* The first four programs are simulations that can be used to supplement laboratory exercises in *Biology in the Laboratory, 3rd ed.*, by Helms *et al. Reaction Time* measures voluntary reaction time and performs statistical analysis on the results. *Reaction Time* is an integral part of several exercises in *Biology in the Laboratory.*
2. **Student exercises**. *Alien, Dueling Alleles, Cycle,* and *Seedling* have the text of complete student exercise on the CD-ROM, including background, suggested experiments, and student worksheets. These student exercises are reprinted in this manual following the comments and information for instructor's (and include several corrections or additions to the information on the CD-ROM). The student exercises that involve *Reaction Time* are used in *Biology in the Laboratory,* Laboratory I, and Chi-square analyses may be used in several other exercises.

USING THE CD-ROM ON A MACINTOSH

System Requirements:

1. System 7.0 or higher.
2. 8 MB RAM.
3. Approximately 4.5 MB of free disk space is required to install the software necessary to access the BioBytes simulations alone. If you want to access the student exercises on the CD-ROM, you need to install Acrobat Reader (included on the CD-ROM) which requires an additional 7 MB of disk space. The student exercises are printed in this manual so it is not necessary to view them on the screen.
4. A color monitor capable of displaying at least 8-bit (256 colors) at 640 × 480 pixels. For best results, set your monitor to "thousands" of colors or more. Refer to the documentation that came with your computer/graphics card for information about setting your monitor's color depth and resolution.
5. A double-speed (2×) or faster CD-ROM drive.

Installing the CD-ROM:

The BioBytes SETUP program will copy onto your computer's hard disk the components necessary to access the materials on the CD-ROM.

1. Place the CD-ROM in the CD drive.
2. Double-click on the "WHFreeman" CD-ROM icon on your desktop.
3. Drag the "BioBytes" folder onto your hard drive.
4. If you wish to install Adobe Acrobat Reader 3.0 for viewing the student exercises on your monitor or for printing them, drag the "ACROREAD" folder to your hard drive. Open it, then open the "MAC" folder, then the "READER" folder, and then another "READER" folder, and find the "Reader 3.01 Installer." Double click on it. After installation is finished, you may discard the "ACROREAD" folder.

5. The memory allocations in the BioBytes programs are adequate for most users, but may not be adequate for computers with many RAM-intensive applications running in the background. To change the memory allocation for a program, click once on the icon to highlight it and press the Command and I keys. This will give you the "Information" dialog on the program. For all the programs except *Dueling Alleles,* set the minimum memory to 800 K and the maximum to 1600 K. For *Dueling Alleles,* a minimum of 400 K and a maximum of 800 K is sufficient.

Accessing the BioBytes Software:

1. To use *Alien, Cycle, Dueling Alleles,* or *Seedling,* double-click on the BioBytes folder to open it. Double-click on BioBytes to access a menu opening any of the programs or open the desired program directly.
2. To use *Reaction Time,* open the BioBytes folder and double-click on its icon. Students may use the program to collect data or use the Chi-square median test to analyze data from other exercises.
3. To view (and print) one of the student manuals on your monitor, use *either* of the following techniques:
   a. Open the Adobe Acrobat 3.0 folder and double-click on the Acrobat Reader 3.0 icon. Then use the Open command to access one of the manuals in the BioBytes folder.
   b. From *Alien, Cycle, Dueling Alleles,* or *Seedling,* click on the Apple icon in the upper left corner of the screen and select Manual. Then quit from both the program and the BioBytes menu screen and the manual will appear on your screen.

Note: The BioBytes student worksheets are reprinted in this *Preparator's Guide.* They can be photocopied and distributed to your students if you wish to do so.

USING THE CD-ROM ON A PC RUNNING WINDOWS

System Requirements:

1. PC with at least an 80386 processor running Windows 3.1 or Windows 95.
2. 8 MB RAM.
3. Approximately 13.9 MB (Win 3.1) or 13.1 MB (Win 95) of free disk space for installing the BioBytes simulations alone. If you want to access the student exercises on your computer and you do not have Acrobat Reader 3.0, its installation will require an additional 5.6 MB (Win 3.1) or 5.3 MB (Win 95). [Removing the Acrobat installation file (AR32E301.EXE or AR16E301.EXE) from the BioBytes folder after installation will regenerate 3.6-3.9 MB.]
4. A color monitor capable of displaying at least 8-bit (256 colors) at 640 × 480 pixels. Refer to the documentation that came with your computer/graphics card for information about setting your monitor's color depth and resolution.
5. A double-speed (2x) or faster CD-ROM drive.

Installing the CD-ROM:

The BioBytes SETUP program will copy onto your computer's hard disk the components necessary to access the materials on the CD-ROM.

Windows 95:

1. Start Windows and place the CD-ROM in the CD drive.

2. From the Start menu, select Run.
3. Type D:\win95\setup, where "D" is the letter of your CD drive (or point the Browse button to your CD-ROM drive and Open "Setup.exe" in the Win95 folder). You can also use My Computer to open the Win95 folder on your CD-ROM drive and double-click on Setup.exe.
4. The setup wizard will appear; follow the instructions presented, and enter your preferred drive and directory when requested (the default location, C;\Program Files\BioBytes, should be used). Note the information provided in the Readme Information window (see below) during installation. The installer will create a new Windows program group called BioBytes.

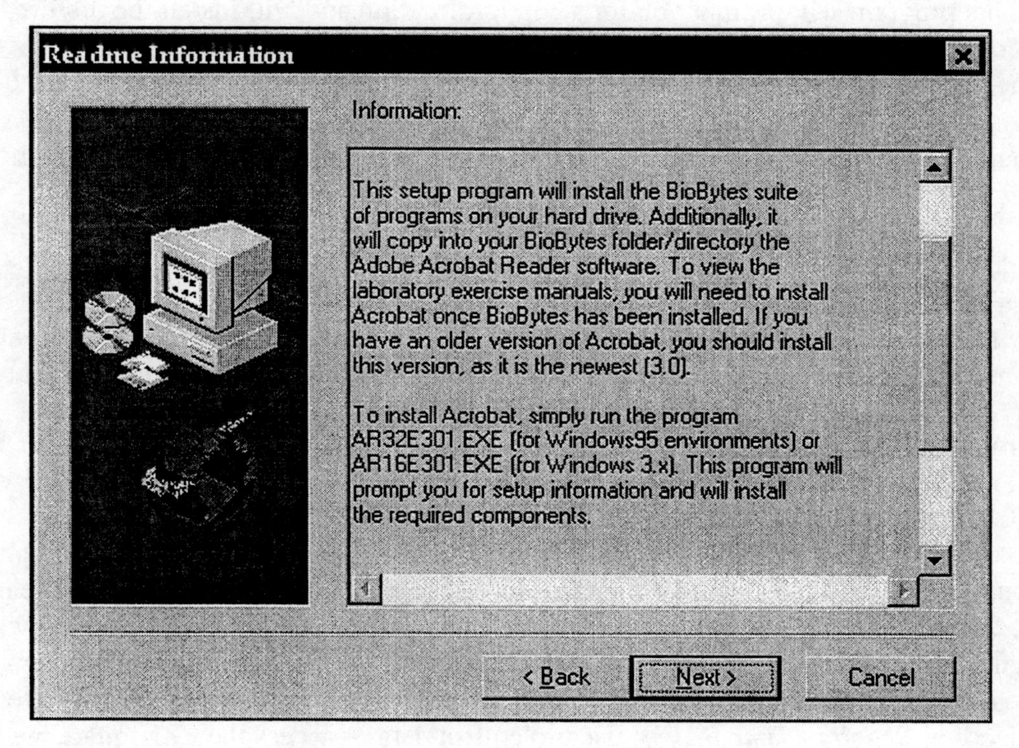

Windows 3.1:

1. Start Windows and place the CD-ROM in the CD drive.
2. From the Program Manager, select Run from the File menu.
3. Type D:\win3x\setup, where "D" is the letter of your CD drive (or point the Browse button to your CD-ROM drive and Open "setup.exe" in the Win3x folder). You can also use My Computer to open the Win3x folder on your CD-ROM drive and double-click on setup.exe.
4. The setup wizard will appear; follow the instructions presented, and enter your preferred drive and directory when requested (the default location, C;\programf\biobytes, should be used). The installer will create a new Windows program group called biobytes (locate the group under the Window menu item in Program Manager).

Accessing The Biobytes Student Worksheets:

Note: The *BioBytes* student worksheets are reprinted in this Preparator's Guide. They can be photocopied and distributed to your students if you wish to do so.

1. To view the student worksheets that accompany the BioBytes simulations, you need to install Acrobat 3.0. The Acrobat installation program is available only after you have installed BioBytes. To install Acrobat 3.0:
   a. Open the folder on your hard drive into which you installed the BioBytes CD. (The default is C:\Program Files\BioBytes (Win 95) or C:\programf\biobytes (Win 3.1).
   b. Double-click on the file AR32E301 (Win 95) or AR16E301.EXE (Win 3.1).
   c. This program will prompt you for setup information and will install the required components.
2. The student worksheets are located in the same directory and folder as the other *BioBytes* programs (C:\Program Files\BioBytes or C:\programf\biobytes). To view a worksheet, double-click on its name. The file names correspond to the following *BioBytes* modules:

Manual_a	*Alien*
Manual_c	*Cycle*
Manual_d	*Dueling Alleles*
Manual_s	*Seedling*

Note that these files can be renamed Alien, Cycle, ... Seedling using the File...Rename command (Win 3.1) or by selecting the name and retyping it (Win 95).

Accessing The Biobytes Software:

1. To use *Alien, Cycle, Dueling Alleles,* or *Seedling,* click on Start, Programs, BioBytes and select the desired program.
2. To use *Reaction Time,* open the BioBytes folder (C:\\Program Files or C:\\programf) and double-click on the *Reaction Time* icon. In Windows 95, a "Shortcut" can be created for *Reaction Time* (and BioBytes) by selecting (not opening) the program, clicking with the right mouse button, and selecting Create Shortcut. The Shortcut can then be dragged onto the desktop or placed in the Start menu of programs—alone or under the BioBytes selection. *Reaction Time* may be used to collect data or access the Chi-square median test to analyze data from other exercises.

HELP WITH THE CD-ROM

If you have any questions about installing the CD-ROM, please send e-mail the publisher at cdrom@whfreeman.com.

LABORATORY I Science—A Process

I. FOREWARD

The exercises in this laboratory will introduce students to the art of making observations, formulating and testing hypotheses, identifying variables, collecting and analyzing data, and forming conclusions. Students will learn the importance of testing the null hypothesis and how to analyze whether to accept or reject the null hypothesis. Exercises A-G should be completed in sequence.

This laboratory can be completed using the *Reaction Time* Program included on the accompanying CD-ROM or by using a simple non-computer based reaction time experiment. Data can be presented and analyzed from simple graphs and charts or by the chi-square median test.

Laboratory I is an important "first step." Unlike other laboratories in which exercises can be treated separately, Exercises A-G in this laboratory all deal with parts of the same investigation and should be completed by students during their course of study. This laboratory "sets the stage" for other investigations in this laboratory manual. Throughout the laboratory manual, additional independent investigations are identified as "Extending Your Investigation" and will provide students with opportunities to reinforce their investigative skills.

During Laboratory I, students will also be asked to construct a Laboratory Report. This can be accomplished by referring to Appendix II, Writing Laboratory Reports.

II. TIME REQUIREMENTS

Exercise A—The Scientific Method (15 minutes)
Exercise B—Reaction Time Experiments: Making Observations (30 minutes)
Exercise C—The Reaction Time Experiments: Formulating a Hypothesis (15 minutes)
Exercise D—The Reaction Time Experiments: Developing an Experimental Design (15 minutes)
Exercise E—The Reaction Time Experiments: Conducting the Test (30 minutes)
Exercise F—Presenting Experimental Data (15 minutes)
Exercise G—Interpretation: Conducting Statistical Analyses and Forming a Conclusion (15 minutes)

III. STUDENT MATERIALS AND EQUIPMENT

	Per Student	Per Pair (2)	Per Group (4)	Per Class (24)
Exercise A				
materials contained in the exercise				
Exercise B (Part 1)				
reaction time program (1)	1			24
Exercise B (Part 2)				
stainless steel ruler (2)		1		12

	Per Student	Per Pair (2)	Per Group (4)	Per Class (24)
Exercise C (Parts 1 and 2)				
see Exercise B				
Exercise D (Parts 1-3)				
materials contained in the exercise		1		12
Exercise E (Parts 1 and 2)				
see Exercise B				
Exercise F				
materials contained in the exercise				
Exercise G				
results from Exercise B, Table AI-2				
χ^2 worksheet (Part 2) **(3)**	1			24
Exercise H				
Appendix II				

IV. PREPARATION OF MATERIALS AND SOLUTIONS

(1) reaction time program

The reaction time program is on the CD-ROM accompanying each manual. Instructions for installing and operating the BioBytes software including *Reaction Time* are included in the section just before this exercise in the *Preparator's Guide* (pages CD-ROM-1 to 4).

(2) stainless steel ruler

A stainless steel ruler with a cork backing (available in most office supply stores) is best for "catching." Wooden or plastic rulers tend to fall at an angle since they are so light, but they can be used if necessary. Stainless steel rulers without a cork backing are so thin that they are difficult to catch. You may want to cut a strip of cork or felt and affix it to the ruler with super-glue or other adhesive.

(2) χ^2 analysis

The program that analyzes χ^2 is also on the CD-ROM. A worksheet for calculating χ^2 manually follows this section of the *Preparator's Guide* and can be duplicated for student use if computers are not available in the laboratory.

V. PREPARATION SUGGESTIONS

Exercise A

This exercise should be completed by students before coming to the laboratory. It can be discussed during the lecture period or during laboratory.

Exercise B

Students should be given enough time to explore the five tests available on the Reaction Time program. The CD-ROM disk contains all of the directions necessary for students to try all experiments. Students using the ruler test should have time to practice. They will need to experiment with the process of dropping the ruler so that all procedures are the same. Most students will find it simplest to rest their arm on the tabletop with the hand extended over the side (you will need to sit at the end of the table for this). Others may want to place an elbow on the table with the hand raised above the tabletop.

Exercise C

Students will need to practice making hypotheses. For most, the concept of the null hypothesis is new. It is the null hypothesis that we can accept or reject after testing since we can always show that either there is or is not a difference. This allows us to support or not support the alternate hypothesis. Make sure that students understand why we say that we can never prove the alternate hypothesis—we can only disprove it.

Exercise D

Students will need to learn how to identify variables and define treatments. Throughout the laboratory manual, students will be asked to identify variables as well as make predictions. It is important for students to understand the difference between paired and unpaired tests since use of the statistics package during other laboratories (osmosis and diffusion, *Sordaria* genetics, mineral nutrition, etc.) is possible.

Exercise E

Conducting the reaction time experiments is straightforward. Students should make sure to standardize all possible variables except what is tested (i.e., make sure room conditions, distance from the computer, etc. are the same). Make sure that students perform enough experimental replicas or that enough students are included per group to yield meaningful conclusions.

Exercise F

Data from experiments in Laboratory I can be presented in tables or charts. They can be analyzed qualitatively or by the use of statistics (see Exercise G).

Exercise G

Two types of statistical tests are used in this laboratory manual. The chi-square test and a special application of this test, the chi-square median test. A discussion of these two tests is presented at the end of this set of laboratory directions (See Chi-Square Tests, p. I-7; and Chi-Square Median Tests, p. I-8).

The chi-square test is used in analyzing data from *Sordaria* crosses (Laboratory 14, Genes and Chromosomes: Chromosome Mapping) where we are interested in the number of observations. The chi-square median test is used for the *Reaction Time* program and for analysis of data in other laboratories (Laboratory 12, Photosynthesis; Laboratory 30, Water Movement and Mineral Nutrition; Laboratory 43, Communities and Ecosystems) where values are analyzed in comparison to expections for distribution above and below the median.

Statistical analysis of data by the chi-square median test can be carried out using the statistics package provided on the CD-ROM. For those using the *Reaction Time* Program, statistical analyses are presented along with the experimental results (but no attempt is made to teach students how to make these calculations).

See Doing Statistical Analyses with the Reaction Time *program at the end of this section.*

For entering data from other laboratories in this manual into the *Reaction Time* program for statistical analysis, *see Manual Entry of Data into the* Reaction Time *program.*

For those using the non-computer options, calculations for the chi-square median test can be carried out using the worksheet that follows this section of the *Preparator's Guide* (or with the assistance of certain statistical calculators, use of statistical functions in computer spreadsheets, etc.).

Other statistical functions are introduced in Appendix I, Presenting and Analyzing Experimental Data: Tables, Graphs, and Statistical Methods.

Doing Statistical Analyses with the *Reaction Time* Program

1. Double-click on the "Reaction Time" icon in the BioBytes folder (Macintosh) or on the "Reaction Time" icon (Windows).
2. Click on the "Collect Reaction Times Data" button.
3. Click on the radio button next to, "Use the reaction time program and then perform an immediate statistical analysis." Then click on the "OK" button.
4. Type in the number of treatments you will use (2-5). Then select the "Yes" radio button if you will be using *exactly* the same reaction time tests for all treatments. If you will be making the smallest change in the test administered (e.g., using low-contrast spot-the-dot for one treatment and high-contrast spot-the-dot for the other treatment), the "No" radio button should be selected. If in doubt, select the "No" button. Then click "OK."
5. Click either on the "Paired" or "Unpaired" analysis buttons. A good rule of thumb is that if you are using the same people for both treatments (e.g., reaction times before and after exercise), you should do a *paired* analysis. Whenever the two treatments are different groups of people, select the *unpaired* analysis. If in doubt, use unpaired analysis. Then click the "OK" button. (See Chi-Square Median Tests.)
6. Select a type of reaction time test.
7. Type in how many trials (2-20) you want per person. Click "OK."
8. Enter the number of the treatment to which this first series of trials belongs. Click "OK."
9. Perform the tests. After the last test, the program will display a table of statistics (including the individual reaction times and the mean reaction time for that series of trials). There is no need to copy these down. The computer has stored them and will perform an analysis on them later. However, if your computer is connected to a printer, you may print this screen.
10. Unless there is some problem with the data, add these observations to treatment 1. Indicate that you are not finished with the experiment yet.
11. You will be returned to the reaction test menu. Repeat steps 6-10 until you have completed all reaction times in all treatments. When the program asks to what treatment the data belong, be sure to enter the right number, or the statistical analysis will be incorrect.
12. When all data are complete, click on the "Yes" button when asked, "Is this the end of your experiment?" The program will analyze the data and show a results screen. At the top of this will be a table of treatment averages, plus the highest and lowest values in each treatment.
13. If your analysis was *unpaired* the middle of the screen will show the number of unpaired observations above and below the median:

Treatment	Below Median	Above Median
1	6	4
2	4	6

14. If your analysis was *paired* the table will show the number of times each treatment had the higher number in each pair of observations:

Treatment 1 Higher	Treatment 2 Higher
6	4

15. The final line of the table will give the Chi-square value and the probability that a chi-square that large would have originated by chance if there were really no difference between the treatments. The null hypothesis cannot be rejected unless this probability is 0.05 (5%) or less.

Manual Entry of Data into the Reaction Time Statistical Program

For entry of data from other laboratory exercises:

1. Double-click on the reaction time program icon in the "BioBytes" folder.
2. Click on the "Perform Statistical Analysis" button.
3. Click on the "Yes" button when asked if you want to perform a median test.
4. Enter the number of treatments in the first text field and click on the "Accept" button. You may only use two treatments in a paired test.
5. Click on either the "Paired" or "Unpaired" radio buttons.
6. Enter a data point in the text field on the lower right of the screen and click the "Accept" button. Continue entering data points until you reach the end of the first treatment. If you make a mistake on a data point, use the "Previous" button to go back and correct the point.
7. When you reach the end of the treatment, click on the "Next Treatment" button. Begin entering the data points for this new treatment.
8. Continue in this fashion until all treatments have been entered. Then press the "Done" button and the statistical analysis screen will be presented.

VI. ORDERING INFORMATION

steel rulers, 12"—Carolina Biological, # 70-2597; Ward's, # 15W4659

Notes:

Chi-Square Tests

Chi-square (χ^2) is a statistic that compares the observed number of observations in different categories with the number expected if some null hypothesis were true. For example:

Experiment	Null Hypothesis	Expected Result If Null Is True
flipping a coin 20 times	equal numbers of heads and h = 10; t = 10 tails	h = 10; t = 10
the numbers of males and females who receive an "A" in General Biology class containing 150 males and 90 females	if 10% of all students get an A,10% of males should get an A and so should 10% of females	15 males get A; 9 females get A
genetic experiment with pea plants; 60 offspring	3:1 ratio of tall to short	tall = 45; short = 15
comparison of reaction times of 8 men and 8 women	Equal distribution of male times above and below the median time; same for female reaction times	4 male above the median and 4 below; 4 females above the median and 4 below

The size of the deviation between the observed and expected is measured by the chi-square statistic:

$$\chi^2 = \sum \frac{(O-E)^2}{E}$$

where O is the observed numbers of observations and E is the expected number of observations. If there is no deviation at all between O and E, $\chi^2 = 0$. The greater the deviation between observed and expected, the larger χ^2 becomes.

Chi-Square Median Tests

Either a paired or unpaired median test is just a special application of chi-square. It works on the same principle: comparison of observed data with the data expected if the null hypothesis were true. Use a median test when you have a series of observations in two or more treatments and you want to evaluate the null hypothesis that the two treatments are equal. Although the median test is insensitive (it cannot detect small differences between treatments), it is robust (it can work in almost any experimental situation).

Unpaired Median Test. Use when the treatments are different groups of people (e.g., men and women, people over 40 and people under 40, different bean seedlings subjected to five different nutrient treatments), or in general whenever there is no reason to associate one observation with another. The test's null hypothesis is that if the treatments are equal, each treatment will have equal numbers of observations above and below the median of all observations.

Paired Median Test. In an experimental design where the treatments are the "before" and "after" responses of the same people, plants, etc., a paired test is appropriate. The paired test avoids being weakened by variability between individuals by pairing each individual's "before" data with the same individual's "after" data. The test focuses only on whether each "before" data point is higher or lower than the "after" data point. The test's null hypothesis is that if there is no difference between the treatments, there will be just as many times when the "before" data is higher than the "after" data as when the opposite occurs.

Worksheet for Manual Computation of Chi-Square Median Tests

UNPAIRED TEST

For a paired test, go to page 3.

1. Arrange all the observations (Obs.) in order *from lowest to highest.* As you do, number them according to which treatment (Trtmt.—1 or 2, for example) they belong. For example:

Lowest →

Obs.	0.194	0.199	0.202	0.212	0.258	0.289	0.305	0.318	0.350	0.360
Trtmt.	1	1	2	1	2	1	1	1	1	1

→ Highest

Obs.	0.372	0.383	0.395	0.405	0.428	0.530	0.591	0.689	0.727	0.779
Trtmt.	2	1	1	2	2	2	2	2	2	2

Your Obs. ______________________________

Trtmt.______________________________

Your Obs. ______________________________

Trtmt.______________________________

2. Determine the median, the point above which and below which there are equal numbers of observations. For the example:

Lowest → **Median** ↓

Obs.	0.194	0.199	0.202	0.212	0.258	0.289	0.305	0.318	0.350	0.360
Trtmt.	1	1	2	1	2	1	1	1	1	1

→ Highest

Obs.	0.372	0.383	0.395	0.405	0.428	0.530	0.591	0.689	0.727	0.779
Trtmt.	2	1	1	2	2	2	2	2	2	2

3. Count up how many observations from each treatment fell above and below the median. Also, sum up the *number of observations* in each treatment. Half of this total is the expected number of observations above and below the median.

Example:

	Above Median	**Below Median**	**Total Number of Observations**	**Expected Value**
Trtmt. 1	2	8	10	5
Trtmt. 2	8	2	10	5

Your data:

	Above Median	Below Median	Total Number of Observations	Expected Value
Trtmt. 1				
Trtmt. 2				
Trtmt. 3				
Trtmt. 4				

Note: you need have only two treatments, and you could have more than four treatments.

4. Compute the chi-square statistic:

$$\chi^2 = \sum \frac{(O - E)^2}{E}$$

where O is the observed numbers of observations and E is the expected number of observations. For the example,

$$\chi 2 = [(2-5)^2]/5 + [(8-5)^2]/5 + [(2-5)^2]/5 + [(8-5)^2/5$$

$$\chi 2 = 9/5 + 9/5 + 9/5 + 9/5 = 7.2$$

Your chi-square:

5. Compare the chi-square statistic with the critical values for $p = 0.05$ listed in Table AI-2. For an unpaired median test, your degrees of freedom = one less than the number of treatments you used.
   For the example, degrees of freedom = 1, and the critical value = 3.84.

6. Conclusion: If your chi-square value is *less than* the critical value, you *cannot reject* the null hypothesis. If your chi-square value is *greater than* the critical value, you *must reject* the null hypothesis.
   Example Chi-square = 7.2 Critical value = 3.84. Conclusion: reject null

Your chi-square __________ Your critical value __________

Your conclusion:

PAIRED TEST

For an unpaired test, go to page 1.

1. Write down the paired observations (usually "before" and "after" results on the *same individuals*). They need not be in any particular order, but pairs of observations must remain together. Write a "+" if treatment 1 is higher than treatment 2 and a "–" if treatment 2 is higher than treatment 1. If the two treatments are equal, write "0."

Example:

Trtmt. 1	0.234	0.455	0.330	0.423	0.289	0.308	0.770	0.310	0.302	1.234
Trtmt. 2	0.199	0.302	0.401	0.440	0.566	0.330	0.288	0.310	0.351	0.245
Result	+	+	–	–	–	–	+	0	–	–

Your results:

Trtmt. 1										
Trtmt. 2										
Result										

Trtmt. 1										
Trtmt. 2										
Result										

2. Count up the number of plus and - signs. Count zeroes as both half a plus sign and half a minus sign. The expected number of + signs and the expected number of – signs is half the number of observation *pairs*.

Example:

	Observed	**Expected**
+	3.5	5
–	6.5	5

Your data:

	Observed	**Expected**
+		
–		

3. Compute the chi-square statistic for your data.

$$\chi^2 = \sum \frac{(O-E)^2}{E}$$

where O is the observed number of observations and E is the expected number of observations (both in the table in step 2). For the example,

$$\chi^2 = [(3.5-5)^2/5 + [(6.5-5)^2]/5$$

$$\chi^2 = 2.25/5 + 2.25/5 = 0.9$$

5. Compare the chi-square statistic with 3.84, the critical value for $p = 0.05$ and 1 degree of freedom (Table AI-2).

6. Conclusion: If your chi-square value is *less than* 3.84, you *cannot reject* the null hypothesis. If your chi-square value is *greater than* the critical value, you *must reject* the null hypothesis.

   Example: Chi-square = 0.9 Critical value = 3.84.
   Conclusion: cannot reject null hypothesis

   Your chi-square __________ Your critical value __________

   Your conclusion:

LABORATORY 1 Observations and Measurements: The Microscope

I. FOREWORD

This laboratory is designed to introduce students to the proper use of both the dissecting and compound microscopes and the correct methods for preparing wet-mount slides. Students will learn that the microscope can be used not only to observe objects but to measure their size. The microscope can be used for both qualitative and quantitative measurements. Extending Your Investigation provides an opportunity for students to form hypotheses and design an experiment using what was learned in Laboratory I. Students work individually in all exercises in this laboratory.

II. TIME REQUIREMENTS

Exercise A— Identifying the Parts of the Compound Microscope (25 minutes)
Exercise B— Using the Compound Microscope (20 minutes)
Exercise C— Preparing a Wet-Mount Slide (45 minutes)
Extending Your Investigation: Measuring Cyclosis (20 minutes)
Exercise D— Measuring the Size of Objects Using the Compound Microscope (20 minutes)
Exercise E— The Stereoscopic Dissecting Microscope (20 minutes)

III. STUDENT MATERIALS AND EQUIPMENT

	Per Student	Per Pair (2)	Per Group (4)	Per Class (24)
Exercise A				
lens paper packet		1		12
compound microscope	1			24
Exercise B				
compound microscope	1			24
letter "e" (prepared slide)	1			24
Oscillatoria (prepared slide)	1			24
Spirogyra (prepared slide)	1			24
Oedogonium (prepared slide)	1			24
Exercise C				
compound microscope	1			24
tap water (dropping bottle)		1		12
Elodea (live piece)				2
forceps (with *Elodea)*				1
microscope slides (75 × 25 mm)	3			72
coverslips (22 × 22 mm)	3			72
10% NaC1 (dropping bottle) **(1)**		1		12
Protoslo (dropping bottle)		1		12
Paramecium culture (live)				2

	Per Student	Per Pair (2)	Per Group (4)	Per Class (24)
Exercise C—continued				
toothpicks	1			24
Extending Your Investigation				
compound microscope	1			24
tap water (dropping bottle)		1		12
Elodea (live piece)				2
forceps (with *Elodea)*				1
microscope slides (75 × 25 mm)	3			72
coverslips (22 × 22 mm)	3			72
Exercise D				
compound microscope	1			24
transparent ruler	1			24
Nostoc (prepared slide)			1	6
ocular micrometer **(2)**			1	6
stage micrometer				2
Exercise E				
dissecting microscope	1			24
three-dimensional specimen **(3)**	1			24

IV. PREPARATION OF MATERIALS AND SOLUTIONS

(1) 10% NaC1

Weigh out 10 g NaCl. Add distilled water to 100 ml and mix.

(2) ocular micrometer

Several demonstration microscopes should be fitted with ocular micrometers and these should be calibrated by the instructor using a stage micrometer. Calibrations should be written on a tag and the tag should be looped around the ocular containing the micrometer. Ocular micrometers should be ordered to fit your specific brand of microscope.

(3) three-dimensional specimen

Preserved insects (dried) mounted on small Styrofoam blocks by straight pins are excellent three dimensional specimens. Pine needles or cones are also useful specimens.

V. PREPARATION SUGGESTIONS

Exercise A

Save on lens paper by cutting small squares from the booklet and placing them in a Petri dish for use by students.

Exercise B

Emphasize care of the microscope to prolong its life and to facilitate its use in visualization of biological specimens.

Exercise C

One box of glass microscope slides and a box of coverslips per laboratory table will suffice. A dirty slide container (such as a cut-off plastic milk bottle) will help in collecting used slides (have students discard used coverslips).

Extending Your Investigation

Be sure that students work rapidly. Excess heat from the microscope bulb will cause cell death.

Exercise D

Be sure to use only ocular micrometers that are specifically made to fit the brand of microscope you are using. Most simply insert into the ocular (after it has been removed from the eyepiece) and are held in place by a small ring of wire. Check to make sure that the scale is not backwards before inserting the ocular micrometer

It is possible for students to calibrate the ocular micrometer rather than having instructors calibrate them prior to the laboratory period. However, stage micrometers are fairly expensive and, if you have a large laboratory section, the process will create a bottleneck. If this option is desired, ask some students to begin with Exercise B while others begin with either Exercises C or D. Be prepared to give one-on-one assistance if students are asked to calibrate the micrometers. While it is an excellent experience in manipulating the metric system, many students find the calibration procedure difficult.

You may want to use transparencies of stage and ocular micrometers to show how to calibrate the ocular micrometer by using a stage micrometer. Show students how one can overlay the other to calibrate the ocular micrometer, and then overlay the ocular micrometer transparency with a third transparency of an object. (Masters for suitable transparencies follow this section of the *Preparator's Guide*.)

You might also want to place an ocular micrometer transparency on your television screen if you are using a video microscope. Give students the calibration for the ocular micrometer and have them determine the size of the object being viewed.

This exercise was included to show students that the microscope can be used to collect quantitative data as well as for observations. This point should be stressed during the laboratory. You should remove the ocular micrometers from the oculars after the laboratory so they will not get broken during general microscope use throughout the year.

Exercise E

Ask students to bring in several three-dimensional objects that they wish to study. Otherwise, preserved insects make excellent objects to study.

VI. ORDERING INFORMATION

letter "e" (prepared slide)—Carolina Biological, # BM1; Ward's, # 94W9111
Oscillatoria (prepared slide)—Carolina Biological, # B10; Ward's, # 91W0110
Nostoc (prepared slide)—Carolina Biological, # B9; Ward's, # 91W0093
Oedogonium (prepared slide)—Carolina Biological, # P7B54; Ward's, # 91W0383
Spirogyra (prepared slide)—Carolina Biological, # B61 or B62; Ward's, # 92W3444
Elodea (live)—Carolina Biological, # 16-2100; Ward's, # 68W7500
microscope slides, 75 × 25 mm—Carolina Biological, # 63-2000; Ward's 14W3500
coverslips, 22 × 22 mm—Carolina Biological, # 63-3015; Ward's, # 14W3251
Protoslo—Carolina Biological, # 88-5141; Ward's, # 37W7950
Paramecium culture (live)—Carolina Biological, # 13-1554; Ward's, # 87W1410
ocular micrometer—contact your microscope manufacturer

stage micrometer, scaled to 0.01 mm—Carolina Biological, # 59-4460; Ward's, # 24W0250

OCULAR MICROMETER

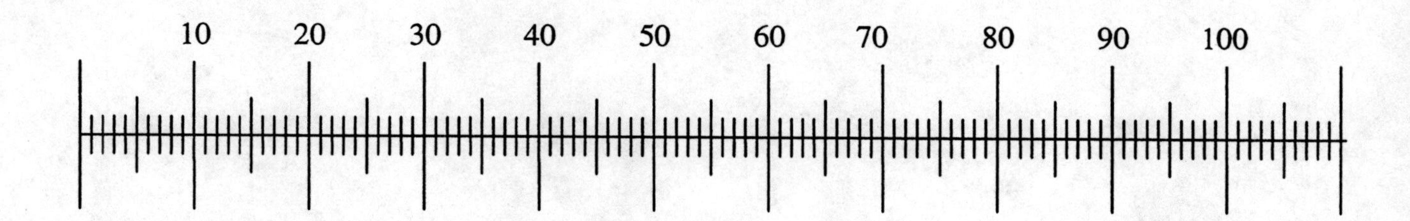

STAGE MICROMETER (mm)

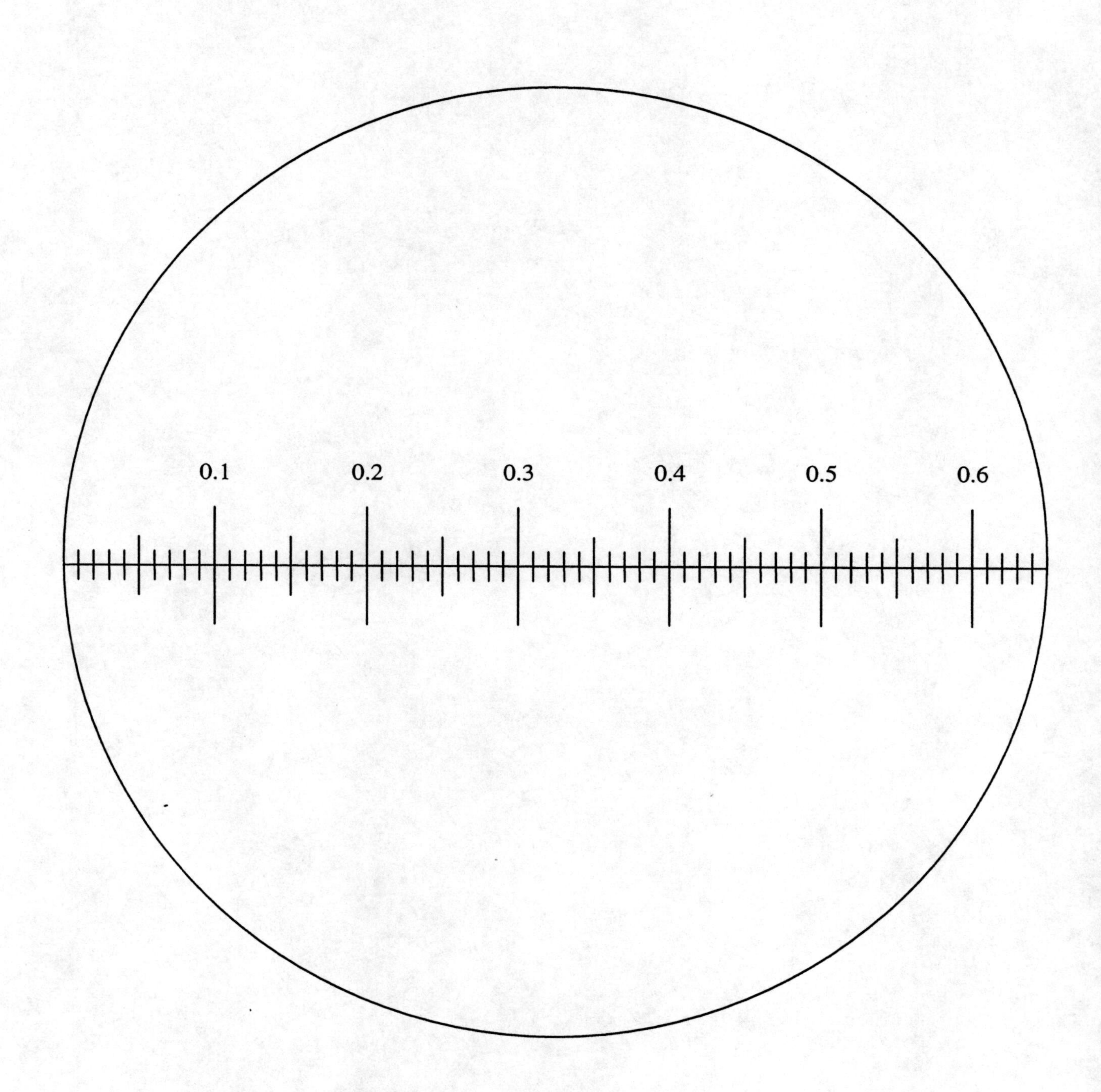

CELL FOR OCULAR MICROMETER MEASUREMENTS

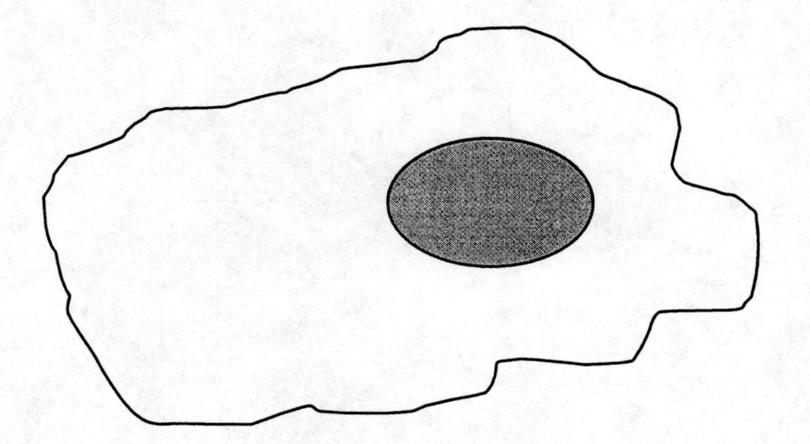

LABORATORY 2 Observations and Measurements: Measuring Techniques

I. FOREWORD

This laboratory is designed to introduce students to the metric system and to give them experience using the metric system while collecting quantitative data. Students should review Appendix I, Presenting and Analyzing Experimental Data: Tables, Graphs, and Statistical Methods, to learn how to report quantitative data in the laboratory. All work in Appendix I can be done outside the laboratory and is best given as a pre-laboratory assignment. Most of Exercise A may also be completed outside the laboratory before coming to class. Exercises B and C focus on being precise and accurate when measuring mass and volume. Exercise D introduces students to guidelines for preparing solutions. Additional information on solutions is included in Appendix III, Preparing Solutions.

II. TIME REQUIREMENTS

Exercise A—The Metric System (25 minutes)
Exercise B—Measuring Mass (30 minutes)
Exercise C—Measuring Volume
  Part 1—The Pipette (20 minutes)
  Part 2—The Graduated Cylinder (20 minutes)
  Extending Your Investigation: Comparing Measuring Devices (20 minutes)
  Part 3—The Volumetric Flask (20 minutes)
  Part 4—Precision and Accuracy of Measuring Devices (30 minutes)
Exercise D—Preparing Solutions (40 minutes)

III. STUDENT MATERIALS AND EQUIPMENT

	Per Student	Per Pair (2)	Per Group (4)	Per Class(24)
Exercise A				
millimeter ruler	1			24
objects to be measured **(1)**		10	60	
Exercise B				
balance (± 0.01 g)			1	6
weighing paper or boat			3	18
objects for weighing—rubber stoppers **(2)**			3	18
Exercise C (Part 1)				
10 ml pipette (blow out), demonstration				1
5 ml pipette (to deliver), demonstration				1
volumetric pipette (100 μl), demonstration				1
10 ml pipette	1			24
red liquid, bottle (25 ml) **(3)**			1	6

	Per Student	Per Pair (2)	Per Group (4)	Per Class(24)
Exercise C (Part 1)—continued				
Propipette	1			24
graduated cylinder (100 ml)			1	6
microcapillary pipete (100 μl)	1			24
digital micropipette (optional)				
(10 μl – 100 μl)				2
Parafilm (square)	1			24
Exercise C (Part 2)				
100 ml graduated cylinder		1		12
10 ml graduated cylinder		1		12
125 ml Erlenmeyer flask		1		12
bottle of water (100 ml)		1		12
Extending Your Investigation				
50 ml graduated cylinder (can use 100 ml cylinder)		1		12
50 ml beaker (can use 100 ml beaker)		1		12
50 ml Erlenmeyer flask (can use 125 ml flask)		1		12
water, bottle (100 ml)			1	6
Exercise C (Part 3)				
100 ml graduated cylinder		1		12
100 ml volumetric flask		1		12
water, bottle (200 ml)		1		12
Exercise C (Part 4)				
100 ml volumetric flask		1		12
Pasteur pipette		1		12
rubber bulb		1		12
100 ml Nalgene graduated cylinder		1		12
150 ml beaker		1		12
balance (± 0.1g)			1	6
Exercise D				
100 ml volumetric flask		1		12
balance (± 0.1 g)				
250 ml beaker		1		12
hydrometer			1	6
bottle of NaCl OR				
2 M NaCl, stock (500 ml) **(4)**				1
test tubes		6		72
test tube racks			1	6
100 ml graduated cylinder			1	6

IV. PREPARATION OF MATERIALS AND SOLUTIONS

(1) objects to be measured

A bag of 10 leaves (do not pick poison ivy!) gives students an excellent set of objects to be measured. Students will have to decide on some rules to employ while collecting data; for example, whether or not to include the petiole. Other objects such as nails, screws, or lengths of yarn could also be used.

(2) objects for weighing

Rubber stoppers provide a ready source of objects to weigh rather than using reagents. Small blocks of wood, small stones or pebbles, fishing weights, or a number of other objects can also be used.

(3) red liquid

Add a few drops of red food coloring to tap water to make practice solutions for pipetting exercises.

(4) 2 M NaCl

Mix 117 g NaCl in 900 ml of distilled water and add water to make 1 liter.
To mix solutions from the 2 M stock, calculate grams needed in the solution.

Example: Prepare 100 ml of a 0.8 M solution
0.8 M NaCl = 46.8 g/l
4.68 g/100 ml

Use the formula:

$$\frac{117g}{1000ml} = \frac{4.68g}{Xml}$$

Pour 40 ml in to a 100 ml volumetric flask and add distilled water to the 100 ml mark to make 100 ml of a 0.8 M NaCl solution from the 2 M stock.

V. PREPARATION SUGGESTIONS

Exercise A

Students usually experience difficulty with scientific notation and the metric system. Assign the written section of Exercise A to be completed before students come to the laboratory. You may wish to give a short quiz at the beginning of the laboratory to identify those students who are still having difficulty (these could be graded by the students themselves if given a sheet of answers). If quiz papers are placed on the laboratory bench so that the instructor can look at the results, each student could be given extra help during the laboratory period.

Objects used for measuring should all be of a size that conversion to different units is possible (this gives students extra practice). You should use objects that are large enough for students to measure in centimeters and then convert to millimeters or meters. Make sure students review Appendix I, Presenting and Analyzing Experimental Data, Tables, Graphs, and Statistical Methods, before attempting to construct their histograms.

Exercise B

It is important to use objects that will show significant changes in mass when added one at a time to the balance pan. Be sure to warn students that the balance pan should always be protected by a sheet of weighing paper or by a weighing boat.

We have used the terms "mass" and "weight" interchangeably for simplicity. You may wish to discuss this usage with your students.

Exercise C (Part 1)

Pipettes to be placed on demonstration should be taped onto pieces of cardboard in sets to show a "to deliver" or blow out pipette, a delivery pipette, a "to contain" pipette, and an example of at least one volumetric pipette. If you do not tape them together, the students will pick them up and scatter them all over the demonstration bench. Consequently, students will have a difficult time identifying the types of pipettes they are supposed to study.

Be sure to announce your laboratory rules for mouth pipetting—(it is not allowed!!). Explain why some pipettes are cotton-plugged. Demonstrate the Propipette or Pi-Pump prior to this exercise.

Exercise C (Parts 2, 3, and 4)

It is difficult to get water out of 10 ml graduated cylinders. If you are running multiple laboratory sections, have students place the cylinders upside down in a test tube rack that has a paper towel covering the mesh at the bottom of the rack. If you are running multiple laboratory sections you may wish to have two sets of volumetric flasks so that one may be dried in a drying oven while the other set is being used.

You may wish to augment Part 4 by collecting class data and computing the mean and standard deviation for class data. To do this, refer to Appendix I, Analyzing Experimental Data: Tables, Graphs, and Statistical Methods, for directions on computing these statistics.

Experiment D

You can set up balances and give students bottles of NaCl to use, but this tends to get messy. Also, to make 100 ml of a 0.1 ml solution only requires 0.585 g/100 ml and this is a small amount to measure accurately. A dilution from a 1 M solution would be more accurate, but this increases the glassware needed. If you want to limit solutions to molarities between 0.01 and 1, then use a 1 M stock solution. If you want to make the calculations more complex so students must "think through" what they are doing, then use a 2 M stock (this also allows students to mix solutions of greater concentration than 1 M).

To make solutions from the 2 M stock, students can use the formula in **(4)** of the Preparation of Materials and Solutions section, p, 2-3.

VI. ORDERING INFORMATION

clear plastic millimeter ruler—Carolina Biological, # 70-2604; Ward's, # 14W0811
weighing paper or dish—Carolina Biological, # 70-2320 (3×3" paper), # 70-2330 (aluminum dish, 43 mm); Ward's. # 15W2000
assorted rubber stoppers—Carolina Biological, # 71-14XX (various sizes); Ward's, # 15W8403
10 ml pipette (blow out)—Fisher, # 13-675 M; Carolina Biological, # 59-6514; Ward's, # 17W1113
5 ml pipette (to deliver)—Fisher, # 13-665 K; Carolina Biological, # 73-6276; Ward's, # 17W1307
100 μl volumetric pipette—Fisher, # 13-157-SS; Carolina Biological, # 72-5116
Propipette (safety pipette filler)—Fisher, # 12-681-50; Carolina Biological, # 73-6868; Ward's, # 15W0510
Pi-Pump (Pipette Pump)—Fisher, # 13-683-50; Carolina Biological, # 73-6875; Ward's, #

15W0532
100 ml volumetric flask—Fisher, # 10-199 D; Carolina Biological, # 72-5116; Ward's, # 17W2812
Pasteur pipettes—Fisher, # 13-678-20C; Carolina Biological, # 73-6062; Ward's, # 17W1146
rubber bulbs—Fisher, # 14-065B
graduated cylinder, 100 ml—Carolina Biological, # 72-1788 (glass); # 72-2266 (Nalgene); Ward's, # 17W0578
graduated cylinder, 50 ml—Carolina Biological, # 72-1786 (glass); 72-2264 (Nalgene); Ward's, # 17W0577
graduated cylinder, 10 ml—Carolina Biological, # 72-1782 (glass); Ward's, # 17W0575
Erlenmeyer flask, 125 ml—Carolina Biological, # 72-6670; Ward's, # 17W2981
Erlenmeyer flask, 50 ml—Carolina Biological, # 72-6684; Ward's, # 17W2971
beaker, 150 ml—Carolina Biological, # 72-1205; Ward's, # 17W4115
beaker, 250 ml—Carolina Biological, # 72-1209; Ward's, # 17W4116
microcapillary pipettes (100 μl)—Carolina Biological, # 21-4517
Accropet—Carolina Biological, # 21-4688
digital micropipette (10 μl – 100 μl)—Carolina Biological, # 21-4636; Ward's, # 15W2083
Parafilm—Carolina Biological, # 21-5600 (dispenser, 21-5602); Ward's, # 15W1940
hydrometer, specific gravity—Carolina Biological, # 72-2580 (1.000-1.220); Ward's, # 15W0879
test tubes, 16 × 125 mm—Carolina Biological, # 73-1418; Ward's, # 17W1403
test tube rack—Carolina Biological, # 73-1890; Ward's, # 18W4213

Notes:

LABORATORY 3 pH and Buffers

I. FOREWORD

This laboratory is designed to introduce students to the concepts of ionization, acid and base relationships, pH, and buffers. The use of the pH meter, a common laboratory instrument, is also explained. The instructor should choose appropriate exercises from those presented, since the combination of all exercises cannot be completed within the laboratory period.

An understanding of scientific notation (see Laboratory 2) is required. Exercise A should be completed before coming to the laboratory. Students may choose from several options in Exercise D. Appendix III, Preparing Solutions, will be used in Exercises E and F and should also be reviewed prior to laboratory.

Throughout this laboratory, students should be working in pairs or groups to save both time and reagents.

II. TIME REQUIREMENTS

Exercise A—Understanding pH (60 minutes; may be completed prior to laboratory)
Exercise B—Using Indicators to Measure pH
  Part 1—Making a pH Indicator (30 minutes)
  Part 2—Measuring pH with Cabbage Indicator (15 minutes)
  Part 3—Using Alkacid Test Paper (15 minutes)
Exercise C—Determining the pH of Some Common Solutions (*Optional*)
  Part 1—pH of Beverages (20 minutes)
  Part 2—pH and Activity of Some Common Medicines (20 minutes)
  Part 3—pH and the Action of Some Cleaning Solutions (20 minutes)
    student pairs may choose to do one of more parts of Exercise C
Exercise D—Soil pH and Plant Growth (*Optional;* 20 minutes)
Exercise E—The pH meter (30 minutes)
Exercise F—Buffers (45 minutes)

III. STUDENT MATERIALS AND EQUIPMENT

	Per Student	Per Pair (2)	Per Group (4)	Per Class (24)
Exercise A				
laboratory manual written exercise				
Exercise B (Part 1)				
red cabbage indicator, bottle (25 ml) **(1)**		1		12
test tubes		9		108
test tube rack		1		12
buffers, pH 1, 2, 4, 6, 8, 10, 12, 14, bottles (300 ml) **(2)**			1	6
5 ml pipettes (1 per buffer)			7	42
Parafilm (small squares)		7		84

	Per Student	Per Pair (2)	Per Group (4)	Per Class (24)
Exercise B (Part 2)				
test tubes	2			48
solution A (10 ml) **(3)**			1	6
solution B (10 ml) **(4)**			1	6
red cabbage indicator, bottle (25 ml) **(1)**		1		1
Exercise B (Part 3)				
test tubes	4			96
solution A (50 ml) **(3)**			1	6
solution B (50 ml) **(4)**			1	6
solution C (50 ml) **(5)**			1	6
solution D (50 ml) **(5)**			1	6
bottle plus stirring rod for each solution (A-D)				
alkacid test paper, 2 inch piece	4			96
forceps	1			24
Exercise C (Part 1)—Optional				
test tubes	4			96
test tube rack			1	6
apple juice (10 ml) **(6)**			1	6
coffee (10 ml) **(6)**			1	6
7-Up (10 ml) **(6)**			1	6
white wine (10 ml) **(6)**			1	6
red cabbage indicator, bottle (25 ml) **(1)**			1	6
alkacid test paper, 2 inch piece	2			48
Exercise C (Part 2)—Optional				
test tubes	4			96
test tube rack			1	6
aspirin solution (10 ml) **(7)**			1	6
Milk of Magnesia (10 ml) **(8)**			1	6
sodium bicarbonate solution (10 ml) **(9)**			1	6
Maalox (10 ml) **(10)**			1	6
red cabbage indicator, bottle (25 ml) **(1)**			1	6
Exercise C (Part 3)—Optional				
test tubes	4			96
test tube rack	1			24
Drano solution (10 ml) **(11)**			1	6
Ivory Liquid solution (10 ml) **(12)**			1	6
Cascade solution (10 ml) **(13)**			1	6
Tide solution (10 ml) **(14)**			1	6
Exercise D—Optional				
potting soil, beaker and stirring rod **(15)**			1	6
clay, beaker and stirring rod **(16)**			1	6
sand, beaker and stirring rod **(17)**			1	6

	Per Student	Per Pair (2)	Per Group (4)	Per Class (24)
Exercise D—continued				
lime, beaker and stirring rod **(18)**			1	6
peat moss, beaker and stirring rod **(19)**			1	6
Exercise E				
pH meter				4
storage beaker for pH meter				4
distilled water, squirt bottle				4
plastic beaker for pH meter rinse water				4
magnetic stirring bar				4
stirring plate				4
apple juice, beaker (50 ml)				4
7 Up, beaker (50 ml)				4
Maalox, beaker (50 ml) **(10)**				4
Tide, beaker (50 ml) **(14)**				4
Ivory Liquid, beaker (50 ml) **(12)**				4
NaOH (bottle)				1
$CaOH_2$ (bottle)				1
KH_2PO_4 (bottle)				1
NaC1 (bottle)				1
balance (± 0.01 g)				4
weighing pan		4		24
distilled water, bottle (1000 ml)			1	6
Exercise F				
pH meter				4
beaker				4
distilled water squirt bottle				4
magnetic stirring bar				4
stirring plate				4
0.1 N NaOH **(20)**				2
0.1 N HC1 **(21)**				2
M K_2HPO_4, bottle **(22)**				4
M KH_2PO_4 (80 ml) **(23)**				4
M KH2PO4/K2HPO4 buffer, bottle **(24)**				4
distilled water, bottle				4
congo red, dropper bottle **(25)**		1	6	
thymolphthalein, dropper bottle **(26)**			1	6
5 ml pipette			2	8

IV. PREPARATION OF MATERIALS AND SOLUTIONS

(1) red cabbage indicator (cabbage extract)

Shave thin slices from a head of red cabbage. Fill a 250 ml beaker half full with the cabbage and add 150 ml of water. Boil until the water turns a medium purple-red in color. If the solution is too light, add more cabbage. Each group of students (or pair) should prepare its own extract. However, to save time, it is possible to make the extract before the laboratory and distribute it in bottles. Simply increase the amounts (fill a 1000 ml beaker half to three-quarters full with shaved cabbage and add 500 ml of water).

(2) **buffers (2000 ml each)**

Buffers can be prepared from chemicals normally found in the chemistry stockroom as given below. Alternatively, buffer concentrates can be ordered in liquid form, as capsules, or dried salts from Fisher or Carolina Biological Supply Company (see ordering information).

pH 1

Use 2000 ml of 1 N HCl (order from Fisher prepared)

pH 2

Part 1. Dissolve 14.91 g KCl in 400 ml of distilled water and dilute up to 500 ml with distilled water

Part 2. Mix the 500 ml of potassium chloride solution prepared in Part 1 with 260 ml of 0.1 N HCl and add distilled water to make 2000 ml.

pH 4

Part 1. Dissolve 20.42 g of potassium hydrogen phthalate in 700 ml of distilled water and dilute up to 1000 ml with distilled water.

Part 2. Mix the 1000 ml of potassium hydrogen phthalate solution (Part 1) with 998 ml of distilled water and 2 ml of 0.1 N HC1 to make 2000 ml.

pH 6

Part 1. Dissolve 13.61 g of potassium phosphate (monobasic) in 700 ml of distilled water and dilute up to 1000 ml with distilled water.

Part 2. Mix the 1000 ml of potassium phosphate (monobasic) solution (Part 1) with 888 ml of distilled water and 112 ml of 0.1 N NaOH to make 2000 ml.

Note: You may wish to double the amount of pH 6 buffer in order to prepare pH 7 buffer if you are teaching large sections.

pH 7

Mix pH 6 solution and pH 8 solution together until you obtain a solution of pH 7.

pH 8

Part 1. Dissolve 13.61 g of potassium phosphate (monobasic) in 700 ml of distilled water and dilute up to 1000 ml with distilled water.

Part 2. Mix the 1000 ml of potassium phosphate (monobasic) solution (Part 1), 922 ml of 0.1 N NaOH, and 78 ml of distilled water to make 2000 ml.

Note: You may wish to double the amount of pH 8 buffer in order to prepare pH 7 buffer if you are teaching large sections.

pH 10

Part 1. Dissolve 7.46 g of potassium chloride and 6.18 g of boric acid in 700 ml of distilled water and dilute to 1000 ml with distilled water (heat to dissolve the boric acid).

Part 2. Mix the 1000 ml of potassium chloride and boric acid solution (Part 1), 874 ml of 0.1 N NaOH, and 126 ml of distilled water to make 2000 ml.

pH 12

Part 1. Dissolve 7.1 g of sodium phosphate (dibasic) in 700 ml of distilled water and dilute to 1000 ml with distilled water.

Part 2. Mix the 1000 ml of sodium phosphate (dibasic) solution, 538 ml of 0.1 N NaOH, and 462 ml of distilled water to make 2000 ml.

pH 14
Use 2000 ml of 2 N NaOH

(3) solution A
Mix 100 ml of 0.1 N HCl and 900 ml of distilled water. Adjust to pH 2 if necessary.

(4) solution B
Mix 0.01 ml of 0.1 N HCl and 999.9 ml of distilled water. Adjust to pH 6 if necessary.

(5) solutions C and D
Choose any two colored solutions or add several drops of different food colorings to samples of solutions A and B.

(6) apple juice, coffee, 7-Up, white wine
Use as bottled

(7) aspirin solution
Dissolve one aspirin in each 100 ml of tap water.

(8) Milk of Magnesia
Use as purchased.

(9) sodium bicarbonate solution
Dissolve 0.5 g of $NaHCO_3$ in each 100 ml of distilled water.

(10) Maalox solution
Use as purchased.

(11) Drano solution
Dissolve 6 g of Drano into each 100 ml of tap water.

(12) Ivory Liquid
Use as purchased.

(13) Cascade solution
Dissolve 6 g of Cascade into each 100 ml of tap water.

(14) Tide solution
Dissolve 6 g of Tide into each 100 ml of tap water.

(15) potting soil mix
Fill a 150 ml beaker with potting soil up to the 40 ml mark and add tap water to the 120 ml mark. The pH should be 6 or 7.

(16) clay
Fill a 150 ml beaker with clay up to the 40 ml mark and add tap water to the 120 ml mark. The pH should be 4.

(17) sand
Fill a 150 ml beaker with sand up to the -40 ml mark and add tap water to the 120 ml mark. The pH should be 7.

(18) lime
Fill a 150 ml beaker with lime up to the 40 ml mark and add tap water to the 120 ml mark. The pH should be 9.

(19) peat moss
Fill a 150 ml beaker with peat moss up to the 40 ml mark and add tap water to the 120 ml mark. The pH should be 8.

(20) 0.1 N NaOH

Purchase already prepared.

(21) 0.1 N HCl

Purchase already prepared.

(22) 0.02 M K_2HPO_4

Dissolve 3.48 g K_2HPO_4 in 700 ml distilled water. Add distilled water to 1000 ml. For 4 liters, dissolve 13.92 g K_2HPO_4 in 3000 ml distilled water. Add distilled water to 4000 ml.

(23) 0.02 M KH_2PO_4

Dissolve 2.72 g KH_2PO_4 in 700 ml distilled water. Add distilled water to 1000 ml. For 4 liters, dissolve 10.88 g KH_2PO_4 in 3000 ml distilled water. Add distilled water to 4000 ml.

(24) 0.02 M KH_2PO_4/K_2HPO_4 buffer (pH 7)

For 1000 ml mix 390 ml KH_2PO_4 with 610 ml K_2HPO_4 ((**22**) and (**23**)).
For 2000 ml mix 780 ml KH_2PO_4 with 1220 ml K_2HPO_4.
For 4000 ml mix 1560 ml KH_2PO_4 with 2440 ml K_2HPO_4.

(25) 0.1% congo red

Add 0.1 g congo red to 100 ml distilled water.

(26) 1% thymolphthalein

Add 1 g thymolphthalein to 100 ml distilled water. (If a small amount does not go into solution, filter through Whatman #1 filter paper.) Make fresh each year.

V. PREPARATION SUGGESTIONS

Exercise A

This exercise should be completed before coming to laboratory. You may find that an additional "help" session is necessary during the week because students generally find the mathematical manipulations associated with determining hydrogen ion concentration or pH difficult. You may need to review methods for using the log and antilog functions on the calculator.

Exercise B (Parts 1 and 2)

Students find this exercise to be fun as well as instructive. To save time, you may wish to boil cabbage ahead of time. To save on glassware, you may wish to use a repipette for each buffer solution (set the dispenser to 5 ml). Place the repipettes on the demonstration bench and tell the students to carry their labeled test tubes (in a test tube rack) to the demonstration bench to fill each with the appropriate buffer. If students do not prepare their own cabbage extract, you may also want to use a repipette for the cabbage extract (set to 3 ml).

Exercise B (Part 3)

Provide antacid test paper in small pieces in a Petri dish. Students can use forceps to place the antacid test paper into the solution.

Exercise C

To save time, split the class into groups and have one group measure the pH of beverages, another group measure the pH of medicine, and a third group measure the pH of common cleaning solutions. Record class data on the board. You might also want to assign students to bring in a sample of a particular substance along with a "guestimate" of its pH.
Examples could include shampoo, a favorite soft drink, Alka-Seltzer, Tums, iced tea, etc.

Exercise D

This exercise is optional. Students should use a glass stirring rod to transfer a small amount of sample to the alkacid test paper. You might want to assign students to bring in a soil sample from somewhere on campus or at home, noting the types of plants living in the area from which the soil was taken. Class data should be placed on the board and pH values compared for plants from similar areas.

Exercise E

If enough pH meters are available, this exercise will flow smoothly. However, to speed things up, you might ask half of the class to complete Exercises C and D while the other half works on Exercises E and F. For the latter group, estimated pH values can be added later to Table 3E-1 when students complete Exercise C. Students should read Appendix III, Preparing Solutions, before working on Exercise E. This can be done prior to the laboratory.

Since students will be using NaOH that is very caustic, make sure that the balance pans and tops of balances are well protected. Have students wear safety glasses. Be sure to indicate which pH meters are to be used for particular pH ranges. Plastic combination electrodes should be used to avoid excessive breakage that is likely if glass electrodes are used.

You will find it most practical if beakers of solutions are labeled and present at the designated balances.

You will also find it helpful if reagents for preparing solutions are located in a separate area of the laboratory. You will need a magnetic stirring plate, magnetic stirring bars, and a retriever. Students should have ready access to running water to rinse out beakers used for mixing solutions. Provide plenty of paper towels and some absorbent fowling for drying glassware. Be sure to indicate which pH meters should be used for measuring the pH of basic, neutral, and acidic solutions.

Exercise F

This exercise can be done in pairs or even in groups of four. By using the indicators thymolphthalein and congo red, students are able to see color changes that indicate whether a solution is buffering against acid only (K_2HPO_4), base only (KH_2PO_4), both acid and base (KH_2PO_4/K_2HPO_4 buffer) or is buffering against neither (water). By visualizing the process, the typical buffer "diagram" begins to make sense. The buffering capacity of the solutions can also be calculated mathematically so that quantitative data supports qualitative observations. Since each student group will use 40 ml of unknown solution, a class of 24 students working in pairs would need approximately 500 ml of each solution. Make enough K_2HPO_4 and KH_2PO_4 to have the amount necessary for making the buffer. Be sure that students rinse glassware well between each test. Place screens in your sink drains so that magnetic bars used to stir solutions are not poured down the drain by students

VI. ORDERING INFORMATION

test tubes, 16 × 125 mm—Fisher, # 14-925H; Carolina Biological, # 73-1418; Ward's, # 17W1403

test tube rack—Carolina Biological, # 73-1896; Ward's, # 18W4213

hydrochloric acid, 1 N—Fisher, # SA48-4; Carolina Biological, # 86-7843; Ward's, # 37W8605

hydrochloric acid, 0.1 N (N/10)—Fisher, # SA54-4; Carolina Biological, # 86-7823; Ward's, # 37W9561

pipette, 5 ml—Fisher, # 13-665K; Carolina Biological, # 73-6512; Ward's, # 17W4854
Parafilm—Fisher, # 13-274-12 (dispenser, # 13-374-18); Carolina Biological, # 21-5600 (dispenser, # 21-5602); Ward's, # 15W1940
potassium chloride (KCl)—Fisher, # P217-500; Carolina Biological, # 88-2900; Ward's, # 37W4634
potassium hydrogen phthalate—Fisher, # P243-500; Carolina Biological, # 88-2900; Ward's, # 37W4634
potassium phosphate (monobasic) (KH_2PO_4)—Fisher, # P285-500; Carolina Biological, # 88-4250; Ward's, # 37W4841
sodium hydroxide, 0.1 N (N/10)—Fisher, # SS276-4; Carolina Biological, # 88-9551; Ward's, # 37W8606
boric acid (H_3BO_3)—Fisher, # A73-1; Carolina Biological, # 84-8440; Ward's, # 37W0864
sodium phosphate (monobasic) (NaH_2PO_4)—Fisher, # S369-1; Carolina Biological, # 89-1350; Ward's, # 37W5655
potassium phosphate (dibasic) (K_2HP0_4)—Fisher, # P288-500; Carolina Biological, # 88-4290; Ward's, # 37W4841
sodium hydroxide, 2 N—Fisher, # SS264-4; Carolina Biological, # 88-9593
sodium chloride (NaCl)—Fisher, # S271-3; Carolina Biological 88-8901; Ward's, # 37W5487
Hydrion buffers in Capsules—Fisher, # 13-640-303 A-J; Carolina Biological, # 849XX (see catalog; numbers vary with pH); Ward's, # 37W5955
Fisher certified buffer concentrate—Fisher, # SB99-SXX (see catalog; numbers vary with pH)
buffer, Chemvelope—Carolina Biological, # 84-9XXX (see catalog; numbers vary with pH); Ward's, # 37W5955
Alkacid (wide-range) test ribbons—Fisher, # A979
stirring plate—Carolina Biological, # 70-1036A; Ward's, # 15W8024
magnetic stirring bar, 2"—Carolina Biological, # 70-1085; Ward's, # 70-1036A
congo red—Fisher, # LC13360-1 (0.1% aqueous); Carolina Biological, # 85-5375; Ward's, # 38W9195
thymolphthalein indicator—Fisher, # LC26070-1 (0.05% in ethanol)

LABORATORY 4 Using the Spectrophotometer

I. FOREWORD

The Milton Roy ("Bausch and Lomb") Spectronic 20 Colorimeter or other spectrophotometer will be used in several of the laboratories presented in this manual. This laboratory is designed to introduce students to the use of the spectrophotometer and to some of the theory involved in its operation.

Students should work in pairs with two or three pairs forming a group to share reagents and the spectrophotometer

II. TIME REQUIREMENTS

Exercise A—How the Spectrophotometer Works
  Part 1—Principles of Spectrometry (10 minutes)
  Part 2—Using the Spectrophotometer (20 minutes)
  Part 3—Determining Transmittance and Absorbance (30 minutes)
Exercise B—Determining the Maximum Absorption Wavelength (45 minutes)
Extending Your Investigation: Absorbance and Transmittance (10 minutes)
Exercise C—Exploring the Relationship between Absobance and Concentration (60 minutes)

III. STUDENT MATERIALS AND EQUIPMENT

	Per Student	Per Pair (2)	Per Group (4)	Per Class (24)
Exercise A (Part 2)				
Spectronic 20 **(1)**			1	6
Spectronic 20 cuvettes, glass tubes **(2)**		1		12
white paper strips		1		12
black paper cylinders		1		12
Exercise A (Part 3)				
Spectronic 20 **(1)**			1	6
Spectronic 20 cuvettes, glass tubes **(2)**		1		12
albumin solution, bottle (50 ml) **(3)**			1	6
coomassie blue solution, bottle (50 ml) **(4)**			1	6
Pasteur pipette and bulb (with albumin)		1		12
1 ml pipette (with distilled water)		1		12
5 m1 pipette (with coomassie blue)		1		12
Exercise B				
(same requirements as Exercise A, Part 3)				
Extending Your Investigation				
Spectronic 20 **(1)**			1	6
Spectronic 20 cuvettes, glass tubes **(2)**			2	12
solution of red dye, bottle **(5)**			1	6
Pasteur pipette			2	12

	Per Student	Per Pair (2)	Per Group (4)	Per Class (24)
Extending Your Investigation—continued				
rubber bulb			2	12
Exercise C				
Spectronic 20 **(1)**			1	6
Spectronic 20 cuvettes (glass tubes) **(2)**		6		72
BSA albumin standard, tube (2 ml) **(6)**		1		12
unknown albumin solution **(7)**		1		12
coomassie blue solution, bottle (50 ml) **(4)**		1		12
test tubes		6		72
test tube rack		1		12
1 ml pipettes		2		24
5 ml pipettes		1		12

IV. PREPARATION OF MATERIALS AND SOLUTIONS

(1) Spectronic 20

Use a wide range phototube with filter. This will cover a range of 400-700 nm without having to change phototubes at 600 nm. Be sure to warn students not to pour liquid into the cuvette holder—**a cuvette must be used**. Also warn students not to pour liquid into the cuvette while it is in place; if they miss, the liquid will get into the machine.

(2) Spectronic 20 cuvettes

To save on cost, Pyrex or Kimax tubes may be used rather than cuvettes (see Ordering Information).

(3) albumin solution

Mix 0.1 g albumin into 100 ml distilled water. Be sure to use a fairly pure source of albumin (see ordering information).

(4) coomassie blue

Dissolve 0.1 g coomassie blue in 50 ml of 95% ethanol. Add 100 ml of phosphoric acid. Add distilled water to make 1000 ml.

(5) red dye

Add 2 drops of red food coloring to 10 ml distilled water.

(6) albumin standard (240 mg/ml)

Mix 0.24 g albumin into 100 ml of distilled water. Dilute 1 ml of this albumin solution to 9 ml with distilled water to obtain a solution of 240 μg/ml. (For larger quantities use 10 ml of albumin solution and dilute it to 90 ml with distilled water.) Dispense to students in tubes of 1-2 ml each.

(7) albumin unknown, "C"

Use dilutions of the albumin standard **(6)** at 1:1, 1:2, 1:5, and 1:10 for a series of unknowns. Number tubes accordingly so that student results can be checked.

V. PREPARATION SUGGESTIONS

Exercise A (Part 2)

Familiarize students with the Spectronic 20. It is fairly simple to tell them that the knob on the right will control what happens to the needle on the scale at the right, and that the knob on the left will control what happens to the needle on the scale at the left. Show them what happens when you place a blank tube into the machine. Students can cup their hands or use a cylinder made from black construction paper to view the changes in light that occur with changes in wavelength. This is an excellent way to get students to understand how the spectrophotometer works. Most students do not understand the relationship between wavelength and color of light until they see this demonstration.

Exercise A (Part 3)

Be sure to explain the use of "blanks" to your students. Explain that a blank contains all of the same components as the "test" sample except for the absorber molecule that allows you to measure concentration. In some cases, the absorber molecule is the solute itself, while in others, it is a chemical that will react colorimetrically with the molecule in question, allowing the concentration of the molecule to be measured. Students should always ask themselves "what" in the mixture would absorb light? Is it what I want to measure, or will it interfere with the measurement so that I should also include it in the blank?

If you do not wish to use coomassie blue, bromophenol blue can also be used. It has a maximum absorbance at 580 nm. Mix 0.1 g bromophenol blue per liter of water to prepare a stock solution. Dilute the stock solution 1:2 (10 ml starch plus 10 ml water), 1:4 (10 ml of the 1:2 mixture plus 10 ml water), and 1:8 (10 ml of the 1:4 mixture plus 10 ml water). Use the 1:8 dilution for this exercise. Save the other dilutions for Exercises B and C.

Exercise B

Students should be shown that substances will absorb light over a fairly wide range of wavelengths, but maximum absorbance occurs at a single wavelength. You can use either coomassie blue or bromophenol blue from Exercise A, Part 3, for this exercise.

Extending Your Investigation

This exercise will help students understand the relationship between absorbance and wavelength. You could extend this investigation by using a green, yellow, orange, or blue solution and ask students to predict the wavelength and color of light for which absorbance is maximum. Why do some materials appear yellow or green? What wavelengths of light are being absorbed? Transmitted? Relate this to perception of leaf color change in the fall when chlorophyll is broken down and other pigments absorb or reflect light. Red leaves absorb? (blue and green) Transmit? (red)

Exercise C

Standard curves are used in several different laboratories in this manual. They provide good graph construction and interpretation experience for the students.

If you wish to use bromophenol blue instead of coomassie blue for this experiment, use the dilutions described for Exercise A, Part 3.

Note: the stock solution contains 0.1 mg/ml of bromophenol blue. You may need to make several more dilutions—1:16, 1:32, 1:64, and so on. Have students figure concentrations of bromophenol blue in each and construct a standard curve from their data and calculations. Read absorbance at 580 nm when constructing the curve. Pick one of the dilutions to serve as an "unknown" or prepare a new solution of 0.005 mg/ml for an unknown.

If you use coomassie blue, you may want to ask the students to think of something they would like to test for protein content—for example, sodas, diet preparations, dilute apple juice, or others.

VI. ORDERING INFORMATION

Spectronic 20 (Milton Roy, formerly B&L)—Fisher, # 07-143-1; Carolina Biological, # 65-3300; Ward's, # 14W5550

wide range phototube and filter—Fisher, # 07-144-10; Carolina Biological, # 65-3312; Ward's, # 14W5553

Spectronic 20 tubes—Fisher, # 14-385-9008; Carolina Biological, # 65-3310; Ward's, # 14W5554 OR

Spectronic 20 tubes, Pyrex (100 × 13 mm)—Fisher, # 14-957C; Carolina Biological, # 73-1408

Spectronic 20 tubes, Kimax (100 × 13 mm)—Fisher, # 14-923D; Carolina Biological, # 73-1408A

albumin—Fisher, # A388-500; Sigma, # A 4378; Carolina Biological, # 84-2250; Ward's, # 39W0197

coomassie brilliant blue G-250—Sigma, # B 5133; Carolina Biological, # 68-9782

phosphoric acid (H_3PO_4)—Fisher, # A280-500; Carolina Biological, # 88-1470; Ward's, # 37W6585

Pasteur pipette, 9"—Fisher, #13-678-20C; Carolina Biological, # 73-6062; Ward's, # 17W1146

rubber bulb—Fisher, # 14-065B

pipette, 1 ml—Fisher, # 13-665F; Carolina Biological, # 73-6270; Ward's, # 17W1305

pipette, 5 ml—Fisher, #13-665K; Carolina Biological, # 73-6276; Ward's, # 17W1307

test tubes, 16 × 125 mm—Fisher, # 14-925H; Carolina Biological, # 73-1418; Ward's, 17W1403

test tube rack—Carolina Biological, # 73-1896; Ward's, # 18W4213

red dye—use food coloring

LABORATORY 5 Organic Molecules

I. FOREWORD

This laboratory is designed to introduce students to some of the structures and terminology used in discussing biochemical macromolecules. Tests are not extensive enough to allow all types of macromolecules to be distinguished but major types are all represented. Testing for unknowns is done with commercial products rather than specific purified molecules in order to make the laboratory a more familiar experience rather than a "chemistry" laboratory. The principles of chromatography are introduced in an optional exercise on paper chromatography of amino acids.

II. TIME REQUIREMENTS

Exercise A—Testing for Carbohydrates
- Part 1—Benedict's Test for Reducing Sugars (20 minutes)
- Part 2—Lugol's Test for Starch (20 minutes)

Exercise B—Testing for Lipids (30 minutes)

Exercise C—Testing for Proteins and Amino Acids
- Part 1—Testing for Protein with Biuret Reagent (20 minutes)
- Part 2—Testing for Amino Acids with Ninhydrin (20 minutes)

Exercise D—Chromatography of Amino Acids (overnight, 60 minutes)

Exercise E—Analyzing Unknowns Qualitatively (60 minutes)

III. STUDENT MATERIAL AND EQUIPMENT

	Per Student	Per Pair (2)	Per Group (4)	Per Class (24)
Exercise A (Parts 1 and 2)				
test tube	16			384
test tube rack	1			24
wax pencil	1			24
distilled water, bottle			1	6
2% potato starch, bottle **(1)**			1	6
6% glucose, bottle **(2)**			1	6
6% maltose, bottle **(3)**			1	6
6% sucrose, bottle **(4)**			1	6
onion juice, bottle **(5)**			1	6
milk, whole, bottle			1	6
potato				1
pipette (5 ml)			7	42
Pi-Pump or Propipette	1			24
Benedict's reagent, dropping bottle **(6)**		1		12
hot plate or Bunsen burner		1		12
boiling chips, jar				1
600 ml beaker		1		12

	Per Student	Per Pair (2)	Per Group (4)	Per Class (24)
Exercise A (Parts 1 and 2)—continued				
test tube holder	1			24
Lugol's solution, dropping bottle **(7)**		1		12
Exercise B				
3 inch square brown wrapping paper, unglazed	1			24
distilled water, dropping bottle			1	6
vegetable oil, dropping bottle			1	6
onion juice, dropping bottle **(5)**			1	6
hamburger juice, dropping bottle **(8)**			1	6
cola (or Sprite), dropping bottle			1	6
Sudan IV, dropping bottle **(9)**			1	6
test tube	5			120
test tube rack (same as Exercise A)				
wax pencil (same as Exercise A)				
Exercise C (Parts 1 and 2)				
distilled water, dropping bottle (same as Exercise A)				
6% egg albumin, bottle **(10)**			1	6
2% potato starch, bottle **(1)**			1	6
6% glucose, bottle **(2)**			1	6
amino acid solution, bottle **(11)**			1	6
test tubes	5			120
test tube rack (same as Exercise A)	1			24
wax pencil (same as exercise A)				
biuret reagent, dropping bottle **(12)**			1	6
filter paper (Whatman #1, 15 cm)	1			24
solution A—proline, dropping bottle **(13)**			1	6
solution B—distilled water, dropping bottle **(13)**			1	6
solution C—amino acid, dropping bottle **(13)**			1	6
solution D—amino acid, dropping bottle **(13)**			1	6
ninhydrin, dropping bottle **(14)**			1	6
hot plate or Bunsen burner (same as Exercise A)				
Exercise D				
paper towels		1		12
Whatman chromatography paper (8.5 × 11 inches) **(15)**		1		12
pencil		1		12
unknown amino acid mixture **(16)**		2		24
wooden applicator sticks		6		72
known amino acid solutions **(17)**		4		48
plastic paper clips		2		24
gallon glass jar		1		12
chromatography solvent, bottle (1200 ml) **(18)**				1
ninhydrin solution, non-aerosol bottle **(19)**		1		12
hair dryer		1		12

	Per Student	Per Pair (2)	Per Group (4)	Per Class (24)
Exercise E				
10 unknown solutions **(20)**				10

IV. PREPARATION OF MATERIALS AND SOLUTIONS

(1) 2% potato starch

Bring 900 ml of distilled water to a boil. Mix 20 g of potato starch in 100 ml of distilled water and slowly pour this mixture into the 900 ml of boiling water. Heat the entire solution to boiling, mix and remove from heat—allow it to cool (cover with aluminum foil while cooling). Some starch will settle out of solution.

(2) 6% glucose

Weigh 60 g glucose (dextrose). Add distilled water to 1000 ml.

(3) 6% maltose

Weigh 60 g maltose. Add distilled water to 1000 ml.

(4) 6% sucrose

Weigh 60 g sucrose. Add distilled water to 1000 ml.

(5) onion juice

Place a peeled, diced onion into a blender. Cover it with distilled water (100-200 ml, depending on the size of the onion) and blend. Strain through cheesecloth.

(6) Benedict's reagent

Order prepared solution (powdered or liquid form can be ordered).

(7) Lugol's solution (I_2KI)

Order prepared solution OR

Dissolve 10 g of potassium iodide in 100 ml of distilled water and add 5 g of iodine. Store in a dark bottle. Prepare 200 ml per class of 24 students.

(8) hamburger juice

Boil hamburger (1/4 of a 250 ml beaker full) with 150 ml water for 5 minutes. Allow hamburger to settle to the bottom of the beaker and use the liquid.

(9) Sudan IV

Weigh 0.2 g Sudan IV. Make up to 100 ml with 95% ethanol.

(10) 6% egg albumin

Weigh 60 g egg albumin. Add distilled water to make 1000 ml.

(11) amino acid solution

Weigh 0.8 g of amino acid (choose any on hand). Add distilled water to 100 ml. If amino acid does not go into solution, adjust pH with 0.1 N NaOH and heat slightly.

(12) biuret reagent

Order prepared solution. Approximately 500 ml per class of 24.

(13) amino acid solutions (A, B. C, and D)

Weigh 0.8 g proline (or other amino acid). Add distilled water to 100 ml. Suggestion: Bottle A—proline; Bottle B—distilled water, Bottle C—methionine; and Bottle D—alanine. If amino acid does not go into solution easily, adjust pH with 0.1 N NaOH and heat slightly.

(14) ninhydrin

Weigh 0.2 g ninhydrin. Dissolve in 100 ml of 95% ethanol. **AVOID BREATHING POISONOUS FUMES!**

(15) Whatman chromatography paper

Using a paper cutter, cut large sheets of chromatography paper into 8.5 × 11" pieces. Do not handle the paper, grease from fingers will interfere with the chromatography process. Wear gloves to facilitate cutting.

(16) and (17) Unknown and known amino acid mixtures

Unknown solutions contain three amino acids. It is best to use one polar (acidic or basic), one non-polar, and one polar uncharged amino acid if possible, but any combination of three amino acids can be used. Suggested solutions include:

Unknowns

(a) proline, cystine, lysine
(b) phenylalanine, proline, lysine
(c) proline, aspartic acid, methionine
(d) leucine, aspartic acid, methionine

Knowns

For unknown solutions a and b, use proline, lysine, cystine, and phenylalanine as knowns.
For unknown solutions c and d, use proline, aspartic acid, leucine, and methionine as knowns.

To prepare unknowns, add 0.05 g of each amino acid to 25 ml of distilled water. Adjust the pH to 5. If you choose an amino acid that is difficult to get into solution, you may have to adjust the pH up or down with 0.1 N NaOH or HC1. (Some amino acids will go into solution more easily if dissolved in a few drops of 95% ethanol before adding them to water.)

To prepare known solutions, add 0.05 g of a single amino acid to 25 ml of distilled water and adjust the pH to 5.

(18) chromatography solvent

Mix 50 ml butanol, 50 ml glacial acetic acid, and 50 ml of distilled water.

(19) ninhydrin solution for spraying chromatograms (0.2% ninhydrin in 95% ethanol)

Add 2 g of ninhydrin to 1000 ml of 95% ethanol. Store in a dark bottle. Dispense the solution into a non-aerosol bottle with a fine spray so that chromatograms can be covered with a fine, dispersed spray and not large spots of ninhydrin solution (use old hair spray or glass cleaner bottles or plant misters).

(20) unknowns

Use the following materials as unknowns: 1) proline (0.8 g/100 ml distilled water); 2) onion (blend with water and filter); 3) potato buds (blend with water and filter); 4) Karo syrup (dilute 1:2 with distilled water); 5) milk (whole milk diluted 1:1 with water); 6) table sugar (6 g sucrose/100 ml distilled water); 7) egg white (blend lightly with water); 8) vegetable oil; 9) hamburger (see **(8)** above); 10) amino acid mix (0.4 g (each)/100 ml distilled water—use two amino acids except proline—see **(13)** above).

V. PREPARATION SUGGESTIONS

Exercise A

It is most convenient to prepare solutions for groups of 4-8 students (depending on the seating arrangement in your laboratory). Materials are suggested for groups of 4, but larger groups are possible. Use 500 ml or liter bottles. Tape a test tube to the side of the bottle (use two pieces of tape—top and bottom of tube). Place a 5 ml pipette into the tube. Label the pipette to match the label on the bottle. Use a Pi-Pump or Propipette to dispense the solution from the bottle. Each student should have a Pi-Pump or Propipette for his or her own use. This helps to prevent mixing of pipettes and solutions and cuts down on the amount of dirty glassware generated by the laboratory.

As an alternative, solutions can be placed in dropping bottles and students can use a dropper full of solution to approximate the 2 ml of liquid required.

A bottle of boiling chips should be available in the laboratory. The 500 ml beaker is to be used as a boiling water bath—add 5-6 boiling chips. A bucket of soapy water and test tube brush should be provided for washing dirty test tubes. Also provide some paper towels to use for removing wax pencil marks. This will help with clean up.

Exercise B

Sudan IV is fairly messy and will get into everything in the laboratory if students fail to wash out test tubes in hot soapy water. When several drops of Sudan IV are added to the test tubes, a red droplet will appear if fat is present. Otherwise the red disperses throughout the solution. This can be done as a demonstration and tubes can be placed in a test tube rack on the demonstration table if necessary. Sudan IV can be purchased in powder or liquid form.

Exercise C

Do not use aerosol spray cans of ninhydrin since FUMES ARE POISONOUS. Store ninhydrin solutions in dark bottles.

Exercise D

Cut 8.5 × 11 inch pieces of chromatography paper from large sheets using a paper cutter. Wear rubber or cloth gloves while doing this to avoid getting grease on the paper. Have students wash their hands before picking up the chromatography paper and have them place the paper on a piece of paper toweling—not on the laboratory bench where wax or other chemicals can get onto the paper. Make sure that students put their name on the upper right hand corner of the paper. This should be done with pencil and all other marks should also be made with a pencil (ink will chromatograph and form a beautiful chromatogram all on its own!).

Be sure to warn students not to punch into the paper with the wooden applicator sticks. Small dots with many applications are best (allow each to dry before adding the next). Ask students to bring their own hair dryers to laboratory—this way, all students will have a dryer and the exercise can be completed in a much shorter period of time. Plastic paper clips are used to join the ends of the chromatography paper to form a cylinder—do not use metal paper clips because they will rust and chromatograph in the solvent.

It is best if you add the chromatography solvent to the chromatography jars before the laboratory so that the air in the jar has a chance to become saturated. Jars can be large glass mustard or pickle jars from local cafeterias.

Have all students start chromatograms at the same time. The instructor should check after four hours to see how far the solvent front has traveled. When the solvent front is within 1 inch of the top of the chromatography paper, remove the chromatograms from the jars, one at a time. Unclip and mark the solvent front on each paper (use a pencil). If you have a hood in a nearby room, hang the papers on string lines using spring-type wooden clothespins.

Otherwise, place chromatograms in a well-ventilated empty room near an open widow until the solvent smell disappears. The chromatograms can be saved until the following week or one student from each student pair can be assigned to pick up the chromatogram.

The student who picks up the chromatogram should also be responsible for treating it with ninhydrin. To do this, use a non-aerosol spray bottle (an old hair spray or liquid cleaner bottle) with a fine spray. Tack the chromatogram to a large, heavy piece of cardboard (the side of a cardboard carton is excellent) with push pins or tape. Spray the entire chromatogram evenly—this is best done in a hood (or outside on the grass). Warn students NOT to inhale fumes. Everything (including floors, hood walls, clothes, hands) that contains amino acids will stain with ninhydrin, so also warn students to be careful where they aim the spray bottle.

Allow the chromatogram to air dry or speed up the process with a blow dryer. To develop the chromatogram, you must heat it. This can be done by placing it in a drying oven for 3-5 minutes, or by holding it approximately one foot above a hot plate while moving it back and forth until colored spots appear or by using a blow-dryer on the hot setting. Since the spots will disappear, have students draw around each spot with a pencil and record the color (yellow, pink-brown, purple, purple-red, etc.).

Student partners should get together to determine R_f values for all amino acids and the identity of unknowns. Chromatograms and student reports can be turned in the following week.

Exercise E

Unknowns can be from commercial sources (as described in Preparation of Materials and Solutions (**20**)) or unknowns can be the same materials as those used during the class period (i.e., vegetable oil, glucose, starch, albumin, amino acid solution, distilled water, and proline solution). Using unknowns from commercial sources is more fun for the students.

Ten or fewer unknowns can be made available. Students can try as many as they can or you can tell them how many to use depending on the length of the class period. A list of possible unknowns should be given to students to develop hypotheses. Note that the first listing is just a list of all the possible unknowns. It might lessen confusion if students just scratch through the existing numbers. You should number the unknowns so that only the instructor knows what is in tube 1 or tube 2, etc. All that students know is the total possibilities, but they do not know what is in the tube marked 1 or the tube marked 2, etc. After unknowns are tested, students can determine the contents. If students work in groups, they can do several tests on each unknown rapidly. Alternatively, have each student choose three numbered tubes from a rack of tubes (you have a master list of 1-10) and do all tests necessary to determine the contents of the three tubes. Once determined, develop a class list to see if all students agree about the contents of each tube.

If unknowns are not dispensed in previously numbered test tubes, then one set of unknowns in 500 ml bottles prepared as in Exercise A (with labeled pipettes in test tubes attached to the bottles) will be sufficient for the entire class. A Pi-Pump should be provided with each bottle if individual students have not been given one to use in earlier exercises.

VI. ORDERING INFORMATION

test tubes, 16 × 125 mm—Fisher, # 14-925H; Carolina Biological, # 73-1418; Ward's, # 17W1403
test tube rack—Carolina Biological, # 73-1896; Ward's, # 18W4213

wax pencil—Carolina Biological, # 65-7730 (red), # 65-7732 (blue), # 65-7734 (black); Ward's, # 15W1159

soluble starch, potato—Fisher, # S516-500 (500 g); Carolina Biological, # 89-2530 (125 g); Ward's, # 39W3275

D-glucose (dextrose) (certified)—Fisher, # D16-500 (500 g); Carolina Biological, # 85-7440 (500 g); Ward's, # 39W1457

maltose (certified)—Fisher, # M75-100 (100 g); Carolina Biological, # 87-3750 (100 g); Ward's, # 39W2285

sucrose (certified)—Fisher, # S5-500 (500 g); Carolina Biological, # 89-2870 (500 g); Ward's, # 39W3280

pipette, 5 ml (to deliver)—Fisher, # 13-665K; Carolina Biological, # 59-6276; Ward's, # 17W1307

Pi-Pump—Fisher, # 13-683C (up to 10 ml); Carolina Biological, # 21-4684 or # 73-6875; Ward's, # 15W0532

Propipette (safety pipette filler)—Fisher, # 12-681-50; Carolina Biological, # 73-6868; Ward's, # 15W0510

Bendict's reagent—Fisher, # LC11650-1 (solution); Carolina Biological, # 84-7091 (powder, sufficient for 1000 ml), # 84-7111 (solution, 500 ml); Ward's, # 37W0700 (powder), # 37W0698 (solution)

hot plate—Carolina Biological, # 70-1010; Ward's, 15W8010

Bunsen burner—Carolina Biological, # 70-6645; Ward's, 15W0604

boiling chips—Carolina Biological, # 84-8278 (100 g) or # 84-8280 (500 g); Ward's, 37W0916

beaker, 600 ml— Fisher, # 02-540M (Pyrex) or # 02-539M (Kimax); Carolina Biological, # 72-1225 (Corning, student grade); Ward's, # 17W4060

test tube holder—Carolina Biological, # 70-2092 or # 70-2900; Ward's, # 15W0841

Lugol's solution—Carolina Biological, # 87-2793 (100 ml) or # 87-2795 (500 ml); Ward's, # 39W1685

dropping bottle, 30 ml— Fisher, # 02-980; Carolina Biological, # 71-6525; Ward's, # 17W6011

biuret reagent—Carolina Biological, # 84-8211 (120 ml); Ward's, # 37W 0790

Whatman filter paper, #1 (15 cm)—Fisher, # 09-805G or Carolina Biological (#201), #71-2712; Ward's, # 15W2811

ninhydrin (powder)—Carolina Biological, # 87-7460 (5 g); Ward's, # 37W3389

amino acids:

L-alanine—Fisher, # BP369-100 (100 g); Eastman Organic Chemicals, # 3125 (25 g); Carolina Biological, # 84-2170 (25 g); Ward's, # 39W4012

L-proline—Fisher, # BP392-100 (100 g); Eastman Organic Chemicals, # 2488 (25 g)

L-methionine—Fisher, # BP388-100 (100 g); Eastman Organic Chemicals, # 5279 (25 g); Carolina Biological, # 87-5030 (200 g)

Whatman chromatography paper, #l—Fisher, #05-714 (package of 100); Carolina Biological, # 21-5610; Ward's, # 15W3708

wooden applicator sticks—Fisher, # 01-340; Carolina Biological, # 70-6865; Ward's, # 14W0105

butanol—Fisher, # A399-1; Carolina Biological, # 84-9792; Ward's, # 39W0462

acetic acid, glacial—Fisher, # A38-212; Carolina Biological, # 841293; Ward's, # 39W0125 albumin—Fisher, # A388-500; Sigma, # A 4378; Carolina Biological, # 84-2252; Ward's, # 39W0198

Sudan IV—Fisher, # S667-25 (25 g); Carolina Biological, # 89-2980 (powder, 25 g) or # 89-2993 (liquid, 120 ml); Ward's, # 38W8789

DL-amino acid kit—Sigma, # DLAA (includes 24 amino acids for greater selection—if specific amino acids are ordered, use L amino acids); Ward's, 38W6800

LABORATORY 6 Prokaryotic Cells

I. FOREWORD

This laboratory is designed as an introduction to prokaryotic cell structure. Students will learn about coacervates and how to recognize bacteria and cyanobacteria. They will learn to classify bacteria as gram negative or gram positive. Students will also have an opportunity to learn some basic techniques in microbiology. Most exercises are fairly short and can be selected for combination with Laboratory 7, Eukaryotic Cells, if desired.

II. TIME REQUIREMENTS

Exercise A—Producing Protobionts (45 minutes)
Exercise B—Examining Bacterial Cells
 Part 1—Observing Bacteria Using the Light Microscope (10 minutes)
 Part 2—Observing Bacteria Using the Transmission Electron Microscope (TEM) (10 minutes)
 Part 3—Using Gram Staining to Study Bacterial Cell Walls (30 minutes)
 Part 4—Examining Cyanobacteria (20 minutes)
Exercise C—Working with Bacteria
 Part 1—Techniques for Transferring Cultures (30 minutes)
 Part 2—Isolating Pure Cultures (20 minutes)

III. STUDENT MATERIALS AND EQUIPMENT

	Per Student	Per Pair (2)	Per Group (4)	Per Class (24)
Exercise A				
1% gelatin (**1**)			1	6
3% gum arabic (**2**)			1	6
5 ml pipette			2	12
Parafilm (pieces)	3			72
test tubes	1			24
glass stirring rod	1			24
pH paper			1	6
microscope slides, box (75 × 25 mm)			1	6
coverslips, box (22 × 22 mm)			1	6
Pasteur pipette	2			48
rubber bulb	1			24
1% HCl, dropping bottle (**3**)			1	6
stain (choice), dropping bottle (**4**)			1	6
5% NaCl, dropping bottle (**5**)			1	6
compound microscope	1			24
Exercise B (Part 1)				
Escherichia coli (prepared slide)				1
Staphylococcus aureus (prepared slide)				1

	Per Student	Per Pair (2)	Per Group (4)	Per Class (24)
Exercise B (Part 1)—continued				
Spirillum volutans (prepared slide)				1
bacterial flagella (prepared slide)				1
compound microscope				3
Exercise B (Part 2)				
laboratory manual				
Exercise B (Part 3)				
microscope slides, box (75 × 25 mm)		1		12
coverslips, box (22 × 22 mm)				
indelible pen		1		12
inoculating loop		1		12
Escherichia coli agar slant culture **(6)**			1	6
Staphylococcus aureus slant culture **(6)**			1	6
distilled water, dropper bottle		1		12
Bunsen burner (or alcohol lamp)		1		12
Petri dish, 60 × 15 mm		1		12
Petri dish, 150 × 15 mm		1		12
Gram's crystal violet stain, bottle **(7)**			1	6
Gram's iodine solution, bottle **(8)**			1	6
Gram's alcohol, bottle **(9)**			1	6
safranin stain, bottle **(10)**			1	6
immersion oil, bottle			1	6
Exercise B (Part 4)				
Oscillatoria (live, class of 30)				1
Nostoc (live, class of 30)				1
Cylindrospermum (live, class of 30)				1
microscope slides, box (75 × 25 mm)			1	6
coverslips, box (22 × 22 mm)			1	6
compound microscope	1			24
Exercise C (Part 1)				
nutrient broth culture tubes **(11)**	1			24
nutrient agar slants **(11)**	1			24
marking pencil	1			24
inoculating loop	1			24
Bunsen burner	1			24
Escherichia coli, nutrient agar slant **(12)**	1			24
Serratia marcescens, nutrient agar slant **(12)**	1			24
E. coli, 24 hour broth culture **(13)**	1			24
S. marcescens, 24 hour broth culture **(13)**	1			24
Exercise C (Part 2)				
E. coli, nutrient agar slant **(12)**	1			24
Micrococcus luteus, nutrient agar slant **(14)**	1			24

	Per Student	Per Pair (2)	Per Group (4)	Per Class (24)
Exercise C (Part 2)—continued				
E. coli/S. marcescens, mixed broth culture **(15)**	1			24
E. coli/M. luteus, mixed broth culture **(16)**	1			24
nutrient agar plate **(17)**	2			48
glass spreading rod	1			24
inoculating loop	1			24
95% ethanol (beaker)	1			24
Bunsen burner	1			24
plaque study plates set (optional) **(18)**				1

IV. PREPARATION OF MATERIALS AND SOLUTIONS

(1) 1% gelatin
Dissolve 10 g gelatin powder in enough distilled water to make 1 liter of solution. Adjust amounts for smaller volumes. Adjust pH to 7 using 0.1 N HaOH or 0.1 N HCl.

(2) 3% gum arabic
Dissolve 30 g of gum arabic powder in enough distilled water to make 1 liter of solution. Adjust amounts for smaller volumes. Adjust pH to 7 using 0.1 N NaOH or 0.1 N HCl.

(3) 1% HCl
Add 1 ml concentrated HCl to 99 ml of distilled water. Wear safety glasses. Remember to always add acid to water.

(4) stain for coacervates
Methylene blue, neutral red, or congo red may be used to stain coacervates:
methylene blue—0.01 g methylene blue stain per 100 ml absolute alcohol
neutral red—0.03 g neutral red stain per 100 ml absolute alcohol
congo red—0.1 g congo red stain per 100 ml absolute alcohol

(5) 5% NaCl
Dissolve 5 g NaCl in enough distilled water to make 100 ml solution. Mix and dispense into dropping bottles.

(6) *Escherichia coli* or *Staphylococcus aureus* agar slant culture
These can be purchased from Carolina Biological Supply Company (# 15-5065 and # 15-555A). *Staphylococcus aureus* is a pathogen and must be handled carefully. As an alternative or for high school use order *S. epidermidis* (# 15-5556).

(7) Gram's crystal violet stain

Solution A

crystal violet (90% dye content)	2.0 g
ethyl alcohol	20 ml

Solution B

ammonium oxalate	0.8 g
distilled water	80 ml

Mix solutions A and B.

(8) Gram's iodine solution

iodine	1.0 g

KI 2.0 g
distilled water 300 ml

Add iodine after KI is dissolved in water.

(9) Gram's alcohol

95% ethyl alcohol 98 ml
acetone 2 ml

(10) safranin stain

safranin O (2.5% solution in 95% ethyl alcohol) 10 ml
distilled water 2 ml

Note: items **(7-10)** can be ordered already prepared—see Ordering Information.

(11) nutrient broth culture tubes and nutrient agar slants

Prepare nutrient broth (see Ordering Information) as follows:

Suspend 0.8 g of dehydrated nutrient broth in 100 ml of distilled water in a 250 ml flask. Prepare a total of 4 flasks. Add 1.5 g of (Bacto)agar to 2 of the 4 flasks and mix well. Cover the flasks with aluminum foil and autoclave at 15 lbs. pressure and 121° C for 15 minutes.

Aseptically dispense 5 ml of the nutrient agar into sterile culture tubes to make approximately 40 broth tubes.

Alternatively, you may purchase materials already prepared from Carolina Biological Supply Company. Tubes of nutrient agar (# 82-6100) can be melted and slanted. Nutrient broth (# 82-6120) is available in bottles at a reduced cost but must be dispensed into sterile culture tubes. (See Ordering Information.)

a. nutrient broth

Add 5.0 g peptone and 3.0 g beef extract to 1000 ml distilled water and adjust to pH 7. Autoclave at 15 lb. pressure for 15 minutes.

b. nutrient agar

Prepare nutrient broth as above, but add 15 g (Bacto)agar to the flask and mix before autoclaving.

(12) *Escherichia coli,* nutrient agar slant, or *Serratia marcescens,* nutrient agar slant

These can be purchased as agar slant cultures from Carolina Biological Supply Company (# 15-5065 and #15-5452, respectively). To prepare additional slants for students, inoculate fresh slants prepared as in **(11)** and incubate for 24 hours at 37° C.

(13) *Escherichia coli* or *Serratia marcescens,* 24 hour broth cultures

Aspetically inoculate 100 ml of nutrient broth, prepared as in **(11)** with a generous loop full of the appropriate bacteria. Incubate for 24 hours at 37° C in an incubator or shaking water bath. *Micrococcus luteus* cultures—see **(14)** below—must be started a day ahead of either *e. coli* or *S. marcescens* cultures.

(14) *Micrococcus luteus,* nutrient agar slant

This can be purchased as a nutrient agar slant culture from Carolina Biological Supply Company (# 15-5155). To prepare additional slants for students, inoculate fresh slants prepared as in **(11)** and incubate for 48 hours at 37° C. (Note: *M. luteus* slants and broth cultures require an additional 24 hours of incubation and must be started one day ahead of *E. coli* or *S. marcescens* cultures.)

(15) *Escherichia coli/Serratia marcescens* mixed broth culture

Prepare a single 24 hour broth culture of *E. coli* and *S. marcescens* as in **(13)**. Aseptically mix 2.0 ml of *E. coli* with 1.0 ml of *S. marcescens* to prepare tubes for use by students.

(16) *Escherichia coli/Micrococcus luteus* mixed broth culture

Prepare a 24 hour broth culture of *E. coli* as in **(13)**. Prepare a 24 hour broth culture of *M. luteus* by inoculating 100 ml of nutrient broth with a generous loop full of the bacteria from the agar slant. (This is the same as the preparation of broth cultures of *E. coli* and *S. marcescens* in **(13)**. The *M. luteus* culture must, however, be started a day ahead of other cultures; i.e., 2 days before the laboratory period.)

Prepare the mixed culture by aseptically mixing 2.0 ml of *M. luteus* with 1.0 ml of *E. coli* in tubes for student use.

(17) nutrient agar plates

Dissolve 23 g of dehydrated nutrient agar medium in 1000 ml of distilled water in a 2 liter flask. Cover with foil or a cloth stopper and autoclave at 15 lbs. pressure for 15 minutes. Cool to 50° C and pour into sterile Petri dishes (yields 40 plates). Plates should be poured at least 2 days ahead to allow moisture to evaporate. Store the plates upside down (inverted) so that condensation on the lid will not drip onto the agar surface.

(18) agar plate demonstration of viral plaques

Carolina Biological Supply Company offers a plaque "study plates" set (# 12-1180) with four plates demonstrating T even bacteriophages and two plates demonstrating φX174. (See Ordering Information.)

Alternatively, plates can be prepared as follows:

1. Prepare a 24 hour nutrient broth culture of *E. coli* B and a 24 hour broth culture of T_2 bacteriophage (coli phage)

   tryptone broth

tryptone	10.0 g
calcium chloride ($CaCl_2$)	0.02 g
sodium chloride (NaCl)	5.0 g

   add distilled water to make 1000 ml—autoclave

2. Prepare tubes of tryptone soft agar as follows:

tryptone	10.0 g
potassium chloride (KCl)	5.0 g
agar	9.0 g

   add distilled water to make 1000 ml—autoclave

3. Prepare Petri plates of tryptone hard agar as follows:

tryptone	10.0 g
calcium chloride ($CaCl_2$)	0.2 g
sodium chloride (NaCl)	5.0 g
agar	11.0 g

   add distilled water to make 1000 ml—autoclave

4. Prepare 10 tubes, each containing 9 ml of sterile tryptone broth. Starting with 10 ml of the 24 hour T_2 bacteriophage culture, make serial dilutions: remove 1 ml from the 10 ml culture and place it in the 9 ml of tryptone broth (10^{-1} dilution) and transfer it to the next tube of 9 ml (10^{-2} dilution); and so on, until dilutions of 10^{-5}, 10^{-6}, 10^{-7}, 10^{-8}, and 10^{-9} have been made.
5. Melt the soft tryptone agar in tubes by placing them in boiling water and cool them to 45° C. To each of 5 tubes, aseptically add 3 drops of the 24 hour *E. coli* B culture and 1 ml of one of the bacteriophage dilutions so that you will have tubes containing 10^{-5}, 10^{-6}, 10^{-7}, 10^{-8}, and 10^{-9} dilutions.

6. Mix and pour each over plates of tryptone agar to form two-layered plates. Incubate plates in an inverted position for 24 hours at 37° C.

V. PREPARATION SUGGESTIONS

Exercise A

The most important aspect of successful coacervate production is a careful adjustment of pH. Students must add only 2 drops of HCl at a time. Be sure to shake the tubes—often the cloudiness disappears. You should try this before your class gets started. Often the pH of the water you start with can exceed the acid requirements. For this reason, adjusting the gum arabic and gelatin solutions to pH 7 is suggested. Adjusting the light for proper viewing with the microscope is also important. Students who have never learned to properly focus the microscope may experience difficulty and will be looking at air bubbles or dirt on the coverslip.

Although the coacervates can be seen without staining, methylene blue, congo red, and neutral red can be used. If you prepare all three stains, different students can compare the results from different stains. These three stains are all "vital" stains and can also be used to stain living material such as protozoans. Apply a drop of stain to a clean glass slide and let the stain dry. Store the slides in a box and, when you want to stain live materials, simply add a drop of your culture (e.g., pond water) to the slide. The stain will dissolve slowly and color the organisms. You may wish to try the same technique with coacervates if students experience difficulty with diluting the material by adding stain.

Exercise B (Parts 1-2)

Slides of bacteria are best observed at 100× with oil. For this reason, demonstration slides, set up ahead of time, are used.

Exercise B (Part 3)

Be careful using Gram stain. Tell students to wear old clothes. Use gallon milk bottles with the top cut off to collect used up stain. Since students must observe the slide using oil immersion, reserve one of the demonstration microscopes for that use. Be sure bacterial smears are dry before heat fixing or the bacteria will "sputter" and spreas everywhere. (Make sure that students do not "cook" the bacteria while heat fixing!) Use *Staphylococcus epidermidis* if obtaining pathogens such as *S. aureus* is not allowed. Make sure that students wash hands after completing this exercise—this is especially important if you use *S. aureus.*

Exercise B (Part 4)

Living *Oscillatoria, Nostoc,* or *Cylindrospermum* are excellent for use in this exercise.

Exercise C

Be sure to order agar slants of bacteria far enough ahead to prepare broth cultures. You may decide to have students simply do a transfer between slants and not do broth cultures if students cannot come back to the laboratory in one to two days to observe cultures.

It is important for students to learn how to make a streak plate. This technique will be used again in Laboratory 18.

All Exercises

Have available a spray bottle of 10% Clorox and have all students spray the surfaces of the laboratory bench before and after their work with bacteria.

VI. ORDERING INFORMATION

gelatin—Fisher, # G8-500; Carolina Biological, # 86-4670; Ward's, # 38W3030
gum arabic (acacia)—Fisher, # G85-500; Carolina Biological, # 86-6108; Ward's, # 37W2190
pipette, 5 ml—Fisher, # 13-665K; Carolina Biological, # 73-1890; Ward's, # 17W1307
Parafilm—Fisher, # 13-274-12 (dispenser, # 13-374-18); Carolina Biological, # 21-5600 (dispenser, # 21-5602); Ward's, # 15W1940
test tubes, 16 × 125 mm—Carolina Biological, # 73-1418; Ward's, # 17W1403
test tube rack, Carolina Biological, # 73-1890; Ward's, # 18W4213
glass stirring rod—Carolina Biological, # 71-1304 (or cut glass rod into desired lengths and fire-polish); Ward's, # 17W6005
pH paper (range 4.5-10)—Carolina Biological, # 89-3942; Ward's, # 15W2558
microscope slides, 75 × 25 mm—Fisher, # 12-550A; Carolina Biological, # 63-2000; Ward's, # 14W3500
coverslips, 22 × 22 mm—Fisher, # 12-542B; Carolina Biological, # 63-3095; Ward's, # 14W3251
hydrochloric acid (HCl)—Fisher, # A144-212; Carolina Biological, # 86-7792; Ward's, # 37W2251
sodium chloride (NaCl)—Fisher, # S271-3; Carolina Biological, # 88-8903; Ward's, # 37W5488
dropping bottle, 30 ml— Fisher, # 02-980; Carolina Biological, # 71-6525; Ward's, # 17W6011
Pasteur pipette, 9"—Fisher, #13-678-20C; Carolina Biological, # 73-6062; Ward's, # 17W1146
rubber bulb—Fisher, # 14-065B
methylene blue stain—Fisher, # M291-25 (powder) or # LC16920-1 (1% solution, ethanol;dilute to 0.1%); Carolina Biological, # 87-5684 (powder) or # 87-5913 (120 ml); Edvotek, # 609; Ward's, # 8W8323
neutral red stain—Fisher, # N129-25 (powder); Carolina Biological, # 87-6830; Ward's, # 38W8385
congo red stain—Carolina Biological, # 85-5348; Fisher, # C580-25; Ward's, # 38W7004
Escherichia coli (prepared slide)—Carolina Biological, # Ba90; Ward's, # 90W2042
Staphylococcus aureus (prepared slide)—Carolina Biological, # Ba235; Ward's, # 90W2080
Spirillum, gram negative (prepared slide)—Carolina Biological, # Ba028; Ward's, # 90W0557
bacterial flagella (prepared slide)—Carolina Biological, # Ba017; Ward's, # 90W7574
microscope slides, 75 × 25 mm—Fisher, # 12-550A; Carolina Biological, # 63-2000
coverslips, 22 × 22 mm—Fisher, # 12-542B; Carolina Biological, # 63-3095
Escherichia coli, nutrient agar slant—Carolina Biological, #15-5065; Ward's, # 85W0400
Staphylococcus aureus—Carolina Biological, # 15-5554A (freeze-dried plus medium); Ward's, # 85W1178
OR *S. epidermidis,* nutrient agar slant—Carolina Biological, # 15-5556; Ward's, # 85W1035
Petri dish, 60 × 15 mm—Carolina Biological, # 74-1346; Ward's, # 18W7099
Petri dishes, sterile disposable, 100 × 15 mm—Fisher, # 08-757-12; Carolina Biological, #74-1350; Ward's, # 17W0700
crystal violet (powder)—Carolina Biological, # 85-6150; Ward's, # 38W7433
ammonium oxalate—Carolina Biological, # 84-4170; Ward's, # 37W0470

OR Gram's crystal violet stain—Carolina Biological, # 86-7483; Ward's, # 85W3983
iodine—Carolina Biological, # 86-8980; Ward's, # 37W2371
potassium iodide (KI)—Carolina Biological, # 88-3808; Ward's, # 37W4771
OR Gram's iodine—Carolina Biological, # 86-5863; Ward's, # 38W9989
acetone—Carolina Biological, # 84-1500; Ward's, # 39W0146
safranin O (powder)—Carolina Biological, # 88-7039; Ward's, # 38W7013
OR Gram's safranin counterstain—Carolina Biological, # 88-6983; Ward's, # 38W7018
bibulous paper—Carolina Biological, # 63-4050; Ward's, # 15#W2500
immersion oil—Ward's, # 14W3326
Oscillatoria (live)—Carolina Biological, # 15-1865; Ward's, # 86W2150
Nostoc (live)—Carolina Biological, # 15-1847; Ward's, # 86W2150
Cylindrospermum (live)—Carolina Biological, # 15-1755; Ward's, # 86W1930
nutrient agar (power form)—Fisher, # B11-472; Carolina Biological, # 78-5300; Ward's, # 38W1200
nutrient broth (power form)—Fisher # B11-479; Carolina Biological, # 78-5360; Ward's, # 38W1230
(Bacto)agar[1]—Fisher, # DF0140-02-9; Carolina Biological, # 21-6720 or 21-6721 (500 g); Ward's, # 38W0015
nutrient agar (tubes)—Carolina Biological, # 77-6360 (or #82-6100, 125 ml bottle); Ward's, # 88W1500
nutrient broth (tubes)—Carolina Biological, # 77-6380 (or # 82-6120, 125 ml bottle); Ward's, # 88W1510
peptone—Fisher, # BP1420-500; Carolina Biological, #79-4260; Ward's, # 39W2751
beef extract—Carolina Biological, # 79-4720; Ward's, # 38W2123
wax pencil—Carolina Biological, # 65-7730 (red), # 65-7732 (blue), # 65-7734 (black); Ward's, # 15W0957
inoculating loop—Carolina Biological, # 70-3060; Ward's, # 14W0957
Bunsen burner—Carolina Biological, # 70-6645; Ward's, # 15W0604
OR alcohol lamp
Serratia marcescens, nutrient agar slant—Carolina Biological, # 15-5452; Ward's, # 85W0400
Micrococcus luteus, nutrient agar slant—Carolina Biological, # 15-5155; Ward's, # 85W0966
tryptone—Fisher, # BP1421-500; Carolina Biological, # 21-6740; Ward's, # 39W3600
calcium chloride ($CaCl_2$)—Fisher, # C79-500; Carolina Biological, # 85-1800; Ward's, # 37W0936
potassium chloride (KCl)—Fisher, # P217-500; Carolina Biological, # 88.2900; Ward's, # 37W4634

[1] Bacto is a designation used on Difco Laboratories products available through Fisher Scientific, Inc. and Carolina Biological Supply Company. Difco Laboratories, PO Box 331058, Detroit, MI 48232-7058; FAX 313-462-8517, Voice, 800-521-0851. Difco maintains a Web page at http://difco.com (consult Reference Guide under Technical Support for information on agar).

LABORATORY 7 Eukaryotic Cells

1. FOREWORD

This laboratory is designed as an introduction to eukaryotic cell structure, Students compare animal cells with plant cells. The use of the microscope for studying both cellular chemistry and cell ultrastructure is emphasized. Cell fractionation and differential centrifugation are used to study cellular organelles. Most exercises are all fairly short and can be reorganized to fit several shorter laboratory periods if necessary.

II. TIME REQUIREMENTS

Exercise A—Examining Plant Cells (45 minutes)
Extending Your Investigation: Cytoplasmic Streaming (15 minutes)
Exercise B—Examining Animal Cells
  Part 1—Studying Animal Cells Using Light Microscopy (15 minutes)
  Part 2—Studying Animal Cells Using Cytochemical Stains (30 minutes)
Exercise C—The Strange Shapes of Cells (20 minutes)
Exercise D—Cell Fractionation: A Study of Eukaryotic Cells (60 minutes)
Exercise E—A Closer Look at Eukaryotic Cells (15 minutes or completed at home)

III. STUDENT MATERIALS AND EQUIPMENT

	Per Student	Per Pair (2)	Per Group (4)	Per Class (24)
Exercise A				
Elodea (live sprig)				4
onion				1
Lugol's solution **(1)**			1	6
potato				1
banana				1
carrot				1
single edge razor blade		1		12
microscope slides, box (75 × 25 mm)			1	6
coverslips, box (22 × 22 mm)			1	6
compound microscope	1			24
Extending Your Investigation				
Elodea (live sprig)				4
microscope slides, box (75 × 25 mm)			1	6
coverslips, box (22 × 22 mm)			1	6
compound microscope	1			24
Exercise B (Part 1)				
frog epithelium (prepared slide)		1		12
salamander epithelium, mitochondria (prepared slide)		1		12
compound microscope	1			24

	Per Student	Per Pair (2)	Per Group (4)	Per Class (24)
Exercise B (Part 2)				
compound microscope				8
DNA in liver (prepared slide)				1
liver, DNAse treated (prepared slide)				1
RNA in liver (prepared slide)				1
liver, RNAse treated (prepared slide)				1
glycogen in liver (prepared slide)				1
liver, amylase treated (prepared slide)				1
fat tissue, osmium tetroxide treated (prepared slide)				1
fat tissue, lipase treated (prepared slide)				1
Exercise C				
neuron (prepared slide)		1		12
Stentor (live, class of 30)				1
Acetabularia (preserved)				1
spermatozoa, human (prepared slide)		1		12
Volvox (live, class of 30)				1
starfish development (prepared slide)		1		12
microscope slides, box (75 × 25 mm)			1	6
coverslips, box (22 × 22 mm)			1	6
compound microscope	1			24
Exercise D				
conical centrifuge tubes for clinical centrifuge			2	12
clinical centrifuge				4
peas (dry, wrinkled), 10 g (in beaker) **(2)**			1	6
cheesecloth (4 layers)			1	6
beaker, 250 ml			1	6
homogenization medium (1 l bottle) **(3)**				1
phosphate buffer, bottle (250 ml) **(4)**			1	6
DPIP (%), bottle (100 ml) **(5)**			1	6
tetrazolium (1 mM), bottle (100 ml) **(6)**			1	6
microscope slides, box (75 × 25 mm)			2	12
coverslips, box (22 × 22 mm)			2	12
Pasteur pipettes			4	24
rubber bulbs			2	12
permanent marker			1	6
light (100 w flood)				1
water bath (35 – 40° C)				1
pipette, 1 ml			1	6
pipette, 5 ml			1	6
Pi-pump			1	6
Exercise E				
student's collection of electron micrographs				

IV. PREPARATION OF MATERIALS AND SOLUTIONS

(1) Lugol's solution

Order prepared or mix 10 g of potassium iodide in 100 ml of distilled water and add 5 g of iodine.

(2) peas

Soak 10 g of peas/student group overnight. Wrinkled peas contain less starch and yield better separation. (Fresh peas can be used but use 20 g of fresh material.) You can homogenize all peas for one laboratory together or you can homogenize each group's 10 g separately.

(3) homogenization medium

Dissolve 170 g sucrose in 900 ml distilled water. Add 6 ml of KH_2PO_4 solution, 4 ml of Na_2HPO_4, and enough distilled water to make 1 liter. Adjust pH to 7 with KH_2PO_4 or Na_2HPO_4 if necessary.

KH_2PO_4

9.08 g KH_2PO_4 (anhydrous) dissolved and diluted to 1 liter with distilled water.

Na_2HPO_4

9.47 g Na_2HPO_4 dissolved and diluted to one liter with distilled water.

(4) phosphate buffer

Same as (3). Sucrose is necessary in the phosphate buffer to avoid osmotic rupture and maintain chloroplast and mitochondria structure.

(5) DPIP (1 mM) (or DCIP, dichlorophenolindophenol)

Dissolve 0.29 g DPIP in enough distilled water to make 1 liter.

(6) tetrazolium (1%)

Dissolve 1 g triphenyl tetrazolium chloride in 100 ml distilled water. Prepare immediately before use.

V. PREPARATION SUGGESTIONS

Exercise A

This exercise is designed to introduce students to the structure of eukaryotic cells. Living organisms should be used where possible.

In Part 3, stress to students that slices of potato and carrot should be very thin and the smear of banana tissue should also be spread out very thinly. The chromoplasts of the carrot are easy to see if sections are made just below the epidermis (removing all of the epidermis makes this easier).

A nice supplement to this exercise is studying cell size using an ocular micrometer (see Laboratory 2).

Extending Your Investigation

One box of glass microscope slides and a box of coverslips per laboratory will suffice. A dirty slide container (such as a cut-off plastic milk bottle) will help in collecting used slides (have students discard used coverslips).

You may wish to ask students to hypothesize about what might happen to cyclosis in *Elodea* if you heat up or cool down the slide or whether light is necessary for cyclosis. Students might also ask if pH affects cyclosis. pH can be varied using small amounts of 0.01 N HCl or NaOH introduced on one side of the coverslip.

Exercise B (Part 1)

Use prepared slides for this exercise. Alternatively, sloughed frog epithelium can be obtained from aquaria if frogs are kept in the laboratory.

Exercise B (Part 2)

Stress to students that a microscope can be used for more than just studying structure. The use of special stains (and illumination techniques) makes the study of cell biochemistry possible. You may wish to order more than one slide for each of the demonstrations. Some slides are better than others. Good methyl-green pyronin slides are particularly difficult to find, often being stained very lightly, making observations problematic.

Exercise C

The human sperm slides will be used a second time (Laboratory 19). The starfish development slide is also needed in Laboratory 19. Rather than using a separate slide of starfish eggs, this slide will suffice. Many developmental stages will be included, but students are intrigued by these and often ask some very interesting questions.

Exercise D

Use 10 g of dry pea seeds soaked overnight for this exercise. If wrinkled pea seeds are used, you will find fewer starch granules and separations are easier. Alternatively you can use fresh peas from the grocery store. Place the seeds into 100 ml *cold* sucrose buffer solution and homogenize for 1 minute at high speed. Make sure the homogenization buffer is cold! Filter the homogenate through 4 layers of cheesecloth into a 250 ml beaker. Pieces of seed coat as well as cell walls and whole cells will be present on the cheesecloth. This should be set aside for examination.

Before filling centrifuge tubes, make sure that you swirl the beaker to suspend all the contents. Keep centrifuge tubes cold—on ice prior to filling. Make sure that students fill tubes to the same level so that opposite positions in the centrifuge will contain tubes that balance.

Obtaining material from the green layer is difficult. Sliding the pipette down the side of the tube helps keep the tip in sight, but you may wish to mark the tip of the pipette with a permanent marker. This makes it easier for students to see the pipette. Alternatively you can remove and save the supernatant and green layer by pouring off *all* material above the pellet. Add 3 ml homogenization medium and recentrifuge this material at 1300× (top speed) in the clinical centrifuge. A small pellet will collect in the bottom of the tube with a larger chloroplast layer above it and mitochondria in the supernatant.

DPIP can be used to test for the presence of both chloroplasts and mitochondria. In the experiment as written, DPIP is used to test for photosynthetic activity and tetrazolium is used to test for mitochondrial activity. Both, however, are electron acceptors. If DPIP is used to test tubes X, Y, and Z for mitochondrial activity, light is not necessary because mitochondria do not need light to run both the Krebs cycle and electron transport. Light is needed for DPIP tests on the chloroplast-containing green layer. The presence of light vs. no light will allow students to distinguish between the layer containing chloroplasts and the layer containing mitochondria.

CAUTION: tetrazolium is a metabolic poison.

If tetrazolium is used, wear gloves when dispensing it. Place all student tubes in one rack in a single water bath and expose them to a 100 w light. If the reaction does not take place in 30 minutes due to lack of sufficient mitochondrial activity, leave for several hours or overnight.

Students can observe demonstration tubes X, Y, and Z to determine results.
Collect tubes and dispose of contents into the proper waste container.

Exercise E

Individually labeled micrographs from a packet of electron micrographs (see Ordering Information) can be placed on the bulletin board in the laboratory. If students cannot identify a particular structure in the micrographs included in the laboratory manual, do not name the structures for them. Rather, send them to the bulletin board or to their textbook to locate the structure by comparison with other micrographs. Students will learn about other structures in addition to the one they are seeking. Those in the laboratory manual include: 1) Golgi body, 2) centrioles, 3) a centriole in cross section, 4) nucleus, 5) mitochondrion, 6) plasmodesmata, and 7) chloroplast.

VI. ORDERING INFORMATION

Elodea—Carolina Biological, #16-2100; Ward's, # 86W7500
Lugol's solution—Carolina Biological, # 87-2793; Ward's, # 39W1685
methylene blue stain—Fisher, # M291-25 (powder) or # LC16920-1 (1% solution, ethanol; dilute to 0.1%); Carolina Biological, # 87-5684 (powder) or # 87-5913 (120 ml); Edvotek, # 609; Ward's, # 38W8323
frog skin (cross section, prepared slide)—Carolina Biological, # H 2060; Ward's, # 92W3643
Amphiuma skin (cross section, prepared slide)—Carolina Biological, # Z 3615
neuron (cow, prepared slide)—Carolina Biological, # H1660; Ward's, # 93W3617
Stentor (live)—Carolina Biological, # L5; Ward's, # 87W1370
Acetabularia (preserved)—Carolina Biological, # PB45; Ward's, # 63W0135
spermatozoa (human, prepared slide)—Carolina Biological, # H4259; Ward's, # 93W7241
Volvox (live)—Carolina Biological, # 15- 1335; Ward's, # 86W0805
starfish development (prepared slide)—Carolina Biological, # E582; Ward's, # 92W8254
DNA in liver (prepared slide)—Carolina Biological, # HC100; Ward's, # 93W2275
liver DNAse treated (prepared slide)—Carolina Biological, # HC102; Ward's, # 93W2351
RNA in liver (prepared slide)—Carolina Biological, # HC105; Ward's, # 93W2354
liver RNAse treated (prepared slide)—Carolina Biological, # HC107; Ward's, # 93W2355
glycogen in liver (prepared slide)—Carolina Biological, # HC110; Ward's, # 93W2731
liver, amylase treated (prepared slide)—Carolina Biological, # HC112; Ward's, # 93W2732
fat tissue, osmium tetroxide treated (prepared slide)—Carolina Biological, # HC115; Ward's, # 93W2374
fat tissue, lipase treated (prepared slide)—Carolina Biological, # HC117; Ward's, # 93W2375
centrifuge tubes, glass (conical)—Fisher, # 05-500 or # 05-501; Carolina Biological, # 73-2014; Ward's, # 17W1330
clinical centifuge—Carolina Biological, # 70-1800
peas (wrinkled), Little Marvel—Carolina Biological, # 15-8883
potassium phosphate (monobasic) (KH_2PO_4)—Fisher, # P285-500; Carolina Biological, # 88-4262; Ward's, #. 37W4845
sodium phosphate (dibasic, heptahydrate) ($Na_2HPO_4 \cdot 7H_20$)—Fisher, # S471-3; Carolina Biological, # 89-1370; Ward's, # 37W4841
DPIP (2,6-dichloroindophenol)—Fisher, # S286-5; Sigma, # D 1878; Carolina Biological, # 86-8600

tetrazolium (2, 3, 5 triphenyl tetrazolium chloride)—Fisher # T413-5; Carolina Biological, # 89-6930
Pasteur pipette, 9"—Fisher, #13-678-20C; Carolina Biological, # 73-6062; Ward's, # 17W1146
rubber bulb—Fisher, # 14-065B
microscope slides, 75 × 25 mm—Fisher, # 12-550A; Carolina Biological, # 63-2000; Ward's, # 14W3500
coverslips, 22 × 22 mm—Fisher, # 12-542B; Carolina Biological, # 63-3095; Ward's, # 13W3251
cheesecloth—Fisher, # 06-665-18 (70 yard bolt); Carolina Biological, # 71-2690 (5 yard package); Ward's, # 15W0015
beaker, 250 ml— Fisher, # 02-540K (Pyrex) or # 02-539K (Kimax); Carolina Biological, # 72-1223 (Corning, student grade); Ward's, # 17W4116
light (flood), 100 w
pipette, 1 ml—Fisher, # 13-665F; Carolina Biological, # 73-6270; Ward's, # 17W1305
pipette, 5 ml—Fisher, #13-665K; Carolina Biological, # 73-6276; Ward's, # 17W1307
Pi-Pump (Pipette Pump)—Fisher, # 13-683C; Carolina Biological, # 73-6875; Ward's, # 15W0532
student's collection of electron micrographs—Carolina Biological, # 45-9800

LABORATORY 8 Osmosis and Diffusion

I. FOREWORD

This laboratory is designed to introduce students to the principles of both diffusion and osmosis and to the concept of water potential. It consists of several separate exercises to be done individually or in pairs.

II. TIME REQUIREMENTS

Exercise A—Brownian Movement (15 minutes)
Exercise B—Diffusion
  Part 1—Diffusion of a Gas in a Cell (15 minutes)
  Part 2—Diffusion of a Liquid in a Liquid (30 minutes)
Extending Your Investigation: Does Temperature Affect the Rate of Diffusion? (30 minutes)
  Part 3—Effect of Molecular Weight on the Rate of Diffusion (15 minutes)
Exercise C—Diffusion Across a Selectively Permeable Membrane (45 minutes)
Exercise D—A Look at Osmosis
  Part 1—Measuring Osmotic Potential (75 minutes, waiting period 60 minutes)
  Part 2—Measuring Pressure Potential: The Osmometer (demonstration)
  Part 3—Measuring the Water Potential of Living Plant Cells (30-minute preparation, waiting period 2 hours or overnight)[1]
Extending Your Investigation: Water Potential of Different Tuber Types (30 minute preparation or question discussion, waiting period 2 hours for experiment)
  Part 4—Observing Osmosis in a Living System (20 minutes)

III. STUDENT MATERIALS AND EQUIPMENT

	Per Student	Per Pair (2)	Per Group (4)	Per Class (24)
Exercise A				
carmine suspension (dropping bottle) **(1)**			1	6
microscope slides, box (75 × 25 mm)	1			24
coverslips, box (22 × 22 mm)	1			24
compound microscope	1			24
Exercise B (Part 1)				
filter paper strips	1			24
phenolphthalein, dropping bottle **(2)**			1	6
ammonium hydroxide (500 ml)				1
500 ml Erlenmeyer flask OR 250 ml graduated cylinder	1			24
brass cup hooks	1			24
corks (# 16 or 18)	1			24

[1] If Exercise D (Part 3) is run for 2 hours, set it up early in the laboratory period and other experiments can be completed while materials in this exercise are allowed to incubate

	Per Student	Per Pair (2)	Per Group (4)	Per Class (24)
Exercise B (Part 2)				
Petri dish, 100 × 15 mm		1		12
water, bottle (30° C)			1	6
plastic mm ruler (white or clear)		1		12
blue dye (food coloring), bottle			1	6
thermometer			1	6
Extending Your Investigation				
see Exercise B (Part 2)				
water, bottle (10° C)			1	6
water, bottle (50° C)			1	6
additional Petri dishes, 100 × 15 mm		2		24
additional plastic mm ruler (white or clear)		2		24
Exercise B (Part 3)				
absorbent cotton, roll				1
cheesecloth, package				1
ammonium hydroxide (500 ml)				1
hydrochloric acid, conc. (500 ml)				1
ring stand				1
utility clamp				1
glass tube (90 cm long × 41 mm diameter)				1
Exercise C				
dialysis tubing, 1 × 12 inch strip **(7)**		1		12
Pasteur pipette		4		48
Pi-Pump or Propipette		2		24
15% glucose, bottle **(3)**		1		12
1% starch **(4)**		1		12
250 ml beaker		1		12
Lugol's reagent **(5)**		1		12
rubber bands		1		12
Benedict's solution **(6)**		1		12
Exercise D (Part 1)				
dialysis tubing, 1 × 12" strip **(7)**			6	36
0.2 M sucrose, bottle **(8)**			1	6
0.4 M sucrose, bottle **(8)**			1	6
0.6 M sucrose, bottle **(8)**			1	6
0.8 M sucrose, bottle **(8)**			1	6
1.0 M sucrose, bottle **(8)**			1	6
plastic cups			6	36
balance (± 0.1 g)			1	6
Exercise D (Part 2)				
40% sucrose with red food coloring, bottle (200 ml) **(9)**				1
small glass tubing (68 cm long × 3 mm ID)				1

	Per Student	Per Pair (2)	Per Group (4)	Per Class (24)
Exercise D (Part 2)—continued				
dialysis tubing (large)				1
one-hole rubber stopper # 7				1
1000 ml beaker				1
ring stand				1
utility clamp				2
Exercise D (Part 3)				
0.2 M sucrose, bottle **(8)**			1	6
0.4 M sucrose, bottle **(8)**			1	6
0.6 M sucrose, bottle **(8)**			1	6
0.8 M sucrose, bottle **(8)**			1	6
1.0 M sucrose, bottle **(8)**			1	6
potato			1	6
cork borer (5 mm inner diameter)			1	6
marking pencil			1	6
balance (± 0.1 g)			1	6
Extending Your Investigation				
see Exercise D (Part 3)				
assorted potatoes, turnips, beets			1	6
Exercise D (Part 4)				
0.5% NaCl, large specimen bowl **(10)**				1
tap water, large specimen bowl				1
Chara (live, class of 12)				1
Elodea (live, per 12)				1
Spirogyra (live, class of 30)				1
tap water, dropping bottle			1	6
10% NaCl, dropping bottle **(11)**			1	6
microscope slides, box (75 x 25 mm)	1			24
coverslips, box (22 x 22 mm)	1			24

IV. PREPARATION OF MATERIALS AND SOLUTIONS

(1) carmine suspension
Add 0.05 g of carmine powder to 100 ml of distilled water.

(2) phenolphthalein solution
Order prepared solution OR weigh 0.5 g phenolphthalein and add 95% ethanol to make 100 ml.

(3) 15% glucose
Mass 75 g of glucose. Add distilled water to 500 ml.

(4) 1% starch
Mass 5 g of potato starch. Add distilled water to 500 ml. Heat on low heat while stirring and increase heat until starch solution becomes clear.

(5) Lugol's solution (I_2KI)
Order prepared solution.

(6) Benedict's solution
Order prepared solution.

(7) dialysis tubing
Soak dialysis tubing for at least one hour in distilled water before use. If teaching multiple laboratory sections, you may cut the necessary number of 12-inch pieces in advance and place them in a beaker to soak—be sure to store the beaker overnight in the refrigerator so mold does not grow. Bags can be formed by twisting the ends and tying double knots in each end of the pieces of tubing (string may also be used). Make sure students fill the dialysis bags only half to two-thirds full. If you use a smaller diameter dialysis tubing, adjust the sucrose volume within the tubing (the 15 ml suggested is not a critical amount).

(8) sucrose solutions
Note that the same sucrose solutions are used for both Parts 1 and 3 of Exercise D.

Approximately 200 ml will be needed by each group to carry out both parts of the exercise. A liter will provide enough for 5 groups (or 1 group at a certain location for 5 laboratory periods).

1.0 M
Dissolve 342.3 g of sucrose in enough distilled water to make 1000 ml of solution.

0.8 M
Dissolve 274.2 g of sucrose in enough distilled water to make 1000 ml of solution.

0.6 M
Dissolve 205.4 g of sucrose in enough distilled water to make 1000 ml of solution.

0.4 M
Dissolve 136.9 g of sucrose in enough distilled water to make 1000 ml of solution.

0.2 M
Dissolve 68.4 g of sucrose in enough distilled water to make 1000 ml of solution.

Alternatively, prepare 3 liters of **1.0 M** sucrose and dilute as follows (retain 1000 ml of 1.0 M sucrose)

0.8 M
800 ml of 1.0 M sucrose plus 200 ml of distilled water.

0.6 M
600 ml of 1.0 M sucrose plus 400 ml of distilled water.

0.4 M
400 ml of 1.0 M sucrose plus 600 ml of distilled water.

0.2 M
200 ml of 1.0 M sucrose plus 800 ml of distilled water.

(9) 40% sucrose with red food coloring
Mass 40 g of sucrose and add distilled water to 100 ml. Mix well. Add red food coloring until solution is dark red.

(10) 0.5% NaC1
Mass 2.5 g NaC1 and add distilled water to 500 ml. Use to fill specimen bowl.

(11) 10% NaC1
Mass 10 g NaCl and add distilled water to 100 ml. Use to fill dropping bottles.

V. PREPARATION SUGGESTIONS

Exercise A

Students will have a difficult time observing Brownian motion, primarily due to lack of microscope skills. This will give you an excellent opportunity to review the use of the microscope. An excellent description of the physics of Brownian motion can be found in Breuer, H. (1975). *Physics for Life Science Students.* Prentice Hall.

Exercise B (Part 1)

Use small brass cup hooks screwed into the bottom of the appropriate size cork (to fit either a 500 ml Erlenmeyer flask (# 16) or a 250 ml graduated cylinder (# 18). If hooks are not available, a paper clip can be pulled apart and cut with wire cutters. Both of the U-shaped pieces can then be inserted into the corks.

Exercise B (Part 2)

Have students practice delivering a drop of blue dye into a Petri dish filled with water before attempting to collect data. If you use a Pasteur pipette and do not have the blue dye at the tip (but close to it), you can release the dye once the pipette tip is beneath the water surface and is resting on the bottom of the Petri dish. Convection currents will create problems. If your bench top is cold, place paper towels below the paper on which you place the Petri dish. You may wish to cover each dish to keep air currents from disturbing the experiment. If enough rulers are available, place a second ruler at right angles to the first. This may help students make more accurate measurements.

Extending Your Investigation

This extension allows you to compare diffusion rates from Exercise B, Part 3, in room temperature conditions to rates in hot and cold water. Be sure to insulate the dishes used and wait for all convection currents to disappear.

Exercise B (Part 3)

The glass tube should be 90 cm long with an inner diameter of 41 mm. Plugs should be made by tying cheesecloth around an appropriately sized plug of cotton. After use, the glass tube will need to be washed with a long-handled brush and hot soapy water.

Exercise C

To reduce the amount of glassware used, tape a test tube to the side of the bottle containing the 15% glucose and 1% starch solutions. Place a Pasteur pipette into the tube. Make sure that the pipette is provided with a label corresponding to the contents of the bottle.

Soak all dialysis tubing overnight in distilled water prior to use in the laboratory. Strips should be cut 8-10 inches long.

Exercise D (Part 1)

Use plastic cups rather than beakers to avoid glassware cleaning. Dialysis bags should be allowed to sit for at least 60 minutes in water before use. Do not leave dialysis bags containing sucrose overnight (some sucrose may eventually be lost from the dialysis bag and results will be difficult to interpret). This experiment can be run for 30 minutes and still give results since the rate of diffusion is proportional to the difference in water potential outside and inside the bag; the greater the difference, the faster water moves into the bag and the faster the mass of the bag increases. However, for accuracy in determining mass, the experiment should be run for at least one hour. The more mass gained over longer periods of time, the easier it is to measure the gain in mass.

You may wish to have students work in pairs or in groups. You might also want to assign certain molarities to different groups to carry this out as a class experiment and reduce solution preparation time or expense. It is recommended, however, that at least two groups have the same molarities so data will be duplicated. Note that sucrose solutions are the same as those used for Exercise D, Part 3, so plan ahead when preparing solutions.

Tie bags by twisting the ends and placing two knots at each end (string may also be used).

To answer question (d), the *Y*-axis of the graph must be extended below zero. In this case, water would flow out of the bag when placed in sucrose solutions of greater than 0.4 M.

Exercise D (Part 2)

Make a groove around the bottom of a # 7 one-hole rubber stopper approximately 1 cm from the bottom surface. Stretch a piece of soaked large dialysis tubing over the bottom of the cork and secure it by tying a piece of string around the cork so that the string fits into the groove. Fill the bag with red sucrose solution through the hole in the stopper until the bag is completely filled and turgid. Use petroleum jelly to grease the end of a piece of glass tubing that has been calibrated (marked) in cm. The tubing should be 65-68 cm long with an ID of 3 mm. Insert the tube in the hole. From a ring stand, suspend the osmometer by the cork in a 1000 ml beaker of distilled water so that the entire bag is below the surface of the water. Use a clamp to hold the cork onto the ring stand and use a second clamp to keep the glass tubing straight. Set a timer for 15-minute intervals during the laboratory period. One student should check the level of fluid in the osmometer at each 15-minute interval, record the value on the blackboard, and reset the timer.

Exercise D (Part 3)

This experiment is set up to be performed in groups of four students. Each group will need to cut four potato cores per solution (total = 24 cores) from the SAME potato (so potatoes should be fairly large). You could also allow two groups to do three solutions each but potato cores must still be from the same potato. Adequate results can be achieved from cores soaked for two hours which means that this experiment must be started at the beginning of the laboratory period. If cores are left overnight, students will need to come back the next day, weigh the cores, and record their data on a class table. Class data of replicates can be averaged and can be distributed during lecture at the end of the week and students should be able to graph data and interpret results prior to the next week's laboratory. Alternatively, if multiple sections of laboratory meet, Tuesday's laboratory sections can weigh cores prepared by Monday's sections, and so on. In this case, all data should be kept on the laboratory blackboard or on overhead transparencies. Class data can be statistically analyzed using the statistics program included with the *Reaction Time* software on the CD-ROM. (See *Instructor's Manual: Preparator's Guide* for installation and operation instructions—pages CD-ROM 1-4 and I-5.)

It would be of interest for some groups of students to use a fresh potato that has been kept in a hydrator while others use a dehydrated potato. Ask students to predict which potato cells would have the higher water potential. You can also use different types of potatoes as well as turnips and beets.

When students calculate water potential, they often try to call it osmotic potential. Make sure students understand that you cannot measure osmotic potential of living plant cells because you do not know how great the pressure potential is inside the plant cells. Since Ψ_p

is not zero, Ψ does not equal Ψ_π. Recall that $\Psi = \Psi_p + \Psi_\pi$.

Exercise D (Part 4)

Any of several live materials can be used to demonstrate plasmolysis and loss of turgor. Lettuce and celery are easily obtained. The green alga *Chara is* ideal. Students can take it from tap water and place it in a 0.5% NaCI solution and watch it wilt, and then place it back into water and watch it "pop" back into shape almost immediately.

VI. ORDERING INFORMATION

carmine aceto-carmine stain—Carolina Biological, # 84-1421
carmine—Fisher, # C579-25; Sigma, # C 1022; Carolina Biological, # 85-3070; Ward's, # 38W7320
dropping bottle, 30 ml— Fisher, # 02-980; Carolina Biological, # 71-6525; Ward's, # 17W6011
microscope slides, 75 × 25 mm—Fisher, # 12-550A; Carolina Biological, # 63-2000; Ward's, # 14W3500
coverslips, 22 × 22 mm—Fisher, # 12-542B; Carolina Biological, # 63-3095; Ward's, # 14W3251
phenolphthalein—Fisher, # P79-100 (powder) or # SP62-500 (1% solution); Carolina Biological, # 87-9963 (120 ml) or # 87-9940 (powder); Ward's, # 39W2873
Erlenmeyer flask, 500 ml—Fisher, # 10-040H (Pyrex) or # 10-039H (Kimax); Carolina Biological, # 72-6676 (Corning, student grade); Ward's, # 17W2974
corks (# 16 or 18)—Fisher, # 07-781X or # 07-792X (various sizes); Carolina Biological, # 71-2000-2020 (various sizes)
graduated cylinder, 250 ml—Fisher, # 08-553F (Pyrex) or # 08-549-5H (Kimax) or # 08-572-6E (Nalgene); Carolina Biological, # 72-1790 (Pyrex) or # 72-1950 (Kimble) or # 72-2268 (Nalgene); Ward's, # 17W0174
Petri dish, 100 × 15 mm—Fisher, # 08-747C (Pyrex) or 08-746C (Kimax); Carolina Biological, # 74-1158; Ward's, # 18W7101
thermometer, – 20 to 110° C—Carolina Biological, # 74-5020; Ward's, # 15W1478
absorbant cotton, roll—obtain locally
cheesecloth—Fisher, # 06-665-18 (70 yard bolt); Carolina Biological, # 71-2690 (5 yard package); Ward's, # 15W0015
ammonium hydroxide—Fisher, # A-669-500 (500 ml); Carolina Biological, # 84-4010 (500 ml); Ward's, # 39W0182
hydrochloric acid (HCl)—Fisher, # A144-212; Carolina Biological, # 86-7790; Ward's, # 37W9561
ring stand, medium—Carolina Biological, # 70-7192; Ward's, # 15W0692
utility clamp—Fisher, # 05-768-10; Carolina Biological, # 70-7376 or 70-7312; Ward's, # 15W0665
glass tubing—Fisher, # 11-365CC; Carolina Biological, # 71-1174
dialysis tubing, 32 mm—Fisher, #08-667D; Carolina Biological, # 68-4212 (10 ft.); Ward's, # 14W4516
Pasteur pipette, 9"—Fisher, #13-678-20C; Carolina Biological, # 73-6062; Ward's, # 17W1145
rubber bulb—Fisher, # 14-065B

Propipette (safety pipette filler)—Fisher, # 13-681-50; Carolina Biological, # 73-6868; Ward's, # 15W0510 OR

Pi-Pump (Pipette Pump)—Fisher, # 13-683C; Carolina Biological, # 73-6875; Ward's, # 15W0532

D-glucose (dextrose) (certified)—Fisher, # D16-500 (500 g); Carolina Biological, # 85-7440 (500 g); Ward's, # 39W1457

soluble starch, potato—Fisher, S516-500; Carolina Biological, # 89-2530; Ward's, # 39W3275

sucrose (certified)—Fisher, # S5-500; Carolina Biological, #89-2870; Ward's, # 39W3280

sodium chloride (NaCl)—Fisher # S271-3; Carolina Biological, # 88-8880; Ward's, # 37W5491

Lugol's solution (I_2KI)—Carolina Biological, # 87-2795; Ward's, # 39W1685

Benedict's reagent—Fisher, # LC11650-1; Carolina Biological, # 84711 (500 ml); Ward's, # 37W0698

Elodea (live)—Carolina Biological, # 16-2100 (per 12 students); Ward's, # 14W4516

Chara (live)—Carolina Biological, # 15-1240 (per 12 students, available May until mid-October)

Spirogyra (live)—Carolina Biological, # 15-2525 (per 12); Ward's, # 86W0650

cork borer—Fisher, # 07-845B; Ward's, # 15W1666

wax pencil—Carolina Biological, # 65-7730 (red), # 65-7732 (blue), # 65-7734 (black); Ward's, # 15W1159

TesTape—Carolina Biological, # 89-3840; Ward's, # 14W4107

LABORATORY 9 Mitosis

I. FOREWORD

A simulation has been chosen as the focus of this exercise that is designed to teach the process of mitosis as well as give examples in living materials.

II. TIME REQUIREMENTS

Exercise A—The Cell Cycle—Interphase and "Getting Ready" (30 minutes, home)
Exercise B—Simulating the Events of Interphase, Mitosis, and Cytokinesis (75 minutes)
Exercise C—Mitosis in Living Tissues—Onion Root Tips (40 minutes)
Exercise D—Phases of the Cell Cycle in the Onion Root Tip (45 minutes)

III. STUDENT MATERIALS AND EQUIPMENT

	Per Student	Per Pair (2)	Per Group (4)	Per Class (24)
Exercise A				
None				
Exercise B				
chromosome simulation kit **(1)**				2
Exercise C				
stained onion root tips **(2)**	1			24
compound microscope	1			24
microscope slides, box (75 × 25 mm)	1			24
coverslips, box (22 × 22 mm)	1			24
tap water, dropping bottle		1		12
Exercise D				
onion root tip, mitosis (prepared slide)	1			24

IV. PREPARATION OF MATERIALS AND SOLUTIONS

(1) chromosome simulation kit

A chromosome simulation Biokit can be purchased from Carolina Biological Supply Company (# P7-17-1100) and all necessary materials will be included. Although the original price is considerable ($129.95/kit), the same kit can be used year after year without additional purchases. The beads are also used in Laboratories 13 and 20.

If you prefer, you can purchase beads only (Carolina Biological, # P7-17-1112). Two sets cost $33.50 (two bags, 900 beads per bag with two colors in a set—this will be enough for a class of 30). Magnetic centromeres can be made from amber rubber tubing (ID 3.2 mm, wall thickness 1.6 mm) with children's magnets placed inside. Centrioles can be made by cutting plastic drinking straws into 10 mm lengths.

Kits for individual students are made by attaching 10 beads to either side of a magnetic centromere. Two yellow "chromosomes" and two red "chromosomes" (four 20-bead strands in all) are then placed in a plastic bag to serve as an individual "kit."

Colored paper clips (vinyl coated) may be substituted for beads. Children's bar magnets can be inserted into narrow diameter amber rubber tubing (ID 4.8 mm, wall thickness 1.6 mm) and chains of paper clips can then be attached to opposite ends of the rubber tubing.

Colored pipe cleaners can also be used as an alternative. Paper clips can be used as centromeres to hold two chromatids together.

(2) stained onion root tips

Gather together the following materials:

camel-hair brushes
plastic Petri dishes (100 mm diameter)
tooth picks 100 ml beakers (10)
onion-starts (10)
single-edge razor blades

Solutions:

ALWAYS ADD ACID SLOWLY TO WATER WHILE STIRRING!
WEAR SAFETY GLASSES

3:1 EtOH-glacial acetic acid

Add 10 ml of glacial acetic acid to 30 ml of 100% (absolute) ethyl alcohol and mix. You will need approximately 40 ml of this solution.

6N HC1

Mix 50 ml of concentrated (12N) hydrochloric acid into 50 ml of distilled water.

Schiff Reagent

Purchase prepared solution.

Procedure:

1. Rooting the onions: first clean the bottom of the onion by rubbing it with your finger. Next insert a round toothpick through the onion, placing the onion in the center of the toothpick. Lower the onion down into a 100-ml beaker of tap water until the toothpick is resting across the top of the beaker. Allow three days for rooting. Alternatively, wrap in a wet paper towel and place in a plastic bag (not sealed). Keep in a warm, dark place overnight.
2. Removing the root tips: after the onions have rooted, harvest the root tips. Cut the root tips off using a razor blade, cutting approximately 5 mm from the tip itself. With a camel-hair brush, gently place the root tips into 3:1 95% ethanol:acetic acid solution. Let the root tips remain in this solution for at least 20 minutes.
3. With the camel-hair brush, transfer the root tips to a Petri dish of distilled water to rinse. Rinse for one minute.
4. With the camel-hair brush, transfer the root tips to a Petri dish of 6N HC1. Incubate for 5 minutes.
5. With the camel-hair brush, transfer the onion tips to a Petri dish of distilled water to rinse. Rinse for one minute.
6. With the camel-hair brush, transfer the onion root tips to a Petri dish of Schiff Reagent for 15 minutes.
7. With the camel-hair brush, transfer the purple-stained onion root tips from the Schiff reagent to a Petri dish of distilled water. Root tips are now ready for use.
8. Onion root tips can be obtained by students from a central location on a demonstration bench. Each student should bring a glass slide with a drop of water on it. Be sure to provide

a camel-hair brush to transfer root tips.

V. PREPARATION SUGGESTIONS

Included above.

VI. ORDERING INFORMATION

Schiff reagent—Carolina Biological, # 88-7263; Ward's, # 39W1850
100% ethanol—Fisher, # A-407
camel-hair brush—Carolina Biological, # 70-6182; Ward's, # 15W3846
chromosome simulation Biokit—Carolina Biological, # P7-17-1100
onion root tip, mitosis (prepared slide)—Carolina Biological, # B551; Ward's, # 93W2145

Notes:

LABORATORY 10 Enzymes

I. FOREWORD

The activities in this laboratory are divided into two major exercises. The first exercise examines the effects of temperature, pH, substrate, and enzyme concentration on enzyme activity. Data can be recorded from qualitative observations or quantitatively using the Spectronic 20. During this laboratory, students work in groups of four (see Preparation Suggestions). The team approach should allow plenty of time for cheese making in the second exercise.

II. TIME REQUIREMENTS

Exercise A—Investigating the Enzymatic Activity of Catecholase
- Part 1—The Effect of Temperature on Enzyme Activity (20 minutes)
- Part 2—The Effect of pH on Enzyme Activity (20 minutes)
- Part 3—The Effect of Enzyme Concentration on Enzyme Activity (30 minutes)
- Part 4—The Effect of Substrate Concentration on Enzyme Activity (40 minutes)

Extending Your Investigation: Making Juices Juicier (15 minutes)

Exercise B—The Essentials of Cheesemaking (60 minutes)

III. STUDENT MATERIALS AND EQUIPMENT

	Per Student	Per Pair (2)	Per Group (4)	Per Class (24)
Exercise A (materials common to Parts 1-4)				
test tube rack	1			24
wax pencil	1			24
masking tape			1	6
catecholase potato extract, dropping bottle **(1)**			2	12
0. 1% catechol, dropping bottle **(2)**			1	6
pH 7 phosphate buffer, bottle (250 ml) **(3)**			1	6
5 ml pipette (with buffer)			1	6
small squares of Parafilm or wax paper (container with 100 pieces)			1	6
5 ml pipette (with buffer)			1	6
Pi-Pump			4	24
Spectronic 20			1	6
Spectronic 20 cuvettes or tubes			29	696
Exercise A (Part 1—individual A)				
test tubes (quantitative)			3	60
test tubes (qualitative)			7	42
150 ml beaker			2	12
hot plate			1	6
thermometer			1	6
ice bucket with shaved ice			1	6

	Per Student	Per Pair (2)	Per Group (4)	Per Class (24)
Exercise A (Part 2—individual B)				
test tubes (quantitative)			10	60
test tubes (qualitative)			5	30
0.1 M phosphate buffer, pH 4, bottle **(4)**			1	6
0.1 M phosphate buffer, pH 6, bottle **(5)**			1	6
0.1 M phosphate buffer, pH 8, bottle **(6)**			1	6
0.1 M phosphate buffer, pH 10, bottle **(7)**			1	6
5 ml pipettes (with buffers)			4	24
Exercise A (Part 3—individual C)				
test tubes (quantitative)			8	48
test tubes (qualitative)			4	24
Exercise A (Part 4)				
test tubes			9	54
Extending Your Investigation				
apple juice (25 ml)			1	6
distilled water, bottle			1	6
pectinase, bottle **(8)**			1	6
spatula			2	12
cheesecloth (4 thicknesses)			2	12
beaker, 250 ml			2	12
graduated cylinder, 100 ml			2	12
Exercise B				
graduated cylinder, 250 ml (sterile)				1
preripened milk (250 ml) **(9)**		1		12
rennilase, bottle (sterile) **(10)**				1
syringe, 1 ml		1		12
sterile glass rod		1		12
cheesecloth (4 thicknesses, sterile)		1		12
water bath				2
sterile Petri dish (100 mm)		1		12
beaker, 1000 ml (sterile)		1		12
beaker, 1000 ml (nonsterile)		1		12
thermometer (clean)		1		12

IV. PREPARATION OF MATERIALS AND SOLUTIONS

(1) catecholase (potato extract)

Put 75 g of diced potato into a blender, add 500 ml of distilled water and blend for 2 minutes at high speed. After blending, strain the solution through four thicknesses of cheesecloth. Immediately dispense the solution into dropping bottles and stopper them tightly since the presence of oxygen will allow the enzyme to work and will lead to darkening of the solution. Prepare this in class immediately prior to use.

(2) 0.1 % catechol

Weigh 0.2 g catechol. Add distilled water to 200 ml. Dispense into dropping bottles.

(3) - (7) buffers

pH 4—0.1 M potassium phthalate-NaOH buffer

Part A—Dissolve 10.21 g of potassium hydrogen phthalate in 300 ml of distilled water. Bring the solution up to 500 ml with distilled water.

Part B—Mix 500 ml of potassium hydrogen phthalate solution, Part A, with 499 ml of distilled water and 1 ml of 0.1 N HCl to make one liter of buffer.

pH 6, 7, and 8—stock solutions (0.1 M sodium phosphate buffer)

Part A—0.1 M solution of sodium phosphate (monobasic), NaH_2PO4. Mass 13.9 g NaH_2PO_4. Make up to 1000 ml with distilled water.

Part B—0.1 M solution of sodium phosphate (dibasic), Na_2HPO_4. Mass 26.825 g $Na_2HPO_47 \cdot H_2O$ or 35.85 g $Na_2HPO_4 \cdot 12H_20$. Make up to 1000 ml with distilled water.

for pH 6, mix 877 ml (Part A) and 123 ml (Part B)
for pH 7, mix 390 ml (Part A) and 610 ml (Part B)
for pH 8, mix 53 ml (Part A) and 947 ml (Part B)

pH 10—0.1 M borax-NaOH buffer

Part A—Dissolve 3.73 g KCl and 3.09 g boric acid into 300 ml of distilled water. Then bring the solution up to 500 ml with distilled water. Heat the solution to dissolve boric acid.

Part B—Mix 500 ml of KCl and boric acid solution (Part A) with 437 ml of 0.1 N NaOH and 63 ml of distilled water to make 1 liter.

(8) pectinase

Order in liquid form. Available from Carolina Biological Supply Company (see Ordering Information).

(9) preripened milk

Twelve to fifteen hours before laboratory, mix 10 ml of buttermilk into each 1000 ml of whole milk. Prepare 3 liters total. Into each of 12 sterile 1000 ml beakers put 250 ml of milk to ripen. Cover each beaker with cheesecloth and leave at room temperature. Do not cover the beakers with an air-tight seal—cheesecloth should remain loose.

(10) rennilase

Mix 1 ml of rennin emporase into 19 ml of cold tap water.

V. PREPARATION SUGGESTIONS

Exercise A

The catecholase solution from potatoes needs to be prepared for each class immediately before use. As soon as it is made, it should be stoppered quickly.

For this exercise, groups of four students work as a team. One student examines the effects of temperature, a second studies pH, a third works with substrate concentration, and a fourth student investigates the effects of enzyme concentration. (You could also work in pairs of students, each pair completing all four parts of Exercise A if time is available.) The tests result in a series of colored solutions that can then be used for discussion of the results among the group members or student pairs. If working in groups, each student should explain the procedure used, demonstrate the results, and explain his or her conclusions.

Exercise A can be completed quantitatively by using the Spectronic 20 if available. It is important to make sure that students realize that their blanks must be made up before they add catechol to their experimental tubes and the timed reactions start. Otherwise, the experimental reactions will be too far along to read results accurately because their blanks are not ready quickly enough.

Blanks use potato juice because the floculence of the potato juice solution tends to scatter light that must be corrected for in the procedure. If your students have completed Laboratory 4, The Spectrophotometer, you should encourage some discussion on the use of blanks and how to decide what components of the experimental tubes should be included in the blanks.

It is best to use a wide range bulb in the spectrophotometer. Tubes are to be read at 420 nm. Note that most reactions can be carried out in regular test tubes and the contents can be transferred to Spectronic 20 or spectrophotometer cuvettes. However, this is messy and students often end up confused with solutions in the wrong tubes. Since the accuracy of matched cuvettes is not necessary for proper interpretation of results from these experiments, it is easier to use Pyrex or Kimax tubes (13 × 100 mm) which are relatively inexpensive and fit into the Spectronic 20. (Always remember to wipe tubes off before inserting them into the machine!) See ordering information.

Class discussion of results is helpful at the end of the laboratory period. This is especially true if you want students to calculate the K_m for catecholase. It is important to get students to realize that each reaction with a different number of drops of catechol generates a typical rate curve over time. To calculate the K_m for the enzyme, students must calculate V_0 for EACH (all eight) of the curves. It is V_0 that is graphed against substrate concentration for each reaction to make the graph from which K_m is determined. If students are working in groups of four, allow one pair of students to do four of the substrate concentrations and the other pair to do the remaining four substrate concentrations. While calculating Km allows for discussion of an enzyme's affinity for substrate and the effects of inhibitors (see laboratory review questions 7 and 8), this step is not necessary to demonstrate the objective of this exercise—demonstration of the effects of increasing substrate concentration.

Exercise B

This exercise is fun for the students and also teaches some practical applications for enzymes. During the preceding week, tell students to bring seasoning (garlic, seasoned salt, pepper, salt, caraway or sesame seeds, etc.) to class. Bring soda and crackers as well.

Make sure that all cheese glassware is either sterile or extremely clean since contamination is a possibility. We keep special glassware just for this laboratory, sterilizing it both after the laboratory is completed and before we do it again the following semester.

By wrapping the cheese in cheesecloth and pressing it into a Petri dish, the consistency is very much like that of a soft, spreadable cheese. You can put a rubber band around the Petri dish and refrigerate it until the following week rather than eating it the day it is made, but this enhances the possibility of contamination and is not recommended.

To make the activity more meaningful, instructors should be prepared to discuss other examples of enzyme activity which students may be aware of— yogurt making, action of meat tenderizer, enzyme detergents, junket making, cottage cheese production, etc.

VI. ORDERING INFORMATION

test tubes, 16 × 125 mm—Fisher, # 14-925H; Carolina Biological, # 73-1418; Ward's, # 17W1403

test tube rack—Carolina Biological, # 73-1896; Ward's, # 18W4213

wax pencil—Carolina Biological, # 65-7730 (red), # 65-7732 (blue), # 65-7734 (black); Ward's, # 15W4213

rennin emporase—Carolina Biological, # 20-2375; Ward's, # 39W2831

dropping bottle, 30 ml— Fisher, # 02-980; Carolina Biological, # 71-6525; Ward's, # 17W6011

catechol—Fisher, # P370-100; Carolina Biological # 88—3540 (100 g)

hydrochloric acid (HCl)—Fisher, # A144-212; Carolina Biological, # 86-7821 (0.1 M) or # 86-7792; Ward's, # 37W9561

potassium chloride (KCl)—Fisher, # P217-500; Carolina Biological, # 88-2910; Ward's, # 37W4631

boric acid (H_3BO_3)—Fisher, # A73-1; Carolina Biological, # 84-8450; Ward's, # 37W0864

sodium hydroxide (NaOH)—Fisher, # S318-1; Carolina Biological, # 88-9470; Ward's, # 37W5560

potassium hydrogen phthalate—Fisher, # P243-500; Carolina Biological, # 88-3470 (500 g); Ward's, # 37W4755

sodium borate, terra (Borax)—Fisher, # S248-500; Carolina Biological, # 84-8440; Ward's, # 37W0851

sodium phosphate (monobasic) (NaH_2PO_4)—Fisher, # S369-1; Carolina Biological, # 89-1350 (125 g); Ward's, # 37W5655

sodium phosphate (dibasic, heptahydrade) ($Na_2HPO_4{\cdot}7H_2O$)—Fisher, # S471-3; Carolina Biological, # 89-1370; Ward's, # 37W5650

sodium hydroxide (NaOH), 0.1 N—Fisher, # S276-4; Carolina Biological, # 88-9SS 1 (500 ml); Ward's, # 37W9562

pipette, 5 ml—Fisher, #13-665K; Carolina Biological, # 73-6276; Ward's, # 17W1307

Parafilm—Fisher, # 13-274-12 (dispenser, # 13-374-18); Carolina Biological, # 21-5600 (dispenser, # 21-5602); Ward's, # 15W1940

Pi-Pump—Fisher, # 13-683C (up to 10 ml); Carolina Biological, # 21-4684 or # 73-6875; Ward's, # 15W0532

beaker, 150 ml— Fisher, # 02-540J (Pyrex) or # 02-539J (Kimax); Carolina Biological, # 72-1222 (Corning, student grade); Ward's, # 17W4030

thermometer, – 20 to 110° C—Carolina Biological, # 74-5020; Ward's, # 15W1478

hot plate—Carolina Biological, # 70-1010; Ward's, # 15W8010

graduated cylinder, 250 ml—Carolina Biological, # 72-1666 or # 72-1685; Ward's, # 17W0174

syringe, 1 ml—Carolina Biological, # 69-7765; Ward's, # 14W1613

glass stirring rod—Carolina Biological, # 71-1304 (or cut glass rod into desired lengths and fire-polish); Ward's, # 17W6005

cheesecloth—Fisher, # 06-665-18 (70 yard bolt); Carolina Biological, # 71-2690 (5 yard package); Ward's, # 15W0015

Petri dish, 100 × 15 mm—Fisher, # 08-747C (Pyrex) or 08-746C (Kimax); Carolina Biological, # 74-1158; Ward's, # 17W0700 beaker, 1000 ml— Fisher, # 02-540P (Pyrex) or # 02-539P (Kimax); Carolina Biological, # 72-1227 (Corning, student grade); Ward's, # 17W4080

pectinase—Carolina Biological, # 20-2380 (100 ml)

Juice Release Biokit—Carolina Biological, # 20-2315

Spectronic 20 (Milton Roy, formerly B&L)—Fisher, # 07-143-1; Carolina Biological, # 65-3300; Ward's, # 14W5550

wide range phototube and filter—Fisher, # 07-144-10; Carolina Biological, # 65-3312; Ward's, # 14W5553

Spectronic 20 tubes—Fisher, # 14-385-9008; Carolina Biological, # 65-3310; Ward's, # 14W1400 OR

Spectronic 20 tubes, Pyrex (100 × 13 mm)—Fisher, # 14-957C; Carolina Biological, # 73-1408

Spectronic 20 tubes, Kimax (100 × 13 mm)—Fisher, # 14-923D; Carolina Biological, # 73-1408A

LABORATORY 11 Energetics, Fermentation, and Respiration

I. FOREWORD

This laboratory introduces students to both fermentation and cellular respiration using two major experiments that could be done on separate days if desired.

II. TIME REQUIREMENTS

Exercise A—Production of Carbon Dioxide and Ethanol by Fermentation (60 minutes)
- Part 1—Examining Yeast Cells
- Part 2—Production of Carbon Dioxide by Fermentation
- Part 3—Requirements for Fermentation in Yeast
- Part 4—Production of Ethanol by Yeast

Exercise B—The Krebs Cycle Reactions in Bean Seeds (90 minutes)
Extending Your Investigation: Studying Inhibition
Exercise C—Heat Production During Respiration in Seedlings (10 minutes)
Exercise D—Respiration by Plant Embryos (20 minutes)

III. STUDENT MATERIALS AND EQUIPMENT

	Per Student	Per Pair (2)	Per Group (4)	Per Class (24)
Exercise A (Part 1)				
stock yeast solution (250 ml Erlenmeyer flask) **(1)**				4
microscope slides, box (75 × 25 mm)			1	6
coverslips, box (22 × 22 mm)			1	6
neutral red, dropping bottle **(2)**			1	6
compound microscope	1			24
incubator (37° C)				1
Exercise A (Part 2)				
sugar and yeast suspension, flask (500 ml) **(3)**				1
10% sucrose, flask (500 ml) **(4)**				1
bromothymol blue, flask or cylinder (500 ml) **(5)**				2
one-hole rubber stopper (# 6 1/2)				4
bent glass tubing ("U")				2
Exercise A (Part 3)				
fermentation apparatus **(6)**		1		12
wax pencil		1		12
yeast suspension, tube (10 ml) **(7)**		1		12
boiled yeast suspension, tube (10 ml) **(8)**		1		12
5% glucose, bottle (25 ml) **(9)**		1		12
5% sucrose, bottle (25 ml) **(10)**				

	Per Student	Per Pair (2)	Per Group (4)	Per Class (24)
Exercise A (Part 4)				
test tubes	1			24
24 hr yeast culture **(11)**				1
10 ml pipette (to deliver with Pi-Pump)				1
10% NaOH **(12)**			1	6
Lugol's reagent **(13)**			1	6
5 ml pipette (to deliver)			2	12
test tube rack		1		12
Exercise B				
test tube			6	36
test tube rack			1	6
Spectronic 20			1	6
Spectronic 20 tubes			6	36
lima bean mitochondria extract **(14)**				1
blender				1
1 mM DPIP, bottle **(15)**			1	6
0.2 M phosphate buffer, pH 7.2, bottle **(16)**			1	6
0.25 M succinate, bottle **(17)**			1	6
0.5 M malonic acid, bottle **(18)**			1	6
Extending Your Investigation				
same requirements as Exercise B				
Exercise C				
vacuum (Thermos) bottle				1
lima bean seeds (100 g)				1
thermometer				1
cotton (roll)				1
one-hole rubber stopper (to fit vacuum bottle)				1
Exercise D				
soaked corn seeds (live) **(19)**				200
soaked corn seeds (boiled) **(19)**				200
tetrazolium (2, 3, 5-triphenyl-tetrazolium), dropping bottle **(20)**			1	6
Petri dish (100 mm)	1			24
dissecting needles			4	24

IV. PREPARATION OF MATERIALS AND SOLUTIONS

(1) stock yeast solution

Prepare 1 liter of 10% molasses or syrup solution (100 ml syrup, 900 ml distilled water) and dispense 200 ml of the solution into each of three 250 ml Erlenmeyer flasks until they are half full. Add 2 g Fleischmann's dried yeast and 0.5 g peptone to each flask. Cotton stopper all flasks and refrigerate.

1. Approximately 12 hr before the start of the laboratory period, place one 250 ml flask in a

37° C oven and incubate for at least 12 hr. Keep stoppered with a cotton stopper. This will be used for Exercise A, Part 1.

2. Approximately 2-3 hr prior to the beginning of the laboratory, place one flask in a 37° C oven and warm it for several hours. This solution will be used for Exercise A, Part 2.
3. Approximately 24 hr before the start of the laboratory, place one 250 ml flask in a 37° C oven and incubate for at least 24 hours. Keep stoppered. This will be used for Exercise A, Part 4.

(2) neutral red

Obtain a mass of 0.2 g neutral red. Add 100% (absolute) ethanol to make 100 ml.

(3) sugar and yeast suspension, flask

Same as stock yeast solution prepared in (1), step 2.

(4) 10% sucrose

Obtain a mass of 10 g sucrose. Add distilled water to make 100 ml.

(5) bromothymol blue

Order prepared (0.04% solution) or grind 0.05 g bromothymol blue powder in 8 ml 0.01 N NaOH and add distilled water to make 125 ml of indicator solution.

(6) fermentation apparatus

The Carolina Biological Supply Company's Basic Fermentation Biokit ($27.20 for a class of 30) can be purchased once and then used year after year with new yeast purchased at the grocery store).

(7) and (8) yeast suspension

Add two packages of dry yeast to 100 ml of warm tap water and mix. Make approximately 5 minutes before use. Boil a 10 ml sample for 5 minutes. Make sure that the yeast suspension itself is actually boiling.

(9) 5% glucose

Dissolve 5 g glucose in 100 ml of distilled water. Each group will need approximately 20-25 ml.

(10) 5% sucrose

Dissolve 5 g sucrose in enough distilled water to make 100 ml of solution. Each group will need approximately 20-25 ml.

(11) 24 hr. yeast culture

See (1), 3.

(12) 10% NaOH

Obtain a mass of 10 g NaOH. Add distilled water to make 100 ml.

(13) Lugol's reagent

Order prepared (Carolina Biological, # 87-2793) or dissolve 10 g potassium iodide in 100 ml distilled water and add 5 g of iodine (handle carefully—do not breathe fumes!).

(14) lima bean mitochondria extract

Materials:

Homogenization medium—prepare 100 ml/class. Obtain a mass of 13.69 g sucrose and add 0.2 M phosphate buffer (pH 7.2) to 100 ml. See (16) for phosphate buffer recipe. Readjust pH to 7.2.

Procedure:

Prepare extract in the classroom immediately prior to use. Soak 50 g white baby lima beans (obtain from grocery store) in tap water overnight. Keep in refrigerator. When ready to use, pour out the water and put the beans into an electric blender. Add 100 ml homogenization medium. Homogenize 3 minutes at top speed. Filter through cheesecloth. Dispense 10-15 ml into each of six 50 ml beakers and give one beaker of extract to each group.

Boil a small amount (approximately 10 ml) for 5 minutes.

(15) 1 mM DPIP (or DCPIP, 2, 4 dichlorophenol-indophenol)

Mass 0.29 g and add distilled water to make 1000 ml.

(16) 0.2 M phosphate buffer, pH 7.2

Solution A. 0.2 M KH_2PO_4

Mass 27.22 g KH_2PO_4. Add distilled water to 1000 ml.

Solution B. 0.2 M K_2HPO_4

Mass 34.84 g K_2HPO_4. Add distilled water to 1000 ml. For 1 liter of buffer, mix 280 ml of solution A plus 720 ml of solution B. Test pH and adjust with either 0.1 N NaOH or 0.1 N HCl if necessary.

(17) 0.25 M succinate

Mass 2.95 g succinate. Add distilled water to 100 ml.

(18) 0.05 M malonic acid

Dissolve 52 g malonic acid and 42 g NaOH in 500 ml distiled water. Add distilled water to 1000 ml and adjust to pH 7.0.

(19) corn seeds

Soak corn seeds overnight in tap water. Mark one set A and the other B. Boil one of the sets of the seeds for 15 minutes to kill corn embryos.

(20) tetrazolium, 0.1%

Mass 1 g tetrazolium. Make up to 1 liter with distilled water.

V. PREPARATION SUGGESTIONS

Exercise A (Part 1)

Yeast cultures for parts 1, 2, and 4 can all be prepared ahead and in the same manner.

Exercise A (Part 2)

In preparing the apparatus for part 2, a piece of glass tubing can be bent into a U to go from the yeast suspension or 10% sucrose flask to the flask of bromothymol blue. If glass tubing is not available, a piece of Tygon tubing will also work. Set-this exercise up as a demonstration only.

Exercise A (Part 3)

The fermentation apparatus gives dependable results. This is a new design for the apparatus. If you have the old style apparatus, simply follow directions included in the kit. For the old style apparatus, it is best to fill the apparatus beneath water in a dishpan.

With the new style apparatus, students may have difficulty discerning the line that marks the top of the yeast solution in the small vial. If this is the case, carefully move the small vial sideways until it is up against the wall of the larger vial.

Exercise A (Part 4)

This reaction is fairly touchy. The yeast suspension must be at least 24 hours old to have enough alcohol in it so that the reaction is positive. Gentle shaking of tubes from side to side is enough for mixing.

Exercise B

Be sure to put directions for using the Spectronic 20 next to the machines. It would be wise to have students practice using the Spectronic 20 before using it in this exercise. If not enough Spectronic 20 machines are available, split the laboratory into two groups of students and have half do Exercise A, Part 3, while the other half completes this exercise.

Results will be more precise if round-bottom cuvettes are used, but we get excellent results with plain Pyrex or Kimax glass tubes (see ordering information).

If the preparation of lima beans it too viscous, dilute it with pH 7 phosphate buffer. This may be particularly necessary with the boiled suspension in which denatured proteins tend to form a viscous mass.

Results will be best if two students sit on either side of the machine. One adds succinate and inverts the tube to mix it. The person reading the Spectronic 20 wipes finger marks off the tube and inserts it into the machine. A person standing behind the reader checks the reading. The person on the other side of the machine records the results.

Malonate acts as a competitive inhibitor of the enzyme. This can be discussed further with the students if desired. K_m values can be compared (see Laboratory 10). Keep in mind that with a competitive inhibitor present, V_{max} can still be obtained if enough substrate is present. For the enzyme in the presence of inhibitor, the rate is slower and K_m is higher than for the enzyme without the inhibitor present.

Extending Your Investigation

Once students have completed Exercise B, the setup for this exercise should be simple. If you desire, several student pairs in the class could do this exercise and compare their results to those for the class. You could also add a tube with excess succinate to Exercise B if desired, although students will not have the opportunity to design their own experiment if this is done. You can increase amounts of succinate 8 to 10 fold for best results.

Exercise C

Soak lima bean seeds overnight and place them in moist cotton in the Thermos. Set up the vacuum bottle at least 3-4 days ahead of the laboratory period and record the temperature at 3-4 hr intervals (skip nights). Place graphed temperature data next to the Thermos when placed on demonstration.

Exercise D

Tetrazolium is a poison and should be handled carefully. The results of this experiment are very obvious and even help students locate the embryo of a corn seed more easily.

If cut seeds are placed cut surface down, the embryos can be viewed by looking through the bottom of the Petri dish if glass or plastic Petri dishes are used.

VI. ORDERING INFORMATION

Erlenmeyer flask, 250 ml—Fisher, # 10-040F (Pyrex) or # 10-039F (Kimax); Carolina Biological, # 72-6672 (Corning, student grade); Ward's, # 17W2982

molasses/syrup—obtain locally

Fleischmann's dry yeast—grocery store; Carolina Biological, # 17-3234; Ward's, # 38W5821

peptone—Fisher, # BP1420-500; Carolina Biological, # 79-4260; Ward's, # 39W2751

microscope slides, 75 × 25 mm—Fisher, # 12-550A; Carolina Biological, # 63-2000; Ward's, #

14W3500

coverslips, 22 × 22 mm—Fisher, # 12-542B; Carolina Biological, # 63-3095; Ward's, # 14W3251

neutral red stain—Fisher, # N129-25 (powder); Carolina Biological, # 87-6853; Ward's, # 38W8383

sucrose (certified)—Fisher, # S5-500; Carolina Biological, #89-2870; Ward's, # 39W3280

bromothymol blue—Fisher, # B388-10; Carolina Biological, # 84-9163 (0.04%); Ward's, # 38W9100

stopper, rubber (#6 ½)—Carolina Biological, # 71-2469; Ward's, # 15W8466

cork borer set—Carolina Biological, # 71-2202; Ward's, # 15W1665

glass tubing, 6 mm—Carolina Biological, # 71-1146; Ward's, # 17W0941

basic fermentation kit—Carolina Biological, # 20-2200

wax pencil—Carolina Biological, # 65-7730 (red), # 65-7732 (blue), # 65-7734 (black); Ward's, # 15W1159

D-glucose (dextrose) (certified)—Fisher, # D16-500 (500 g); Carolina Biological, # 85-7440 (500 g); Ward's, # 39W1457

sodium hydroxide (NaOH)—Fisher, # S318-1; Carolina Biological, # 88-9468; Ward's, # 39W1685

Lugol's reagent—Carolina Biological, # 87-2793; Ward's, # 39W1685

lima beans—obtain locally

blender—Carolina Biological, # 70-1905 (glass) or # 70-1908 (stainless); Ward's, # 15W2675

cheesecloth—Fisher, # 06-665-18 (70 yard bolt); Carolina Biological, # 71-2690 (5 yard package); Ward's, # 15W0015

beaker, 50 ml—Fisher, # 02-540G (Pyrex) or # 02-539G (Kimax); Carolina Biological, # 72-1220 (Corning, student grade); Ward's, # 17W4010

test tubes, 16 × 125 mm—Carolina Biological, # 73-1418; Ward's, # 17W1403

test tube rack—Carolina Biological, # 73-1890; Ward's, # 18W4213

Spectronic 20 (Milton Roy, formerly B&L)—Fisher, # 07-143-1; Carolina Biological, # 65-3300; Ward's, # 14W5550

wide range phototube and filter—Fisher, # 07-144-10; Carolina Biological, # 65-3312; Ward's, # 14W5553

Spectronic 20 tubes—Fisher, # 14-385-9008; Carolina Biological, # 65-3310; Ward's, # 17W1400 OR

Spectronic 20 tubes, Pyrex (100 • 13 mm)—Fisher, # 14-957C; Carolina Biological, # 73-1408

Spectronic 20 tubes, Kimax (100 • 13 mm)—Fisher, # 14-923D; Carolina Biological, # 73-1408A

DPIP (2,6-dichloroindophenol)—Fisher, # S286-5; Sigma, # D 1878; Carolina Biological, # 86-8600; Ward's, # 39W1670

sodium phosphate (monobasic) (NaH_2PO_4)—Fisher, # S369-1; Carolina Biological, # 89-1350; Ward's, # 37W5655

potassium phosphate (dibasic) (K_2HP0_4)—Fisher, # P288-500; Carolina Biological, # 88-4290; Ward's, # 37W4841

hydrochloric acid (HCl)—Fisher, # A144-212; Carolina Biological, # 86-7790; Ward's, # 37W2250

succinate (succinic acid)—Fisher, # A294-500; Carolina Biological, # 89-2830 (500 g)

malonic acid—Fisher, # A170-100; Carolina Biological, # 87-3720; Ward's, # 37W2869
corn seeds—obtain locally; Carolina Biological, # 15-9243; Ward's, # 86W8080
tetrazolium (2, 3, 5 triphenyl tetrazolium chloride)—Fisher # T413-5; Carolina Biological, # 89-6930; Ward's, # 39W3534
dropping bottle, 30 ml— Fisher, # 02-980; Carolina Biological, # 71-6525; Ward's, # 17W6011
Petri dish, 100 × 15 mm—Fisher, # 08-747C (Pyrex) or 08-746C (Kimax); Carolina Biological, # 74-1158; Ward's, # 17W0700
dissecting needles—Carolina Biological, # 62-7200; Ward's, # 14W0650

Notes:

LABORATORY 12 Photosynthesis

I. FOREWORD

The progression of exercises in this laboratory allows students to first examine the effects of light intensity and light quality on photosynthesis. After studying the absorption of light by different pigments, an action spectrum is generated. The final exercise investigates how plants use trapped light energy for synthesis of photosynthetic products. Exercises can be organized to be completed on separate days if desired.

BioBytes *Seedling* can be used to supplement this laboratory. *Seedling* provides students with the opportunity to investigate the relationship between photosynthesis and transpiration.

II. TIME REQUIREMENTS

Exercise A—The Energy-Capturing Reactions
- Part 1—Determining the Effect of Light Intensity (45 minutes)
- Part 2—The Spectrum of Visible Light (15 minutes)
- Part 3—Determining the Effects of Light Quality—The Action Spectrum (45 minutes)

Exercise B—The Pigments of Chloroplasts
- Part 1—How Plant Pigments Use the Light Spectrum (30 minutes)
- Part 2—Absorption Spectra of Chloroplast Pigments (45 minutes)
- Part 3—The Role of Light in Chlorophyll Synthesis (10 minutes)

Exercise C—The Light-Independent Reactions of Photosynthesis (30 minutes)

Extending Your Investigation: Do All Plants Store Starch? (30 minutes)

III. STUDENT MATERIALS AND EQUIPMENT

	Per Student	Per Pair (2)	Per Group (4)	Per Class (24)
Exercise A (Part 1—as class exercise)				
0.2% $NaHCO_3$ (2 liters) **(1)**			1	6
2000 ml beaker		1		12
ring stand		1		12
3-prong clamp		1		12
light assembly		1		12
light bulbs (40, 60, 100, or 150 w)	1			24
Petri dish (clear)		1		12
Petri dish (covered with black tape) **(2)**		1		12
side-arm flask, 250 ml				1
rubber stopper, one-hole (size 6 ½)				1
fresh spinach (bag)				1
cork borer, #3		1		12
paper towels or Styrofoam board		1		12
vacuum tubing				1
vacuum pump (or aspirator, Nalgene)				1

	Per Student	Per Pair (2)	Per Group (4)	Per Class (24)
Exercise A (Part 2)				
equilateral or 90° right angle glass prism				1
slide projector				1
slit-slide **(3)**				1
Exercise A (Part 3)				
ring stand			1	6
photoflood lamp (150 w)			1	6
meter stick			1	6
2 liter beaker			1	6
Petri dish covered with colored filter **(4)**			1	6
250 ml side-arm flask				1
1 hole rubber stopper (size 6 ½)				1
fresh spinach, bag				1
cork borer, #3			1	6
paper towels or Styrofoam board			1	6
vacuum tubing				1
vacuum pump (or aspirator, Nalgene)				1
Exercise B (Part 1)				
chromatography paper **(5)**				1
capillary pipette	1			24
stoppered test tube, large			1	6
chromatography solvent, bottle **(6)**				1
chloroplast pigment extract **(7)**				1
test tube rack				1
Exercise B (Part 2)				
xanthophyll pigments, bottle **(8)**				1
chlorophyll a, bottle **(8)**				1
chlorophyll b, bottle **(8)**				1
carotenoids, bottle **(8)**				1
solvent blank for chlorophyll b and xanthophylls **(8)**				1
solvent blank for carotenoids and chlorophyll a **(8)**				1
Spectronic 20			1	6
Spectronic 20 tubes			1	6
Exercise B (Part 3)				
2 flats wheat seedlings (grown in dark and light) **(9)**				1
Exercise C				
variegated *Coleus* plant **(10)**				2
hot plate			1	6
250 ml beaker			1	6
hot 70% ethanol, bottle			1	6
Lugol's solution, dropping bottle **(11)**			1	6

	Per Student	Per Pair (2)	Per Group (4)	Per Class (24)
Extending Your Investigation				
potato				1
onion				1
Lugol's solution (I_2KI), dropping bottle **(12)**			1	6
test tube	3			72
Benedict's reagent, bottle **(13)**			1	6
pipette, 5 ml			1	6

IV. PREPARATION OF MATERIALS AND SOLUTIONS

(1) 0.2% $NaHCO_3$

Mass 2 g $NaHCO_3$. Add distilled water to make 1 liter.

(2) Petri dish (dark)

Cover the top and bottom of a Petri dish with black electrician's tape.

(3) slit slide

Take a 35 × 38 mm piece of black paper or an exposed and developed piece of 35 mm film and make a slit in it, 1-2 mm wide. Mount this in a 2 × 2 slide mount.

(4) Petri dishes covered with colored filters

Materials:

Filters—daylight blue # 851 medium green # 874 medium red (2 layers) #823.

Procedure:

Cover the bottom and sides of the Petri dishes with black plastic (electrician's) tape. Cut circles from the large sheets of filters and cover the top of the Petri dish, taping the filter on with black tape around the sides. Be sure to use two layers of red filters.

(5) chromatography paper

Use Whatman # 1 filter paper for chromatography. Cut strips 90 mm long for 100 mm tubes. This should give enough of an end to bend over and tack to the cork. Cut one end into a point so that the tapered portion is approximately 15-20 mm of the length of the strip.

(6) chromatography solvent

For paper chromatography, use 4 parts petroleum ether to 1 part acetone.

(7) chloroplast pigment extract

Use fresh spinach. Place 10-20 g of spinach in 50 ml of 100% acetone. Crush the leaves using a mortar and pestle until the acetone solution is a dark green. (**Caution**: this is a flammable mixture so do not use a blender.)

(8) chlorophyll extract for chromatography and for action spectrum

Chlorophyll extracts can be ordered from Sigma (see Ordering Information). These include chlorophyll a, chlorophyll b, carotene, and xanthophyll.

When new pigment containers are opened, they should be rinsed out with acetone to remove the pigment from the vial. Store pigments in the freezer. Pigments are prepared by putting a small amount of pigment stock in a spectrophotometer tube or cuvette to which a larger amount of acetone is added. Each pigment solution will need to be standardized to an absorbance range at a set wavelength by adding either acetone or stock pigment. Cuvettes should be tightly stoppered with green rubber #000 stoppers (these are general instructions because everything relies on pigment stock solution that varies with orders). The following table indicates wavelength settings for the Spectronic 20 and the absorbance range to be obtained for each pigment.

Tube #	Pigment	Wavelength	Absorbency
1	whole extract	450/660	0.8-0.9/0.3-0.4
2	chlorophyll a	660	0.3-0.4
3	chlorophyll b	660	0.3-0.4
4	xanthophyll	450	0.8-0.9
5	carotene	450	0.8-0.9

Pigments should be checked before use (each day if multiple lab sections are taught). Tubes should be numbered near the top so that tape does not interfere with absorption readings. Pigments and corresponding numbered blanks (acetone only) should be placed at spectrophotometer stations for students. When work with pigments is completed, acetone can be evaporated under a fume hood and dry pigments can be stored in the freezer.

To extract pigments from spinach or parsley, use the following directions: [1]

WORK UNDER A FUME HOOD

Gather together the following materials:

Chemicals and Solutions

acetone
$CaCO_3$
petroleum ether
methanol
diethyl ether
30% KOH in methanol (30 g KOH: add methanol to 100 ml)

Materials and Equipment

spinach (fresh) or parsley (dried)
blender
Buchner funnel and filter paper
side arm filter flask a
aspirator or vacuum pump
separatory funnel
2 125 ml Erlenmeyer flasks
1 100 ml graduated cylinders
4 screw top test tubes

Procedure:

1. Place 3 g dried parsley in 40 ml 80% acetone or 10 g fresh spinach in 50 ml 100% acetone (latter preferred).
2. Add a pinch of $CaCO_3$ to prevent Mg^{2+} loss from chloroplasts.
3. Homogenize in blender for 3 minutes, top speed. (Be careful to avoid sparks!)
4. Vacuum filter through Buchner funnel to remove debris.
5. Readjust volume of filtrate to 40 ml.
6. Place in separatory funnel containing 60 ml petroleum ether.
7. Add 70 ml water to the pigment mixture by pouring the water down the side of the

[1] Acknowledgement: Ross, Cleon W. *Plant Physiology Laboratory Manual.* Wadsworth Publishing Co.

funnel.

8. Stopper and rotate slowly until the upper layer contains nearly all of the chlorophyll. Gas pressure will rise in the funnel—unstopper and vent carefully.
9. Permit the layers to separate.
10. Drain off the lower layer (acetone) and discard. The upper layer (petroleum ether) now contains the chlorophyll (all chloroplast pigments).
11. Add 20 ml distilled water to wash the petroleum ether and remove any traces of acetone. Do this twice.
12. Remove 5 ml of the petroleum ether-chlorophyll layer and put it in a test tube. Allow this to evaporate down to get a very concentrated extract (several hours or less). Stopper when three-fourths of the fluid has evaporated. This chlorophyll extract can be used for chromatography.
13. Add 50 ml of 92% methanol (92 ml methanol, 8 ml distilled water) to the petroleum ether extract and mix. Do not breathe the solvents.
14. Chlorophyll b and xanthophylls are polar enough to dissolve in the methanol while carotenes and chlorophyll a will remain in the petroleum ether (upper layer).
15. Draw off the two layers into two separate 125 ml flasks.
16. Place 50 ml of the methanol layer (chlorophyll b and xanthophylls) back into the funnel and add 50 ml of diethyl ether.
17. Add approximately 25 ml of distilled water, 5 ml at a time down the side of the funnel. Mix each 5 ml portion by inverting the funnel. This will remove the methanol, but the chlorophyll a and xanthophylls will remain in the upper ethyl-ether layer.
18. Discard the lower layer and place 30 ml of the diethyl ether layer into a 125 ml Erlenmeyer flask.
19. Now return the petroleum ether layer (steps 14 and 15) which is already in a similar 125 ml flask.
20. Add 15 ml of 30% KOH in methanol to this flask and to the flask of the diethyl ether extract (18).
21. You are hydrolyzing the phytol tail off the chlorophyll molecule so that the chlorophyll pigments will dissolve in a more polar solvent. Swirl the flasks frequently (for at least 10 minutes or until the yellow upper layer is free of any green color in each flask.
22. Now add 30 ml of distilled water to each flask and mix by gently swirling.
23. Pour the contents of each of the flasks into separate graduated cylinders and allow the phases to separate.
24. The petroleum ether extract will contain carotenes in the upper layer and chlorophyll a in the lower layer. The diethyl ether extract will contain xanthophylls in the upper layer and chlorophyll b in the lower layer.
25. Separate the four layers with a pipette and dispense into four separate test tubes with plastic screw tops. Refrigerate.

For blanks to use with spectrophotometry, you must use petroleum ether with carotenes and chlorophyll a and diethyl ether with xanthophylls and chlorophyll b.

You may need to dilute the pigments with the appropriate solvents for use with the Spectronic 20.

If you leave the pigments under the hood and significant evaporation occurs, simply add more solvent (this is not quantitative).

(9) wheat seedling

Plant two flats of wheat seedlings approximately 4-5 days before the laboratory. Put one flat in the dark and leave the other in the light.

(10) *Coleus* plant

Place a *Coleus* plant in the dark 48 hours prior to use.

(11) Lugol's solution

Order prepared or dissolve 10 g of potassium iodide in 100 ml of distilled water and add 5 g of iodine.

V. PREPARATION SUGGESTIONS

Exercise A (Part 1)

Students work in pairs during this exercise. Each pair should cut approximately 50 spinach disks using the cork borer. If you have access to a metal shop, have metal tubing (5 mm) cut and sharpened on one end. Add a handle to the other end. Attach a water aspirator to the sink faucet. Put all spinach disks (from the entire class) into the side-arm flask containing $NaHCO_3$. Pull a vacuum for 1 minute. Release the vacuum and check to see if disks have sunken. (Do not overdo the "degassing" procedure because you might rupture the leaf cells.) If some disks remain floating after 1 minute, repeat the procedure. Turn the lights off in the room before trying to dispense degassed disks to students (otherwise photosynthesis may produce oxygen and cause the disks to float). Tell students to make sure that all disks they use settle to the bottom of the Petri dish to begin with. Each student pair should expose the disks to one light source (either 40, 60, 100, or 150 w) and should have one dish covered with black tape to keep disks in the dark. A 2000 ml beaker should be used as a heat sink on top of the clear Petri dish. Collect class data for all light intensities.

Exercise A (Part 2)

Shine the light of a slide projector through a slit slide. Prop the prism up so that the light goes through it and rotate the prism until the light spectrum can be seen on a wall or piece of white poster board in the room. When you are ready to look at the effects of different wavelengths of light (Exercise A, Part 3), simply place the Petri dish covered by the filter in front of the projector and see which colors of light disappear from the spectrum. Discuss with the students what a green filter absorbs and what it transmits or reflects. Do the same with other colors of filters.

Exercise A (Part 3)

Have students cut out spinach disks. If you assign one filter color to a group of students (so there is a no filter group, a red group, a green-group, a blue group, and a dark group),have all students place their disks into one flask and aspirate all disks at the same time. Students can then obtain disks from one common source. (The degree of aspiration will have an effect on the amount of time it takes for disks to float.) If a side arm flask is used, place a piece of masking tape over the hole in the one-hole stopper. While aspirating, peel off the tape from time to time to see if the disks will sink. If they do not, replace the tape. Remember to release the vacuum in this manner before turning off the water. If using a two-hole stopper with glass tubing for making the connection to an aspirator, then use the second hole as described.

Be sure to use filters only from Edmund Scientific (see Suppliers). Using another red or blue filter will not give you the proper results since distances and amount of light have been calculated for these filters. If you use different filters, adjust the height for each 100 w flood light using a light meter under the proper filter.

Preparation materials are written to assign one group to a filter color each to save on the number of lights and the preparation of Petri dishes with filters.

Exercise B (Part 1)

Cut strips of Whatman #1 filter paper to fit into large test tubes. Cut one end to form a point (like an arrow). Put approximately 0.5" of solvent in the bottom of the test tube and use a rubber cork to close the tube tightly. If you cut pointed ends on the paper strips, you do not need to tack the strips to the cork. Simply lower them gently into the solvent (do not drop the paper strips into the tube because solvent can splash onto the paper). While the air in the tube is saturating with solvent, apply a line (or dot) of chlorophyll extract to the paper strip above the pointed end. (A simpler method requires that a leaf of ivy or spinach be placed bottom side on the paper and a dime can be rolled over the leaf to make the line of chlorophyll extract on the paper.) Place the paper strip into the solvent and allow it the solvent to ascend until it is almost at the top. Remove the paper and mark the solvent front with a pencil. The order of pigments from top (solvent front) to bottom (pointed end): carotenes (orange-yellow pigments), xanthophylls (yellow pigments), chlorophyll a (bright green to blue-green pigment), chlorophyll b (yellow-green to olive-green). Some grayish breakdown products may also be associated with the xanthophylls. The R_f can be calculated for each pigment if paper is used:

$$R_f = \text{(distance pigment migrated)/(distance solvent front migrated)}$$

Exercise B (Part 2)

Separate tubes of extracts with matching blanks are provided for student use. Tell students not to throw material away—it can be used again and again. Number tubes and blanks with matching numbers or make sure proper blanks are used.

Have students use different colors of chalk or markers to put all action spectra on one graph on the blackboard or on a transparency for class discussion.

Exercise B (Part 3)

Grow wheat seedlings for 4-5 days in Styrofoam pots. Place one pot in the dark and one in the light.

Exercise C

Coleus or geraniums can be used in this exercise. Do not let students boil the alcohol because it could start a fire. Place a beaker of alcohol inside a second beaker of water (similar to a double-boiler).

Place the variegated *Coleus* in the dark for two days prior to this experiment to make sure all starch has been translocated from the white or pink areas.

If variegated *Coleus is* not available, use plain green *Coleus* and cover parts of the leaves with foil for a week before use. Starch will be present in the uncovered leaves, but not in the areas covered by foil.

Extending Your Investigation

This investigation gives students the opportunity to further improve their hypothesis formation and experimental design skills. Potatoes store starch in their roots while onions store sugar. Onion will turn greenish and finally brick red when boiled with Benedict's reagent, indicating the presence of glucose. Is sucrose present too? This can be determined by mixing chopped onions and their juice with 10 ml distilled water. Add 1 ml of 5% cobalt nitrate solution (5 g cobalt nitrate dissolved in distilled water to make 100 ml). Then add 1 ml of a 10 N NaOH solution (40 g NaOH dissolved in distilled water to make 100 ml). A violet color indicates the presence of sucrose. Students might also ask what type of carbohydrate is made in the leaves. Allow the potatoes and onions to sprout and use the first green leaves to repeat the test. Most dicots (including potatoes) support starch production in the leaves, while most monocots (including onions) support glucose production and storage.

VI. ORDERING INFORMATION

clamp, 3 prong—Fisher, # 05-768-10; Carolina Biological, # 70-7376 or 70-7312; Ward's, # 15W0665
ring stand, medium—Carolina Biological, # 70-7192; Ward's, # 15W0692
photoflood lamp and bulb—local hardware store or Carolina Biological, # 68-7000 or 68-7100
Parafilm—Fisher, # 13-274-12 (dispenser, # 13-374-18); Carolina Biological, # 21-5600 (dispenser, # 21-5602); Ward's, # 15W1940
meter stick—Carolina Biological, # 70-2620; Ward's, # 15W4065
prism, 90 deg., variety of sizes—Edmund Scientific; Carolina Biological, # 79-4956; Ward's, # 25W2007
Petri dish, 100 × 15 mm—Fisher, # 08-747C (Pyrex) or 08-746C (Kimax); Carolina Biological, # 74-1158; Ward's, # 17W0700
side-arm flask, 250 ml—Carolina Biological, # 72-6346
Buchner funnel, 71 mm—Carolina Biological, # 73-4626
Whatman filter paper, 70 mm—Fisher, # 09-805C
vacuum pump (aspirator), Nalgene —Fisher, # 09-960-2; Carolina Biological, # 71-1972; Ward's, # 18W1570
beaker, 2000 ml— Fisher, # 02-540R (Pyrex) or # 02-539R (Kimax); Carolina Biological, # 72-1215 (Pyrex) or # 72-1215A (Kimax); Ward's, # 17W4122
rubber stopper, 1 hole (size 6 ½)—Fisher, # 14-130K; Carolina Biological, # 71-2439
cork borer set—Carolina Biological, # 71-2202; Ward's, # 15W1666
vacuum tubing—Carolina Biological, # 71-1544
vacuum pump, Nalgene —Fisher, # 09-960-2; Carolina Biological, # 71-1972
meter stick—Carolina Biological, # 70-2620 or 70-2624 (metal ends); Ward's, # 15W4065
Whatman chromatography paper, #l—Fisher, #05-714 (package of 100); Carolina Biological, # 21-5610; Ward's, # 15W3708
daylight blue filter—Edmund Scientific, # H82,031[2]
medium green filter—Edmund Scientific, # H82,041
medium red filter—Edmund Scientific, # H82,015
capillary pipette, 50-100 μl—Carolina Biological, # 21-4517
test tube, large (25 × 200 mm)—Carolina Biological, # 73-1210
Lugol's solution—Carolina Biological, # 87-2793; Ward's, # 39W1685

[2] The three recommended filters are no longer listed in the Edmund catalog but are available at the time of publication (but future availability is not guaranteed). The following filters are not identical but are suggested by Edmund as suitable replacements: # 35,135 (blue), # 35,136 (green), and # 35,137 (red). Contact Edmund Scientific (see Suppliers) for further information.

Benedict's reagent—Fisher, # LC11650-1; Carolina Biological, # 84-711 (500 ml); Ward's, # 37W0698

Spectronic 20 (Milton Roy, formerly B&L)—Fisher, # 07-143-1; Carolina Biological, # 65-3300; Ward's, # 14W5550

wide range phototube and filter—Fisher, # 07-144-10; Carolina Biological, # 65-3312; Ward's, # 14W5553

Spectronic 20 tubes—Fisher, # 14-385-9008; Carolina Biological, # 65-3310; Ward's, # 17W1400 OR

Spectronic 20 tubes, Pyrex (100 × 13 mm)—Fisher, # 14-957C; Carolina Biological, # 73-1408

Spectronic 20 tubes, Kimax (100 × 13 mm)—Fisher, # 14-923D; Carolina Biological, # 73-1408A

pipette, 5 ml—Fisher, #13-665K; Carolina Biological, # 73-6276; Ward's, # 17W1307

hot plate—Carolina Biological, # 70-1010; Ward's, # 15W8010

Erlenmeyer flask, 250 ml—Fisher, # 10-040F (Pyrex) or # 10-039F (Kimax); Carolina Biological, # 72-6672 (Corning, student grade); Ward's, # 17W2982

sodium bicarbonate ($NaHCO_3$)—Fisher, # S233-500; Carolina Biological, # 88-8380; Ward's, # 37W5450

chlorophyll a—Sigma, # C 5753

chlorophyll b—Sigma, # C 5878

carotene—Sigma, # C 9750

xanthophyll—Sigma, # X 7250

petroleum ether—Fisher, # E139-500; Carolina Biological, # 87-9580

acetone—Fisher, # A18-500; Carolina Biological, # 84-1500

calcium carbonate ($CaCO_3$)— Fisher, # C64-500; Carolina Biological, # 85- 1760; Ward's, # 37W0921

methanol—Fisher, # A408-1; Carolina Biological, # 87-4913

diethyl ether—Carolina Biological, # 86-1408

potassium hydroxide (KOH)—Fisher, # P250-1; Carolina biological, # 88-3490

blender—Carolina Biological, # 70-1905 (glass) or # 70-1908 (stainless); Ward's, # 15W2675

separatory funnel—Carolina Biological, # 73-4585 (250 ml) or # 73-4587 (500 ml)

Erlenmeyer flask, 125 ml—Fisher, # 10-040D (Pyrex) or # 10-039D (Kimax); Carolina Biological, # 72-6670 (Corning, student grade); Ward's, # 17W2972

graduated cylinder, 100 ml—Fisher, # 08-553B (Pyrex) or # 08-549-5E (Kimax) or # 08-572-6D (Nalgene); Carolina Biological, # 72-1788 (Pyrex) or # 72-1948 (Kimble) or # 72-2266 (Nalgene); Ward's, # 17W0578

screw top test tubes, 16 × 125 mm—Carolina Biological, # 73-1518

Coleus plant, live—buy locally; Carolina Biological, # 15-7310; Ward's, # 96W6800

Notes:

LABORATORY 13 Meiosis: Independent Assortment and Segregation

I. FOREWORD

The exercises included in this laboratory are designed to help students understand the chromosomal events in the process of meiosis. A simulation has been chosen because it provides the instructor with an opportunity to check each student's understanding of the process by simply observing the student's work displayed on the table.

Segregation and independent assortment are brought into the exercises to provide an introduction to the genetic consequences of meiosis.

Supplementing the laboratory simulations with slides of meiosis in lily or *Ascaris* is possible if slides are available, but we have found that such exercises are of little help to the students because they do not fully understand the biology of the systems they are observing.

Crossing over and its consequences are also examined in the fungus, *Sordaria fimicola.* Students prepare crosses for examination of asci during the following week. Varied arrangements of spores within the asci allow students to map one of the genes for spore color.

II. TIME REQUIREMENTS

Exercise A—Simulation of Chromosomal Events During Meiosis (30 minutes)
Exercise B—Mendel's First Law: Alleles Segregate During Meiosis (60 minutes)
Exercise C—Mendel's Second Law: Alleles of Unlinked Genes Assort Independently (45 minutes)
Extending Your Investigation: Meiosis and Linked Genes (15 minutes)
Exercise D—Meiosis and Crossing-Over in *Sordaria (Optional)* (60 minutes)
Part 1—Crossing *Sordaria* Strains (Week 1)
Part 2—Analyzing Hybrids and Crossovers (Week 2)

III. STUDENT MATERIALS AND EQUIPMENT

	Per Student	Per Pair (2)	Per Group (4)	Per Class (24)
Exercise A				
chromosome simulation kit **(1)**				2
Exercise B				
chromosome simulation kit **(1)**				2
labels for alleles, container **(2)**			1	6
colored pencils (red, yellow, 1 each)	2			48
monohybrid cross (corn), display box **(3)**			1	6
monohybrid cross (corn), ears **(4)**		1		12
Exercise C				
chromosome simulation kit				2
dihybrid cross (corn), display box **(3)**			1	6
dihybrid cross (corn), ears **(4)**		1		12

	Per Student	Per Pair (2)	Per Group (4)	Per Class (24)
Exercise C—continued				
colored pencils (red, yellow, 1 each)	1			24
labels for alleles, container **(2)**			1	6
Extending Your Investigation				
chromosome simulation kit **(1)**				2
Exercise D (optional)				
Sordari fimicola, wild type, agar plate **(5)**				1
Sordari fimicola, tan mutant, agar plate **(5)**				1
crossing agar plate **(6)**	1			24
flat weighing spatula (small)				2
95% ethanol, beaker				2
alcohol burner (or Bunsen burner)				2
Sordaria + × *t* cross plates (hybrid black/tan) **(7)**		1		12
microscope slides (75 × 25 mm)	5			120
coverslips (22 × 22 mm)	5			120
compound microscope	1			24

IV. PREPARATION OF MATERIALS AND SOLUTIONS

(1) chromosome simulation kit

If you completed Laboratory 9, Mitosis, you have all of the materials necessary for this laboratory. Student simulation kits should be assembled using the two large BioKits. Each student chromosome simulation kit should consist of 2 yellow strands of beads and two red strands of beads. Four centriole bodies will be necessary for meiosis. See Preparation of Materials and Solutions, section (1) in Laboratory 9 for a more detailed description of the chromosome simulation kit and possible modifications.

(2) labels for alleles

Do not let the students wrap tape around the beads—it is impossible to remove. We use sheets of white envelope labels from which we have cut small circles using a hole punch. Students can put a letter on a circle and peel the back off to attach the label to the beads.

(3) plastic display boxes of corn

Plastic display boxes of monohybrid and dihybrid corn—order from Carolina Biological Supply Company or Ward's. (See Ordering Information.)

(4) demonstration ears of monohybrid and dihybrid corn

Order from Carolina Biological Supply Company or Ward's. (See Ordering Information.)

(5) *Sordaria fimicola,* wild type or tan mutant (agar plates)

Cultures of wild type or mutant tan *Sordaria* can be ordered from Carolina Biological Supply Company or Ward's. (See Ordering Information.) Two plates of each should be enough for a class of 24. If more plates are needed, it is easy to propagate additional plates from the original as follows.

Prepare corn meal agar plates—these are not the same as the crossing agar plates indicated in **(6)**.

corn meal agar	17 g
dextrose	2 g
yeast extract	1 g

Add distilled water to 1 liter and autoclave 15 minutes at 15 lbs. pressure and 121° C. Cool and pour approximately 30 ml into each sterile Petri dish.

Aseptically, remove a small block of the agar containing mycelia of the desired strain. Cut the block using a small metal spatula (see Ordering Information) sterilized by dipping it in ethanol and then flaming it. Cool the spatula (it should not sizzle when it touches the agar) before cutting. Place the agar block upside down on the surface of the corn meal agar plate. Incubate in the dark at room temperature for approximately 5 days. At this time, the dish should be covered with a mycelial mat.

To prepare plates for student use, take a sterile spatula and cut crosswise, forming a grid of small blocks approximately 5 mm square (they can be as large as 1 cm square). Replace the lid on the plate.

Students will only have to sterilize a spatula and lift two squares of culture from the plate of the wild type and two squares from the plate of mutant tan *Sordaria.* These are to be placed on a plate of crossing agar **(6)**. Crosses will be ready for analysis after incubating for approximately 7 days in the dark (room temperature, 22-24° C). Cooler temperatures can be used if you wish to delay formation of perithecia for 14-15 days. If cultures develop too quickly, you may slow them down by placing them in the refrigerator.

(6) crossing agar plates

Using a 1000 ml Erlenmeyer flask, dissolve the following in 500 ml of distilled water:

corn meal agar	8.5 g
sucrose	5.0 g
dextrose	3.5 g
potassium phosphate (monobasic)	0.05 g
yeast extract	0.5 g

Autoclave for 15 minutes at 15 lb. pressure (121° C). After autoclaving, let the agar cool and then pour enough agar (20-30 ml) solution into each Petri dish to cover the bottom.

(7) *Sordaria* + × *t* cross plates (hybrid black/tan)

Crosses of black (+) and tan (*t*) *Sordaria* can be made as above ((**5**) and (**6**)) a week ahead of this laboratory, or crosses can be ordered from Carolina Biological Supply (# 15-5846) but ORDERS MUST BE PLACED 3 WEEKS IN ADVANCE for crosses. Crosses will be needed for Laboratory 14, Exercise A, so they should be made during this laboratory or should have been ordered. (Laboratory 14, Exercise A can be completed in combination with Laboratory 14, Exercise D, Part 2 to avoid needing hybrid plates during two different weeks.) Plates can be refrigerated (upside down) if they need to be kept for a week but condensation upon warming may create problems so be prepared with some sterile lids so you can change lids if necessary.

V. PREPARATION SUGGESTIONS

Exercise A

Check the work on the table in front of each student to make sure that he or she has properly separated homologous pairs. Go around the room and place different combinations of chromosomes on each student's workplace. For instance, a stranded red long and 2 stranded red short. Ask for mitosis. The student should tell you that he or she cannot do it because that is a haploid cell. Different combinations of colors and lengths can be used to test understanding.

Exercise B

A discussion of how corn kernels are formed is advisable. Most students are not very knowledgeable concerning the biology of corn and, in order to understand the relationships of gametes to kernel color, some review is necessary. This will enhance the value of the exercise.

Exercise C

Check the work on the table in front of each student to make sure that the homologous chromosomes have been separated. Students tend to put two yellow chromosomes together and two red chromosomes together. Size is the difference which is important in identifying different chromosome types in this exercise. Color simply identifies whether chromosomes are maternal or paternal in origin.

Check all drawings and paperwork to make sure that students understand the concept of independent assortment and can relate it to the meiotic process.

Extending Your Investigation

This exercise can be completed at home as a problem assignment. It is simple but will test your students' understanding of meiosis and independent assortment.

Exercise D

Be sure to use wild type *Sordaria* crossed with the tan mutant for this exercise. The allele for tan color maps at approximately 26 crossover units from the centromere, while the allele for gray spore color maps at about 60 crossover units from the centromere. Because of the occurrence of double crossovers, distances greater than 33.3 map units cannot be calculated, and if you try to map the gray allele, you will obtain a value close to 33.3 rather than the correct 60. If tan mutants are crossed with gray mutants, four spore colors are possible and make mapping difficult. (It might, however, be desirable to maintain a single plate of gray mutant *Sordaria* or to make a gray × tan cross for demonstration purposes.) A prepared demonstration slide (Carolina Biological, # B 2615) is also useful for explaining what students are to look for on their slides.

If you desire to maintain wild type or mutant *Sordaria* cultures in your laboratory, new plates can be inoculated using the spores on the lid of a Petri dish in which perithecia have formed (you can also use mature perithecia). Inoculate opposite sides of the plates, approximately 2 cm from the edge. Allow mycelia to begin growth and store the plates in the refrigerator. Make new plates, using spores, at six month intervals.

If you have never worked with *Sordaria* before, it is also possible to purchase a *Sordaria* genetics BioKit (#15-5847) from Carolina Biological Supply Company. The kit comes with all necessary materials and instructions.

VI. ORDERING INFORMATION

chromosome simulation BioKit—Carolina Biological, # 17- 1100 (order 2 kits if they are not available from Laboratory 9)

monohybrid cross (corn, display box)—Carolina Biological, # 17-6810; Ward's, # 67W1160

dihybrid cross (corn, display box)—Carolina Biological, # 17-6900; Ward's, # 67W1161
genetic corn ears, monohybrid (*RR* × *rr*)—Carolina Biological, # 17-6500; Ward's, # 86W8900
genetic corn ears, dihybrid (*R/R Su/Su* × *r/r su/su*)—Carolina Biological, # 17-6600; Ward's, # 86W8917
Sordaria, wild type (black)—Carolina Biological, # 15-6291; Ward's, # 85W8010
Sordaria, tan mutant—Carolina Biological, # 15-6295; Ward's, # 85W9102
Sordaria, + × *t* cross—Carolina Biological, # P7-15-5846 (order 3 weeks in advance with specific Friday ship date); Ward's, # 85W9103
corn meal agar—Carolina Biological, #78-2461; Ward's, # 38W0321
sucrose (certified)—Fisher, # S5-500; Carolina Biological, # 89-2870; Ward's, # 39W3280
dextrose—Fisher, # D16-500; Carolina Biological, # 85-7440; Ward's, # 39W1457
potassium phosphate (monobasic) (KH_2PO_4)—Fisher, # P285-500; Carolina Biological, # 88-4260 (125 g); Ward's, # 37W4845
yeast extract—Sigma, # Y 4000; Carolina Biological, # 79-4780 (100 g) or # 79-4781 (500 g); Ward's, # 38W5800
weighing spatula (small)—Fisher, # 14-373 or 21-401-10; Carolina Biological, # 70-2706; Ward's, # 15W4330
Bunsen burner—Carolina Biological, # 70-6645; Ward's, # 15W0604
OR alcohol lamp with cap—Carolina Biological, # 70-6604; Ward's, # 17W0211
microscope slides, 75 × 25 mm—Fisher, # 12-550A; Carolina Biological, # 63-2000; Ward's, # 14W3500
coverslips, 22 × 22 mm—Fisher, # 12-542B; Carolina Biological, # 63-3095; Ward's, # 14W3212

Notes:

LABORATORY 14 Genes and Chromosomes: Chromosome Mapping

I. FOREWORD

In this laboratory, students are introduced to several methods of mapping chromosomes. Students map the distance between the centromere and the allele for tan spore color in the fungus *Sordaria fimicola* by observing the effects of crossing over on the distribution of wild type and tan ascospores within asci. Crosses must be set up a week before the laboratory (see Laboratory 13). Students will also have the opportunity to visualize the banding pattern on the polytene chromosomes of *Drosophila.* This will help students to understand the banded conformation of human chromosomes (Laboratory 15). Students can then map several genes on the circular chromosome of *Escherichia coli* by interrupting the process of conjugation—a classic technique now augmented by recombinant DNA methodology or by using restriction enzymes and electrophoresis of DNA fragments to establish a restriction map.

II. TIME REQUIREMENTS

Exercise A—Mapping the Genes of *Sordaria fimicola* (60 minutes)
  Part 1—Mapping *Sordaria* Chromosomes (Week 2; see Laboratory 13, Exercise D)
  Part 2—The Chi-Square Test (optional)
Exercise B—Examining the Giant Chromosomes of *Drosophila*—(60 minutes)
Exercise C—Mapping the Chromosome of *Escherichia coli* (45 minutes)
Exercise D—Restriction Endonucleases: Mapping Bacteriophage Lambda (2 hours, gel destaining overnight)
  Part 1—Estimating DNA Fragment Size Using Gel Electrophoresis
  Part 2—Constructing Restriction Maps

III. STUDENT MATERIALS AND EQUIPMENT

	Per Student	Per Pair (2)	Per Group (4)	Per Class (24)
Exercise A				
Sordaria fiimicola, wild × tan cross on agar plate **(1)**		1		12
Sordaria simulation **(2)**				1
microscope slides, box (75 × 25 mm)			1	6
coverslips, box (22 × 22 mm)			1	6
small cork	1			24
compound microscope	1			24
Exercise B				
Drosophila mojavensis, third instar larvae, vial **(3)**		1		12
aceto-orcein or aceto-carmine stain, dropper bottle **(4)**		1		12
microscope slides, box (75 × 25 mm)		1		12
coverslips, box (22 × 22 mm, No. 1)		1		12
dissecting needles		2		24

	Per Student	Per Pair (2)	Per Group (4)	Per Class (24)
Exercise B—continued				
bibulous paper, book		1		12
insect Ringer's solution, bottle **(5)**		1		12
Exercise C				
Escherichia coli, recipient (K12 $pro^-thr^-leu^-thi^-str^r$ F^-) broth culture, tube **(6)**		1		12
E. coli, donor (K12 $pro^+thr^+leu^+thi^+str^s$ Hfr) broth culture, tube **(7)**		1		12
agar plates, minimal medium supplemented with streptomycin and thiamine (STR/THI) **(8)**		4		48
agar plates, minimal medium supplemented with amino acids (pro, leu, thr) and thiamine **(9)**		2		24
conjugation tube, sterile, capped **(10)**		1		12
capped tubes containing 9.0 ml sterile distilled water **(11)**		5		60
95 % ethanol, beaker		1		12
glass spreading rod		1		12
1 ml pipettes, sterile		10		120
marking pencil		1		12
Exercise D				
kits available **(12)**				
electrophoresis apparatus **(13)**		1		12
agarose, bottle (60 ml) **(14)**		1		12
Pasteur pipette		1		12
electrophoresis running buffer (200 ml) **(15)**		1		12
micropipette (50 μl)		1		12
Lambda DNA (*Eco*RI digest), tube (50-100 μl) **(16)**		1		12
Lambda DNA (*Hind*III digest), tube (50-100 μl) **(16)**		1		12
Lambda DNA (undigested), tube (50-100 μl) **(16)**		1		12
bromophenol blue dye **(17)**				1
methylene blue stain **(18)**				1
semi-log paper (sheet)		1		12

IV. PREPARATION OF MATERIALS AND SOLUTIONS

(1) *Sordaria fimicola,* wild × tan cross on agar plate (1)

Directions for preparation of *Sordaria* crosses are presented in Laboratory 13—see **(6)** and **(7)**. Plates are ready for use when perithecia are visible. If black spores begin to appear on the undersides of the lids, refrigerate the cultures until they are analyzed by the students.

(2) Sordaria simulation

Use the asci pattern at the back of this section. Make 50 copies and use a paper cutter to make each column of 8 ascospores into an ascus (a rectangular strip of paper with 8 dots on it). Group 10 asci MI + MII together and place in a small brown paper bag (the virtual "perithecium"). Asci should be grouped in combinations of MI and MII for each bag as shown below:

Bag	MI	MII	Bag	MI	MII	Bag	MI	MII
1	5	5	**11**	6	4	**21**	5	5
2	2	8	**12**	3	7	**22**	3	7
3	8	2	**13**	5	5	**23**	4	6
4	7	3	**14**	6	4	**24**	8	2
5	6	4	**15**	7	3	**25**	7	3
6	6	4	**16**	6	4	**26**	2	8
7	8	2	**17**	6	4	**27**	2	8
8	7	3	**18**	5	5	**28**	3	7
9	4	6	**19**	3	7	**29**	3	7
10	4	6	**20**	2	8	**Total**	143	147

Note that nonhybrid asci are also included. Make several bags of all 10 nonhybrid asci. Students will wonder whether to count these. It is a good place to stop and explain why they are not counted.

If bags are put together as given above, there should be a total of 143 MI and 147 MII asci in all 29 bags. The map distance for this simulation is 25.3 map units. This matches the data in the *Answer Guide*. The chi-square data for this simulation is also included in the *Answer Guide*.

(3) *Drosophila mojavensis*, third instar larvae

D. mojavensis larvae are larger than those of *D. melanogaster* and are ideal for studying polytene chromosomes. Larvae are usually larger if reared at 18° C and if 1-2 drops of a thick yeast suspension are added daily to the culture.

Fresh *Drosophila* medium can be prepared as follows:

To 3 liters of boiling distilled water add:

corn meal	300 g
baker's yeast	76 g
agar	60 g

Add to the above a mixture of 800 ml of distilled water and 400 ml of molasses.

Let the suspension simmer for 10 minutes. Cool for 10 minutes. Add 26 ml proprionic acid and mix. Pour into vials and refrigerate (sufficient for about 200 vials with 20 ml/vial).

Drosophila instant medium can also be purchased through Carolina Biological Supply Company (see Ordering Information).

(4) aceto-orcin and aceto-carmine stains

aceto-orcein stain:

Gather the following materials

aceto-orcein	2 g
glacial acetic acid	50 ml

Procedure:

Mix and simmer gently for 10 minutes. Bring to a boil for 1 minute and add 50 ml of 85% lactic acid. Cool and filter. May also be purchased already made (see Ordering Information).

aceto-carmine stain:

This stain can be prepared in the same manner. It can also be purchased already made (see Ordering Information).

Both stains are intensified by small amounts of iron. Using non-stainless steel dissecting needles or insect pins (mounted in balsa wood for handles) is helpful.

(5) insect Ringer's solution

Add 7.5 g NaCl, 0.35 g KCl, and 0.21 g $CaCl_2$ to 750 ml distilled water and add distilled water to 1000 ml.

(6) *Escherichia coli*, recipient (K12 $pro^-thr^-leu^-thi^-str^r$ F^-) broth culture

Order an agar slant of this *E. coli* strain from the American Type Culture Collection (see Ordering Information). Strain ATCC 23744.

1. Prepare a broth as follows:
   Suspend 30 g of dehydrated tryptic soy broth in 1000 ml distilled water and dispense into two 500 ml sidearm (cotton plugged) flasks and cover with aluminum foil. To a third flask (not sidearm), add 1.5 g (Bacto) agar.
   Autoclave for 15 minutes at 15 lbs pressure and 121° C. Cool.
   Aseptically dispense 5 ml amounts of the agar into sterile 17 mm capped tubes. Immediately slant the tubes containing the agar.
   Aspetically, using a sterile pipette, remove two 5 ml aliquots of tryptic soy broth from each sidearm flask and dispense into four 17 mm sterile capped tubes, 5 ml per tube.
   Alternatively, tryptic soy broth and tryptic soy agar can be purchased in dehydrated form or in bottles. (See Ordering Information.)
2. Prepare several fresh agar slants of the recipient bacteria by inoculating agar slants. Do this three days prior to the laboratory. Incubate at 37° C.
3. Inoculate each of the four tubes containing 5 ml of tryptic soy broth with a generous loop full of bacteria from one of the fresh slant cultures (step 2) after they have incubated for 24 hours. Do this 24 hours before the laboratory period. Incubate 24 hours at 37° C.
4. Early in the morning of the day of the laboratory, inoculate each of the two sidearm flasks containing tryptic soy broth with 1 ml of fresh tube culture from step 3.
5. Incubate in a shaking water bath until cultures reach an OD of 0.6 at 550 nm (approximately 5×10^7 cells/ml). This can be checked by aseptically removing a sample of culture from the flask and using the Spectronic 20 (it usually takes 3-4 hours to reach the required density of cells). Chill on ice until use in laboratory if necessary.
6. Aseptically dispense 2.5 ml of the bacterial culture into small, 13 mm sterile tubes (see Ordering Information) for use by students.

(7) *E. coli* donor (K12 $pro^+leu^+thr^+thi^+str^s$ Hfr) broth culture

Order an agar slant of this *E. coli* strain from the American Type Culture Collection (see Ordering Information). Strain ATCC 23729.

Prepare tubes of broth culture as done for the recipient strain described in **(5)**, steps 1 through 6.

(8) agar plates, minimal medium supplemented with streptomycin and thiamine (STR/THI)

Prepare the following stock solution: streptomycin sulfate (50 mg/ml)

Dissolve 0.5 g streptomycin sulfate in 10 ml distilled water. Filter sterilize using a 0.45 µm syringe filter (see Ordering Information). The final concentration in the medium will be 50 µg/ml.

Prepare minimal medium plus streptomycin and thiamine as follows:

1. Suspend 10.6 g of dehydrated Davis minimal broth (see Ordering Information) in 900 ml of distilled water in a 2l flask. Stir to dissolve. Add 15 g of (Bacto) agar and 0.001 g thiamine prior to autoclaving. Cover with aluminum foil and autoclave for 15 minutes at

15 lb. pressure and 121° C.

2. Dissolve 10 g glucose in 100 ml of distilled water in a 250 ml flask and autoclave as above. Cool to 50° C and then add the 10 ml of sterile glucose solution to the minimal medium.
3. Add 1 ml of sterile streptomycin solution to the agar when almost cool (but still liquid enough to pour plates) for a final concentration of 50 μg/ml and pour plates.

Dispense into sterile plastic Petri dishes with enough agar (20 ml) to cover the bottom of the dish (yields approximately 40 dishes). Let dishes cool and invert to store. Pour dishes 3 days ahead to allow moisture to evaporate.

Alternatively, prepare minimal medium from basic ingredients as follows (you will combine solutions A and B):

Solution A (adjust to pH 7.0)

potassium phosphate monobasic (KH_2PO_4)	3.0 g
potassium phosphate dibasic (K_2HPO_4)	6.0 g
ammonium chloride (NH_4Cl)	2.0 g
sodium chloride (NaCl)	5.0 g

Add distilled water to yield 800 ml and adjust the pH using 0.1 N NaOH or HC1.

Solution B

glucose	8.0 g
magnesium sulfate ($MgSO_4 \cdot 7H20$)	0.1 g
(Bacto)agar	15.0 g
thiamine	0.001 g

Add distilled water to make 200 ml

Autoclave solutions A and B for 15 minutes at 15 pounds pressure and 121° C, and combine. After combining, cool until the flask can be picked up and add 0.05 g streptomycin. Alternatively, add 1 ml of a 50 mg/ml sterile streptomycin sulfate solution as described above.

Dispense into sterile plastic Petri dishes (approximately 20 ml each). Let cool and invert to store. This is best done 1-3 days ahead to allow moisture to evaporate.

(9) agar plates; minimal medium supplemented with amino acids (pro, leu, thr) and the vitamin thiamine (M/PCTT)

Prepare the following solutions:

1. proline (15 mg/ml)

   Dissolve 0.15 g of proline in 10 ml distilled water and filter sterilize, using a syringe filter (0.45 μm). Add 2 ml/liter of agar for a final concentration of 30 μg/ml.
2. leucine(l0 mg/ml)

   Dissolve 0.l g of leucine in 10 ml of distilled water and filter sterilize, using a syringe filter (0.45 um). Add 2 ml/liter of agar for a final concentration of 20 μg/ml.
3. threonine (40 mg/ml)

Dissolve 0.4 g of threonine in 10 ml of distilled water and filter sterilize, using a syringe filter (0.45 μm). Add 2 ml/liter of agar for a final concentration of 80 μg/ml.

Prepare minimal medium agar plates as described in **(8)**, including the thiamine, but NOT the streptomycin. Add the amount of each amino acid to 1000 ml of agar to give the final concentrations designated for **(9)** 1, 2, and 3 (above).

(10) conjugation tubes

It is simplest to use sterile, capped polypropylene tubes such as those available from Carolina Biological, # 21-5080. You can also use other types of sterile capped glass tubes if available (Carolina Biological—# 19-8910 tubes and # 19-9042 caps), or other types of disposable tubes (17 × 100 mm) with snap tops (such as Fisher, # 14-956-6B). You may wish to order extra caps for glass tubes in case students drop caps (Fisher, # 14-956-11).

(11) capped tubes containing 9.0 ml sterile distilled water

Use glass tubes (25 × 100 mm) with caps (see Ordering Information). Fill each tube with 9 ml of distilled water and autoclave for 15 minutes at 15 lbs. pressure (slow exhaust). Alternatively, autoclave a flask of distilled water and dispense it aseptically into sterile, disposable plastic tubes such as those used for conjugation.

(12) general information

For **Exercise C**, the following kits are available:

Lambda DNA cleaved with *Eco*RI for electrophoresis with *Hin*dIII fragments (Carolina Biological Supply Company, kit # 21-1149). Methylene blue is included.

DNA restriction analysis kit (Carolina Biological Supply Company, kit # 21-1151). This kit allows students to cut thei own DNA with restriction enzymes *Eco*RI, *Bam*HI, and *Hind*III.

Lambda DNA predigested with *Eco*RI endonuclease (Edvotek, kit # 112). This kit contains undigested *Lambda* DNA, *Eco*RI digested *Lambda* DNA, and standard *Hin*dIII digested *Lambda* DNA fragments ready for electrophoresis. Agarose and micropipettes for loading samples and methylene blue stain are all included. Enough material is provided to run five gels. Data can be distributed to each student for analysis.

Cleaving of *Lambda* DNA with *Eco*RI endonuclease (Edvotek, kit # 212). An introduction to restriction enzymes. This kit provides *Lambda* DNA, restriction enzymes, and reaction buffers. Students prepare their own digests for electrophoresis. Agarose and micropipettes for loading samples are included. Enough materials are provided for 10 restriction digests.

(13) electrophoresis apparatus

Any small electrophoresis apparatus can be used. If using one with many lanes, make sure that students write down lane numbers.

Gel electrophoresis apparatus (Carolina Biological Supply Company, electrophoresis apparatus, # 21-3668 or # 21-3654, two-gel capacity and power supply, # 21-3672 can be ordered separately). Edvotek electrophoresis chamber, # 502, and power packs #503 or # 508 (dual) or together as minilab station # 501.

(14) agarose

In Carolina Biological Supply Company and Edvotek kits, agarose is supplied. If a kit is not used, a 1% agarose gel is made from a 1% agarose solution in 10× Tris borate buffer as follows:

1. Prepare 10× Tris borate buffer

Tris base	100 g
boric acid	55 g
EDTA (ethylenediamine tetraacetate disodium salt)	8.35 g

Add distilled water to make 1000 ml. Adjust to pH 8.0-8.2 by adding 1 M NaOH. The

EDTA will dissolve only after the pH has been raised.

2. Prepare 1% agarose solution

agarose	1 g
10× Tris borate buffer	10 ml
distilled water	90 ml

This solution can be prepared immediately before use or it can be kept at room temperature and liquefied using a microwave oven.

(15) electrophoresis buffer

Buffer for running agarose gels is prepared as follows:

10× Tris borate buffer	100 ml **(13)**
distilled water	900 ml

Alternatively, this buffer can be purchased already mixed from Edvotek or Carolina Biological Supply Company (see Ordering Information).

(16) Lambda DNA

Lambda DNA, undigested and digested with *Eco*RI and *Hind*III, can be purchased from Carolina Biological Supply Company or Edvotek (see Ordering Information). These are also included in appropriate kits (see General Information).

If you order *Lambda* DNA and predigested *Lambda* DNA, you will want to use approximately 2 μg of DNA per gel if staining with methylene blue.

Lambda DNA—80 μg (200 μl, 0.4 μg/μl)

Mix 4 μl *Lambda* DNA and 4 μl loading dye plus 12 μl TBE buffer in a microcentrifuge tube. This will give a total of 20 μl. Load entire amount into a single lane.

Lambda DNA digested with *Eco*RI—50 μg (200 μl; 0.25 μg/μl)

Mix 8 μl DNA plus 8 μl TBE buffer plus 4 μl loading dye in a microcentrifuge tube. This will give a total of 20 μl. Load entire amount into a single lane.

Lambda DNA digested with *Hind*III—50 μg (200 μl; 0.25 μg/μl)

Mix 8 μl DNA plus 8 μl TBE buffer plus 4 μl loading dye in a microcentrifuge tube. This will give a total of 20 μl. Load entire amount into a single lane.

(17) bromophenol blue loading dye

Bromophenol blue loading dye is available from Edvotek or Carolina Biological Supply company (see Ordering Information).

(18) methylene blue stain

Methylene blue stain is available from Edvotek as Methylene Blue Plus™ (# 609) or Carolina Biological Supply Company as Carolina Blu™ (#21-7300). (See Ordering Information.) If Carolina Blu™ DNA stain is used, some must also be mixed into agar.

V. PREPARATION SUGGESTIONS

Exercise A

The *Sordaria* simulation is an excellent way to make sure that students understand how crossing-over can be used to map genes before they actually start to collect data from *Sordaria* plates. A pattern for asci to be used in the *Sorderia* simulation is included at the end of this section. Asci can be cut out in strips and brown "lunch bags" are filled with 10 asci—a combination of MI and MII asci. Numbers for asci combinations are given in the Preparation of Materials and Solutions Section **(2)**. Using the simulation helps students to understand what to count—to recognize MI and MII and also some non-hybrid peritheca (include some bags of all black or all tan). Students will want to count them, but will learn that they should NOT include non-crossover asci. Use of the simulation also gives students some data if crosses do not work.

Be sure that students have familiarized themselves with Exercise D in Laboratory 13. If students have prepared their own crosses, they will have a greater understanding of this exercise. If not, you may need to review the life cycle and the events of meiosis, including crossing over, to show students how the various arrangements of ascospores of different colors are possible. Be sure to read the Preparation Suggestions for Exercise D in Laboratory 13 to learn more about the care and crossing of *Sordaria.*

The chi-square test is used to see how closely student data agrees with the known map distance of 26 crossover units for the tan allele. Students tend to "hunt" for crossover asci. If your data do not agree with the known 26 map units, it is usually because students counted MII asci preferentially (the difference between observed and expected, in this case, is Not due to chance alone. Tell students to move from one field to the next and count everything in the field until approximately 50 fields have been counted. Be sure to have students read Appendix I, Presenting and Analyzing Experimental Data: Tables, Graphs, and Statistical Methods, before attempting the chi-square test outlined in this exercise. An introduction to the chi-square test is presented in Laboratory I.

The tan mutant is used rather than the gray mutant because the allele for the latter is approximately 60 units from the centromere. In such cases, double crossovers can occur and interfere with gathering data. ALSO, be sure to use wild × tan and not gray × tan crosses.

Students can work singly, in pairs, or in groups in Exercise A, according to how plates were set up during Laboratory 13.

Exercise B

Third instar larvae should be used for this experiment. *Drosophila mojavensis* is recommended because of its larger size (and the larger size of its salivary glands). If *D. melanogaster is* used, larvae should be fed with extra yeast solution (see Preparation of Materials section). Generally, 8 day old larvae should be used (if grown at 21° C). These will be the oldest and largest larvae available, since larvae will form pupae at about day 9.

The lighting used for dissecting glands is crucial. If using transmitted light, the glands will appear to be more translucent (we prefer to use transmitted light reflected off a mirror below a clear glass stage). If a black stage and reflected light are used, the glands will appear whitish. If glands are filled with saliva, individual cells will be swollen and the glands will appear grape-like. If students are having difficulty identifying the salivary glands, gastric caeca, or Malpighian tubules, have them look for the piece of tissue in which they can actually see the individual cells (and even nuclei) using the dissecting microscope.

Use dissecting needles (non-aluminum) if possible to remove as much fat body tissue as possible (fat cells interfere with the production of good squashes). It is possible to place insect pins or steel needles into the ends of balsa wood strips or dowels to make dissecting needles. Some students can manipulate insect pins easily without the aid of handles.

It is important that the initial separation of the head from the body be done quickly, but then the students should NOT continue to try to pull the intestine out of the body. This leads to breakage and a general mess. Simply let the internal parts of the larvae flow out of the larval cuticle.

Be sure to keep an eye on the drop of stain on the glands while staining. It tends to dry up quickly (students should not leave the slide under the lights on the microscope stage). Stain for at least 10 minutes—the longer the better and the darker the bands. Save stain—the older the better. Simply filter the stain each year before use. If students have used too much stain, transfer the glands to a drop of 45% acetic acid to make the squash (only do this if the glands have stained darkly). If glands are in one piece, you may wish to cut them up with the dissecting needles before squashing, taking care that no pieces of gland stick to the needles. When squashing, use several thicknesses of bibulous paper or use a small cork (or pencil eraser). Do not get discouraged if the first slide does not turn out well—it takes some time to learn how hard to press on the slide. If chromosomes are shattered into small pieces, press more lightly next time.

To map *Drosophila* chromosomes, you may wish to use Carolina BioKit # 17-1964. Students can study crossing and distance between genes as with *Sordaria* (Exercise A). Three mutants (white, miniature, and forked) are used to prepare crosses.

Exercise C

ATCC cultures are fairly expensive ($12.00/tube), so, if you have access to any E. *coli* K12 mutants other than those suggested, you may wish to substitute and adjust amino acid requirements in the medium as necessary. Directions can be followed as they are written.

Be sure to leave enough time (3 days) to make cultures from the ATCC slants to be used for producing broth cultures for the conjugation experiment. Glass tubes with caps may be used and reused, but sterile, disposable tubes make the preparation and clean up much easier. Have students work in pairs during this experiment. Students should plate out all dilutions 1/10 and 1/100 to ensure room for error or poor technique. After the laboratory is completed, dispose of all materials by autoclaving or using a 10% bleach solution. Have students at one table start the conjugation experiment at one time and put all conjugation tubes in one rack. This way, other students will not bump the tubes when trying to reclaim their own tubes. Gently place the rack into the incubator. At the end of 30 minutes, remove the rack and shake it to disrupt conjugation.

Exercise D

Kits (described under Preparation of Materials and Solutions) include all materials for this exercise. Alternatively, solutions can be prepared separately (see Preparation of Materials and Colutions **(12)** and **(16)**) or purchased already made (see Ordering Information).

To cast agarose gels follow these suggestions:

1. Place agarose, 50 ml/bottle, in sterile bottles (if using a Carolina Biological apparatus). These can be liquefied in a water bath or in the microwave (loosen caps!). Be sure ALL lumps are gone before pouring. If using Carolina Blu™ stain, you will need to add a small amount (2 drops/bottle) to the liquefied agarose. Make up only enough agarose for use because, once the stain has been added, the agarose cannot be stored for later use.
2. Put absorbent paper under the gel box—usually some leaking occurs and cleaning solidified gel off lab tables and floors is no fun.
3. Care should be taken by the instructor to ensure that the comb teeth are set completely down on their "shoulders" when placed on the gel box, and covered to the correct depth with the agarose gel.

   Wells that are too shallow will result in some DNA splashing out due to vigorous unloading of the micropipette by students with shaking hands. Wells should be able to accommodate 50 μl samples.

Too deep a well may have a "shouldered connection" to the next well, allowing mixing with DNA cut by other restriction enzymes—not to mention the controls— and the resulting bands will be blurred and indistinct.

4. Pour the gels on unmovable counters—inevitably students bump tables while the gel is setting.
5. Recovering the comb after gel solidification must also be done very gently. Slowly lifting one end at a time reduces the number of ripped gel beds. This can be done when the gel is placed in the gel box and buffer has been added which helps the comb to slip out.

Follow these suggestions for loading gel beds:

1. You may wish to prepare some agarose in Petri dishes with wells made by the combs in order for students to practice loading their samples. (You could also use a 20% agar solution for practice in order to cut down on the expense of agarose.) Simply put some loading dye in a small container for practice
2. Be sure that students have identified their well sequence as starting from a particular side and have recorded this information in their lab books, as well as what they put in each well. Remind them to change pipette tips after loading each well.
3. Encourage students to "get down" and "eyeball" each well as closely as they can—a dark background may heighten the contrast.
4. If a digital micropipette is used for loading, two hands on the pipette will help, but not together (one should be close to the base and the other on the handle with a thumb on the button). Slowly expel the contents of the pipette, being sure that the tip is not stabbing through the bottom of the well and making a hole in the gel.

In order to electrophorese the gel, follow these directions:

Place the gel bed into the gel box and cover with running buffer. Just cover the gel; do not add excess buffer. When pouring buffer into the box, pour to the side and not on top of the gel. Remove the comb (if not done earlier) and check wells for bubbles. If bubbles are found, use a Pasteur pipette to blow buffer across the wells to try to dislodge the bubbles. If using Carolina Blu™ stain, you must add stain to the running bugger (12 drops/liter).

1. Be sure to cover the tank. It is essential, to ensure safety during this procedure.
2. With the power supply turned off and the plug not in the outlet, connect the red lead to the positive pole (anode—red) and the black lead to the negative pole (cathode—black). Then plug in and turn the power on. If using an Edvotek chamber, set the voltage to 50 volts. For the Carolina apparatus, use 80 volts and check for a current reading of 50-100 milliamperes.
3. Be sure to watch for dye movement as the experiment begins to ensure that the apparatus is working properly.
4. You can tell when to turn off the gel by watching the movement of the purple bromophenol blue band (from the loading dye). When it reaches the opposite end of the gel, you can turn off the power supply. When finished (about 1 ½ - 2 hours), turn the power off first; unplug the power supply from the outlet, and then unplug the leads to the electrophoresis chamber.
5. Gingerly move the gel to a disposable plastic tray for staining. You can use a kitchen spatula. Make sure that students wear gloves. If a kit is not used, gels can be stained

with 0.25% methylene blue and destained with water. Remember, gloves should always be worn when handling the staining and destaining of gels.

Staining: stain for 30 minutes, rinse, and let stand overnight.

Once gels are stained, wrap them in plastic wrap or overlay them with a 1/4th sheet of plastic transparency film. Mark the bands using a waterproof felt tip marker. Also mark the location of the corners of the gels and the wells. Tape the plastic wrap or transparency to a piece of white paper so it can be copied on a copying machine. (If the electrophoresis is done as a demonstration, copies can be distributed to all students.) If some students are unsuccessful, copies of data from other experiments can be shared.

If students were not introduced to the concept of standard curves covered in Laboratory 4, have them review Exercise C.

You may wish to do the electrophoresis experiment as a separate laboratory or have students carry out the experiment as an extra credit project. Since the running time does not require student involvement, students can be doing other activities or can leave and return later.

VI. ORDERING INFORMATION

Sordaria fimicola, wild type (plate)—Carolina Biological, # 15-6291; Ward's, # 85W8010
S. fimicola, tan mutant (plate)—Carolina Biological, # 15-6295; Ward's, # 85W9102
microscope slides, 75 × 25 mm—Fisher, # 12-550A; Carolina Biological, # 63-2000
coverslips, 22 × 22 mm—Fisher, # 12-542B; Carolina Biological, # 63-3095
Drosophila mojavensis—Carolina Biological, # 17-2870; Ward's, # 87W6651
corn meal—purchase locally
baker's yeast—obtain locally; Carolina Biological, # 17-3234; Ward's, # 38W5821
agar (power form)—Fisher, # B11-472; Carolina Biological, # 78-5300; Ward's, # 38W1201
instant *Drosophila* medium:
plain—Carolina Biological, # 17-3200; Ward's, # 38W0597
blue—Carolina Biological, # 17-3210; Ward's, # 38W0594
aceto-orcein stain—Carolina Biological, # 84- 1451; Ward's, # 38W9050
orcein—Sigma, # 09004; Carolina Biological, # 87-8471; Ward's, # 38W8417
aceto-carmine stain—Carolina Biological, # 84-1421; Ward's, # 38W9020
carmine—Sigma, # C-6752; Carolina Biological, # 85-3070; Ward's, # 38W7320
lactic acid—Carolina Biological, # 87-1690; Ward's, # 39W2095
dissecting needles—Carolina Biological, # 62-7200; Ward's, # 14W0650
bibulous paper—Carolina Biological, # 63-4050; Ward's, # 15W2500
sodium chloride (NaCl)—Fisher, # S271-3; Carolina Biological, # 88-8880; Ward's, # 37W5487
potassium chloride (KCl)—Fisher, # P217-500; Carolina Biological, # 88-2910; Ward's, # 37W4634
calcium chloride ($CaCl_2$)—Fisher, # C79-500; Carolina Biological, # 85-1810; Ward's, # 38W0937
tryptic soy broth—Fisher, # DF0370-01-1; Carolina Biological, # 78-8440 (dehydrated) or # 77-6840 (bottle); Ward's, # 88W0817
tryptic soy agar—Carolina Biological, # 78-8420 (dehydrated) or # 77-6800 (bottle); Ward's, # 88W1815

Escherichia coli, K12 $pro^- thr^- leu^- thi^- str^r$ F^-—ATCC 23744[1]
Escherichia coli, K12 $pro^+ thr^+ leu^+ thi^+ str^s$ Hfr—ATCC 23739
side-arm flask, 500 ml—Carolina Biological, # 72-6388
Erlenmeyer flask, 500 ml—Fisher, # 10-040H (Pyrex) or # 10-039H (Kimax); Carolina Biological, # 72-6676 (Corning, student grade); Ward's, # 17W2974
(Bacto) agar—Fisher, # DF0140-02-9; Carolina Biological, # 21-6720; Ward's, # 38W0015
sterile tubes, snap cap (12 × 75 mm)—Fisher, # 14-956-3D
sterile tubes, snap cap (12 × 100 mm)—Fisher, # 14-956-6B; Carolina Biological, # 19-8940
disposable culture tubes (borosilicate glass, 25 × 150 mm)—Carolina Biological, # 19-8910; Ward's, # 17W1405
caps for glass culture tubes—Carolina Biological, # 19-9042; Ward's, # 18W7186
inoculating loops—Carolina Biological, # 70-3036 (disposable) or # 70-3050 (Nichrome); Ward's, # 15W0954
sterile pipettes, 5 ml—Carolina Biological, # 21-4624; Ward's, # 17W4854
sterile pipettes, 1 ml—Fisher, # 13-678-1 lb.; Carolina Biological, # 21-4620; Ward's, # 18W7196
Petri dishes, sterile disposable—Carolina Biological, #74-1350; Ward's, # 18W7101
streptomycin sulfate—Sigma, # S6501
thiamine—Sigma, # T-1270
L-proline—Sigma, # P-0380
L-leucine—Sigma, # L-8000
L-threonine—Sigma, #T1645
syringe for sterile filtration—Carolina Biological, # 19-9606; Ward's, # 14W1620
disposable syringe filters (45 gm)—Carolina Biological, # 19-9592; Ward's, # 18W1581
Davis minimal broth—Difco, # 0756-01-5[2]
potassium phosphate (monobasic) (KH_2PO_4)—Fisher, # P 285-500; Carolina Biological, # 88-4260; Ward's, # 37W4845
potassium phosphate (dibasic) (K_2HPO_4)—Fisher, # P 288-100; Carolina Biological, # 88-4298; Ward's, # 37W4841
ammonium chloride (NH_4Cl)—Fisher, # A661-500; Carolina Biological, # 84-3810; Ward's, # 37W1965
sodium chloride (NaCl)—Fisher, # S271-500; Carolina Biological, # 88-8880; Ward's, # 37W5480
magnesium sulfate ($MgSO_4 \cdot 7H_20$)—Fisher, # M63-500; Carolina Biological, # 87-3376; Ward's, # 37W2861
glucose (dextrose)—Fisher, # D16-500; Carolina Biological, # 85-7440; Ward's, # 39W1457
tubes, polypropylene (sterile)—Carolina Biological, # 21-5080
conjugation tubes, 25 × 100 mm—Carolina Biological, # 19-3910 (# 19-9042, caps); 17 × 100 mm—Fisher, # 14-956-6B (caps, # 14-956-11)
Lambda DNA cleaved with *Eco*RI for electrophoresis with *Hind*III fragments (kit)—Carolina

[1] ATCC cultures can be purchased from the American Type Culture Collection. If this is your first order, you must make a request by purchase order or on "letterhead" stationary, c/o Sales and Marketing Department, 12301 Parklawn Dr., Rockville, MD 20852, to establish an account for your institution. Once established, FAX orders (FAX 301-231-5816) or phone orders (800-638-6597) can be placed. (See Suppliers.)
[2] Difco Laboratories, PO Box 331058. Detroit, MI 48232-7058 (800-521-0851). (See Suppliers.)

Biological, # 21-1149
DNA restriction analysis (kit)—Carolina Biological, # 21-1151
analysis of *Eco*RI cleavage patterns of *Lambda* DNA (kit)—Edvotek, # 112
cleavage of *Lambda* DNA with *Eco*RI endonuclease (kit)—Edvotek, # 212
mini-lab station for electrophoresis—Edvotek, # 501 OR
gel electrophoresis apparatus—Carolina Biological, # 21-3668 or 21-3654 (two-gel capacity); Edvotek, # 502
electrophoresis power supply—Carolina Biological, # 21-3672; Edvotek, # 503 or # 508 (dual)
Tris base (Trizma base)—Sigma, # T- 1503
boric acid—Carolina Biological, # 84-8440; Fisher # A74-500
EDTA, disodium salt—Carolina Biological, # 86- 1790; Fisher 5311- 100
agarose (low EEO)—Edvotek, # 605, Carolina Biological, # 21-7080; Sigma, # A-3768
electrophoresis buffer—Edvotek, # 607; Carolina Biological, # 21-9024
microcapillary pipete, 50 µl—Carolina Biological, # 21-4516
phage *Lambda* DNA—Carolina Biological, # 21-1410
phage *Lambda* DNA, *Eco*RI fragments—Carolina Biological, # 21-1474
phage *Lambda* DNA, *Hind*III fragments—Carolina Biological, # 21- 473
bromophenol blue loading dye—Edvotek, # 606, or Carolina Biological, # 21-8200
methylene blue stain—Edvotek, # 609, or Carolina Biological, # 21-8290 or # 21-7300 (Carolina Blue™)

Notes:

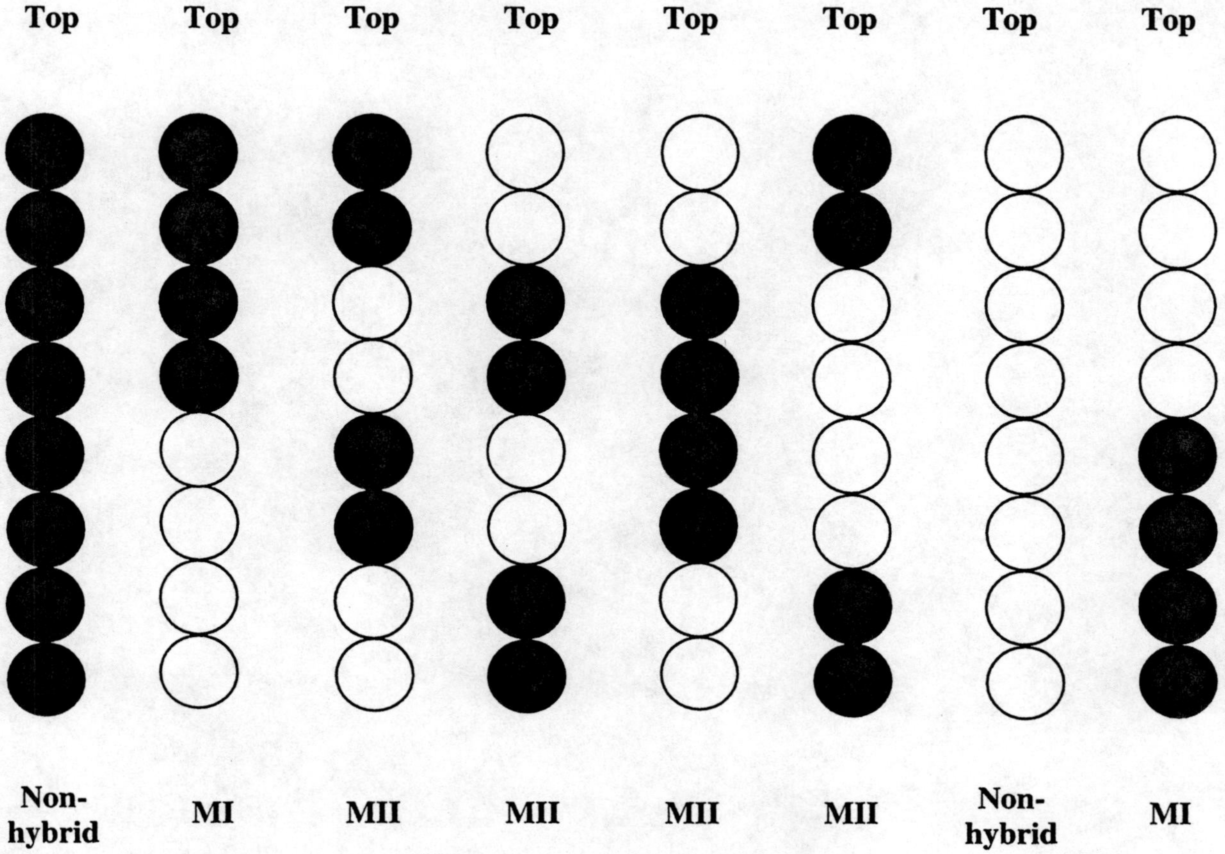

Top	Top	Top	Top	Top	Top	Top	Top
Non-hybrid	MI	MII	MII	MII	MII	Non-hybrid	MI

LABORATORY 15 Human Genetic Traits

I. FOREWORD

Basic concepts in human genetics are covered in this laboratory. Work with human chromosomes in emphasized. DNA forensics is also introduced. This represents another application of restriction enzyme techniques. Students work independently in all exercises.

II. TIME REQUIREMENTS

Exercise A—Human Cytogenetics (home,15 minutes discussion in laboratory)
Exercise B—*X* and *Y* Chromosomes (15 minutes)
Exercise C—Mendelian Inheritance in Humans (45 minutes)
  Part 1—Monohybrid Crosses in Humans
  Part 2—How Individual Is Each Individual?
Exercise D—Chromosomal Abnormalities—Nondisjunction and Translocation (60 minutes)
  Part 1—Nondisjunction: Sex Chromosomes
  Part 2—Nondisjunction: Autosomes
Exercise E—Constructing a Human Pedigree (30 minutes)
Exercise F—Forensic Science: DNA Fingerprinting (60 minutes)

III. STUDENT MATERIALS AND EQUIPMENT

	Per Student	Per Pair (2)	Per Group (4)	Per Class (24)
Exercise A				
completed at home				
Exercise B				
cheek scrapings, male (prepared slide)				1
cheek scrapings, female (prepared slide)				1
compound microscope				2
Exercise C (Parts 1 and 2)				
no preparation or materials needed				
Exercise D (Parts 1, 2, and 3)				
blank karyotype	1			24
metaphase chromosome spread (photographs) (1)	1			24
scissors	1			24
tape or glue			1	6
Exercise E				
no preparation or materials needed				
Exercise F				
PCR forensics simulation kit				1

IV. PREPARATION OF MATERIALS AND SOLUTIONS

(1) metaphase chromosome spread (photographs)

Specially prepared reproducible photographs and kayrotypes are included at the end of this section in the *Preparator's Guide* as follows:

metaphase spread, normal male
karyotype, normal male
metaphase spread, Down's syndrome, male
karyotype, Down's syndrome, male
metaphase spread, Down's syndrome, female
karyotype, Down's syndrome, female
metaphase spread, trisomy 18, female
karyotype, trisomy 18, female

A special note of thanks should be given to Dr. Judy Capra and Dr. Arthur Robinson of the Genetics Unit, Health Sciences Center, University of Colorado, for sharing their excellent photographs with users of this laboratory manual.

V. PREPARATION SUGGESTIONS

Exercise A

Karyotyping during Exercise E will proceed much more rapidly if students learn the technique during Exercise A. This should be tied to a discussion of amniocentesis techniques to make the exercise more meaningful. For additional studies of chromosome banding, the banding patterns of polytene chromosomes can be examined (Laboratory 13).

Exercise B

This exercise is introduced as a contrast to Exercise A where sex is determined by the tedious process of karyotyping. It is possible for students to make their own Barr body slides from cheek scrapings if desired.

Exercise C (Parts 1 and 2)

This exercise is designed to be a simple introduction to some human genetic traits. The exercise in Part 2 is designed to investigate the unique genetic nature of individuals.

Exercise D

Human chromosome abnormalities can be studied by karyotyping. If time does not permit students to complete this exercise, it can easily be done at home. It is also possible (and sometimes more fun) for students to work in pairs. This exercise can be supplemented by a study of human pedigrees and completion of pedigree forms as prepared by genetic counselors (Exercise E).

Exercise E

The family history presented in this exercise is adapted from an actual family history. You may wish to use other pedigree examples as take home problems for students to work on after familiarizing themselves with pedigree notations.

Exercise E

Use Carolina Biological Supply Company Kit # K3-21-1210 (the footnote reference on p.15-16 in *Biology in the Laboratory, 3rd ed.* is incomplete). Directions are included with the kit. Be sure to store the kit in a refrigerator prior to use.

Laboratory Review Questions and Problems

Note that a discussion of blood types and several problems have been included in this section. To avoid student and instructor contact with bodily fluids (in this case, blood), traditional blood typing exercises have been eliminated from this human genetics laboratory.

VI. ORDERING INFORMATION

cheek scrapings, male (prepared slide)—Carolina Biological, # G506; Ward's, # 93W8111
cheek scrapings, female (prepared slide)—Carolina Biological, # G504; Ward's, # 93W8110
PCR forensics simulation kit—Carolina Biological, # 21-1210

Notes:

metaphase spread, normal female

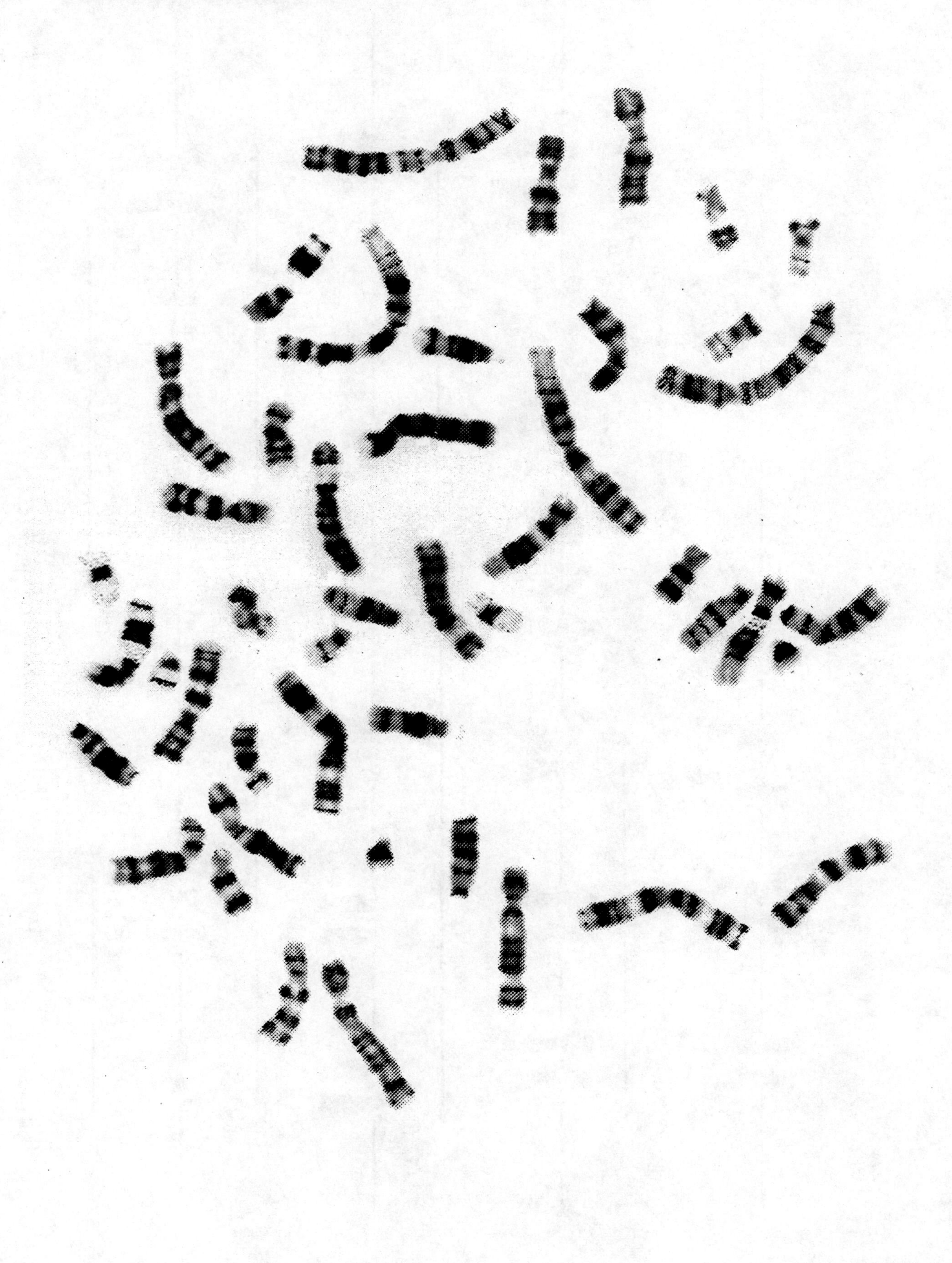

karyotype, normal female

metaphase spread, Down's syndrome, male

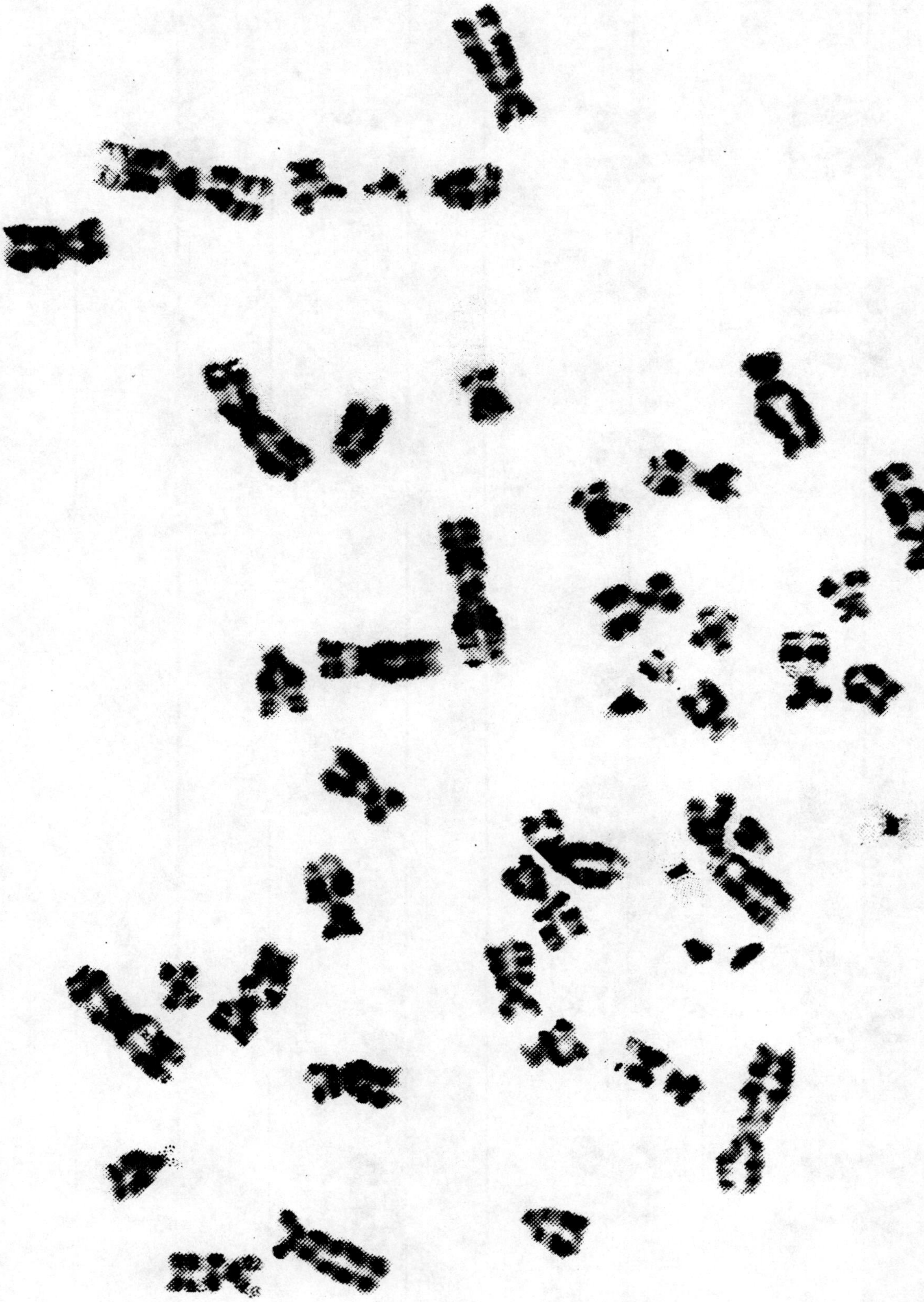

karyotype, Down's syndrome, male

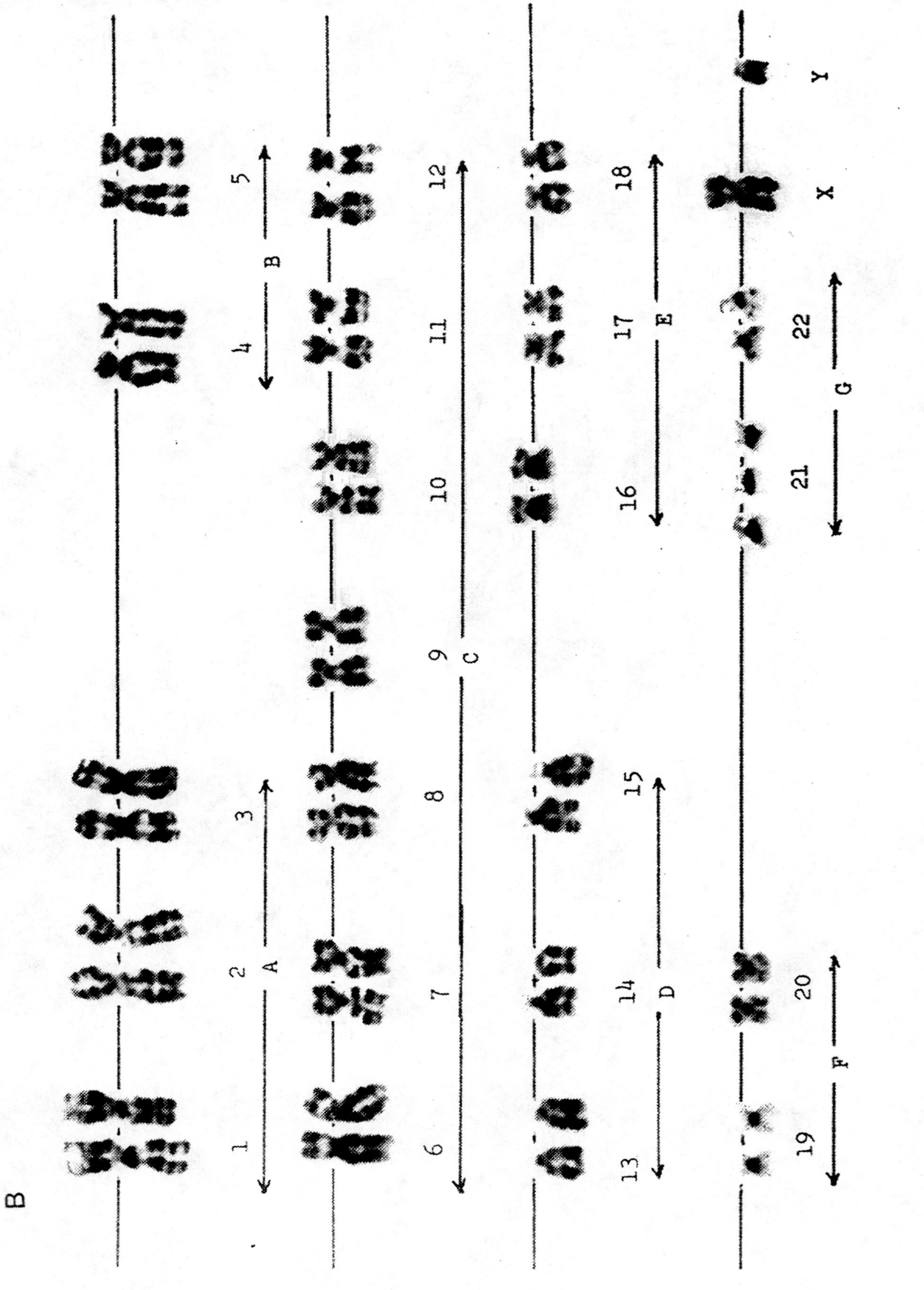

B

1 2 3 A

4 5 B

6 7 8 9 10 11 12 C

13 14 15 D

16 17 18 E

19 20 F

21 22 G

X Y

metaphase spread, Down's syndrome, female

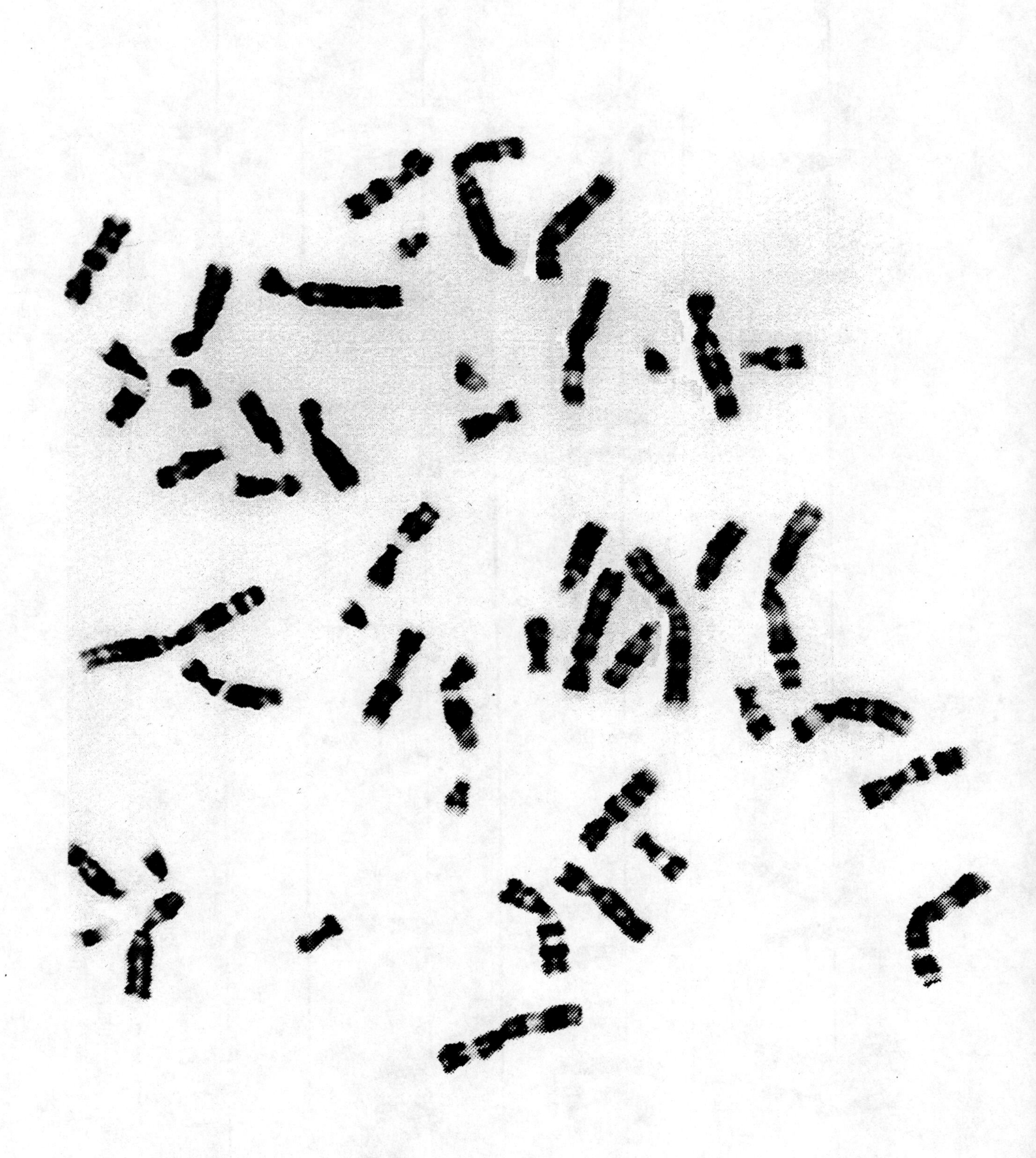

karyotope, Down's syndrome, female

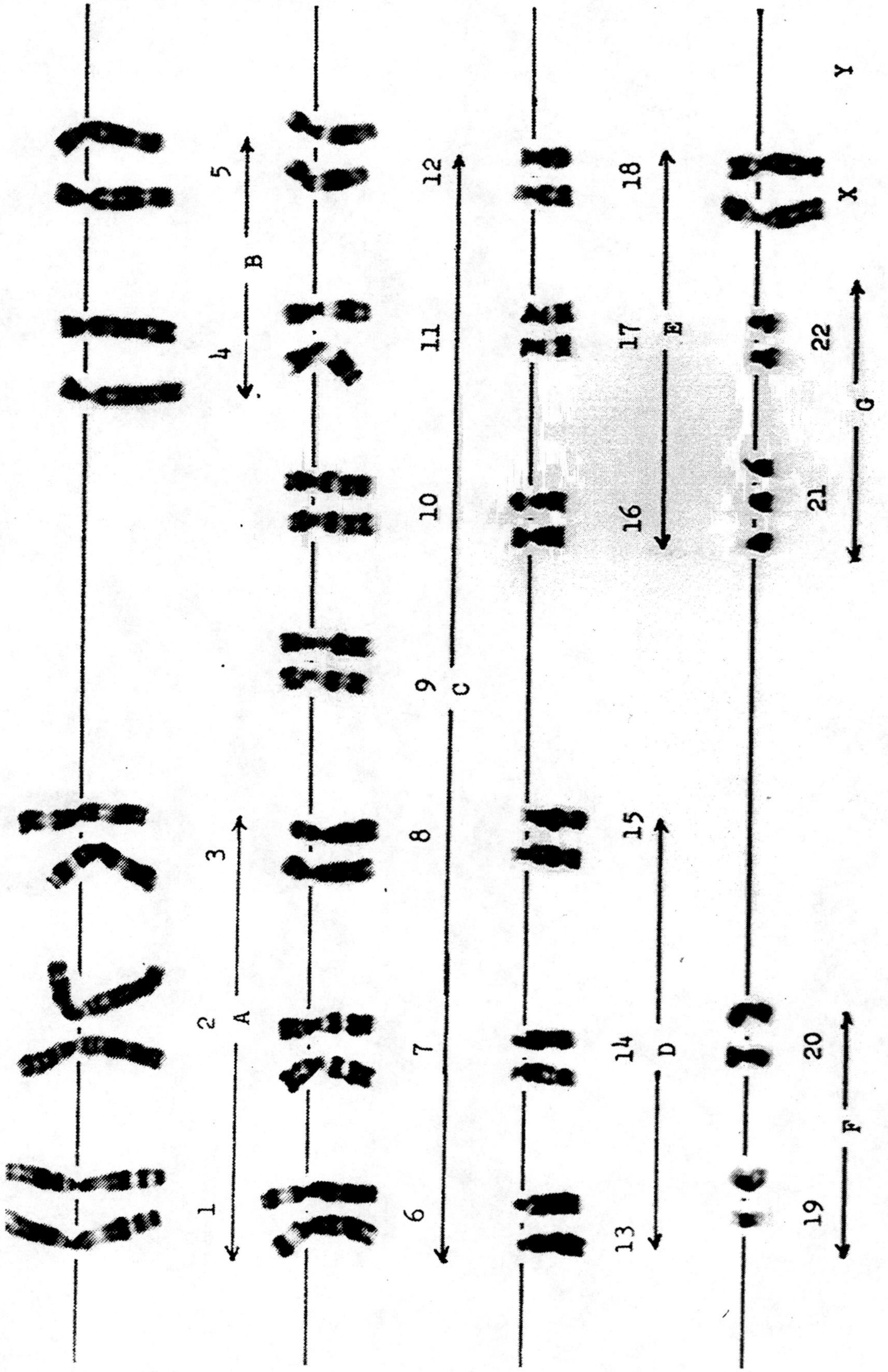

WORTH *175* 0-87901-688-4

metaphase spread, trisomy 18, male

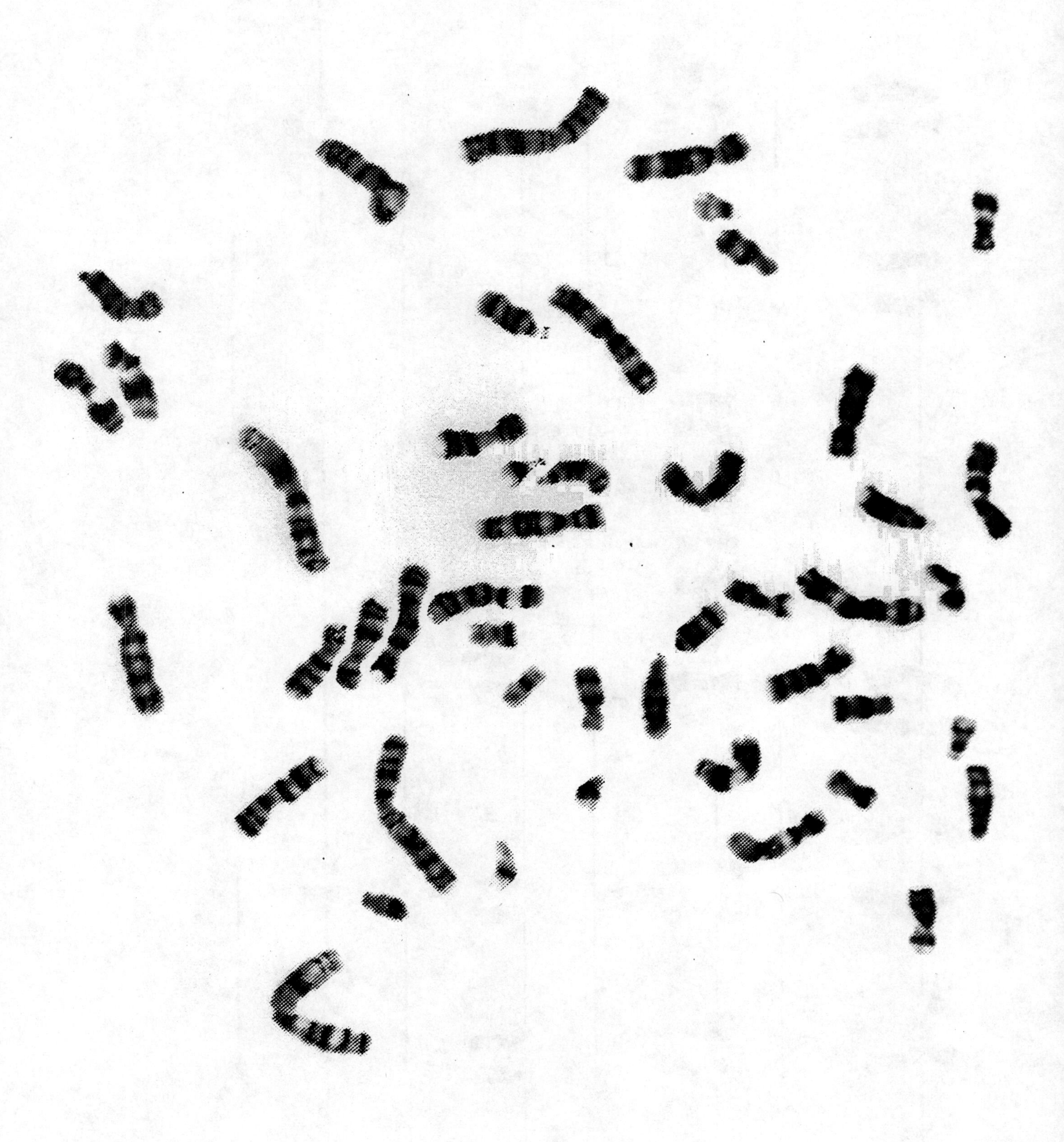

karyotype, trisomy 18, male

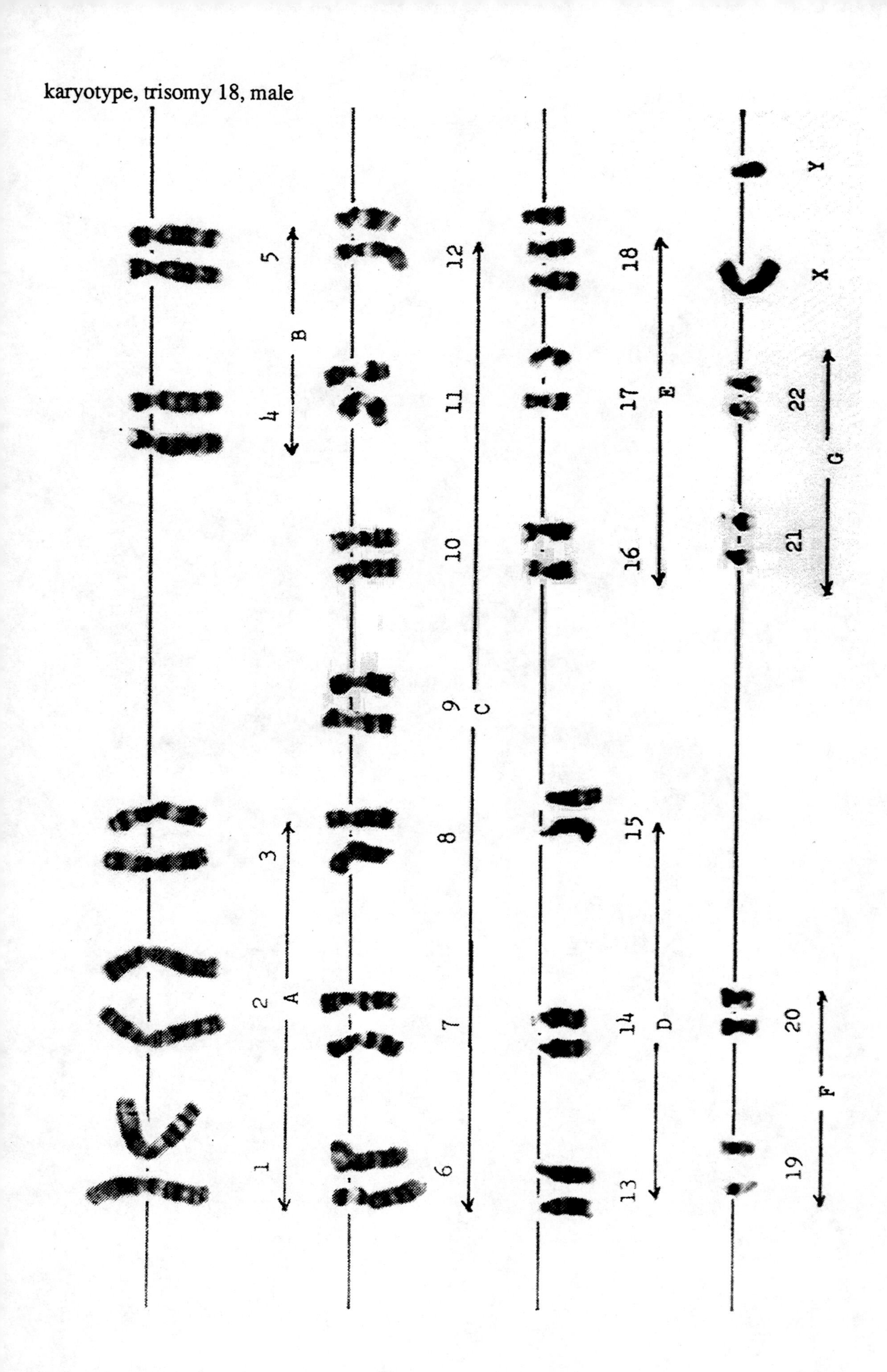

LABORATORY 16 DNA Isolation

I. FOREWORD

This laboratory includes three exercises, each designed to isolate DNA from fresh material. You may use bacteria, plant or animal tissues, or you may wish to assign a different group of students to each tissue and compare results. Students will work in pairs during all exercises.

II. TIME REQUIREMENTS

Exercise A—DNA Isolation Procedure
- Part 1—DNA Isolation *(Chloroform)* (150 minutes)
- Part 2—DNA Isolation *(No chloroform)* (30 minutes)

Exercise B—Isolation of DNA from Animal Cells *(Optional)*
- Part 1—Isolation of DNA (150 minutes)
- Part 2—Colorimetric Detection of DNA (25 minutes)

Exercise C—Isolation of DNA from Bacteria *(Optional)*
- Part 1—Isolation of DNA (150 minutes)
- Part 2—Measuring Absorbance of DNA (20 minutes)
- Part 3—Colorimetric Detection of DNA (20 minutes)
- Part 4—Preparing a Standard Curve (30 minutes)

III. STUDENT MATERIALS AND EQUIPMENT

	Per Student	Per Pair (2)	Per Group (4)	Per Class (24)
Exercise A (Part 1)				
blender				1
60° C water bath				1
ice bucket		1		12
95% ethanol, cold (freezer, 1000 ml) **(1)**				1
chloroform (10 ml)		1		12
homogenization medium (100 ml) **(2)**		1		12
cutting board		1		12
onion (medium)		1		12
knife		1		12
weighing boat		1		12
cheesecloth (4 thicknesses) **(3)**		1		12
100 ml cylinder		1		12
1000 ml beaker		1		12
600 ml beaker		1		12
250 ml beaker		2		24
250 ml Erlenmeyer flask		2		24
5 ml pipette (to deliver)		1		12
conical centrifuge tubes (15 ml)		6		72
table top centrifuge				2

	Per Student	Per Pair (2)	Per Group (4)	Per Class (24)
Exercise A (Part 1)—continued				
glass rod for spooling **(4)**		1		12
balance (± 0.01 g)				1
Exercise A (Part 2)				
onion (yellow)				1
blender				1
homogenization buffer, chilled (100 ml) **(5)**				1
cheesecloth (25 cm square **(3)**				1
meat tenderizer, 6% solution (50 ml) **(6)**				1
test tubes	1			24
Pasteur pipette	1			24
95% ethanol, ice cold (500 ml) **(1)**				1
spooling pipette or glass rod **(4)**	1			24
Exercise B (Part 1)				
testis tissue (2-3 mm piece) **(7)**		1		12
homogenization buffer (5 ml) **(8)**		1		12
SDS, 25% solution (1 ml) **(9)**		1		12
chloroform or 2 M NaC1 (5 ml) **(10)**		1		12
95% ethanol (5 ml) **(11)**		1		12
tissue homogenizer **(12)**		1		12
variable speed drill				1
ice bucket with ice			1	6
centrifuge tube (conical, stoppered) **(13)**		2		24
clinical centrifuge				2
Pasteur pipette		4		48
glass rod		1		12
pipette, 1 ml		1		12
Exercise B (Part 2)				
4% NaCl, tube (3 ml) **(14)**		1		12
wooden applicator stick		1		12
DNA standard solution, tube (3 ml) **(15)**		1		12
diphenylamine solution (repipette) **(16)**				1
test tubes		3		36
boiling water bath (or hot plate with beaker of water containing boiling stones)		1		12
marbles		3		36
Exercise C (Part 1)				
Escherichia coli paste (2 g block) **(17)**		1		12
homogenization medium, bottle (25 ml) **(18)**		1		12
conical centrifuge tube (large)		1		12
lysozyme solution, tube (1 ml) **(19)**		1		12
25% sodium dodecyl sulfate solution, tube (2 ml) **(9)**		1		12
5 M $NaC1O_4$, tube (8 ml) **(20)**		1		12

	Per Student	Per Pair (2)	Per Group (4)	Per Class (24)
Exercise C (Part 1—continued				
chloroform/isoamyl alcohol 50:1, tube (10 ml) **(21)**		1		12
water bath, 37° C				1
water bath, 50° C				1
centrifuge, clinical or refrigerated				2
85% ethanol (25 ml) **(22)**		1		12
ice bucket with ice		1		12
pipette, 1 ml		1		12
pipette, 5 ml		1		12
Exercise C (Part 2)				
spectrophotometer, UV/visible				1
saline citrate buffer, tube (5 ml) **(23)**		2		24
wooden applicator stick		1		12
Exercise C (Part 3)				
see Exercise B (Part 2)				
Exercise C (Part 4)				
Spectronic 20, wide range bulb				6
diphenylamine, tube (20 ml) **(16)**		1		12
DNA standard, tube (20 ml) **(17)**		1		12
test tubes		6		36
saline citrate buffer, tube (5 ml) (23)		1		12

IV. PREPARATION OF MATERIALS AND SOLUTIONS

(1) cold 95% ethanol

Ethanol must be cold for this procedure to work. Place a bottle of ethanol in a freezer overnight. Be sure that the cap is loose and the bottle is not completely full. Place the bottle on paper towels. It will not freeze.

(2) homogenization medium

5% SDS; 0.15 M NaCl; 0.15 M sodium citrate; 0.001 M EDTA—Prepare 2 liters

	for 1 liter	for 2 liters
sodium dodecyl sulfate (SDS)	50.0 g	100.0 g
NaC1	8.77 g	17.54 g
sodium citrate	4.41 g	8.82 g
ethylenediamine tetraacetic acid (EDTA)	0.292 g	0.584 g
add distilled water	up to 1 liter	2 liters

DO NOT place in refrigerator or the SDS will turn the solution an opaque white color.

(3) cheesecloth

Cut 8 inch squares—four thicknesses will be adequate.

(4) spooling pipette or glass rod

To collect ("spool") the DNA, it is best to use a glass rod or pipette that has been scored with a file or a diamond pencil. The rough surface helps the DNA strands adhere to the rod so they can be more easily "spooled."

(5) homogenization buffer (100 ml)

Add 1.5 g non-iodized salt (NaCl) to 50 ml distilled water. Dissolve. Add 10 ml liquid *dishwasher* detergent. Mix gently. Bring up to 100 ml with distilled water. Chill in refrigerator.

(6) meat tenderizer (6% solution)

Add 3 g meat tenderizer to 50 ml distilled water. Make sure that the brand of meat tenderizer that you use contains papain to digest proteins (check the label).

(7) testis tissue

It is best to use tissue with a high nucleus/cytoplasm ratio for any DNA extraction. Also, tissue should contain as little lipid or glycogen as possible. Testis tissue contains a high concentration of DNA and has a high nucleus/cytoplasm ratio. Testes are readily available from local spay or neuter clinics. Dog, horse, or bull testes are most commonly used. You may obtain the material up to 6 months ahead of laboratory and store it frozen. We have found that it is best to cut the testis material into small pieces that can be stored individually. Each pair of students then receives a packet of tissue. Thawing proceeds quickly since the pieces are small and, if packets are left on the laboratory bench top while students gather their materials together, the tissue will be ready for use.

(8) homogenization buffer for animal tissue

Prepare a 0.1 M EDTA (ethylenediamine tetraacetic acid, disodium salt) by dissolving 3.7 g of EDTA (disodium salt) in 100 ml of distilled water. Add 0.87 g of NaC1 to 10 ml of the 0.1 M EDTA and dilute to 100 ml with distilled water. Keep on ice; the homogenization buffer should be kept cold. Dispense the buffer to students in tubes of 5 ml.

(9) SDS (25% solution)

Obtain a mass of 25 g of sodium dodecyl sulfate (SDS), also called sodium lauryl sulfate (SLS), and add distilled water to make 100 ml. Dispense to students in tubes of 1-2 ml.

(10) chloroform or 2 M NaC1

Either chloroform or 2 M NaC1 can be used to denature and precipitate proteins as a white, middle layer if chloroform is used or as a white precipitate if 2 M NaC1 is used. If chloroform cannot be safely used in your classroom, the use of 2 M NaC1 in this exercise allows students to isolate DNA that cannot be done as easily with other tissues. With the use of 2 M NaCl, however, the isolated DNA is not as "clean" and contains associated protein materials.

To prepare 2 M NaCl, obtain a mass of 11.7 g NaCl and add distilled water to 100 ml. Dispense to students in tubes of 5 ml.

(11) 95% ethanol

Ethanol must be cold. Place a bottle of ethanol in the freezer a day ahead of the laboratory. Dispense to students in tubes of 5 ml and have students place on ice. (If you forget to chill the alcohol, simply dispense tubes at the beginning of the period and it will be cold enough for use by the time students have completed the DNA extraction.

(12) tissue homogenizer

Use a tissue homogenizer (Potter-Elvehjem type) that has a Teflon plunger attached to a metal rod so that the metal rod can be attached directly to the chuck of a variable speed drill. Start out on low speed, carefully moving the plunger up and down as it rotates, increasing the volume of each stroke little by little. Keep the homogenizer in ice so it remains cold. Be sure to wash and dry the homogenizer before the next pair of students use it. Caution students not to touch the Teflon plunger with their fingers because this will contaminate the plunger with nucleases (enzymes that can break down DNA).

(13) centrifuge tubes

Use conical glass tubes (15 ml) without a stopper if using 2 M NaCl, but use a stopper or use disposable screw cap conical tubes if using chloroform for DNA extraction (see Ordering Information.)

(14) 4% NaCl

Obtain a mass of 4 g NaC1 and add distilled water to 100 ml. Alternatively, DNA can be redissolved in 0.1 M EDTA (3.7 g EDTA, disodium salt, in distilled water to make 100 ml).

(15) DNA standard solution

Add 0.1 g DNA (see Ordering Information) to 200 ml distilled water. Add 1 drop of glacial acetic acid (heat may be required to get the DNA into solution). Store frozen—will keep for several years if not contaminated. Dispense to students in tubes of 3 ml.

(16) diphenylamine solution

Purchase diphenylamine reagent already prepared from Carolina Biological Supply Company (see Ordering Information), or prepare as follows:

1. Dissolve 15 g diphenylamine in 1000 ml of glacial acetic acid (use new bottle).
2. Add 15 ml concentrated H_2SO_4 (use new bottle).
3. Store in a dark bottle.
4. On the day of use, add 1 ml acetaldehyde solution (1 ml acetaldehyde to 500 ml distilled water) per every 100 ml diphenylamine.

Since this is a strongly acidic solution, make sure that students use Propipettes and wear safety glasses while dispensing it. We dispense diphenylamine using a repipette (see Ordering Information) set to 3 ml. This allows you to use one repipette for the entire class and safety problems are reduced.

(17) *Escherichia coli* paste

Prepare a bacterial cell paste by centrifuging cells (grown in tryptic soy broth, 30 g/1), in late log growth phase at 6,000 rpm for 10 minutes. You may freeze the paste for use as needed.

You may purchase *E. coli* from Carolina Biological Supply Company, # 15-5075—suspend with enough tryptic soy broth to make a turbid mixture. Use 1 ml to inoculate 100 ml of medium. Grow overnight. Use this to inoculate 1000 ml until slightly cloudy. In approximately 4 hours, the cells should be in late log phase (just when you cannot see through the medium or with absorbance set at 660 = 0.5). Generation time is approximately 0.5 hours at 37° C. Bacterial paste may also be purchased from Grain Processing Corporation (see Ordering Information).

(18) homogenization buffer for bacteria

NaCl	8.8 g
EDTA (disodium)	37.2 g

Add distilled water to 1000 ml. Dispense to students in bottles of 25-30 ml.

(19) lysozyme solution (10 mg/ml)

Dissolve 0.5 g lysozyme (see Ordering Information) in 50 ml distilled water. Dispense to students in tubes of 1 ml.

(20) 5 M $NaClO_4$

Obtain a mass of 140.5 g of $NaClO_4$ and add distilled water to make 100 ml (CAUTION—strong oxidizer, handle with care). Dispense to students in tubes of 7.5 ml.

(21) chloroform/isoamyl alcohol 50:1

Add 2 ml isoamyl alcohol to 98 ml of chloroform. Do this under a hood and use safety glasses.

(22) 85% ethanol

Dilute 447.5 ml ethanol to 500 ml with distilled water. This must be cold when used. Store it in a freezer and dispense to students in bottles of 25 ml. Have students put bottles on ice immediately.

(23) saline citrate buffer

Prepare solutions A and B as follows:

A. Dissolve 2.1 g of citric acid in 100 ml distilled water

B. Dissolve 2.94 g of sodium citrate (dihydrate) in 100 ml of distilled water. Add 7.2 ml of A to 42.8 ml of B and dilute to 100 ml.

V. PREPARATION SUGGESTIONS

Exercise A (Part 1)

Use small individual Styrofoam ice buckets. Some water should be added to the ice to make a slush bath which works better.

Pipetting chloroform should be done under a hood. Do not inhale fumes. Use a Pi-Pump or Propipette. DO NOT PIPETTE BY MOUTH.

If homogenization medium gets cold at any time, it will turn white but this will not affect its function. All steps except the first heat treatment and the deproteinization should be done in the cold (ice bucket).

Scoring the end of the glass rods with a diamond pencil will help the DNA to adhere to the rod while spooling. Do not touch the end of the rod with your fingers.

Exercise A (Part 2)

This exercise allows students to isolate DNA without using chloroform or phenol or other substances prohibited in many classrooms. The DNA will not be very pure, but the strands can be seen and spooled from the mixture very easily. The major thing to remember is that everything must be kept cold. When students add meat tenderizer (also cold), warn them to mix *gently.* Otherwise, they may shear the DNA into small pieces.

Adding ice-cold ethanol must be done slowly in order to create two layers. The DNA can be seen precipitating as long threads at the interface of the alcohol and the buffer. Spooling should be done as if one is winding yarn onto a knitting needle—spooling is different from mixing and must be done gently.

Exercise B (Part 1)

Testis tissue is an excellent source of material for DNA isolation since sperm cells are primarily nucleic acid with little accompanying cytoplasm. DNA yields are high.

Testis tissue can be obtained from your local veterinarian who neuters dogs and cats. You can also use testis tissue from bulls slaughtered at your local abattoir. You only need a small amount of tissue—about the size of the nail on your smallest finger.

This exercise allows students to isolate DNA without using chloroform. The buffer stabilizes the DNA as Na^+ ions associate with the negatively charged phosphate groups of the DNA. EDTA in the buffer chelates divalent cations such as magnesium, often a cofactor for DNases that break down the DNA. SDS dissolves cell membranes and denatures proteins during the homogenization. The proteins are then precipitated by chloroform or 2M NaCl. Cold ethanol is used to precipitate the DNA which can then be wound out of solution using a glass rod. Be sure that the glass rod is clean (acid washed is best) and that students do not touch the end used to wind out the DNA since DNases (nucleases) are present on the fingers. Scoring the glass rod to make it "rough" also helps in winding. Isolated DNA can be dried and stored frozen.

If buffer is supplied in tubes to students, only 1 ml of buffer is needed and 100 ml will be enough for the entire class. Keep all materials and solutions cold during the procedure except for the 25% SDS which will precipitate out at cold temperatures to form an opaque, white solution.

Exercise B (Part 2)

DNA from Exercise B (Part 1) can be tested with diphenylamine. DNA standards turn blue when boiled with diphenylamine. Since protein and RNA are still associated with the DNA isolated by the students, do not expect the same color blue.

Use a repipette to dispense diphenylamine. This cuts down on safety problems and on glassware. Use any size repipette and set for 3 ml. If you prepare the DNA standards and water tubes with 3 ml in them, then students will not have to use pipettes and the exercise will proceed quickly. Be sure that students wear safety glasses during the exercise.

Diphenylamine fumes which are noxious and corrosive can be reduced by placing glass marbles on the tops of test tubes while heating. These act as condensers. Be sure to wash marbles after use. Also be sure that some boiling stones are present in the beaker of water in which you heat the tubes.

Exercise C (Part 1)

DNA isolated from *E. coli is* fairly clean and is not as contaminated with protein as that from animal cells which is associated with both histone and acidic proteins.

Prepare *E. coli* paste up to 6 months or a year ahead and freeze it in a single mass. Once frozen, cut it into 2 g blocks and wrap each individually in aluminum foil. Dispense to students in packets and tell them to immediately put the E. *coli* paste into the centrifuge tube. If it melts in the foil, it will be difficult to handle.

Although the separation is best if done using a centrifuge, it can be completed without a centrifuge. Materials will settle out into layers if tubes are allowed to stand for 15-20 minutes. Make sure to remove all chloroform/isoamyl alcohol using a Pasteur pipette.

You will need to use a large, capped tube for this isolation and the appropriate head for the centrifuge, or else the homogenate should be divided into two tubes before the chloroform/isoamyl alcohol is added. Use additional chloroform/isoamyl alcohol to balance the tubes before centrifuging.

Alcohol used to precipitate the DNA should be cold. Store some in the freezer before the lab and dispense in bottles. Use a clean glass rod (preferably acid washed) to wind out the DNA. Remind students not to touch the rod with their fingers due to the presence of DNAases (nucleases) on their fingers. Score the end of the rod to make it "rough." This will make it easier to wind out or "spool" the DNA.

Exercise C (Part 2)

DNA gives a characteristic absorption spectrum with a maximum at 260 nm. If a UV/visible spectrophotometer is present in the laboratory, measure the 280/260 ratio of the DNA. If a scanner is attached to the machine, demonstrate the spectrum for DNA by scanning from

230 nm to 300 nm as a demonstration.

Exercise C (Part 3)

DNA from *E. coli* can be redissolved in 4% NaC1 or sodium citrate (saline citrate) buffer and a diphenylamine test can be done as described for Exercise B (Part 2).

Exercise C (Part 4)

Once isolated, DNA is dissolved in a known amount (3 ml) of buffer and its concentration per ml can be calculated using a standard curve. The DNA standard used for the diphenylamine test contains 0.5 mg of DNA per ml (500 μg/ml). Make 1:10 dilutions of the standard and use 3 ml of each dilution plus 3 ml of diphenylamine to make a standard curve (see Laboratory 4). Since diphenylamine turns blue with DNA, read absorbance at 600 nm using the Spectronic 20. Dissolve all isolated DNA in 3 ml of 4% NaC1 or saline citrate buffer, and add 3 ml of diphenylamine. Read absorbance and calculate concentration per ml by using the standard curve. This amount multiplied by 3 will give you a rough estimate of the total amount of DNA isolated (in ma) from 2 g of *E. coli.*

VI. ORDERING INFORMATION

blender—Carolina Biological, # 70-1905 (glass) or # 70-1908 (stainless); Ward's, # 15W2675

chloroform—Fisher, # C605-1; Carolina Biological, # 85-4000 (500 ml) or 85-4002 (4 l); Ward's, # 39W0920

sodium dodecyl (lauryl) sulfate (SDS or SLS, certified)—Fisher S529-500 (500 g); Carolina Biological, # 85-8740 (100 g) or # 21-8820 (50 g); Ward's, # 37W3001

sodium chloride (NaCl)—Fisher, # S271-3; Carolina Biological, # 88-9060; Ward's, # 37W5481

sodium citrate—Fisher S279-500 (500 g); Carolina Biological, # 88-9060; Ward's, # 38W5491

ethylenediamine tetraacetic acid, disodium salt (EDTA)—Sigma, # E 4884; Fisher, # S311-100; Carolina Biological, # 86-1790; Ward's, # 39W1255

cheesecloth—Fisher, # 06-665-18 (70 yard bolt); Carolina Biological, # 71-2690 (5 yard package); Ward's, # 15W0015

graduated cylinder, 100 ml—Fisher, # 08-553B (Pyrex) or # 08-549-5E (Kimax) or # 08-572-6D (Nalgene); Carolina Biological, # 72-1788 (Pyrex) or # 72-1948 (Kimble) or # 72-2266 (Nalgene); Ward's, # 17W0173

beaker, 250 ml— Fisher, # 02-540K (Pyrex) or # 02-539K (Kimax); Carolina Biological, # 72-1223 (Corning, student grade); Ward's, # 17W4040

beaker, 600 ml— Fisher, # 02-540M (Pyrex) or # 02-539M (Kimax); Carolina Biological, # 72-1225 (Corning, student grade); Ward's, # 17W4060

beaker, 1000 ml— Fisher, # 02-540P (Pyrex) or # 02-539P (Kimax); Carolina Biological, # 72-1227 (Corning, student grade); Ward's, # 17W4080

Erlenmeyer flask, 250 ml—Fisher, # 10-040F (Pyrex) or # 10-039F (Kimax); Carolina Biological, # 72-6672 (Corning, student grade); Ward's, # 17W2982

dishwasher detergent—local grocery store

sodium chloride (NaCl)—Fisher, # S271-3; Carolina Biological, # 88-8870; Ward's, # 36W5481

meat tenderizer (Adolphs)—local grocery store

test tubes, 16 × 125 mm—Fisher, # 14-925H; Carolina Biological, # 73-1418; Ward's, # 17W1403

test tube rack—Carolina Biological, # 73-1896; Ward's, # 18W4213
95% ethanol—Fisher, # A407-500; Ward's, # 39W0277
glass rods—Fisher, # 11-380C; Ward's, # 17W0928
tissue homogenizer (Potter-Elvehjem), 15 ml—Fisher, # 08-414-14C; Carolina Biological, 08-414-166
Pasteur pipette, 9"—Fisher, #13-678-20C; Carolina Biological, # 73-6062; Ward's, # 17W1146
rubber bulb—Fisher, # 14-065B
pipettes, 1 ml—Fisher, # 13-675-9B; Carolina Biological, # 73-6055; Ward's, # 17W1305
pipettes, 5 ml—Fisher, # 13-675-28F; Carolina Biological, # 73-6057; Ward's, # 17W1307
pipettes, 1 ml—Fisher, # 13-675-28G; Carolina Biological, # 73-6058
centrifuge tubes, glass (conical)—Fisher, # 05-500 or # 05-501; Carolina Biological, # 73-2014; Ward's, # 17W1330
centrifuge tubes, polypropylene (conical, capped)—Fisher, # 05-538-53D; Carolina Biological, # 18-8460
large centrifuge tubes, polypropylene (conical, capped)—Fisher, # 05-538-49; Carolina Biological, # 21-5100
DNA—Sigma, # D 1626
diphenylamine reagent, 100 ml bottle—Fisher, # LC13640-7 (0.3% solution); Carolina Biological, # 85-8371; Ward's, # 39W1032
diphenylamine (powder)—Fisher, # 02611-100; Carolina Biological, # 85-8330; Ward's, # 39W1031
acetic acid, glacial—Fisher, A38-212; Carolina Biological, # 84-1293; Ward's, # 39W0125
sulfuric acid—Fisher, # A300-212; Carolina Biological, # 89-3302; Ward's, # 37W5770
acetaldehyde—Fisher, # 0-1004; Carolina Biological, # 84-4480 (500 ml); Carolina Biological, # 84-1200; Ward's, # 39W0150
Propipette (safety pipette filler)—Fisher, # 13-681-50; Carolina Biological, # 73-6868; Ward's, # 15W0510
repipette—Fisher, # 13-687-62B
E. coli paste—Grain Processing Corporation., 1600 Oregon St., Muscatine, Iowa 52671 (319-264-4265)
lysozmye—Sigma, # L 6876 or # L 7001
sodium perchlorate ($NaC10_4{\cdot}H20$)—Fisher, # S490-500
chloroform—Fisher, # C298-1; Carolina Biological, # 85-4000; Ward's, # 39W0920
isoamyl alcohol—Fisher, # A 393-500; Carolina Biological, # 84-4480 (500 ml); Ward's, # 39W0230
Spectronic 20 (Milton Roy, formerly B&L)—Fisher, # 07-143-1; Carolina Biological, # 65-3300; Ward's, # 14W5550
wide range phototube and filter—Fisher, # 07-144-10; Carolina Biological, # 65-3312; Ward's, # 14W55532
Spectronic 20 tubes—Fisher, # 14-385-9008; Carolina Biological, # 65-3310 OR
Spectronic 20 tubes, Pyrex (100 × 13 mm)—Fisher, # 14-957C; Carolina Biological, # 73-1408
Spectronic 20 tubes, Kimax (100 × 13 mm)—Fisher, # 14-923D; Carolina Biological, # 73-1408A

wooden applicator sticks—Fisher, # 01-340; Carolina Biological, # 70-6865; Ward's, # 14W0105

citric acid—Fisher, # A 940-500; Carolina Biological, # 85-4700; Ward's, # 39W0932

sodium citrate (dihydrate)—Fisher, # S 279-500; Carolina Biological, # 88-9050; Ward's, # 37W5492

LABORATORY 17 DNA—The Genetic Material: Replication, Transcription, and Translation

I. FOREWORD

In this laboratory, students use a cardboard and paper model to learn about the processes of transcription and translation. Model pieces are cut from poster board and paper, using patterns found at the end of this section of the *Preparator's Guide*. The model is inexpensive and can be used over and over again by simply reproducing the paper parts on a copying machine.

Students will learn about semi-conservative replication and will see that polymerization takes place in a 5' to 3' direction. They will learn how amino acids are activated and how tRNA molecules are charged, and will investigate the breakage and formation of the chemical bonds involved. Students will also learn how to use the genetic code and how specific codons in mRNA can specify the correct amino acids in a peptide. The processes of initiation, elongation, translocation, and termination, and the factors involved in each are covered in this laboratory using the hands-on model. Students should work individually with one model kit per student.

II. TIME REQUIREMENTS

Exercise A—Replication (45 minutes)
Exercise B—Transcription and Translation (60 minutes)
  Part 1—Transcription—RNA Synthesis
  Part 2—Translation—Protein Synthesis
Exercise C—Point Mutations in DNA (45 minutes)
  Part 1—Base Substitutions—Possible Effects
  Part 2—Base Substitution Resulting in Sickle-Cell Anemia
  Part 3—Frame-Shift mutations: Base Additions and Deletions

III. STUDENT MATERIALS AND EQUIPMENT

	Per Student	Per Pair (2)	Per Group (4)	Per Class (24)
Exercises A, B. and C				
DNA model kit, (**1**) including	1			24
DNA molecule (white paper)				
deoxyribonucleotides (blue paper)				
ribonucleotides (green paper)				
amino acids (yellow paper)				
ribosome (black cardboard)				
4 amino-acyl tRNA molecules (green cardboard)				
1 ATP molecule (orange cardboard)				
transparency model (**2**)				**1**

IV. PREPARATION OF MATERIALS AND SOLUTIONS

(1) DNA model kit

Cut out all cardboard model parts and place in a brown envelope for distribution in the laboratory. If you make several cardboard patterns, a group of students or teaching assistants can trace around the patterns to make as many model kits as needed—one for each student.

Hand out green and blue sheets of nucleotides and yellow amino acids and ask students to cut these out, collect them in an envelope, and bring them to the laboratory. Distribute the white DNA molecule in the laboratory and explain how to cut it out into two wide pieces to be taped together, forming one long strand.

(2) transparency model

You may wish to use reduced size model parts to make transparencies. These can be cut out to use on an overhead projector to help demonstrate how the model works. (If you have color transparencies available, use them to make the pieces in different colors. Patterns for the transparency model are also found at the end of this section of the *Preparator's Guide*.)

V. PREPARATION SUGGESTIONS

Exercise A

In this exercise, students study the process of semi-conservative replication. They will learn about the difference in nucleotides found in DNA and RNA, the role of hydrogen bonds and phosphodiester bonds, the direction of polymerization, the relation of template and complement, and the antiparallel structure of the DNA molecule.

The white DNA molecule is in two pieces and must be taped together to form one long strip. The 5' ends of the molecule are labeled I and II. Make sure that students understand that the zig-zag line between bases represents the location of hydrogen bonds between bases and that bases within each single strand are held together by phosphodiester bonds. To replicate the DNA, students must cut the two strands apart along the zig-zag line (this simulates the breakage of hydrogen bonds).

You will note that both the green and blue sheets of nucleotides contain all five nitrogen bases, A, T. G. C, and U, used in the synthesis of both DNA and RNA. Blue nucleotides (pattern copied onto blue paper) are used to synthesize DNA. The blue U nucleotides are included to make students choose the four nucleotides (A, T. G. C, and not U) to be used. Walk around the classroom and look for Us being placed in the new DNA strands and call this mistake to each student's attention. Next, look to see that students are synthesizing the new strand in a 5' to 3' direction on the 3' 5' template. This means that they will be forming new strands in opposite directions on the two templates. Understanding the fact that DNA polymerization takes place in a particular direction is basic to understanding the nature of replication and the need for forming short Okazaki fragments. Walk around the room and check to see in which directions students are working. Many students do not understand the 5' to 3' concept until after using this hands-on model. Be sure to clear up this point with students who are having trouble.

Students should turn the ends of a long piece of tape under (so that the sticky side is up). After laying the template next to the strip of tape, the complementary nucleotides are put into place in a 5' to 3' direction. Alternatively, stick one end of a strip of tape to the table opposite the 3' end of the template and, holding the other end of the tape in the air, line up the complementary nucleotides by sticking them, face up, to the sticky underside of the tape (this produces a single smooth surface along the strip of nucleotides). Some students will try to tape the nucleotide pairs together, blue to white, forming hydrogen bonds. While this does indeed happen, explain to students that polymerization comes first, and, in this model, we will keep the blue strand separate from the white (forming only imaginary hydrogen bonds) so that we do not have to cut the template and complement apart when we are ready to transcribe messenger RNA.

After completing this exercise, students should have two double stranded blue and white DNA molecules. Check to see that the 3' and 5' ends of each blue and white molecule are labeled to demonstrate the antiparallel orientation of the two polynucleotide strands in the DNA.

Exercise B (Part 1)

Students should use the green nucleotides to synthesize mRNA. Make sure that they use the correct strip of DNA (the newly synthesized I' blue DNA made from the I white strip). Be sure that the mRNA is being synthesized in a 5' to 3' direction. It is easy to check on this by wandering around the room to watch the students at work—the first three nucleotides should be AUG (the initiation codon). If you do not see these as the first nucleotides to be put down, then check to see what the students have done incorrectly.

Make sure that the polynucleotide chain has 1 long strip of tape covering its length on the upper surface (surface with letters). This will allow the mRNA to be pulled through the ribosome more easily.

Be sure to have students label the 5' and 3' ends of the message they have synthesized.

Exercise B (Part 2)

In this exercise, students will learn about the process of translation, including the steps involved in initiation, elongation, translocation, and termination.

Have students remove all four green amino acyl tRNA synthetases from their kits and place them on the table in front of them.

Have them find the four blue tRNA molecules and the ATP molecule, also included in the kit, and place them on the table along with the four yellow paper amino acids that they cut out and brought to class. Show students the three binding sites on the enzyme, one for the R-group of the amino acid, one for the AMP, and one for the anticodon of the tRNA (you may need to interject a short discussion of tRNA structure if this has not been covered already in lecture).

Use the transparency model to show them how ATP is used to activate an amino acid, making sure that they understand the nature of the bond between AMP and the amino acid. Then, use the transparency model to demonstrate how the amino acid gets hooked to the tRNA molecule. Use your own paper model to show them how to cut off the -OH on the amino acid and review the nature of the ester bond formed between the amino acid and the tRNA.

After demonstrating how one tRNA is charged, ask students to charge the remaining tRNA molecules and then return the enzyme molecules and ATP molecule to the kit's envelope.

The black ribosome is in one large piece to make the model easier to handle, but explain to students that, in the cell, subunits are separate and initiation of translation begins when an mRNA molecule attaches to the small ribosomal subunit. The large subunit only joins after the first tRNA molecule pairs with the initiation codon.

Use the transparency model to demonstrate how to position the mRNA on the ribosome and how the anticodons of the tRNAs match the codons on mRNA in both the P and A sites.

You may wish to review the fact that codon and anticodon are antiparallel (left to right; the codon is 5' to 3' while the anticodon is 3' to 5' left to right—whenever two polynucleotide strands are complementary to one another, they must be antiparallel.) Use a paper model or extra transparency pieces to show how to make a peptide bond between amino acids. As students do this, walk around the class and it will be easy to identify those students who do not understand the peptide bond structure. Also, make sure students write the name of each amino acid on each yellow paper amino acid as it is put in place in the growing peptide chain. Again, if you walk around the classroom, it will be easy to identify students who are looking up the anticodon rather than codon (a common mistake) in the "genetic code chart" in order to identify amino acids. Once complete, students should have a chain of four amino acids, each identified. Be sure to review the use of protein factors and the fact that this is an energy demanding process, using GTP.

Once complete, have students return the ribosome to the kit's envelope and save kits for another time. We keep several kits on hand, including paper pieces, and allow students to check them out for review. Upper class students also use kits for review. Transfer students who have not completed this laboratory and seem to have difficulty with the specifics of the process are also referred to the kits by professors as a way of reviewing the concepts involved.

Exercise C (Parts 1, 2, and 3)

By re-examining the model, students have a chance to observe the effects of base substitution and frame shift mutations. This requires only pieces of the model already on hand and can be done as an at-home exercise, if preferred.

VI. ORDERING INFORMATION

All parts of this kit can be made from patterns included at the back of this section of the *Preparator's Guide*. A video tape is available from the author and can be used to guide students through this model, step-by-step. This also allows the model to be placed in a resource center or library (cost #15 including postage; email BIOL110@clemson.edu).

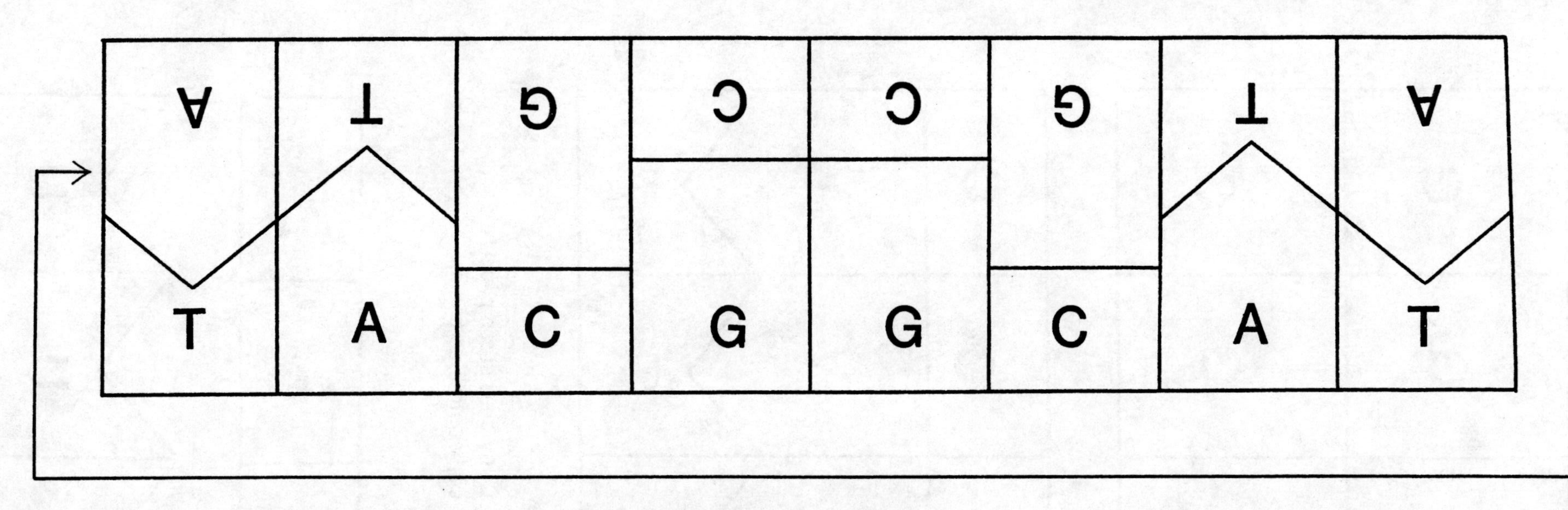

Attach the two ends as indicated

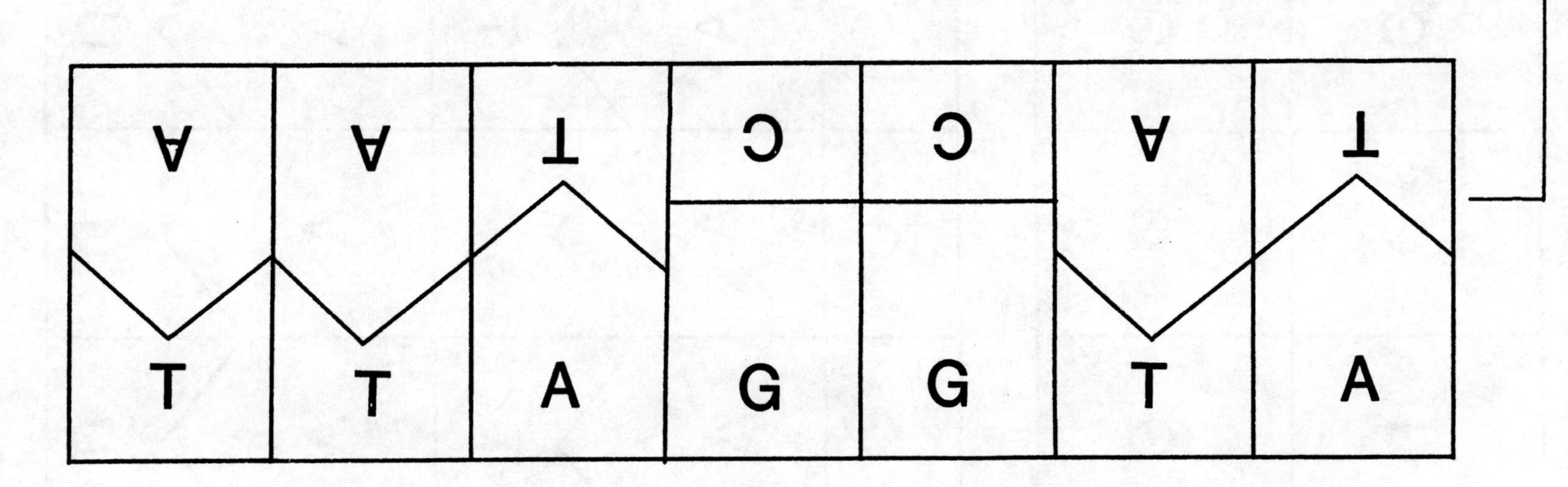

Copy onto white paper

G	G	G	G	G	G
G	G	G	G	G	G

C	C	C	C	C	C
A	A	A	A	A	A
T	T	T	T	T	T

A	A	A	
U	U	U	C C C

Copy onto green paper and
copy onto blue paper

U	U	U	U	U	U

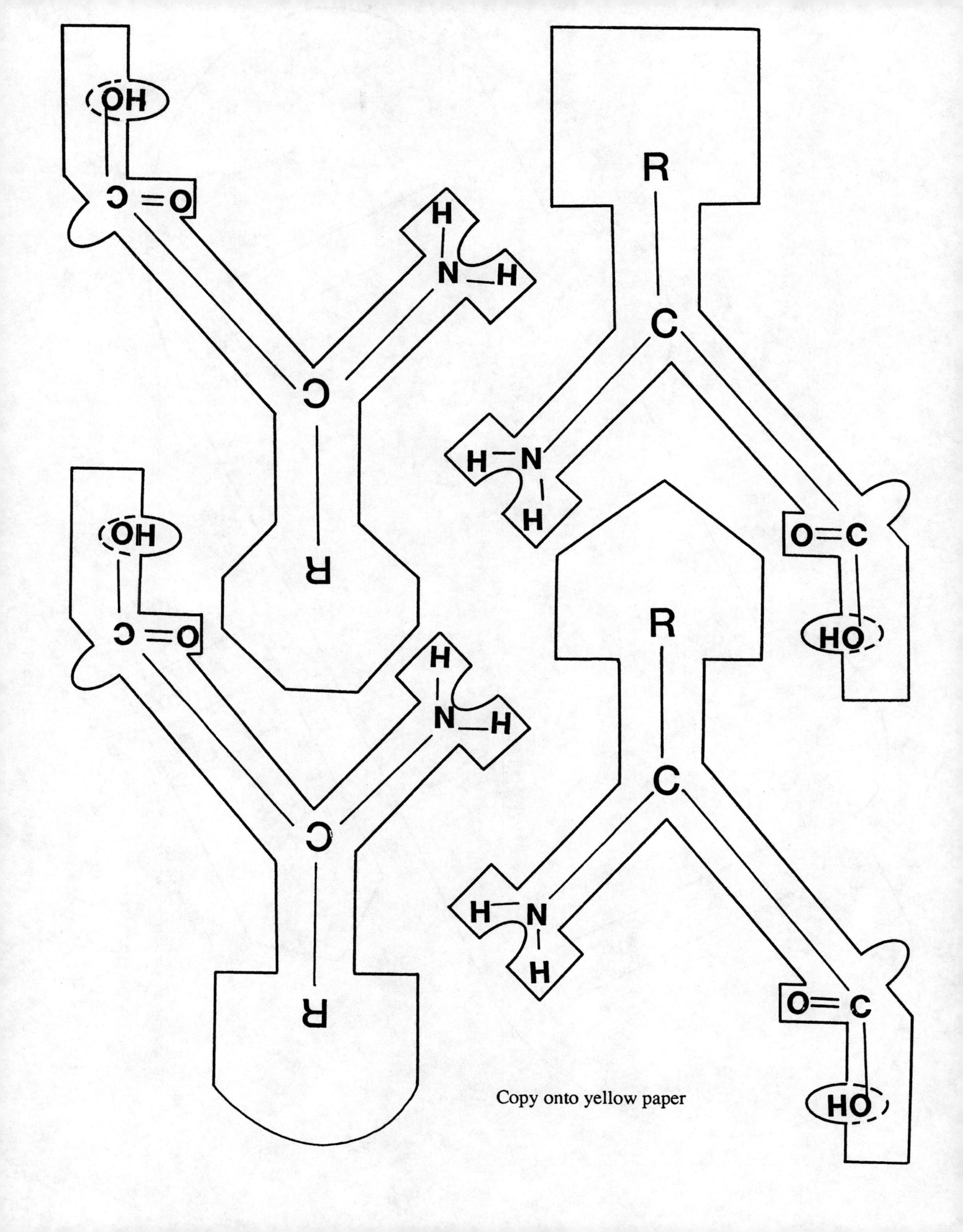

Copy onto yellow paper

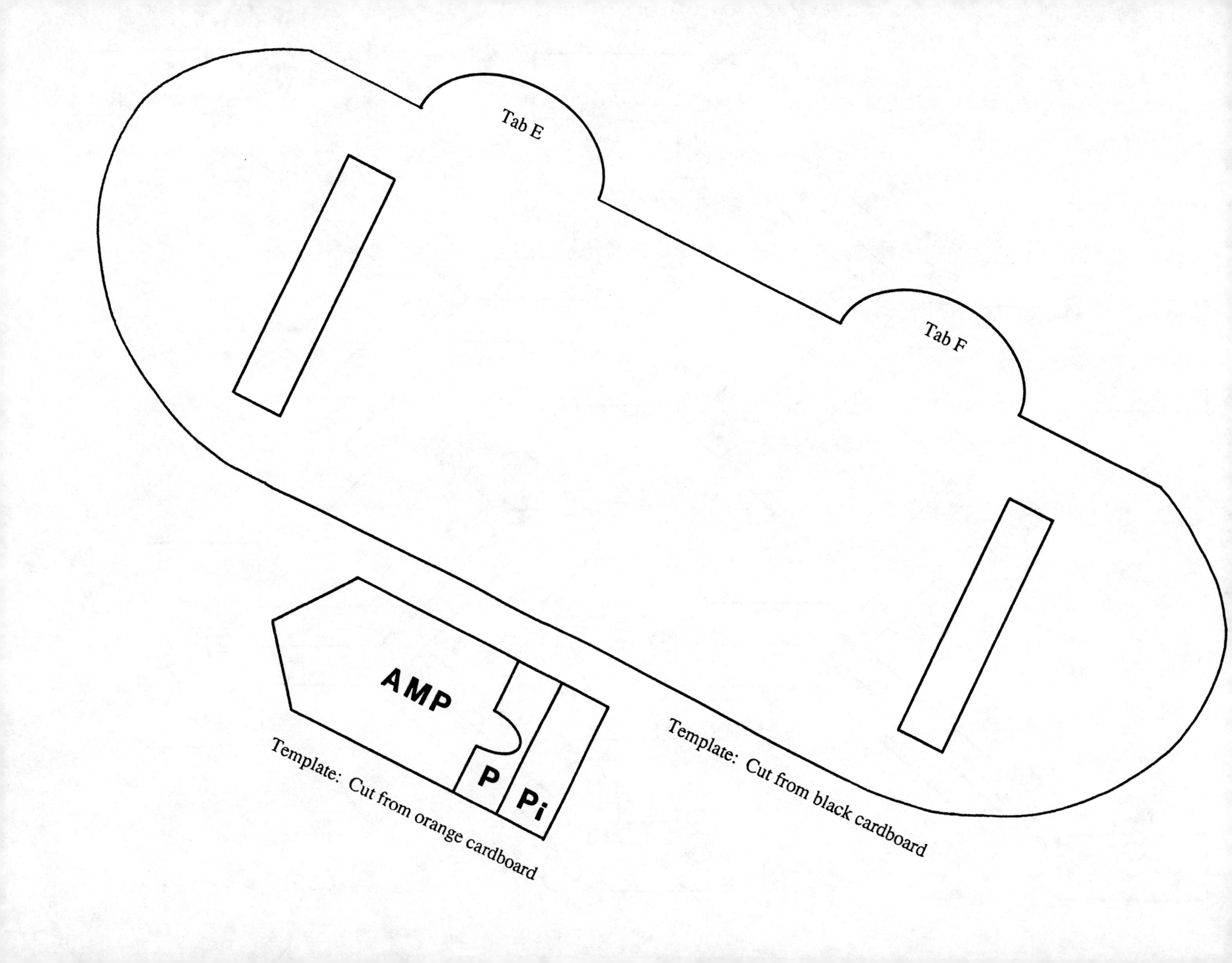

Tab E
Tab F
AMP
P
Pi
Template: Cut from orange cardboard
Template: Cut from black cardboard

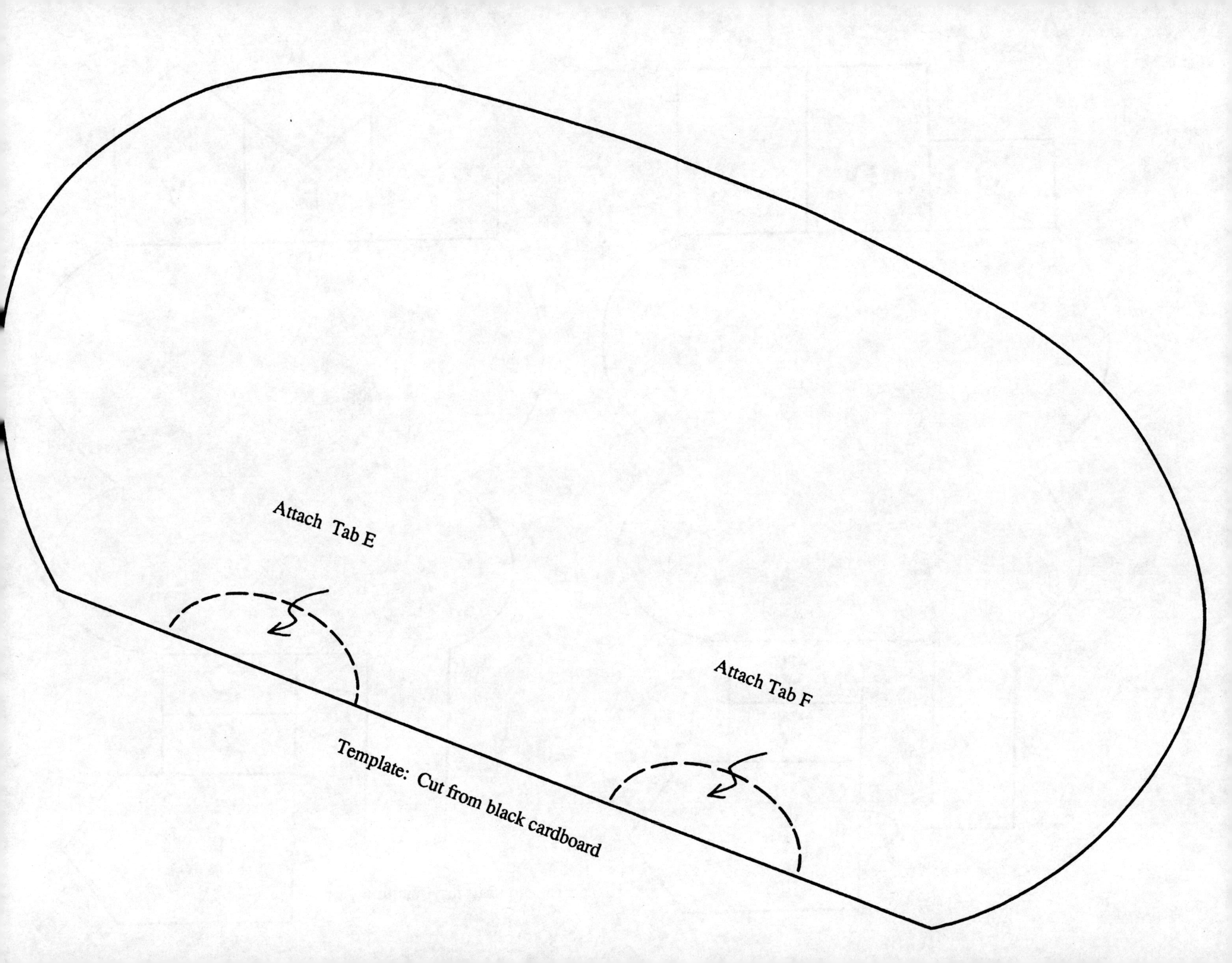
Attach Tab E
Attach Tab F
Template: Cut from black cardboard

Template: Cut from blue cardboard

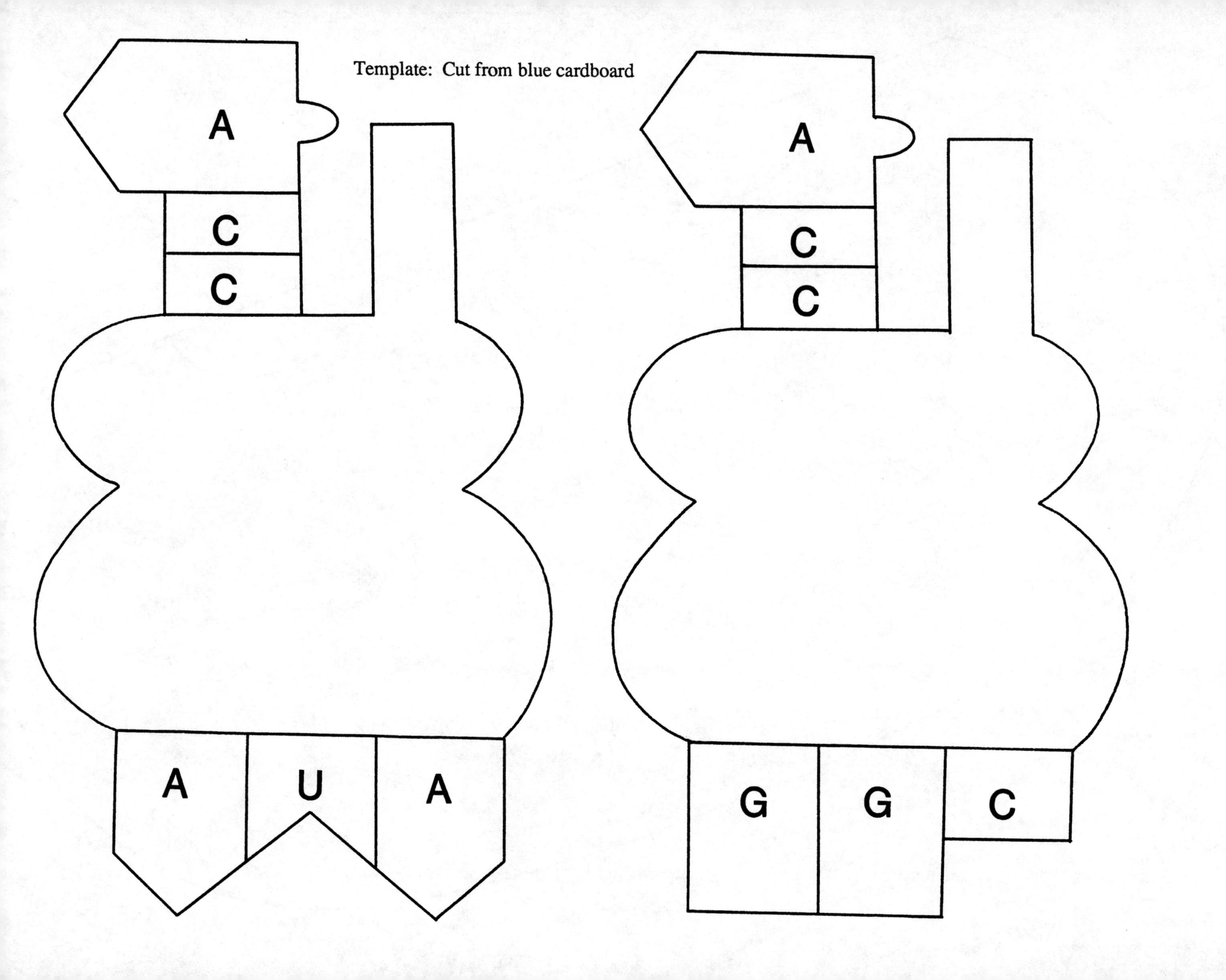

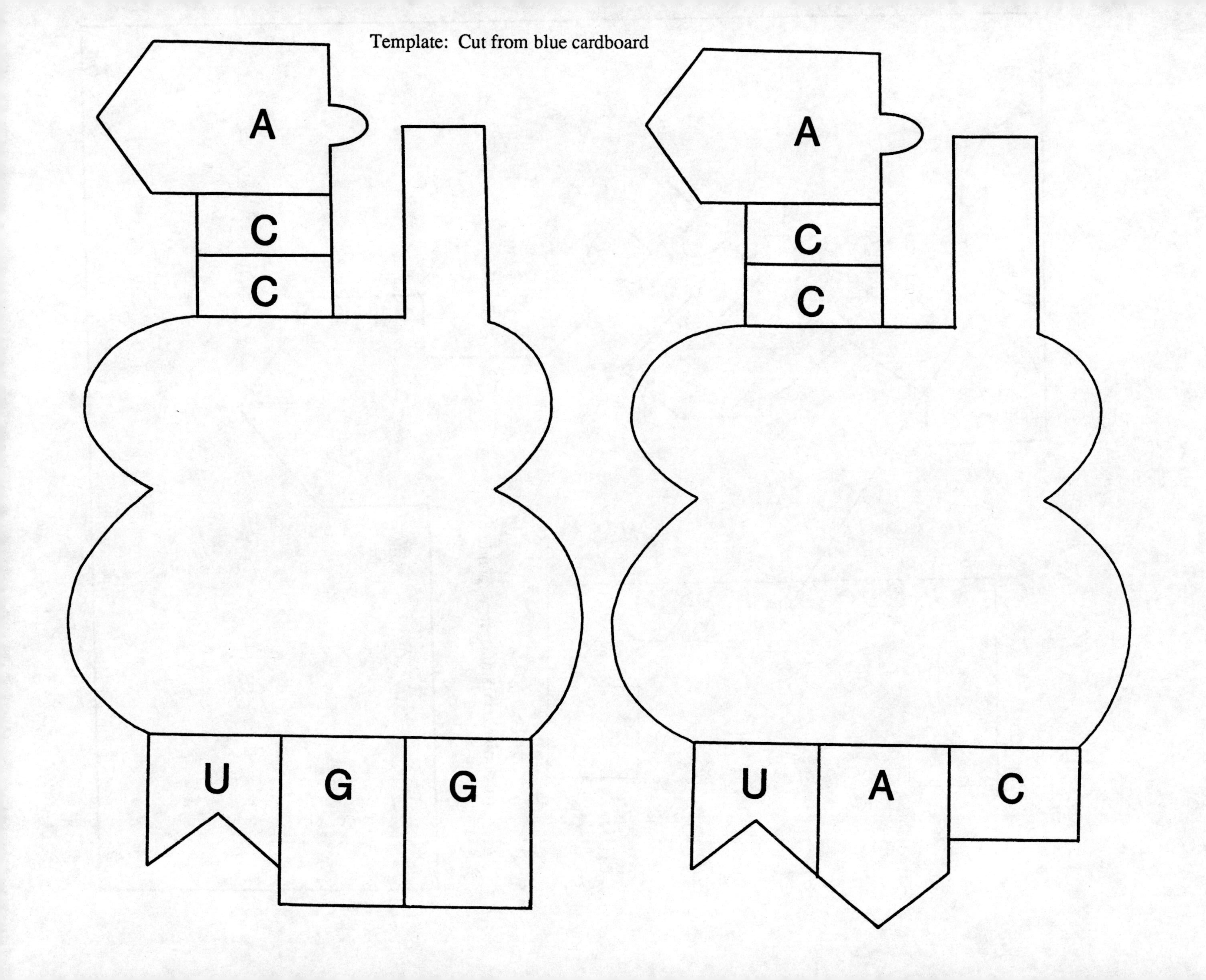

Template: Cut from blue cardboard

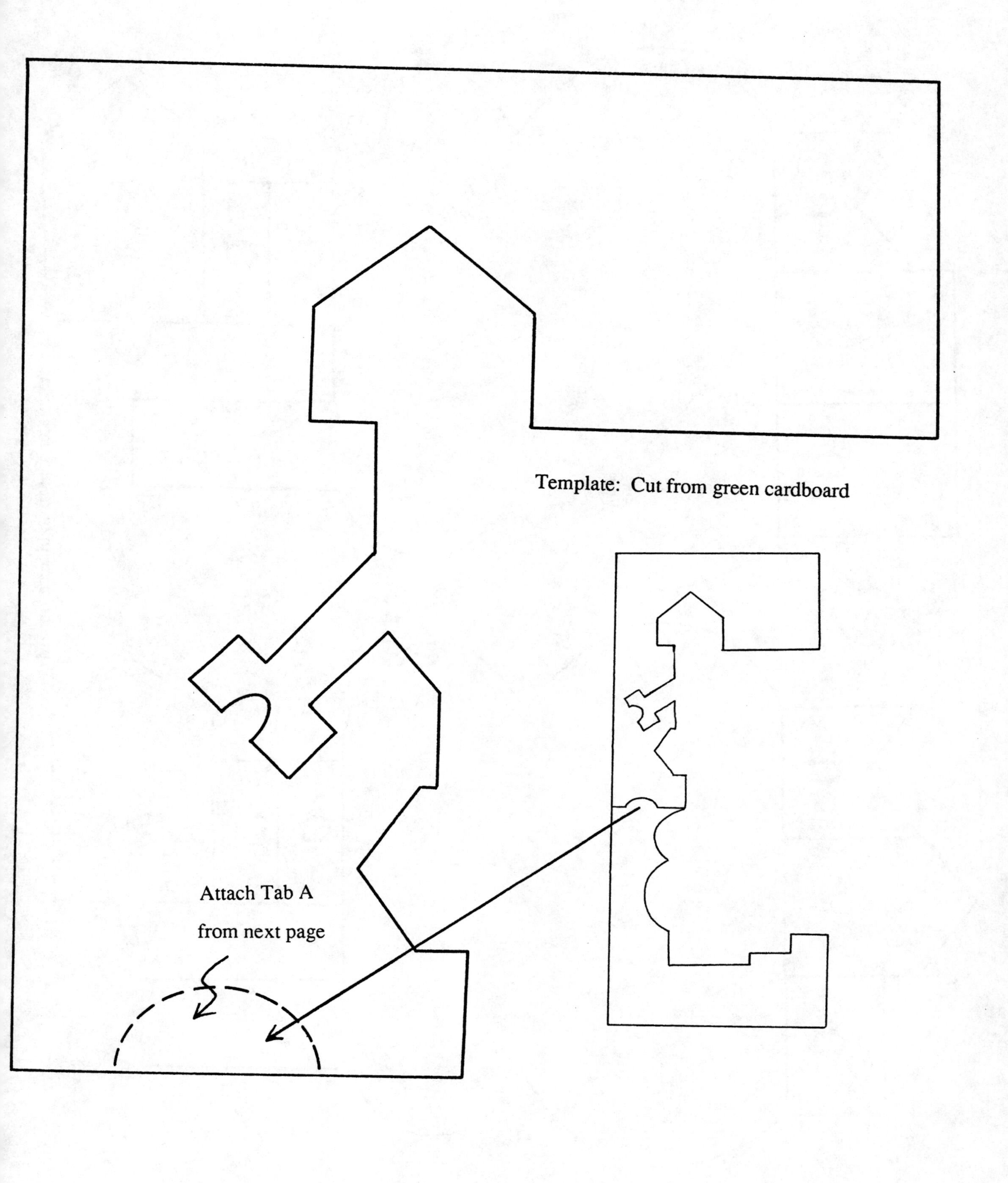
Template: Cut from green cardboard
Attach Tab A
from next page

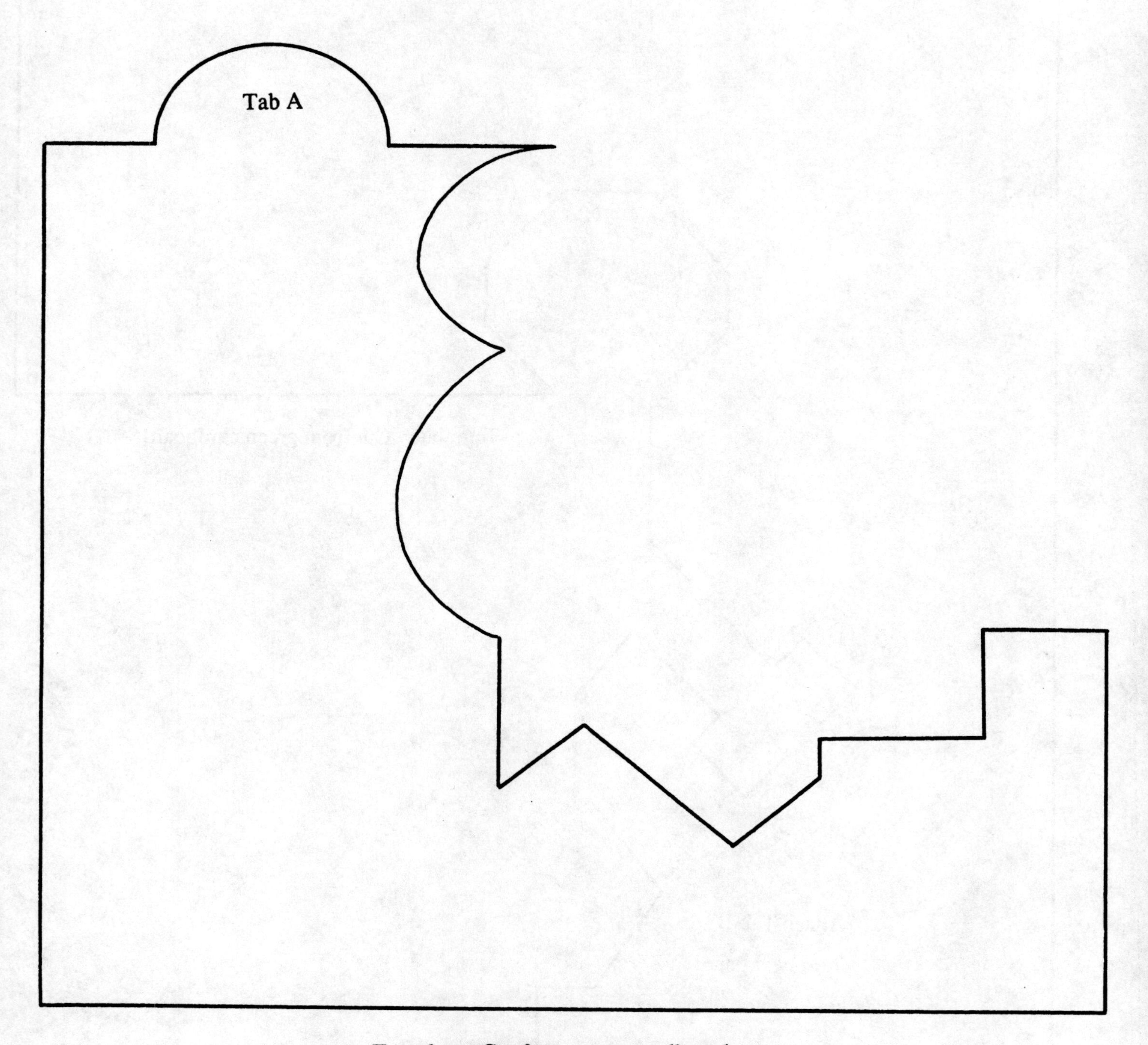

Template: Cut from green cardboard

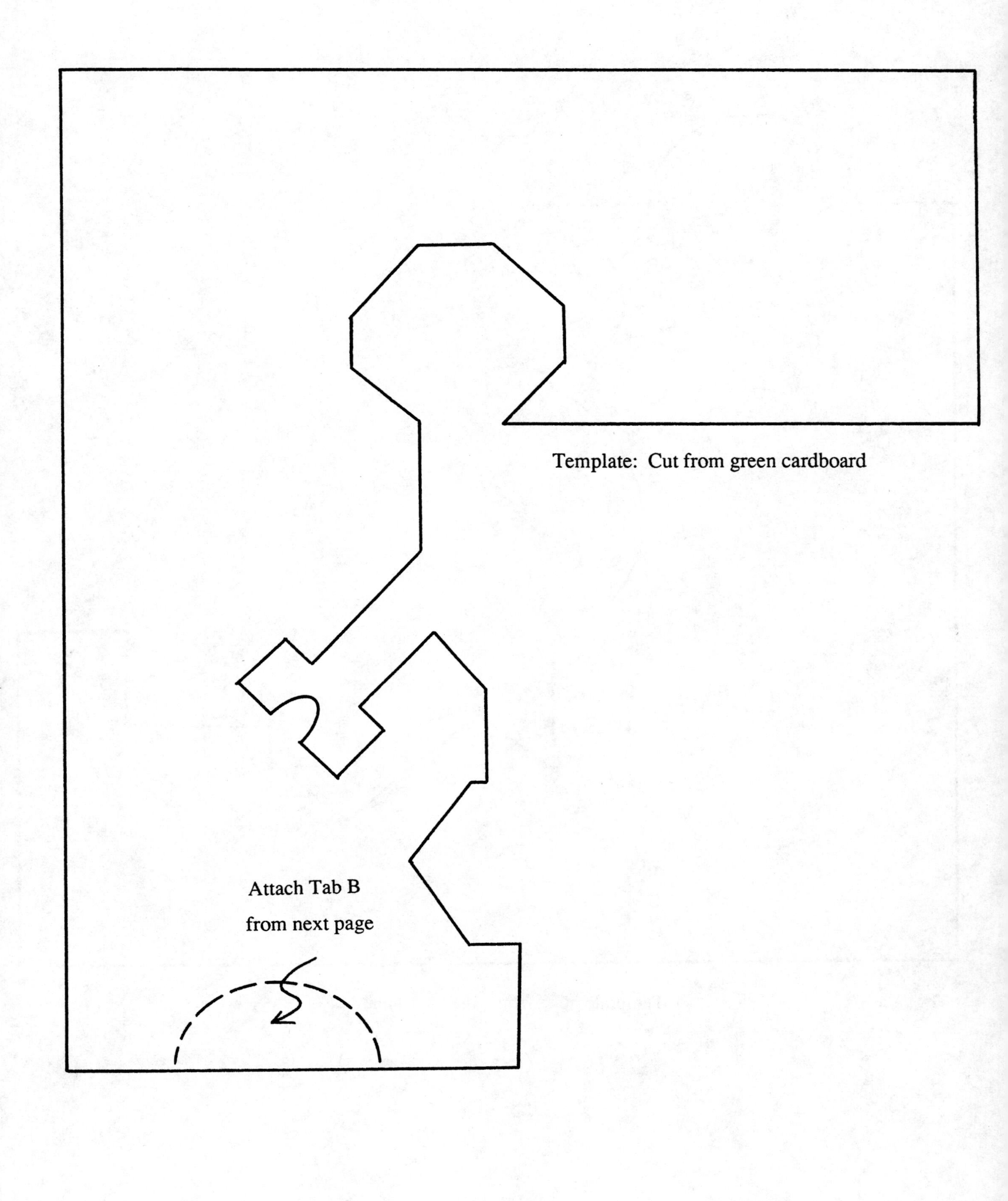
Template: Cut from green cardboard
Attach Tab B
from next page

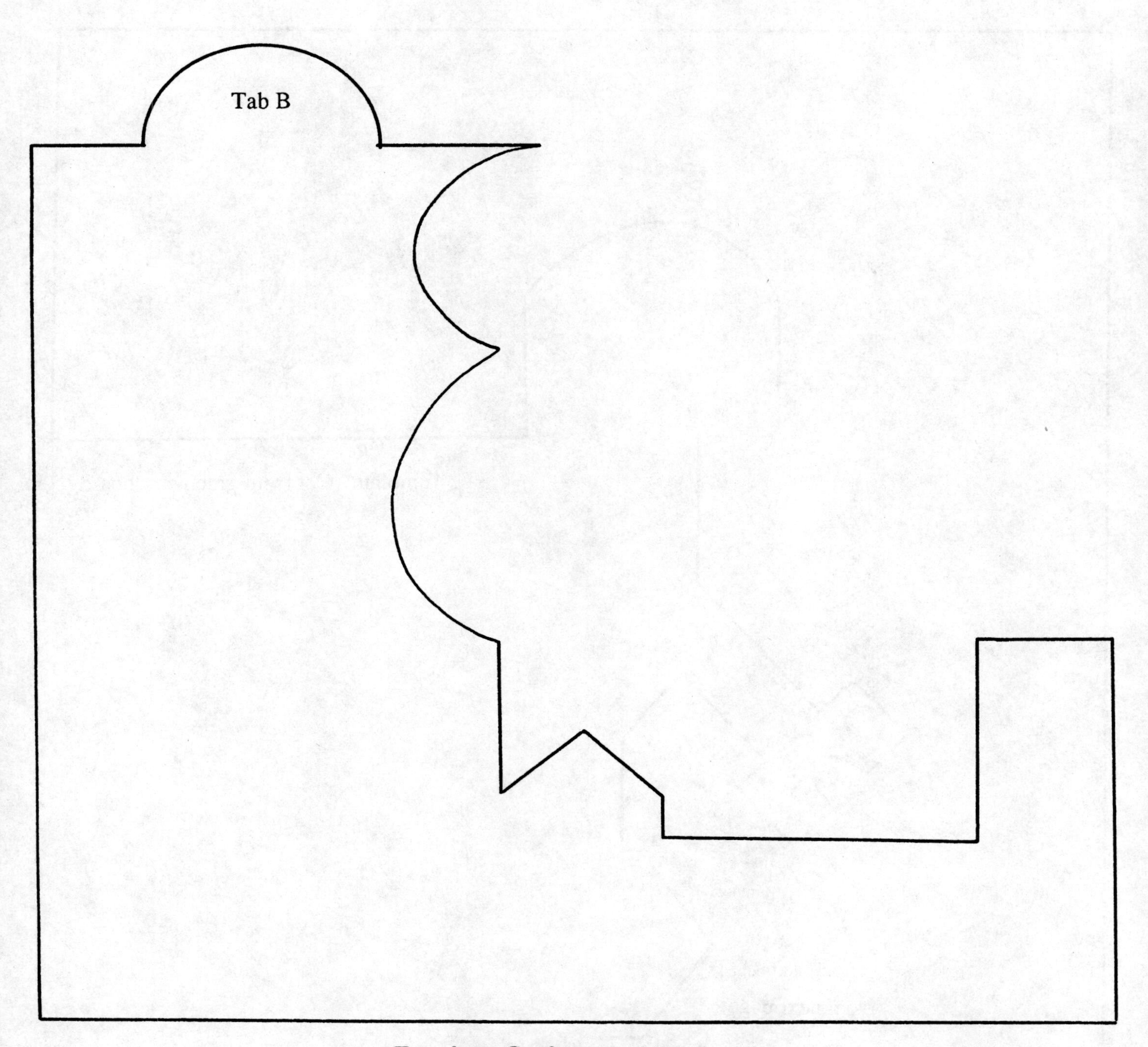

Template: Cut from green cardboard

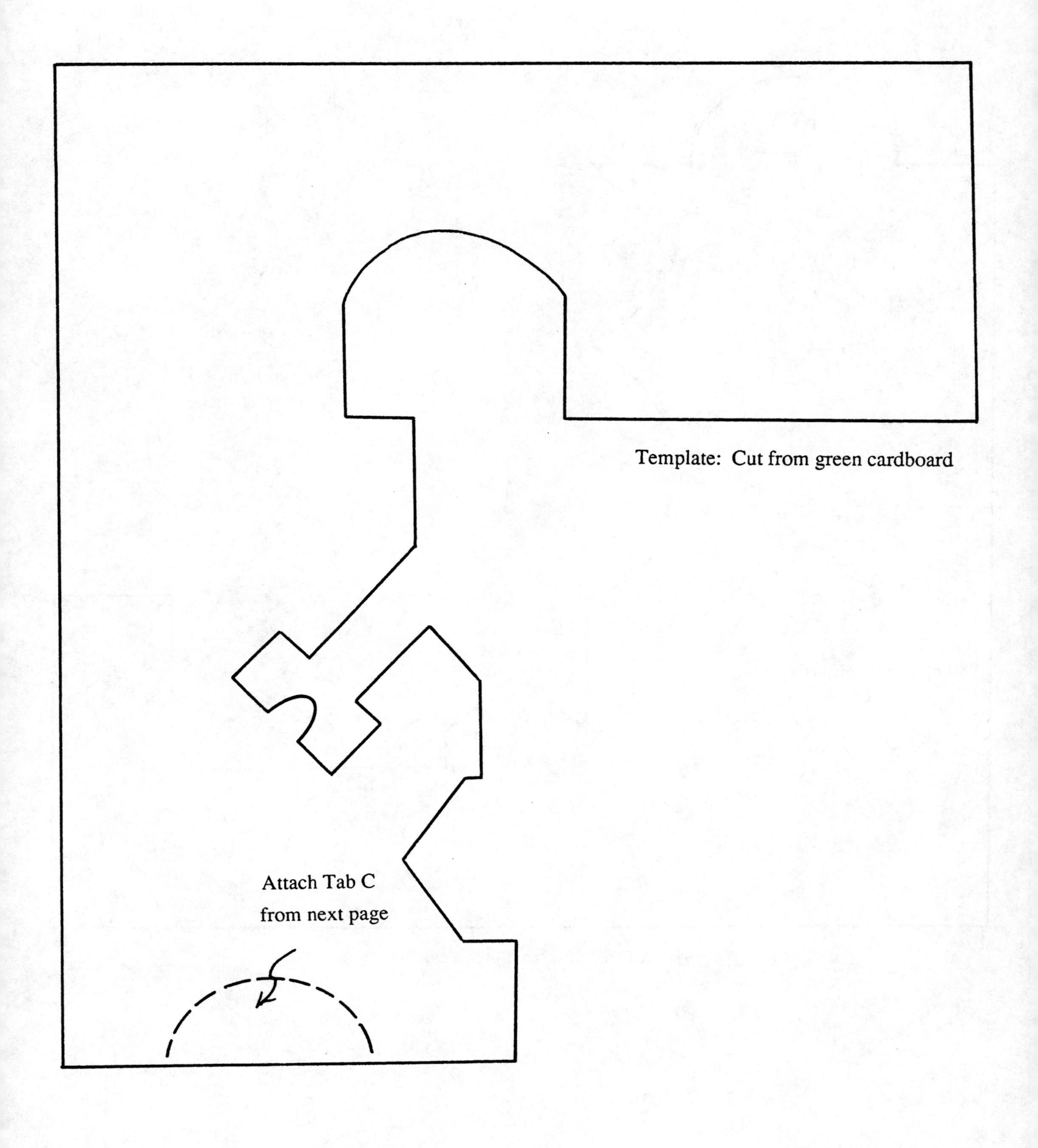
Template: Cut from green cardboard
Attach Tab C
from next page

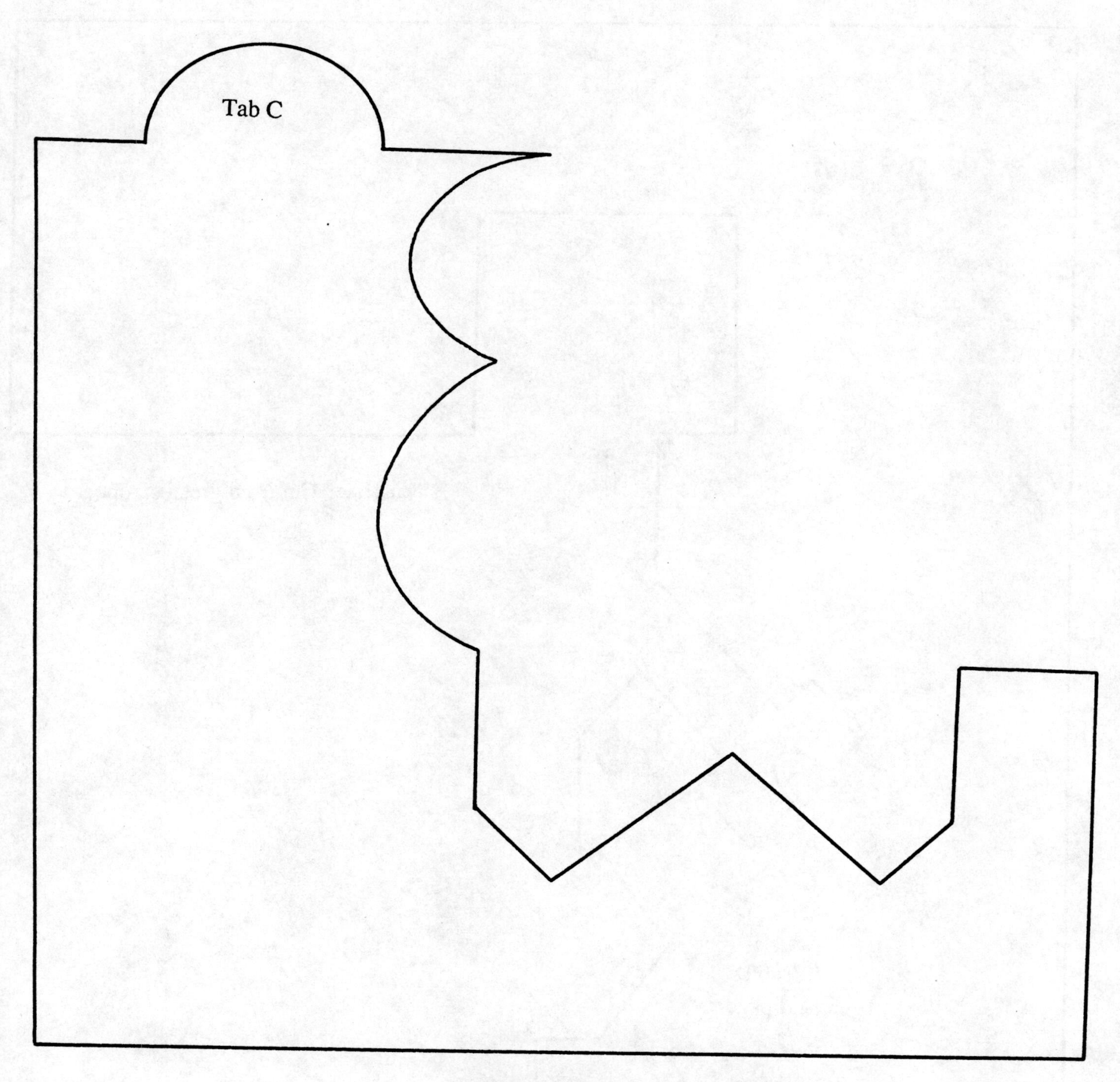

Template: Cut from green cardboard

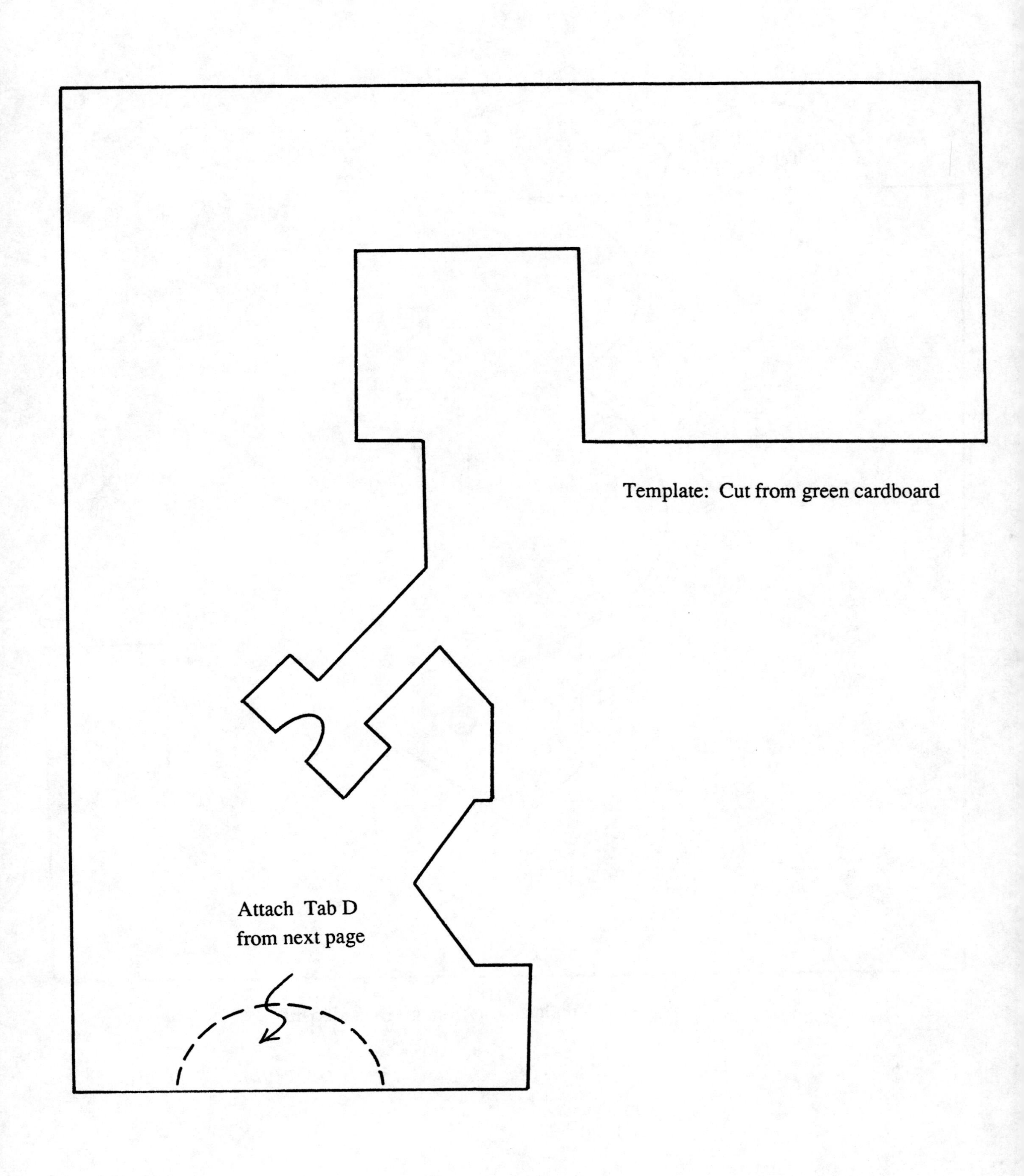
Template: Cut from green cardboard
Attach Tab D
from next page

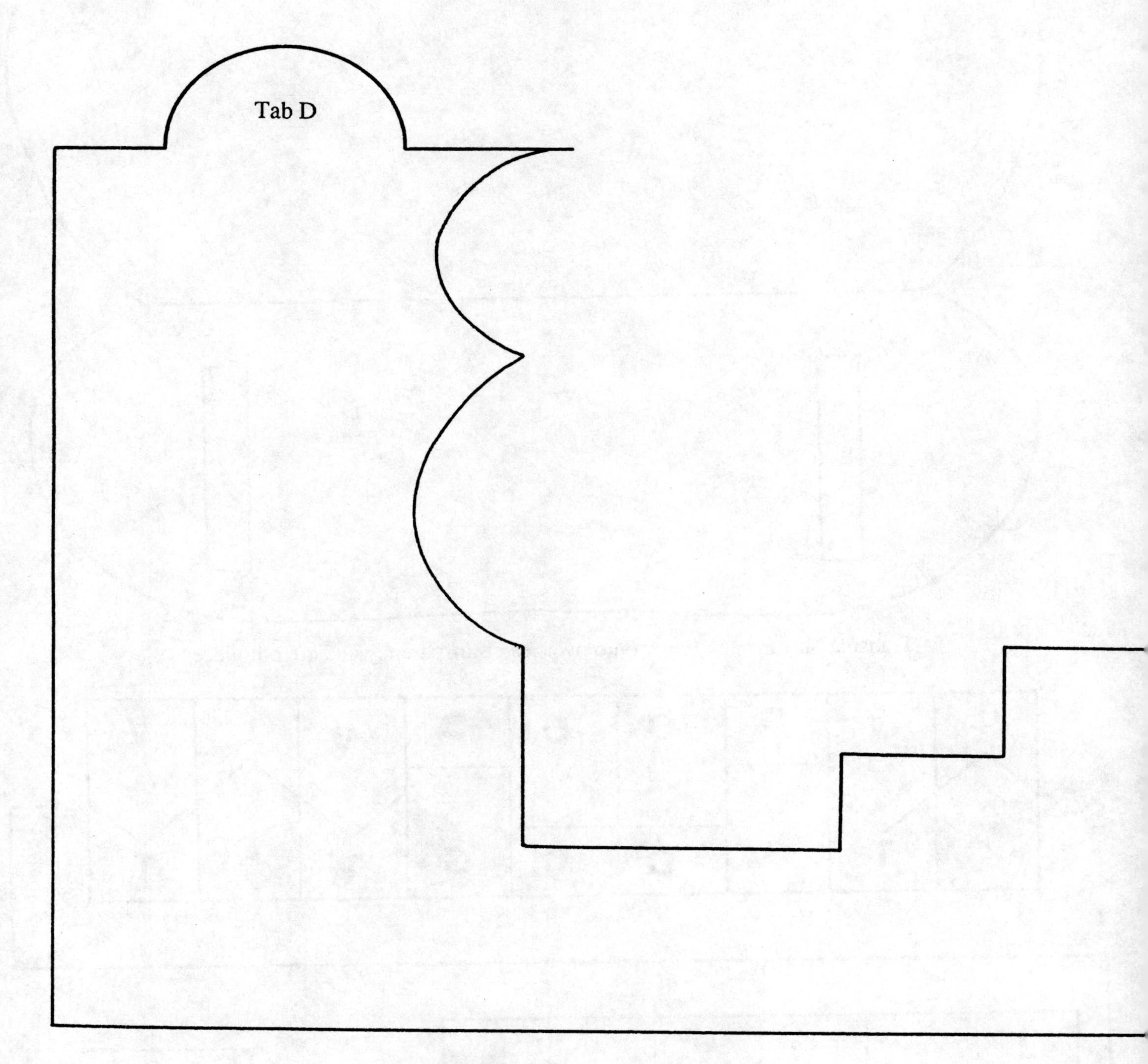

Template: Cut from green cardboard

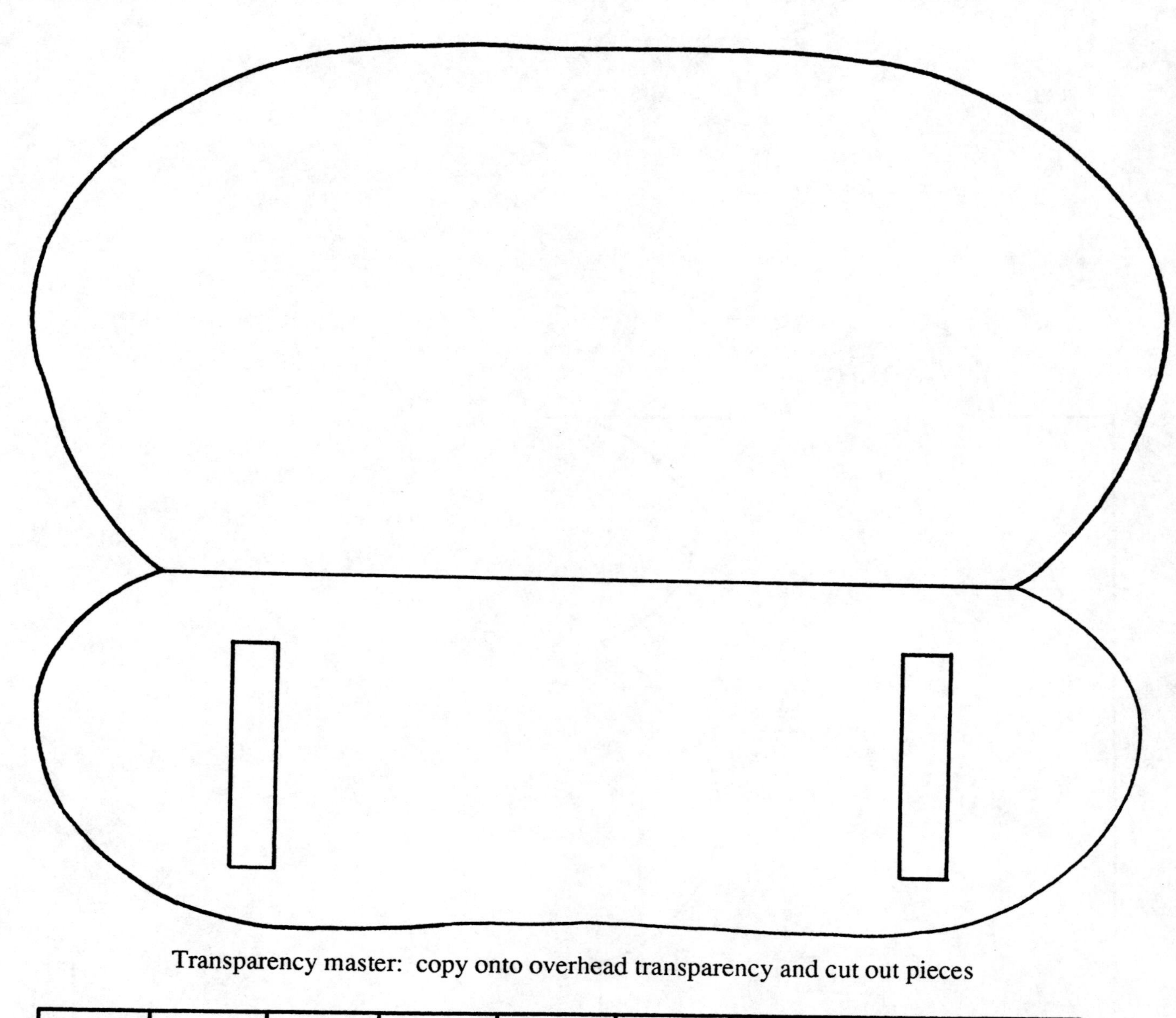

Transparency master: copy onto overhead transparency and cut out pieces

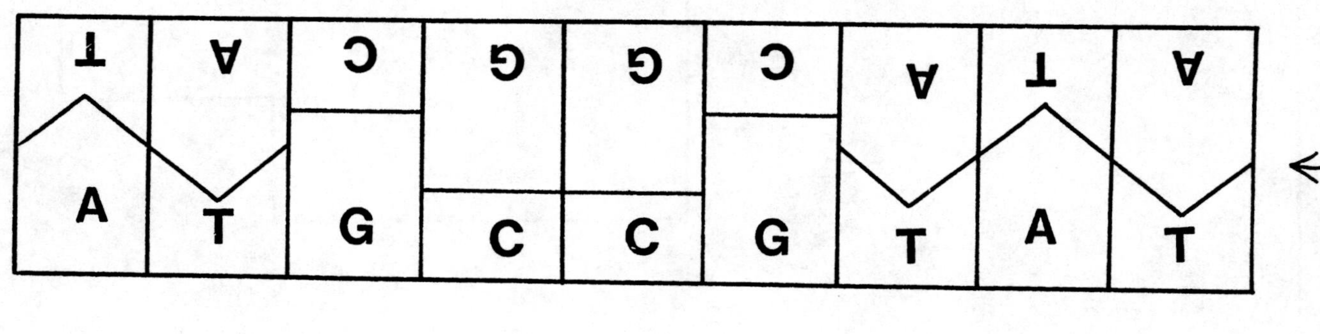

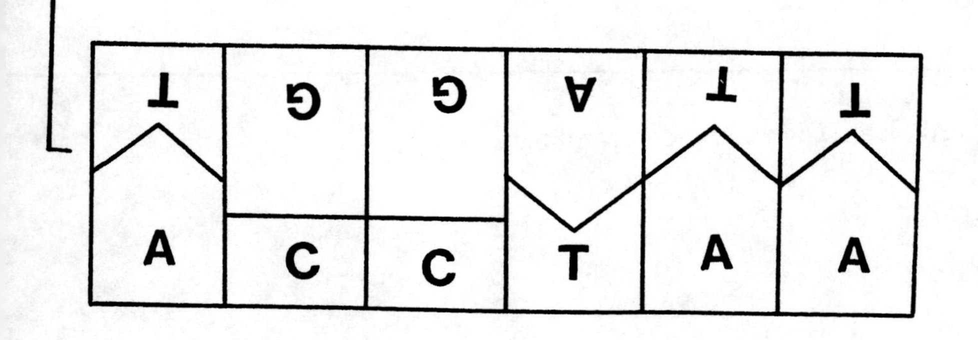

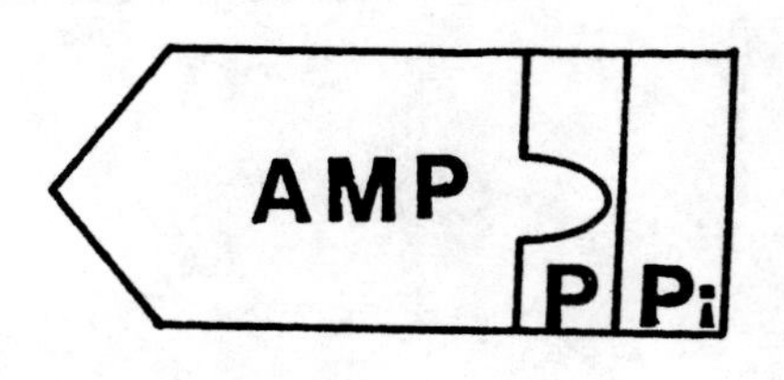

Transparency master: copy onto overhead transparency and cut out pieces

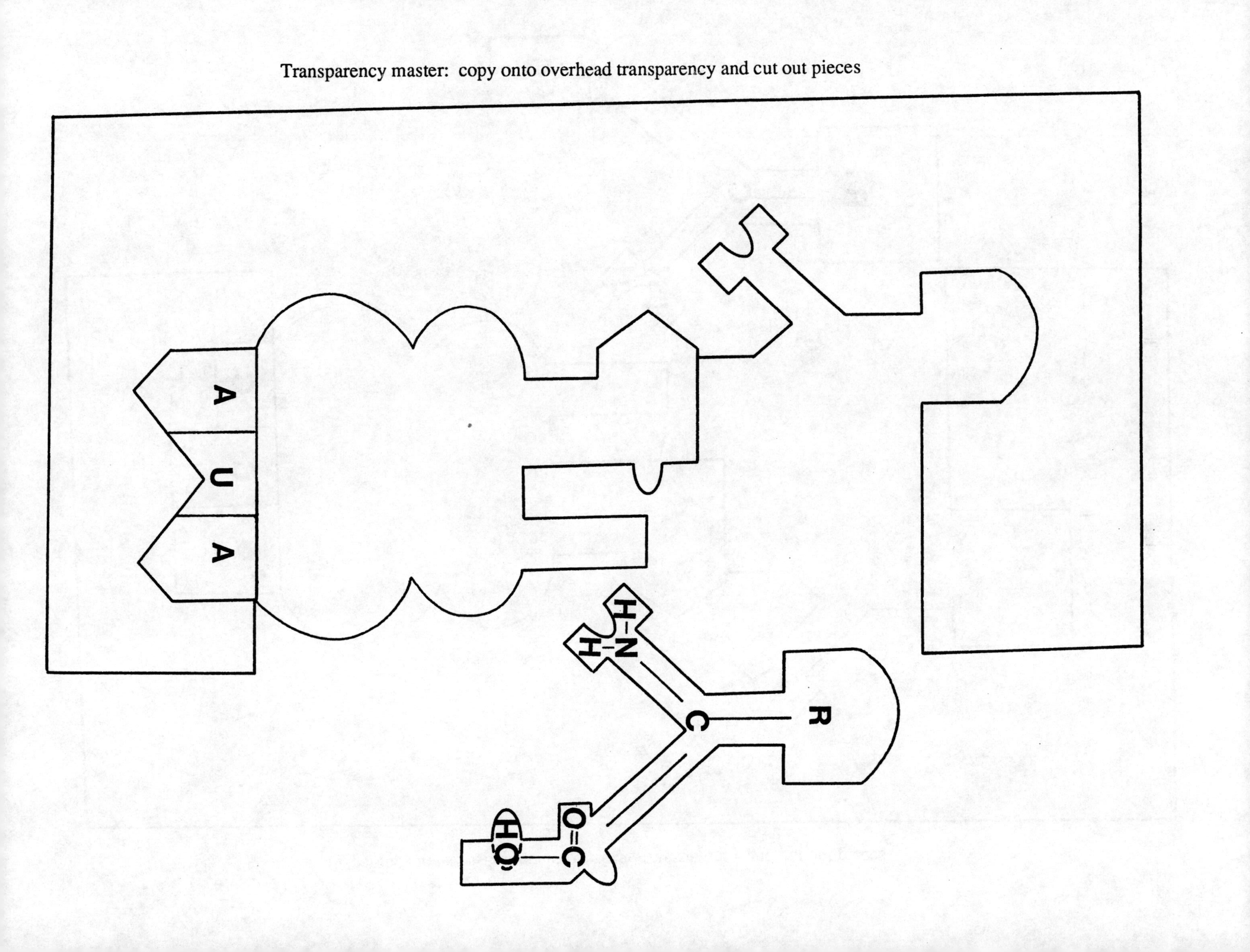

Transparency master: copy onto overhead transparency and cut out pieces

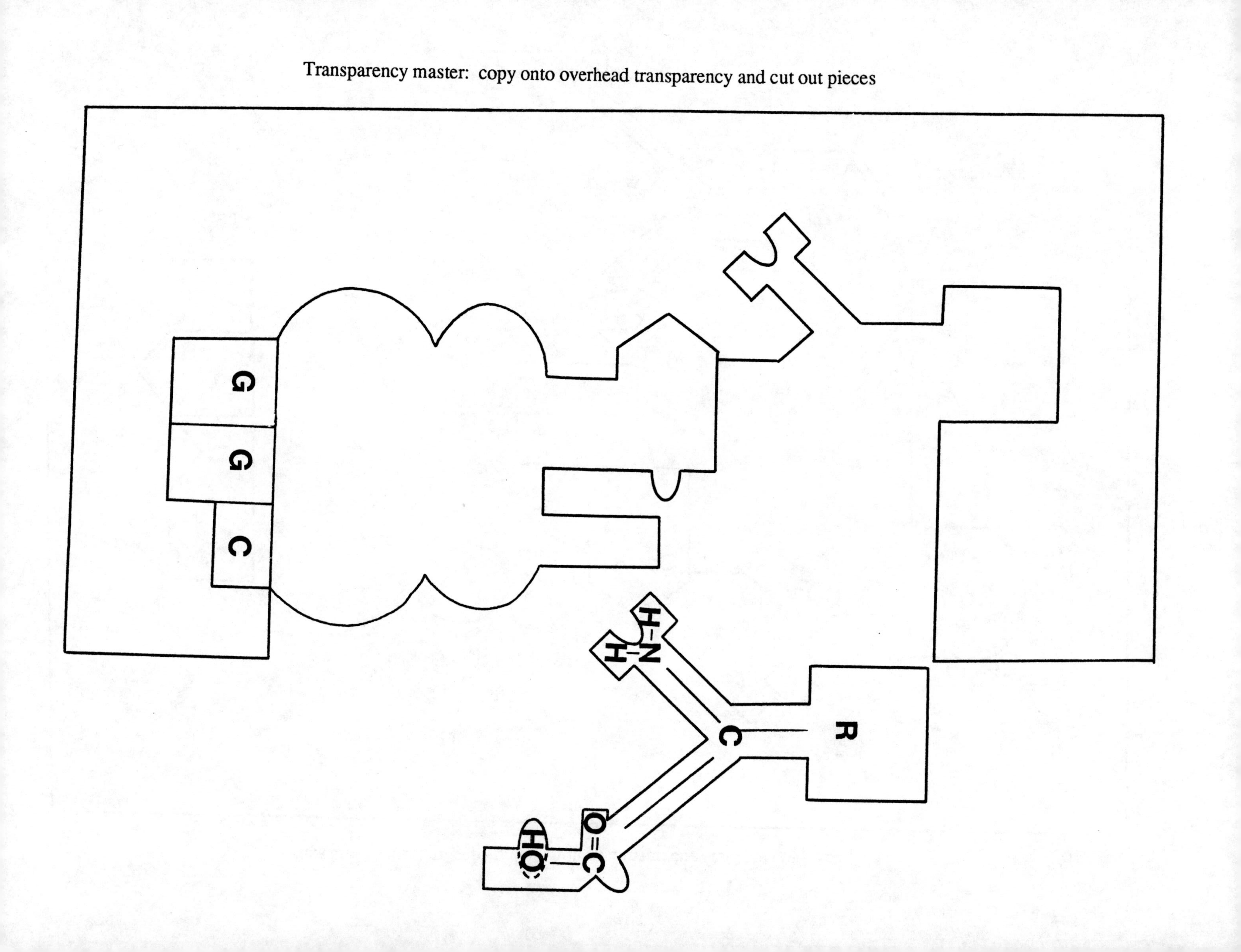

Transparency master: copy onto overhead transparency and cut out pieces

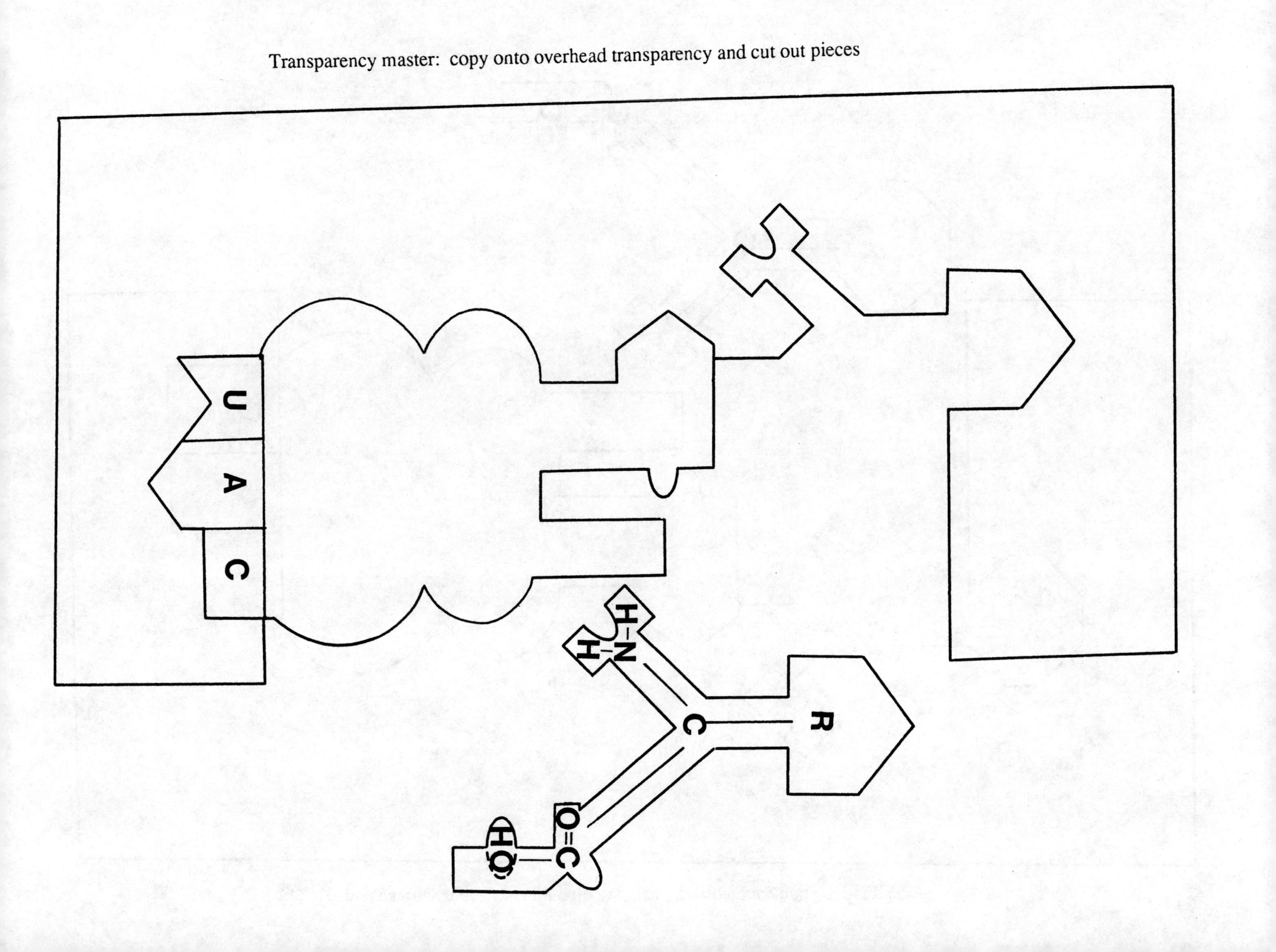

Transparency master: copy onto overhead transparency and cut out pieces

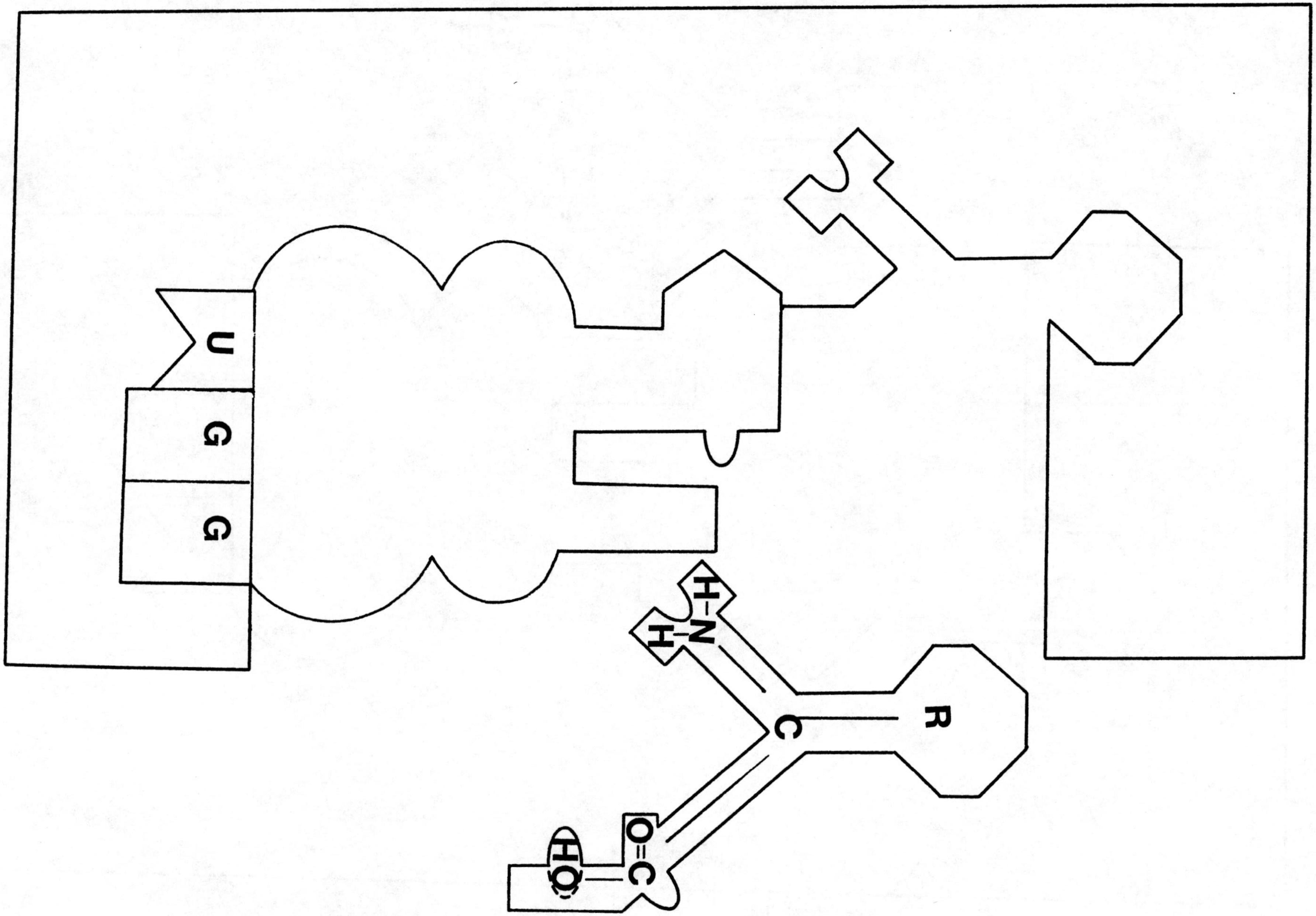

U G G

R

C

H–N

H

O=C

HO

LABORATORY 18 Molecular Genetics: Recombinant DNA

I. FOREWORD

In this laboratory, students are introduced to some common techniques in recombinant DNA technology. Knowledge of aseptic technique (Laboratory 6) is necessary, and students should be encouraged to review rules for working with bacteria. Exercise A can be completed prior to the laboratory. Students may work individually or in pairs on Exercises B and C. Both Exercises B and C demonstrate the transformation of ampicillin-sensitive bacteria and selection for recombinants. In addition, Exercise C demonstrates how a phenotypic marker can be used to identify transformants.

II. TIME REQUIREMENTS

Exercise A—Constructing Recombinant Plasmids (30 minutes, can be completed prior to the laboratory)
Exercise B—Rapid Colony Transformation with pAmp (60 minutes)
Exercise C—Transformation of *E. coli* with pBLU (60 minutes)

III. STUDENT MATERIALS AND EQUIPMENT

	Per Student	Per Pair (2)	Per Group (4)	Per Class (24)
Exercise A				
scissors	1			24
tape (roll)	1			24
Exercise B				
microcentrifuge tubes	2			48
inoculating loop	1			24
Escherichia coli MM294 plate **(1)**		1		12
0.05 M $CaCl_2$, sterile tube (1 ml) **(2)**	1			24
plasmid pAmp (10 µl) **(3)**	1			24
ice bucket with ice		1		12
marking pen		1		12
water bath (42° C) and beaker with Styrofoam float **(4)**		1		12
micropipette, 10 µl **(5)**	1			24
micropipette, 100 µl **(5)**	1			24
micropipette, 250 µl **(5)**	1			24
Luria broth, tube (1 ml) **(6)**	1			24
LB agar plate **(7)**	2			48
LB/Amp agar plate **(8)**	2			48
Bunsen burner		1		12
spreading rod	1			24
95% ethanol, beaker		1		12

	Per Student	Per Pair (2)	Per Group (4)	Per Class (24)
Exercise B—continued				
masking tape, roll				1
Exercise C				
plasmid, pBLU(10 μl) **(9)**		1		12
microcentrifuge (Micro-Test) tubes		1		12
Escherichia coli JM101, plate **(10)**		1		12
competent *E. coli,* JM101 cells, tube (200 μl) **(10)**		1		12
micropipette (0-10 μl) **(5)**		1		12
micropipette (100-1000 μl) **(5)**		1		12
ice bucket with ice		1		12
water bath (42° C) and beaker with Styrofoam float **(4)**		1		12
Luria broth, tube (1 ml) **(6)**		1		12
X-gal/Amp agar plate **(11)**	1		12	
spreading rod		1		12
95% ethanol, beaker		1		12
marking pen		1		12
masking tape				1

IV. PREPARATION OF MATERIALS AND SOLUTIONS

General[1, 2]

Kits are available from Carolina Biological Supply Company and from Edvotek for both Exercises B and C. If you are inexperienced with recombinant DNA, we suggest that you work with kits the first time through this laboratory. These kits come with all components, including cells, plasmids, agar, pipettes, and plates for transformation.

If a kit is used, follow these hints:

When you receive a kit, place cells, plasmid, ampicillin, and X-gal in the refrigerator. Agar can remain in bottles at room temperature until used.

Two to three days before the laboratory, make agar plates using the sterile agar. It is best to heat the bottles of agar slowly in a water bath (loosen, but do NOT remove caps). A microwave is possible but agar usually spills out the top. Make sure *all* agar is melted and no cloudy lumps remain in the bottle. Add ampicillin or ampicillin and X-gal (both come in vials with the kit and should be refrigerated until used) to the agar after it has cooled to the

---

[1] Catalogs for companies that offer a variety of supplies, kits, and biologicals for DNA work can be requested from the following sources (see Suppliers for added information including URLs):

Carolina Biological Supply Company, 2700 York Road, Burlington, NC 27215 or Box 187, Gladstone, OR 97027 (800-334-5551)

Edvotek, Inc., PO Box 1232, West Bethesda, MD 20827-1232 (800-338-6835)

Fotodyne Corporation, 950 Walnut Ridge Dr., Hartland, WI 53029 (800-362-3686)

Life Technologies (formerly Bethesda Research Laboratories, 3175 Staley Rd., Grand Island, NY 14072 (800-344-3100)

Wards Natural Science Establishment, Inc., Box 92912, Rochester, NY 14692-9012 (800-962-2660)

We strongly suggest that you obtain catalogs from these companies before proceeding with these laboratories.

[2] Development of this laboratory was completed with the assistance of Dr. Jack Chirikjian, Edvotek, Inc., and Drs. David Micklos, Mark Brow, and Greg Freyer of the DNA Learning Center, Cold Spring Harbor.

point where you can hold it (but it is still hot—be careful not to let it begin to resolidify!). Pour plates by lifting lids slightly. Pour them thin to make enough plates. After cooling, store plates upside down.

Make streak plates (see Laboratory 6 or **(1)** below) 24 hours ahead. Bacterial slants should be kept in the refrigerator but should be warmed up to room temperature before plating.

For **Exercise B**, the following kits are available:

Colony transformation kit. (Carolina Biological Supply Company, kit # 21-1142). This kit uses the rapid colony transformation technique developed at Cold Spring Harbor as described in the directions of Exercise B. The plasmid used for transformation is pAmp.

Transformation of *E. coli* cells with plasmid pBR322 kit. (Edvotek, kit # 201). This kit provides the reagents and directions for the transformation of *E. coli* cells with plasmid DNA that carries genes for antibiotic resistance. The plasmid used for transformation is pBR322. The protocol for its use differs slightly from that in Exercise B.

For **Exercise C**, the following kits are available:

pBLU colony transformation kit. (Carolina Biological Supply Company, kit # 21-1146. Transformation of E. *coli* strain JM 101 with pBLU™ plasmid DNA. Contains all materials and plastic-ware necessary to complete this exercise. The kit contains enough material for 20 students. (JM101 cells used in Exercise C must be maintained on minimal medium and are relatively difficult to propagate. For this reason, a kit is recommended. See Ordering Information.)

transformation of *E. coli* with pGAL™ kit. (Edvotek, kit # 221). This kit contains the biologicals and labware necessary for Exercise B.

If a kit is not used, materials can be prepared as follows:

(1) *E. coli* MM294 cells, plate

From an agar slant of *E. coli* MM294 bacteria (see Ordering Information), prepare a fresh streak plate (see Laboratory 6) to isolate single colonies. (Alternatively, plates of *E. coli* are provided in both the Carolina and Edvotek kits.)

(2) 0.05 M calcium chloride ($CaCl_2$), sterile

Dissolve 5.5 g anhydrous calcium chloride ($CaCl_2$) or 7.25 g $CaCl_2 \cdot 2H_2O$ in distilled water to make a solution of 100 ml. Autoclave for 15 minutes, 15 pounds pressure, at 121° C in a loosely capped bottle covered with aluminum foil. Tighten cap after cooling. Alternatively, dispense into tubes of 1-2 ml each, cover with aluminum foil, and autoclave as above. This will allow each student to have his or her own sterile $CaCl_2$ solution.

The rapid colony transformation technique can be used. For this, prepare a fresh streak plate for each laboratory section (see Laboratory 6). Use a sterile transfer loop to remove a small amount of bacteria from a slant tube culture. Spread this in a zig-zag motion over the surface of the plate, to one side. Turn the plate 90° and cross the first zig-zag once, making a new zig-zag pattern down the surface of the plate. Repeat twice more, never crossing an original zig-zag more than once. This dilutes the bacteria. Incubate overnight at 37° C. Single colonies should be isolated for the transformation. Each pair of students should use a colony about the size of the head of a round-headed straight pin. Caution students NOT to use too much cell material or the plasmid will be "overwhelmed" and transformation efficiency will be low. Also, caution students to avoid getting agar in the loop when they harvest a colony for transformation.

You can also make cells competent beforehand and ensure the density of cells for transformation. If you choose to do this, prepare cells the day before (remember that there is a possibility of damaging cells during this procedure if you are not careful). See directions in **(10)** for JM101 cells and follow the same directions for MM294 cells.

(3) plasmid pAMP

Plasmid pAMP (200 μl), 0.005 μg/ml, may be purchased separately from Carolina Biological Supply Company. (Plasmid pAMP is provided with the Carolina kit, 0.005 μg/μl; 0.2 ml.) This can be aseptically dispensed into microcentrifuge tubes for students, but we suggest that the instructor dispense the plasmid to avoid pipetting errors. Students should use 10 μl of the purchased plasmid pAMP (0.005 μg/μl) or 0.05 μg in the 250 μl of $CaCl_2$ and cells. (If this is the case, the transfer loop can be used to transfer 10 μl of plasmid from stock solution to the student's tube containing $CaCl_2$.)

If you wish to give each student a microcentrifuge tube containing plasmid, we suggest that you dilute the plasmid 1:1 with TE buffer and give each student 20 μl rather than 10 μl to be transferred to the $CaCl_2$. TE buffer can be purchased in prepared form from Carolina Biological Supply Company (see Ordering Information), or can be-prepared as follows:

1 M Tris (pH 8.0)	1.0 ml
0.5 M EDTA	0.2 ml
distilled water	200.0 ml

(Adjust to pH 8.O, see below.)

1. For 1 M Tris (pH 8.0), dissolve 12.1 g Tris base in 80 ml distilled water. Add concentrated HCl to adjust the pH (about 4-4.5 ml). Add distilled water to 100 ml. **Be sure to wear a mask and cover your nose and mouth when working with Tris powder.** (1 M Tris can also be purchased in prepared form from the Carolina Biological Supply Company.)
2. For 0.5 M EDTA, add 18.6 g ethylenediamine tetraacetate (EDTA, disodium salt) to 80 ml of distilled water. Adjust the pH to 8.0 by adding approximately 2 g NaOH.

(4) water bath with Styrofoam float

A beaker of water can serve as a water bath for an individual or a pair of students. Use a thin piece of Styrofoam and make a hole in it large enough to accommodate the microcentrifuge tube (or Micro-Test tube). Float the Styrofoam holding the tube of cells on the water bath to heat shock the cells. This can also be done in a large water bath regulated to 42° C, but make sure students label their floats with tape.

If you conduct the experiment in larger sterile transformation tubes (Carolina Biological, # 21-508Q), be sure that students do not just put the tubes in the water expecting them to float—they leak!

(5) micropipettes (10 μl, 100 μl, and 250 μl)

There are many different types of micropipettes available. The Carolina and Edvotek kits come with special capillary-type micropipettes. You can also purchase glass capillary pipettes calibrated in μl amounts. The Ultramicro Accropet (see Ordering Information) is made for use with capillary pipettes and is like a miniature Pi-Pump. By turning the Accropet, you can pull up or deliver accurate amounts of solution. Be sure to sterilize capillary pipettes in covered glass test tubes before use.

More expensive digital micropipettes are also available with special autoclavable, disposable tips. Groups of students can use a single pipette by changing tips. Pipettes are available from 2 to 10 μl, 10 to 100 μl, or 100 to 1000 μl (remember that 100 μl is 0.1 ml and sterile glass serological pipettes can be used for such larger amounts).

(6) **Luria broth** (provided in both kits)

To make Luria broth for growing *E. coli,* use:

bacterotryptone	10 g
yeast extract	5 g
NaCl	10 g

1. Add ingredients to a 2 liter flask and add 1000 ml of distilled water. Swirl to dissolve.
2. Dispense into five 200 ml bottles. Cap loosely and cover the top with aluminum foil.
3. Autoclave 15 minutes, 15 pounds pressure, at 121° C.
4. Cool and tighten lids. Store at room temperature. (Cloudiness indicated contamination.)
5. Dispense 1 ml amounts into sterile, capped test tubes (enough for each student or pair of students) for use in recovery of transformed cells. Use disposable transformation tubes (Carolina Biological, # 21-5080).

(7) **LB agar plates** (plates and agar provided in kits; agar ready to pour)

To make Luria broth agar (LB agar) plates, use:

bacterotryptone	10 g
yeast extract	5 g
NaC1	10 g
agar	15 g

1. Add ingredients to a 2 liter flask and add 1000 ml of distilled water to the flask. Swirl to dissolve lumps. Cover with aluminum foil
2. Autoclave for 15 minutes, 15 pounds pressure, at 121° C. Let cool until you can pick up the flask with your bare hands. (If lumps are present, autoclave again.)
3. Remove the cover from a sterile Petri dish, but hold the cover at an angle above the dish and pour in the agar. Replace the cover immediately (never set the cover down on the counter top). Repeat for the rest of the plates.
4. After the agar has solidified, invert the plates and let them sit until condensation disappears. Label the plates "LB" and store in plastic bags in the refrigerator. Store inverted.

Alternatively, prepoured plates or ready-to-pour LB agar is available from Carolina Biological Supply Company (see Ordering Information). Plates can be poured *very* thin.

(8) **LB/Amp agar plates** (plates, agar, and ampicillin provided in kits; agar ready-to-pour)

To make LB/Amp agar plates:

1. Add 0.03 g ampicillin to 1000 ml of autoclaved LB agar **(7)**. Add the ampicillin only after the agar has cooled enough to hold the flask in your hand.
2. Swirl and pour plates as described above.
3. Be sure to mark the Plates "LB-Amp."

Alternatively, prepoured plates or ready-to-pour LB agar containing ampicillin is available from Carolina Biological Supply Company (see Ordering Information).

(9) **plasmid pBLU**

Plasmid pBLU™ (0.005 μg/μl) is included as part of the Carolina Biological Supply Company or as pGAL™ in Edvotek kits. If a kit is not used, the plasmid (1 μg) (200 μl; 0.005 μg/μl) can be purchased from Edvotek or Carolina Biological Supply Company (see Ordering Information).

You will need 0.03-0.05 μg of plasmid per every 100 μl of cells to be transformed. If you order 200 μl; 0.005 μg/μl, this means that each student will use 10 μl. This amount (10 μl) is approximately what a transfer loop will hold and it can be dispensed by allowing each pair of students to obtain a loopful from the original vial. Thus, a vial (1 μg) (200 μl; 0.005 μg/μl) will be good for 18-20 pairs of students. Depending on the size of your class, order accordingly.

If you prefer to provide each student with a vial containing plasmid, dilute 10 μl of plasmid in 20 μl of sterile TE buffer.

(10) JM101 *E. coli* cells, plate

JM101 cells are provided with both Edvotek and Carolina kits. If a kit is not used, the cells can be purchased from Carolina Biological Supply Company or from Life Technologies (see Ordering Information). Purchase a slant culture of JM101.

The rapid colony transformation technique can be used. For this, prepare a fresh streak plate for each laboratory section (see Laboratory 6). Use a sterile transfer loop to remove a small amount of bacteria from a slant tube culture. Spread this in a zig-zag motion over the surface of the plate, to one side. Turn the plate 90° and cross the first zig-zag once, making a new zig-zag pattern down the surface of the plate. Repeat twice more, never crossing an original zig-zag more than once. This dilutes the bacteria. incubate overnight at 37° C. Single colonies should be isolated for the transformation. Each pair of students should use a colony about the size of the head of a round-headed straight pin. Caution students NOT to use too much cell material or the plasmid will be "overwhelmed" and transformation efficiency will be low. Also, caution students to avoid getting agar in the loop when they harvest a colony for transformation.

You can also make cells competent beforehand and ensure the density of cells for transformation. If you choose to do this, prepare cells the day before (remember that there is a possibility of damaging cells during this procedure if you are not careful).

Approximately 24 hours ahead of the experiment:

Prepare 1000 ml of Luria broth (see **(6)**).

Prepare 100 ml of sterile 0.1 M $CaCl_2$:

$CaCl_2 \cdot 2H_2O$	14.7 g
distilled water	100 ml

Prepare 100 ml of 0.1 M $CaCl_2$/15% glycerol:

$CaC1_2 \cdot H_2$	14.7 g
distilled water	85 ml
glycerol	15 ml

1. Add glycerol to water and dissolve $CaC1_2$ in the mixture.
2. Autoclave for 15 minutes at 121° C in a loosely capped bottle covered with aluminum foil. Tighten cap after cooling.
3. Store in a refrigerator.

Procedure:

1. Grow *E. coli* (JM101) in 100 ml of Luria broth, shaking at about 550 rpm at 37° C until absorbance on the Spectronic 20 reads 0.2-0.3 at 660 nm. At 37° C, this will take about 2-3 hours. If you have a shaking water bath, the process will be faster. Otherwise, place the cells in an incubator or a 37° C water bath and simply shake from time-to-time.
2. Pour 15 ml of culture into each of 6 sterile conical centrifuge tubes and pellet cells at top speed in a clinical centrifuge (approximately 4-6 K or 5,000 rpm) for 10 minutes.

3. Pour off supernatant and pipette to resuspend cells in each tube in 5 ml of cold (4° C) 0.1 M $CaCl_2$. Combine cells from three tubes to give two tubes of 15 ml each. **Keep on ice at all times.** Let cells sit for 60 minutes.
4. Recentrifuge for 10 minutes as in Step 2. Pellets should be spread out with a hole in the middle, resembling a doughnut. If pellets are compact, resuspend and repeat the $CaCl_2$ treatment for 25 minutes.
5. Pour off-the supernatant and resuspend in 0.1 M $CaCl_2$/15% glycerol using approximately 1.5 ml in each tube.
6. Store suspended competent cells overnight on ice or in the freezer. If using a low temperature (–80° C) freezer, cells will last for several months. Otherwise use frozen cells within 2-3 days.

Transformation with plasmid pBLU can be carried out as described in Preparation Suggestions. Use 250 μl of competent cells for transformation.

(11) LB/Amp/X-gal agar plates

To make LB/Amp/X-gal agar plates:

1. Prepare 1000 ml of LB-Amp agar as above **(8)**.
2. After the agar has cooled enough to hold the flask, add 2.5 ml of 2% X-gal (see below). (In kits, the X-gal is combined with ampicillin. N,N' dimethyformamide is also in this mixture—this mixture is hazardous, so handle carefully.)

To make 2% X-gal:

1. Dissolve 0.16 g X-gal in 8 ml of N,N' dimethylformamide (DMF) in the hood. Please note: Because DMF is **toxic**, it must be used in a hood.
2. Pour agar into plates and mark LB/Amp/X-gal. Invert and store.

V. PREPARATION SUGGESTIONS

Exercise A

All materials for this exercise are included in the laboratory manual. This exercise can be done at home prior to the laboratory in order to save time for experimentation. If the remainder of the laboratory is not used, this exercise will serve as an excellent supplement to lectures or reading on plasmids and recombinant DNA techniques. As a supplement, the text *DNA Science,* Micklos, David A., and Greg A. Freyer, Cold Spring Harbor Press, is available through Carolina Biological Supply Company. It is an excellent teacher supplement. The test, *Recombinant DNA and Biotechnology: A Guide for Teachers,* Kreuzer, H. and A. Massey, ASM Press, is also excellent and is available through Carolina Biological Supply Company.

Exercise B

The easiest way to prepare for this exercise if you have small numbers of students is to order one of the kits described in the General Information section (see Preparation of Materials and Solutions). For large numbers, purchase MM294 bacteria, pAMP plasmid separately, and make Luria broth, LB agar, and LB/Amp agar plates according to directions (see **(6)**, **(7)**, and **(8)** in Preparation of Materials and Solutions). A solution of 0.05 M $CaCl_2$ will also be needed (see **(2)** in Preparation of Materials and Solutions).

It is important to use a single colony of bacteria for transformation. The colony should be fairly large (but this does not mean that you should use a large amount of bacteria if colonies have grown together and are indistinguishable). This can be isolated from a fresh 24-hour streak plate. When the bacteria are introduced into the 0.05 M $CaCl_2$, be sure to agitate the inoculating loop to break up the colony before introducing the plasmid. Otherwise, transformation will not be as effective. The heat shock must be fast and clean, moving from ice to heat to ice very quickly. Do not let students wander around the laboratory with tubes in hand. Do not place plastic transformation tubes into the water without making sure they are securely held. The float will tip over and the tube may leak.

Exercise B demonstrates the processes of transformation and selection, allowing for selection of ampicillin resistance, since only cells resistant to ampicillin can grow on agar containing this antibiotic. The laboratory is designed so that each student can carry out the exercise. Students can also work in pairs or groups of 4 since there are 4 different plates to be inoculated at the end of the experiment. You will need 2 LB plates and 2 LB/Amp plates for each "group." See **(7)** and **(8)** in Preparation of Materials and Solutions for directions for making agar plates.

Exercise C

In this exercise, the processes of transformation and selection for antibiotic resistance are, once again, investigated as in Exercise B. However, a second gene that confers the Lac+ phenotype to recipient bacteria is also present on the plasmid, pBLU (or pGAL), used in this exercise. (Note: this is a different plasmid from pAmp used in Exercise B.) The recipient strain of bacteria is JM101 (note: this is also a different strain of bacteria from those used in Exercise B, so do not mix materials from the two parts of this exercise). JM101 is a mutant strain and is not capable of making the α peptide of β galactosidase, but transformed bacteria can make the α peptide of β galactosidase and can, therefore, utilize lactose. However, this is not being selected for (as we select for ampicillin resistance) when bacteria are plated out. (If we wanted to select for this, we would need to use a medium containing only lactose as a food source.) In this experiment, X-gal is used as a histochemical substrate that substitutes for lactose. When cleaved by β galactosidase, a blue precipitate is formed and cells capable of using X-gal turn blue. Only transformed bacteria will turn blue. Other bacteria, assuming they are resistant to Amp, could grow on LB X-gal/Amp plates, but would not have the ability to utilize X-gal and would not turn blue. Thus, this exercise demonstrates a change in genotype by recognizing an alteration in phenotype (blue). Note, however, in this specific exercise, that white colonies should not grow independently on X-gal plates. Ampicillin is included in the plates but only transformed bacteria picking up the pBLU plasmid can grow—both genes for Lac^+ and ampicillin resistance are on the same pBLU plasmid. Thus, ampicillin resistant cells will also turn blue.You may find small white colonies surrounding blue colonies, especially if growth is dense. These are "feeder colonies." The main colony (blue) has destroyed the ampicillin in the agar surrounding the colony so that untransformed JM101 cells can grow in the "halo" area.

As in Exercise B, the simplest way to prepare for this exercise is to order one of the kits described in the General Information section (see Preparation of Materials and Solutions). For large numbers of students, JM101 cells and plasmid pBLU can be ordered separately. (Because JM101 cells must be maintained on minimal medium and are difficult to culture for any length of time, you will probably not want to try to maintain cultures from year-to-year.)

You can use the rapid colony transformation technique to make cells competent at the same time they are transformed, or you can make them competent prior to the transformation kit (see **(10)**).

IF A KIT IS NOT USED, the competent cells can be transformed with plasmid pBLU as follows:

1. Place a 200 μl sample of competent cells into a microcentrifuge tube and add 10.0 μl (approximately 0.05 μg of DNA) of pBLU for transformation. (Since this is such a small amount, the plasmid can be diluted using sterile TE buffer **(3)**. The amount of plasmid added to competent cells should be increased by the same ratio used to dilute the plasmid; i.e., if 20 μl of plasmid is diluted by the addition of 40 μl of TE buffer, use 20 μ1 instead of 10 μl for the transformation of 200 μl of bacterial cells.)
2. Place the tube containing the competent cells and plasmid on ice for 60 minutes.
3. Heat shock the cells for 2 minutes in a 42° C water bath.
4. Return the cells to ice for 5 minutes.
5. Add 200 μl Luria broth to the cells and allow them to recover on ice for 15 minutes.
6. Plate out 200 μl of cells onto an LB agar plate and 200 μl onto an X-gal/Amp plate.

If the Edvotek or Carolina Biological Supply kits are used, the instructor is responsible for preparing the control plates. This is done to save on the cost of kits. Before the day of the laboratory, the teacher must prepare sterile agar plates. They may be stored in the refrigerator for several days prior to lab.

Each group of students will need 2 LB plates (unless the instructor prepares control plates), plus 2 LB/Amp plates and 2 LB/Amp/X-gal plates. Make plates according to preparation directions **(7)**, **(8)**, and **(11)**.

24 hours before the experiment:

If using the rapid colony transformation technique, prepare streak plates and incubate for 24 hours at 27° C.

On the day of the experiment:

If using frozen cells that you have made competent, carefully thaw (on ice) the frozen competent JM101 cells. Set up two control LB plates. Competent cells (100 μl) streaked on these plates should grow normally and produce a lawn. Also set up two control Amp-only plates. Competent cells (100 μl) streaked on these plates should not grow. Each student group should be provided with 10 μl of the dissolved pBLU plasmid solution. Students will transform the competent cells with pBLU (this may be diluted with TE buffer to make the plasmid easier to handle—see Preparation of Materials and Solutions **(9)**) and plate these cells onto the LB/Amp/X-gal plates according to the student directions.

Disposal of culture materials:

At the end of the laboratory period, the teacher should collect all bacterial cultures, tubes, pipettes, and other instruments that have come into contact with the cultures. Likewise, 12-24 hours is long enough to culture cells and check for recombinants. Dispose of these plates after 24 hours or other unwanted contaminating organisms may begin to grow on the plates.

Disinfect materials and glassware using a 10% bleach solution. This can be sprayed onto the surface of the plates using a plastic spray bottle. Soak all glassware in the same solution. Tape culture plates together in groups of three or four and close all tubes. Collect materials in a heavy brown bag to autoclave. For non-disposables (glassware and pipettes), disinfect with a 10% bleach solution.

VI. ORDERING INFORMATION

colony transformation, kit—Carolina Biological, # 21- 1142
transformation of *E. coli* cells with plasmid pBR322, kit—Edvotek, # 201
pBLU colony transformation, kit—Carolina Biological, # 21-1146
transformation of *E. coli* strain JM 101 with pGAL, kit—Edvotek, # 221
E. coli MM294, slant culture—Carolina Biological, # 21-1530
calcium chloride ($CaCl_2$), anhydrous—Fisher, # C77-500; Carolina Biological, # 85-1840
OR calcium chloride ($CaCl_2{\cdot}H_2O$)—Fisher, # C 70-500
plasmid pAmp (0.005 μg/μl)—Carolina Biological, # 21-1430 (1 μg; 200 μl)
Tris base (Trizma base)—Sigma, # T 1503; Carolina Biological, # 21-9020' Ward's, # 37W5935
ethylenediamine tetraacetic acid, disodium salt (EDTA)—Sigma, # E 4884; Fisher, # S311-100; Carolina Biological, # 86-1790; Ward's, # 39W1255
microcapillary pipette, 10 ul—Carolina Biological, # 21-4510 (or 4512)
microcapillary pipette, 100 μl—Carolina Biological, # 21-4517
Ultramicro Accropet—Carolina Biological, # 21-4699
serological pipete, 1 × 0.01 ml (disposable)—Carolina Biological, # 73-6055
Luria broth (powdered)—Edvotek, # 611
Luria broth agar plates (LB agar), prepoured—Carolina Biological, # 21-6610
Luria broth agar, ready-to-pour—Carolina Biological, # 21-6620
bacterotryptone—Carolina Biological, # 79-4420; Ward's, # 39W3600
yeast extract—Carolina Biological, # 79-4780; Ward's, # 38W5800
sodium chloride (NaCl)—Fisher, # S271-3; Carolina Biological, # 88-8903; Ward's, # 37W5488
(Bacto)agar[3]—Fisher, # DF0140-02-9; Carolina Biological, # 21-6720 or 21-6721 (500 g); Ward's, # 38W0015
Bunsen burner—Carolina Biological, # 70-6645; Ward's, # 15W0604
OR alcohol lamp with cap—Carolina Biological, # 70-6604; Ward's, # 17W0211
ampicillin—Sigma, # A-9393; Carolina Biological, # 21-6880
Luria broth agar + ampicillin (LB-Amp) plates, prepoured—Carolina Biological, # 21-6611
plasmid pBLU (0.05 μg/μl)—Carolina Biological, # 21-1427 (1 μg; 200 μl)
E. coli JM101 cells—Edvotek, #405;[4] Carolina Biological, # 21-1561 (slant) or # 21-1562 (plate)
Luria broth agar + ampicillin, ready to pour—Carolina Biological, # 21-6621
Luria broth agar + ampicillin + X-gal, ready-to-pour—Carolina Biological, # 21-6624
X-gal—Edvotek, # 614; Carolina Biological, # 21-7190; Life Technologies, # 5520UB
N,N'dimethylformamide (DMF)—Edvotek, # 614; Carolina Biological, # 21-7350
glycerol (redistilled)—Carolina Biological, # 21-7730, or Life Technologies, # 5514UA
Petri dishes, sterile disposable—Carolina Biological, # 74- 1350

---

[3] Bacto is a designation used on Difco Laboratories products available through Fisher Scientific, Inc. and Carolina Biological Supply Company. Difco Laboratories, PO Box 331058, Detroit, MI 48232-7058; FAX 313-462-8517, Voice, 800-521-0851. Difco maintains a Web page at http://difco.com (consult Reference Guide under Technical Support for information on agar).

[4] Edvotek, Inc., PO Box 1232, West Bethesda, MD 20817-1232 (800-338-6835)—see Suppliers.

LABORATORY 19 Genetic Control of Development and Immune Defenses

I. FOREWORD

This laboratory is designed to introduce students to the fundamentals of early developmental processes that are under the control of both maternal and zygotic genes. While slides and preserved materials are used as a basis for the laboratory, the instructor should try to enhance the learning experience by using living organisms: sea urchin, frog, and chick, if possible. The feasibility of doing this will vary with location and time of year. Supplemental film loops and films might be used as an alternative, as suggested.

Students will also explore the nature of the antigen-antibody response using the precipitin ring test. Both the control and development and variable structure of immunoglobulins are under unique types of genetic control. Homeotic genes and DNA splicing are discussed as genetic mechanisms.

II. TIME REQUIREMENTS

PART I—Development
- Exercise A—Fertilization and Early Development in Sea Urchins (60 minutes)
- Exercise B—Cleavage (30 minutes)
- Exercise C—Formation of the Blastula (30 minutes)
- Exercise D—Gastrulation (30 minutes)
- Exercise E—Neurulation (15 minutes)
- Exercise F—Development of the Chick (15 minutes)
- Exercise G—Formation of Extraembryonic Membranes in the Chick (30 minutes or longer, optional)

PART II—Immune Responses
- Exercise H—Demonstrating the Immune Response by the Precipitin Ring Test (30 minutes)

III. STUDENT MATERIALS AND EQUIPMENT

	Per Student	Per Pair (2)	Per Group (4)	Per Class (24)
Exercise A				
sea urchins (live) **(1)**				10
0.5 M KC1, bottle (100 ml) **(2)**				1
tuberculin syringe (26 or 30 gauge needle)				1
beaker of sea water **(3)**				1
finger bowl or Petri dish of sea water				1
depression slide	1			24
Pasteur pipette	1			24
toothpick	1			24
microscope slides, box (75 × 25 mm)			1	6
coverslips, box (22 × 22 mm)			1	6

	Per Student	Per Pair (2)	Per Group (4)	Per Class (24)
Exercise B				
starfish development (prepared slide)		1		12
fertilized eggs (frog, preserved)				1
two cell stage (cleavage, frog, preserved)		1		12
early cleavage (frog, prepared slide)				2
depression slide		1		12
compound microscope	1			24
dissecting microscope				2
Exercise C				
starfish development (prepared slide) (from Exercise A)		1		12
blastula (frog, sagittal section, prepared slide)		1		12
compound microscope	1			24
Exercise D				
starfish development (prepared slide) (from Exercise A)		1		12
dorsal lip, early gastrula (frog, preserved, per 12)				1
lateral lip, mid-gastrula (frog, preserved, per 12)				1
yolk plug, late gastrula (frog, preserved, per 12)				1
yolk plug (frog, sagittal section, prepared slide)		1		12
depression slide				2
compound microscope	1			24
dissecting microscope				2
Exercise E				
late neurula (frog, cross section, prepared slide)		1		12
neural groove (frog, preserved, per 12)				1
neural tube (frog, preserved, per 12)				1
tail bud (frog, preserved, per 12)				1
external gills (frog, preserved, per 12)				1
depression slide				2
compound microscope	1			24
dissecting microscope				2
Exercise F				
chick, 16 hour (whole mount, prepared slide)		1		12
chick, 33 hour (whole mount, prepared slide)				1
chick, 48 hour (whole- mount, prepared slide)				1
chick, 72 hour (plastic mount)				1
compound microscope	1			24
dissecting microscope				2
Exercise G				
chicken egg, incubated 96 hours		1		12
chicken egg, incubated 5-6 days		1		12
chicken Ringer's solution (bottle) (**4**)			1	6
finger bowl	1			24

	Per Student	Per Pair (2)	Per Group (4)	Per Class (24)
Exercise G—continued				
dissecting microscope	1			24
forceps	1			24
Exercise H				
microculture (Durham) tubes			5	30
rack for Durham tubes			1	6
Pasteur pipettes			4	24
rubber bulbs			1	6
antiserum, rabbit (vial) **(5)**				1
normal rabbit serum (vial) **(6)**				1
bovine serum albumin (0.05%, vial) **(7)**				1
PBS buffer (5 ml), bottle ((**8**)			1	6
light (100 w bulb)			1	6

IV. PREPARATION OF MATERIALS AND SOLUTIONS

(1) see urchins (live)

Sea urchins are available from several marine biological laboratories (see Preparation Suggestions for this exercise).

(2) O.5 M KCl

Dissolve 37.28 g KC1 in enough distilled water to make 1000 ml of solution.

(3) sea water (beaker)

Sea water can be purchased as artificial sea water in 1 to 5 gallon amounts or can be mixed from artificial salt mixtures according to directions on the bag. (See Preparation Suggestions for this exercise.)

(4) chick Ringer's solution

Mix 7.2 g NaCl, 0.23 g $CaC1_2 \cdot 2H_20$, and 0.37 g KC1. Add distilled water to make 1000 ml.

(5) rabbit antiserum

Use Sigma Immunochemicals (B-7276) against bovine serum albumin (anti-BSA), containing at least 50 to 100 μg antibody protein per milliliter. You will add 2 ml deionized water to the contents of one vial. Rotate until dissolved. Follow accompanying instructions carefully. Make fresh before laboratory and keep refrigerated. One vial will provide enough material for 6 groups.

(6) normal rabbit serum

Use Sigma Immunochemicals (R-9133), available in 5 ml amounts (enough for 12 groups). Store at – 20° C (avoid frost-free freezers).

(7) bovine serum albumin (0.05%)

Use Sigma Immunochemicals (A-9647), Fraction V powder, 96-99% albumin, available in 5 g amounts. Dilute 0.05 g to 100 ml with PBS buffer (see **(8)**).

(8) PBS buffer

Phosphate buffered saline can be mixed by the following recipe given below:

0.23 g NaH_2PO_4 (anhydrous) (1.9 mM)
1.15 g Na_2HPO_4 (anhydrous) (8.1 mM)

9.00 g NaCl (154 mM)
Add distilled water to 900 ml.
Adjust to pH 7.2 to 7.4 using 1 M NaOH or 1 M HCl.
Add distilled water to 1 liter.

V. PREPARATION SUGGESTIONS

Exercise A

Live sea urchins are available from several marine laboratories or may be collected if you have easy access to the coast. Urchins can be held in marine aquaria in a well air-conditioned room. *Arbacia* and *Lytechinus* urchins prefer temperatures of 18-23° C, while *Stronglyocentrotus* urchins do better at colder temperatures of 10° C. Urchins (usually *Lytechinus)* are available from Carolina Biological Supply Company. A sea urchin embryology kit is also available from the same vendor (see Ordering Information).

Seawater can be made from synthetic mixes (for example, Instant Ocean), or can be purchased in 5 gallon buckets of water collected at the coast (see Ordering Information). If you live near the coast, take several carboys to the beach and fill them with sea water (make sure that the carboys have never been washed out with soap—preferably they have never had materials other than distilled or seawater in them). Depending on the source of your urchins, you may need to adjust salinity. Test salinity in the shipping bags when the urchins arrive and compare to that in your tank. Be sure to use only calcium carbonate marine gravel in your holding tank. Set up the aquarium at least two weeks in advance.

Depending on the time of year, the species you must work with will be different and eggs will be different colors. If ordered from a biological supply house, be sure to tell the person you contact that you will be using the urchins to study embryology. Most suppliers ship their animals by airfreight or express mail. Try to procure urchins as quickly as possible once they have reached their final destination. Equilibrate the temperature by floating the bags of urchins in your aquarium. Urchins should be fed after arrival if they are to be kept for more than a week. Use small pieces of frozen shrimp (uncooked) or kelp (fresh or dried). Feed sparingly—do not foul your tank.

Be sure that all glassware is clean. It is best to use glassware that has not been exposed to soap or any toxic chemical. Avoid overcrowding embryos and overheating them on the microscope stage. While students wait for the first cleavage to occur, have them remove their depression slides from the microscope stage. Be sure to have students check for evaporation during their wait. If the seawater is evaporating, introduce fresh seawater beneath the coverslip using a Pasteur pipette.

When injecting sea urchins with 0.5 M KCl, use only 0. l to l ml. Inject this using a tuberculin syringe with a 26 or 30 gauge needle attached. Injections should be made in- the soft tissue on the oral surface. Shake the urchin gently after injecting. Some urchins do not respond to KCl injection, but most respond to electrical stimulation as suggested in the student laboratory directions.

If you are using *Arbacia punctulata* or *Stronglyocentrotus purpuratus,* eggs can be kept for several hours in the refrigerator or in an ice bath. Likewise, semen, if collected "dry" (not in seawater), can be stored on ice for several hours.

Once urchins have stopped shedding their gametes, return them to a DIFFERENT aquarium. If placed back into the aquarium from which they were taken, they may trigger other animals in the aquarium to spawn.

After students have prepared their slides, you may wish to fertilize the remaining eggs. Change the seawater (the eggs will have settled to the bottom of the beaker) and add several drops of dry sperm. Swirl to mix the eggs and sperm. After 10-15 minutes, when the eggs have again settled, pour off the seawater and add fresh seawater. Allow the eggs to settle and change the water a second time. Eggs develop best if stirred slowly with a magnetic stirring bar on a stir plate. They should be maintained at 20-25° C. If you teach multiple sections on consecutive days, have students take samples from beakers and try to identify stages of development by comparing with prepared slides. (See Ordering Information.)

Exercises B-E

Two organisms, the starfish and frog have been chosen to study the process of development. The starfish (and sea urchin) gastrulate by a simple pushing inward (imagination) of the cells on one side of the blastula while in the frog, complicated epibolic and embolic movements allow cells to migrate across the surface of the blastula and to invaginate to the inside of the area of the blastopore. Try to help students see how these two processes manage to move some cells from the outside of the hollow blastula to the inside where they will contribute to the formation of internal organs.

If you use live sea urchins to study echinoderm development, substitute *Arbacia* slides for *Asterias*—the explanation is correct except that the larva is a pluteus rather than a bipinnaria.

Use depression slides to place preserved frog embryos on demonstration. Check periodically to make sure that the embryos remain covered with preservative and do not dry out.

Depending on the season of the year, live frogs should be used if possible. Note that your use of these organisms requires the approval of your institution's animal care and use committee (IACUC)—be sure to submit your protocol in advance. Females can be artificially induced to ovulate (most easily from April through December) by injecting pituitaries from male or other female frogs. This provides sufficient LH to cause ovulation. Fresh pituitaries are best for use in this procedure. Pith chilled frogs (see Preparation Suggestions in the directions for Laboratory 32), destroying only the spinal cord. Cut the head off from the corner of the mouth, slanting posteriority as far as possible (behind the eyes and through the tympanic membrane area). On the ventral side of the head (the roof of the mouth or palate), make two incisions directed anteriorly from the foramen magnum (the posterior opening of the cranial cavity leading to the spinal cord) and slightly outward from the midline of the skull. Lift up the flap of bone produced and the pituitary will be found, usually in the piece of bone removed. It is a small pink body 1-2 mm in size and often surrounded by white-shiny connective tissue (the dura mater). Place each pituitary in a small vial with about 1 ml of amphibian Ringer's solution (see Ordering Information). Collect the pituitary glands from about three females or six to eight males (depending on size). Inject all of the pituitaries collected (in Ringer's) into the abdominal cavity of a large female. Ovulation will occur in one to three days. Grasp the female firmly and push posteriorly on the abdomen using firm but gentle pressure. Once egg laying begins, maintain a gentle pressure to assist in the process.

Gonadotropin can also be supplied using pituitary extract (Carolina Biological Supply Company, pituitary kit, # L 1494K), but we have found that it is not as effective as fresh pituitaries in inducing ovulation.

Development is a process and students should be able to view it as such.

If live materials are not available, the following films are excellent:

echinoderm development—Richard A. Cloney, Kalmia Science Series 4011, 4012, 4013, and 4014. Kalmia Company, Inc., Department B6, 21 West Circle, Concord, Massachusetts 01742

frog development—Eugene Bell, Kalmia Science Series 4001, 4002 (address above). also in movie form—*Development of the Leopard Frog*.

Exercise F

Stress that gastrulation movements in the chick are like those in the frog, but on a flat surface. Show students how you can take the two ends of the chick blastodisc after gastrulation is complete and bring them together to form a cylinder that looks exactly like the cross section of a frog gastrula.

Exercise G

Have students place embryos in a large plastic container. Dispose of this waste after the laboratory period.

The allantois will be most easily seen in 96 hour embryos, but the other membranes are more noticeable in older embryos. At this stage, the allantois is fairly extensive and is fused to the chorion and cannot be distinguished.

Exercise H

This exercise is designed so that the antibody is always on the bottom of the tube and antigen on the top (except when buffer is substituted for one or the other). Make sure that students understand the following relationships:

Ab = antiserum (antibody) = rabbit antiserum
(sensitized against bovine serum albumin—contains antibodies against the antigen BSA)

N = normal rabbit serum
(does not contain antibodies against BSA)

AG = bovine serum albumin (BSA), the antigen

Only tube 5 will form a ring since it contains the antigen (BSA) and the antibodies against this antigen contained in Ab (rabbit antiserum). Use a light behind the tubes to make the ring easier to see.

In tubes 1, 3, 4, and 5, the bottom layer is already heavier. You will experience some mixing in tube 2. If you wish to increase the density of the bottom layer, you can mix each with a few grains of sucrose. Mix by tapping gently against your finger before adding the materials for the upper layer.

The Durham tubes are small. They will fit in the bottom part of a test tube rack, but be sure to put a paper towel under them on the rack so they do not fall through.

VI. ORDERING INFORMATION

sea urchin development (prepared slide)—Carolina Biological, # E465; Ward's, # 92W8330

potassium chloride (KCl)—Fisher, # P217-500; Carolina Biological, # 88-2900; Ward's, # 37W4634

hypodermic syringe, 1 ml—Carolina Biological, # 69-7765; Ward's, # 14W1613

hypodermic needles, sterile-Carolina Biological, # 69-7808 (26 gauge), # 69-7812 (30 gauge)

beaker, 1000 ml— Fisher, # 02-540P (Pyrex) or # 02-539P (Kimax); Carolina Biological, # 72-1227 (Corning, student grade); Ward's, # 17W4080

depression slide—Carolina Biological, # 63-2230; Ward's, # 14W3510

Pasteur pipettes—Fisher, # 13-678-20C; Carolina Biological, # 73-6062; Ward's, # 17W1146
rubber bulbs—Fisher, # 14-065B
microscope slides, 75 × 25 mm—Carolina Biological, # 63-2000; Ward's, # 14W3500
coverslips, 22 × 22 mm—Carolina Biological, # 63-3015; Ward's, # 14W3251
sea urchin larvae (prepared slide)—Carolina Biological, # E467; Ward's, # 92W8331
sea urchins (live)—Carolina Biological, # 16-2500; Ward's, # 87W7340[1]
sea urchin embryology kit—Carolina Biological, # 16-2505; Ward's, # 87W9030
sea water, artificial (5 gal. carboy)—Carolina Biological, # 16-3393
sea water, Gulf Coast, collected (5 gal. carboy)—Carolina Biological, # 16-3390
Instant Ocean—local marine aquarium store; Carolina Biological, # 16-3390 or # 16-3993; Ward's, # 21W7354
marine gravel (10 lbs.)—Carolina Biological, # 16-3240; Ward's, # 21W1802
starfish development (prepared slide)—Carolina Biological, # E580; Ward's, # 92W8255
fertilized egg (frog, preserved, per 12)—Carolina Biological, # P1530AF; Ward's, # 69W2208
two cell cleavage (frog, preserved, per 12)—Carolina Biological, # P1530BF; Ward's, # 69W2209
early cleavage (frog, sagittal section, prepared slide)—Carolina Biological, E1630; Ward's, # 92W8811
blastula (frog, sagittal section, prepared slide—Carolina Biological, # E1650; Ward's, # 92W8813
dorsal lip, early gastrula (frog, preserved, per 12)—Carolina Biological, # P1530KF; Ward's, # 69W2229
lateral lip, mid-gastrula (frog, preserved, per 12)—Carolina Biological, # P1530MF
yolk plug, late gastrula (frog, preserved, per 12)—Carolina Biological, # P1530NF; Ward's, # 69W2238
yolk plug (frog, sagittal section, prepared slide)—Carolina Biological, # E1670; Ward's, # 92W8815
late neural tube (frog, cross section, prepared slide)—Carolina Biological, # E1691; Ward's, # 92W8817
neural groove (frog, preserved, per 12)—Carolina Biological, # P1530QF; Ward's, # 69W2239
neural tube (frog, preserved, per 12)—Carolina Biological, # P1530RF; Ward's, # 69W2248
tail bud (frog, preserved, per 12)—Carolina Biological, # P1530SF
external gills (frog, preserved, per 12)—Carolina Biological, # P1530UF; Ward's, # 69W2258
chick, 16 hour (prepared slide)—Carolina Biological, # E2020; Ward's, # 92W9010
chick, 33 hour (prepared slide)—Carolina Biological, # E2080; Ward's, # 02W9040
chick, 48 hour (prepared slide)—Carolina Biological, # E2120; Ward's, # 92W9055
chick, 72 hour (prepared slide)—Carolina Biological, # E2200; Ward's, # 92W9080
sodium chloride—Carolina Biological, # 88-8903; Ward's, # 37W5480
calcium chloride—Carolina Biological, # 85-1800; Ward's, # 38W0937
sodium phosphate (monobasic) (NaH_2PO_4)—Fisher, # S374-500; Carolina Biological, # 89-1350; Ward's, # 37W5655
potassium phosphate (dibasic) (K_2HP0_4)—Fisher, # P288-500; Carolina Biological, # 88-4290; Ward's, # 37W4841

---

[1] Suppliers of live sea urchins: Gulf Specimen Company, PO Box 237, Panacea, FL 32346 (*Arabacia punctulata*) and Pacific Bio-Marine Laboratories, Inc., Box 536, Venice, CA 90291 (*Lytechinus pictus* and *Stronglyocentrotus purpuratus*). [See Suppliers.]

microculture tubes (6 × 50 mm)—Fisher, # 14-958A
tube rack—Fisher # 14-754-15
rabbit antiserum—Sigma Immunochemicals, # B-7276
normal rabbit serum—Sigma Immunochemicals, # R-9133
bovine serum albumin—Sigma Immunochemicals, # A-9647
Ringer's solution, amphibian—Carolina Biological, # 88-6503
Ringer's solution, chicken—Carolina Biological, 3 88-7507

LABORATORY 20 The Genetic Basis of Evolution I—Populations

I. FOREWORD

This laboratory demonstrates various aspects of quantitative genetic variation and sexual dimorphism. It also demonstrates the use of the Hardy-Weinberg equation to illustrate the effects of population isolation and drift. The effects of heterozygote advantage and the recessiveness of deleterious alleles are studied as examples of how changes in allele frequencies are the foundation of evolutionary change. Concepts of sample size and sampling error can be emphasized during this laboratory.

BioBytes *Dueling Alleles*, extends and augments this laboratory. *Dueling Alleles* is a microcomputer simulation of the effect of genetic drift and selection on the frequency of an allele. Optimally, it should be used with Exercise E to allow enough generations for analysis of the effects in shifts in allele frequencies.

II. TIME REQUIREMENTS

Exercise A—Understanding Variation
  Part 1—Measuring Cephalic Index (20 minutes)
  Part 2—Measuring Relative Sitting Height (20 minutes)
Exercise B—Estimating Allelic Frequency from a Population Sample (30 minutes)
Exercise C—The Founder Effect as an Example of Genetic Drift (50 minutes)
Exercise D—The Role of Gene Flow in Similarity Between Two Populations (50 minutes)
Exercise E—The Effect of Selection on the Loss of an Allele from a Population
  Part 1—The Effects of Recessiveness on Deleterious Alleles (25 minutes)
  Part 2—Heterozygote Advantage (50 minutes)
Extending Your Investigation: Selection Pressure (60 minutes)

III. STUDENT MATERIALS AND EQUIPMENT

	Per Student	Per Pair (2)	Per Group (4)	Per Class (24)
Exercise A (Part 1)				
calipers			1	6
metric ruler (30 cm)			1	6
Exercise A (Part 2)				
meter stick (1 m)			1	6
Exercise B				
PTC taste test papers, roll			1	6
Exercise C				
beads from chromosome simulation kit				
OR beans (container) **(1)**			1	6
pint jar or 500 ml beaker			1	6

	Per Student	Per Pair (2)	Per Group (4)	Per Class (24)
Exercise C (continued)				
plastic cups		10	20	120
Exercise D				
beads from chromosome simulation kit				
OR beans (container) **(2)**				2
Styrofoam plate or dissecting pan			2	12
pint fruit jar or 500 ml beaker			1	6
Exercise E (Part 1, A and B)				
beads from chromosome simulation kit				
OR beans (container) **(3)**			1	6
plastic cups			2	12
Exercise E (Part 2)				
beads from chromosome sululation kit				
OR beans (container) **(3)**			1	6
plastic cups			2	12
Extending Your Investigation				
yarn "wooly worms" **(4)**				100

IV. PREPARATION OF MATERIALS AND SOLUTIONS

(1) beads or beans for Exercise C

The chromosome simulation kits used in Laboratories 9 and 13 contain red and yellow beads that can be separated and used in these exercises. Otherwise beans of two different colors and approximately the same size can be substituted (e.g., Navy and Pinto beans). Use 360 of one color (common) and 40 of the second color (rare). Each group should start with a population of 400 individuals.

(2) beads or beans for Exercise D

Two trays (or plates) of beans should contain 35 common and 15 rare beads (beans) in one tray (TRAY A) and 45 common and 5 rare beads (beans) in the other tray (TRAY B).

(3) beads or beans for Exercise E (Parts 1 and 2)

Provide a container each with 200 beads (beans) of a color; one container of common "alleles" and one container of rare "alleles" for each group of 4 students.

(4) yarn "wooly worms"

For a class of 15-30, you will need at least 100 pieces of yarn for each of 16 different colors (1,600 pieces). Yarn pieces should be 2-3 inches long. Wrap yarn around 4 fingers and then cut to make 3 inch lengths. Distribute the yarn on a grassy lawn, 50-100 feet on a side, depending on class size.

V. PREPARATION SUGGESTIONS

Exercise A

List all measurements for males in a column on the left side of a page and all measurements for females in a column in the middle of the page. Total all measurements for each sample (to give the sum of *X*s or ΣX). Divide each total by the number of observations ($\Sigma X/n$) to give the arithmetic average or mean. Subtract the smaller mean from the larger and record the result or difference (d).

There are a variety of ways to test this difference to see if it is "significant." (See Appendix I—the following directions use the shorter method for calculating these statistics using a calculator.)

One standard test involves the calculation of the sums of squares (SS). Square each measurement in each sample and total the squares (ΣX^2) separately for males and for females. Square the sum of the original observations, divide the sum by the number of observations, and subtract this result from the total of the squares to give the sums of squares—thus

$$SS = \sum X^2 - (\sum X)^2 / n$$

Do this for both groups or samples. Add the two sums of squares, divide the total by the sum of each sample size minus 1, multiply the result by the sum of the two sample sizes divided by their product, and take the square root of the result. This is the standard error of the difference (s_d). Expressed as an equation, this series of operations is as follows:

$$s_d = \sqrt{((SS_1 + SS_2)/((n_1 - 1)(n_2 - 1)) \times ((n_1 + n_2)/n_1 n_2)}$$

While this equation looks formidable, it is easy to solve if the calculations are performed inside each level of parentheses in order. To test the differences, calculate Student's-*t* by dividing the difference between the two sample means and the standard error of the difference

$$t = d / s_d$$

(See Appendix I.) For a class of 24, a value equal to or greater than 2.074 indicates a 1-in-20 chance (or less) that the difference observed is due to chance. A value greater than 2.819 indicates a 1-in-100 chance (or less) that the difference is due to chance. For other sample sizes, consult the *t*-table in Appendix I or any statistics text (add the two numbers of observations in each group and subtract 2 to determine the degrees of freedom).

This calculation of t assumes that variances are equal. Other versions of the *t*-test can be used under certain conditions and other tests are available. Consult standard statistical references. Sampling, estimates of central tendency, and testing of hypotheses (along with types of errors) can be introduced as useful tools using these calculations if desired.

Exercise B

Students use PTC paper for this exercise. Although carcinogenic in large amounts, the papers are considered safe. If you do not wish to use the test paper, simply write "T" or "NT" on small pieces of paper (approximately 25 of each type) and allow students to draw their phenotype. In this exercise, emphasize that even knowing the number with the recessive phenotype, we cannot calculate allelic frequencies for certain because our population, being small, violates the tenets of the Hardy-Weinberg theorem and we cannot assume that the class is in Hardy-Weinberg equilibrium. Differences in the class population and North American population (large) can be compared to emphasize this point.

Exercise C

For each group, start with 400 beads (pull beads in the chromosome simulation kit apart) in a pint fruit jar or 500 ml beaker. Place 360 of one color ($p = 0.9$) and 40 of the other ($q = 0.1$). The bead of the rarer color is to represent the rarer allele. To show a more dramatic founder effect, increase the number of beads in the parent population relative to the number in the founder population. If beads are not available, you can use beans of different colors, but make sure that these are the same size and shape so that student sampling techniques will not be affected.

Exercise D

Students may work alone if desired, but the number of set-ups will need to be increased. The starting frequencies of beads in the two trays are different: 35 common beads and 15 rare in one tray; 45 common and 5 rare in the other tray. Individual beads represent genes (alleles). Trays A and B are experimental populations where gene flow takes place without regard to genotype as students exchange beads. For each gene pool immigration equals emigration so that the size of the gene pool stays the same. Each repetition or iteration is equivalent to one generation in which one individual is lost from the population due to natural selection and one individual joins the population as a potential reproducer. Gene flow as a result of emigration and immigration should result in allele frequencies in populations becoming more similar.

Exercise E

In some cases, alleles can be selected against (if they are deleterious) or can enjoy a selective advantage. When an allele gets rare enough, it can actually "hide" because it remains in the heterozygote population. Also, its frequency in the population will not decrease as rapidly if the dominant gene in the heterozygote masks its presence. If heterozygotes experience an advantage (as in the case of sickle cell anemia), the deleterious recessive allele can actually increase in number.

Students may not get to this part of the laboratory since the laboratory tends to be long as students struggle with the concepts. If you have not divided the laboratory into two parts, you can assign Exercise E to be done using *Duelling Alleles* on the CD-ROM that accompanies *Biology in the Laboratory*

The BioBytes user's guide contains pedagogical suggestions on how to present *Dueling Alleles* with various numbers of computers and with different instructional objectives.

Extending Your Investigation

The purpose of this exercise is to introduce students to the concepts of variation, adaptation, selection pressure, and cryptic coloration (a specific adaptation). Pieces of colored yarn represent "wooly worms" distributed throughout the ecosystem. Students represent predators (birds) that feed on the worms. Depending on the colors of wool pieces used, some will be harder to find than others. When the colored yarn pieces that have been found are tallied as data, a Chi square test is used to determine if the wool pieces are collected randomly or by a differential selection process.

Allow students to work in teams of 3 or 4. Students should "feed" on the wooly worms for 10 minutes (make sure that at least 10 pieces of the hardest to find colored yarn are collected). It is best to conduct this activity outside. If you have a rainy day, a quick substitute can be made by using "holes" from a paper hole punch. We have successfully used a room with a mottled pattern floor tile. Different colored holes can be produced quickly. Transparencies make excellent "hard to find" holes. Students can pick up the "holes" with a wet finger. On occasion, more pieces of wool are collected than originally distributed. Check to see if some have become unraveled so that one piece becomes two.

If the experiment is repeated using only those pieces of yarn that were collected, then gene frequencies will change from one generation to the next with a subsequent reduction in variation within the species in time. Drastic changes in the environment can lead to extinction, but this is the exception rather than the rule.

VI. ORDERING INFORMATION

anthropometric calipers—Carolina Biological, # 696245
metric ruler—Carolina Biological, # 70-2616
meter stick—Carolina Biological, # 702620
PTC taste test papers—Carolina Biological, # 17-4010
chromosome simulation Biokit—Carolina Biological, # 17-1100
colored yarn—local store

Notes:

LABORATORY 21 Genetic Basis of Evolution II—Diversity

I. FOREWARD

In this laboratory, students are introduced to systematics and some of the different ways in which phylogenetic relationships among organisms are determined. A study of LDH isoenzymes demonstrates how information at the molecular level can be used in the study of phylogeny.

The study of LDH isoenzymes can be carried out using a kit developed by John Anderson of Modern Biology, Inc. (see Preparation Suggestions), or by the following preparation directions given in the manual.

II. TIME REQUIREMENTS

Exercise A—Understanding Evolutionary Classification (30 minutes)
Exercise B—Electrophoretic Analysis of LDH in Ungulate mammals (120 minutes)

III. STUDENT MATERIALS AND EQUIPMENT

	Per Student	Per Pair (2)	Per Group (4)	Per Class (24)
Exercise A				
laboratory manual and textbook	1			24
Exercise B				
electrophoresis chamber			1	6
power supply			1	6
gel comb			1	6
tape (roll)			1	6
tris citrate gel buffer, 1 liter **(1)**			1	6
1.2% agarose, 40 ml **(2)**			1	6
loading dye, 0.2 ml **(3)**				1
cow serum, 15 μl **(4)**			1	6
goat serum, 15 μl **(4)**			1	6
horse serum, 15 μl **(4)**			1	6
sheep serum, 15 μl **(4)**			1	6
microcapillary pipette (20 μl)			4	24
tris-HCl buffer, 200 ml bottle **(5)**			4	24
Li lactate, 50 ml bottle **(6)**				1
NAD, 25 ml bottle **(7)**				1
MTT, 10 ml bottle **(8)**				1
PMS, 10 ml bottle **(9)**				1
staining dish			1	6
graduated cylinder, 50 ml			1	6
pipette, 10 ml (1 with bottle of tris HCl)				1

	Per Student	Per Pair (2)	Per Group (4)	Per Class (24)
Exercise B--continued				
pipette, 5 ml (1 with bottle of Li lactate)			1	
pipette, 1 ml (1 each with NAD, MTT, PMS)			3	
water bath (37° C)			1	
10% acetic acid, 500 ml bottle **(10)**				1

IV. PREPARATION OF MATERIALS AND SOLUTIONS

(1) tris-citrate gel buffer, pH 8.0

Dissolve 41.6 g trizma and 16.5 g of citric acid in 450 ml distilled water and bring to 500 ml with distilled water.This is a stock solution that is diluted with distilled water (30 ml gel buffer, 870 ml water) for making and running gels. Adjust pH with 2 N NaOH (10 × or 50 × concentrates can be ordered from Fisher or Sigma—see Ordering Information).

[To make 100 ml of stock solution, use 8.32 g trizma and 3.3 g citric acid.]

(2) 1.2% agarose

Mix 1.2 g agarose with 100 ml tris-citrate gel buffer. Heat to dissolve agarose. Store in bottles of 50 ml—enough agarose to make 1 gel.

(3) loading dye

Mix 30 ml glycerine with 70 ml water to make 30% glycerol. Add 0.25 g bromophenol blue to this solution.

(4) serum with loading dye

Mix 720 μl (0.72 ml) serum (horse, sheep, goat, cow) with 144 μl (0.144 ml) loading dye. Dispense 15 μl aliquots into Eppendorfs for student use.

(5) tris-HCl stain buffer, 0.2 M at pH 8.0

Mix 2.42 trizma with 100 ml distilled water. Adjust pH to 8.0 with 2 N HCl. Each gel will need 30 ml of this buffer.

(6) Li lactate, 0.5 ml

Mix 0.48 lactic acid, lithium salt, with 10 ml water. Each gel will need 6.0 ml.

(7) NAD (nicotinamide dinucleotide)

Mix 0.1 g NAD with 10 ml water. Each gel will need 1.3 ml.

(8) MTT

Mix 0.1 g MTT tetrazolium salt with 10 ml water. Each gel will need 0.3 ml.

(9) phenazine methosulfate (PMS)

Mix 0.1 g PMS (phenazine methosulfate) with 10 ml water. Each gel will need 0.5 ml.

(5) – (9) gel stain solution

Each student makes stain (enough for 1 gel) immediately before staining.

Mix:
- 30 ml 0.2 M tris-HCl buffer
- 6.0 ml 0.5 M Li lactate
- 1.3 ml NAD
- 0.3 ml MTT
- 0.5 ml PMS

(10) acetic acid, 10%

Mix: 10 ml conc. acetic acid (glacial) and 90 ml tap water
50 ml conc. acetic acid (glacial) and 450 ml tap water

V. PREPARATION SUGGESTIONS

Exercise A

Exercise A can be completed before coming to laboratory. There are many ways to interpret the phylogenetic trees illustrated in this exercise. The major point is that different taxonomic classifications result in different grouping of organisms—there is no "right" or "one" way to classify.

Exercise B

This laboratory exercise was adapted from the work of John N. Anderson and is available as a kit from Modern Biology, Inc., 111 North 500 West, West Lafayette, IN 47906. To place orders, call 1-800-733-6544 or FAX 1-317-449-4577. Request Experiment 6(106) Protein Fingerprinting. If you have not worked with protein fingerprinting before, *you should use this kit for best results.*

Electrophoresis techniques are similar to those used in Laboratories 15 and 18. Refer to complete directions in the laboratories and the *Preparator's Guide* for these laboratories if this is your first experience with electrophoresis.

Since proteins tend to have a negative charge, make sure that students place the comb used to make your gels on the end of the gel bed with the black (negative) stripe. When you place the gel in the gel box, make sure that the black line lines up with the black dot on the gel box so proteins will migrate to the opposite (positive; anode) end.

Stain for gels must be prepared *immediately* before use. Each student group should prepare enough stain for just one gel as indicated in the laboratory manual. The stain *cannot* be poured down the sink. Decant the stain from gels into a waste container and dispose of it according to institutional guidelines. Gels need to stain for 30 minutes.

You can use large weighing boats or Petri dishes for staining. Place these on top of test tube racks in the water bath during incubation. Incubation should be in the dark. Do not let students lift the water bath lid many times to determine progress.

VI. ORDERING INFORMATION

gel electrophoresis apparatus—Edvotek, # 502; Carolina Biological, # 21-3668; Ward's, # 36W5160
electrophoresis power supply—Edvotek, # 503; Carolina Biological, # 21-3673; Ward's, # 36W5112
agarose (low EEO)—Edvotek, # 605; Carolina Biological, # 21-7080; Sigma, # A-3768; Ward's, # 38W0006
glycerine—Carolina Biological, # 86-5560; Ward's, # 39W1435
bromophenol blue loading dye—Edvotek, # 606; Carolina Biological, # 21-8200; Ward's, # 38W9115
bromophenol blue—Fisher, # B392-5
horse serum—Sigma, S 6380; Gibco, # 16050114[1]

[1] See Life Technologies in the list of Suppliers for Gibco products.

sheep serum—Sigma, S 2382; Gibco, # 16070096 (lamb)
goat serum—Fisher, # S 6898; Gibco, # 16210072
cow serum—Fisher, # S 1507; Gibco, # 16170078 (calf)
microcapillary pipette, 20 µl—Carolina Biological, # 21-4513
Tris base (Trizma base)—Sigma, # T 1503; Carolina Biological, # 21-9020; Ward's, # 37W5935 OR 10× or 50× concentrates
citric acid—Fisher, # A 940-500; Carolina Biological, # 85-4700; Ward's, # 39W0932
hydrochloric acid (HCl), 2 N—Fisher, # SA431-500
Li lactate (lactic acid, lithium salt)—Sigma, # L 2250
NAD (β-nicotinamide adenine dinucleotide)—Sigma, # N 3014
MTT tetrazolium salt (3-[4,5-dimethylthiazol-2-yl]-2,5-diphenyltetrazolium bromide; Thiazolyl blue)—Sigma, # M 2128
phenazine methosulfate (PMS)—Sigma, # P 9625; Fisher, # 13-601-55
weighing dish (for staining gel), 5½ × 5½"—Carolina Biological, # 70-2334; Ward's, # 18W3003
graduated cylinder, 50 ml—Fisher, # 08-553A (Pyrex) or # 08-549-5D (Kimax) or # 08-572-6C (Nalgene); Carolina Biological, # 72-1786 (Pyrex) or # 72-1946 (Kimble) or # 72-2264 (Nalgene)
pipette, 1 ml—Fisher, # 13-665F; Carolina Biological, # 73-6270; Ward's, # 17W4852
pipette, 5 ml—Fisher, #13-665K; Carolina Biological, # 73-6276; Ward's, # 17W4854
pipette, 10 ml—Fisher, # 13-665M; Carolina Biological, # 73-6278; Ward's, # 17W4853
acetic acid, glacial—Fisher, A38-212; Carolina Biological, # 84-1293; Ward's, # 39W0125

LABORATORY 22 Diversity—Kingdoms Eubacteria, Archaebacteria, and Protista

I. FOREWORD

This is the first of three laboratories designed to study diversity of three kingdoms: Eubacteria, Protista, and Fungi. Live material should be used as often as possible. By selecting parts of exercises, the laboratory can be shortened and combined with one or both of the other diversity laboratories.

II. TIME REQUIREMENTS

PART I—DOMAINS BACTERIA AND ARCHAEA
- Exercise A—Morphology of Bacteria (10 minutes)
- Exercise B—Characteristics of Bacteria (30 minutes)
- Extending Your Investigation: How Effective Is Your Soap? (30 minutes, overnight incubation)
- Exercise C—Nitrogen-Fixing Bacteria (10 minutes)
- Exercise D—Bioluminescent Bacteria (10 minutes)
- Exercise E—Diversity and Structure of Cyanobacteria (40 minutes)

PART II—KINGDOM PROTISTA
- Exercise F—Identifying Protozoans (30 minutes)
- Exercise G—Symbiosis in the Termite: A Study of Flagellates (15 minutes)
- Exercise H—Plasmodial Slime Molds (10 minutes)
- Exercise I—Water Molds (10 minutes)
- Exercise J—Studying and Classifying Algae (30 minutes)
- Exercise K—Diversity Among the Green Algae: Phylum Chlorophyta (30 minutes)
- Exercise L—Recognizing Protists Among the Plankton (15 minutes)

III. STUDENT MATERIALS AND EQUIPMENT

	Per Student	Per Pair (2)	Per Group (4)	Per Class (24)
Exercise A				
bacilli bacteria (prepared slide)				1
cocci bacteria (prepared slide)				1
spirilla bacteria (prepared slide)				1
compound microscope (demonstration)				3
nutrient agar, Petri dish **(1)**	1			24
Exercise B				
antibiotic agar Petri dish containing *Escherichia coli* bacteria and five different antibiotic sensitivity disks **(2)**				1
antibiotic agar Petri dish containing *Staphylococcus aureus* bacteria and five different antibiotic sensitivity disks **(3)**				1

	Per Student	Per Pair (2)	Per Group (4)	Per Class (24)
Exercise B—continued				
Escherichia coli, gram stain (prepared slide)	1			24
Staphylococcus aureus, gram stain (prepared slide)	1			24
Extending Your Investigation				
S. aureus broth culture **(4)**			1	6
E. coli broth culture **(5)**			1	6
nutrient agar, Petri dish **(1)**	1			24
paper disks **(6)**	2			48
cleaning solutions (student selected), bottles **(6)**			1	6
forceps	1			24
wax pencil		1		12
ruler		1		12
cotton swabs (Q-Tips)	2			48
Exercise C				
soybean plants, pot (inoculated with *Rhizobium*) **(7)**				1
soybean plants, pot (not inoculated with *Rhizobium*) **(7)**				1
Rhizobium (prepared slide)				1
compound microscope (demonstration)				1
Exercise D				
Vibrio fischeri, tube of broth culture **(8)**				4
sterile rubber stoppers	1			24
Exercise E				
microscope slides, box (75 × 25 mm)			1	6
coverslips, box 22 × 22 mm)			1	6
Nostoc (live culture)				1
Cylindrospermurn (live culture)				1
Oscillatoria (live culture)				1
Anabaena (live culture)				1
Gleocapsa (live culture)				1
compound microscope	1			24
Exercise F				
Trypanosoma (prepared slide)	1			24
Amoeba proteus (live culture)				1
Parameciurn caudaturn (live culture)				1
Plasmodiurn vivax (prepared slide)	1			24
microscope slides, box (75 × 25 mm)			1	6
coverslips, box (22 × 22 mm)			1	6
compound microscope	1			24
Exercise G				
insect Ringer's solution, dropping bottle **(9)**			1	6
termites (larvae) **(10)**	3-5			up to 120

	Per Student	Per Pair (2)	Per Group (4)	Per Class (24)
Exercise G—continued				
microscope slides, box (75 × 25 mm)			1	6
coverslips (22 × 22)			1	6
dissecting needle	2			48
Giemsa stain, dropping bottle **(11)**			1	6
glycerol, dropping bottle **(12)**		1	6	
Exercise H				
agar Petri dishes **(13)**	1			24
Physarum on agar plate **(14)**	1			24
Physarum (dried plasmodium starts) **(15)**	1			24
oat flakes, box				1
Exercise I				
water mold (live, demonstration) **(16)**				1
water mold (wet mount of mycelium, demonstration **(17)**				1
Saprolegnia (prepared slide)	1			24
compound microscope	1			24
Exercise J				
Euglena (live culture)				1
diatomaceous earth (bottle)				1
diatoms (prepared slide)	1			24
Peridinium (prepared slide)	1			24
Fucus (preserved, complete plant)				1
Laminaria (preserved, complete plant)				1
Ectocarpus (live culture)				1
Polysiphonia (live culture)				1
Corallopsis (coralline algae, live culture)				1
Dasya (preserved)				1
microscope slides, box (75 × 25 mm)			1	6
coverslips, box (22 × 22 mm)			1	6
compound microscope	1			24
Exercise K				
Chlamydomonas (live culture)				1
Spirogyra (live culture)				1
Gonium (live culture)				1
Volvox (live culture)				1
Zygnema (live culture)				1
Stigeoclonium (live culture)				1
Ulva (preserved)				1
Ulothrix (live culture)				1
desmids (live culture)				1
microscope slides, box (75 × 25 mm)			1	6
coverslips, box (22 × 22 mm)			1	6

	Per Student	Per Pair (2)	Per Group (4)	Per Class (24)
Exercise K—continued				
compound microscope	1			24
Exercise L				
plankton sample **(18)**				1
microscope slides, box (75 × 25 mm)			1	6
coverslips, box (22 × 22 mm)			1	6
compound microscope	1			24
Ward's dichotomous key to free-living protists	1			24

IV. PREPARATION OF MATERIALS AND SOLUTIONS

(1) nutrient agar Petri dishes

Using a 500 ml Erlenmeyer flask, dissolve 3.1 g of nutrient agar (see Ordering Information) in 100 ml of distilled water. Heat to a boil to dissolve nutrient agar. Autoclave for 15 minutes at 15 pounds pressure (121° C). Place a cotton plug covered with cheesecloth in the flask before autoclaving. After autoclaving, let the agar solution cool (but not enough to gel), and then pour enough solution to cover the bottom of each Petri dish.

(2) and (3) Petri dishes for antibiotic sensitivity tests

Gather together the following materials:

Obtain the following:

antibiotic agar
Difco antibiotic disks
tryptic soy broth
sterile Petri dishes (100 mm)
500 ml Erlenmeyer flask (cotton plugged)
two 250 ml Erlenmeyer flasks (cotton plugged)
inoculating loop
sterile spreader
Bunsen burner
incubator (37° C)

Procedure:

1. Antibiotic agar Petri dishes. Using a 500 ml Erlenmeyer flask, dissolve 3.05 g of antibiotic agar (see Ordering Information) in 100 ml of distilled water. Heat to a boil to dissolve the agar before autoclaving at 121° C for 15 minutes at 15 pounds pressure. After autoclaving, let the solution cool, then pour enough antibiotic agar solution to cover the bottom of each Petri dish. (See alternative method described below.)
2. Tryptic soy broth solution. Using two 250 ml Erlenmeyer flasks, each flask containing 100 ml of distilled water, dissolve 3 g of tryptic soy broth in each. Heat slightly to dissolve the tryptic soy broth before autoclaving for 15 minutes at 15 pounds pressure (121° C). After autoclaving, let both flasks cool.
3. After the flasks of tryptic soy broth have cooled, flame and break open a vial of E. *coli* freeze-dried bacteria, and empty the vial into a flask of tryptic soy broth after flaming the top of the flask. Do the same with *S. aureus.* If inoculating loops are not available, a sterile Pasteur pipette can be used to transfer a few drops of culture. The drops can then

be spread. Use a sterile bent glass rod to spread the bacteria in a film to cover the entire Petri dish. Keep the lid over the culture dish while spreading to insure sterility. With sterile forceps, place five appropriate antibiotic disks (your choice) on the surface of plates of *E. coli* and *S. aureus*. Use the same kinds of disks on both plates. Allow 24-28 hours of incubation (37° C).

Alternate method:

Antibiotic agar can be poured into sterile capped tubes instead of Petri dishes. When ready to inoculate, the agar can be melted at 45-48° C and can then be inoculated with bacteria (approximately 0.3 ml or several loops full). Pour the inoculated agar onto a plate and let it harden. Place antibiotic disks on the surface of the agar as above.

(4) *S. aureus* broth culture

Follow steps 2 and 3 in **(2)** to prepare a tryptic soy broth culture of *S. aureus*. Use freeze-dried *S. aureus* or a loopful of bacteria from a prepared slant (see Ordering Information). Spread the bacteria onto the surface of a nutrient agar Petri dish, prepared as in **(1)**. Place approximately 0.1 ml of a 24 hour tryptic soy broth culture onto the center of the agar plate and use a glass spreading rod to spread the bacteria. Alternatively, students can use a sterile Q-Tip to spread the bacteria. You may also wish to use *S. epidermidis* (living culture), a non-pathogenic staphylococcus found on the skin instead of *S. aureus*.

(5) *E. coli* broth mixture

Prepare as in **(4)** above, but use freeze-dried *E. coli* (see Ordering Information).

(6) paper disks with cleaning solutions

Use blank paper disks for preparing commercial cleaning or soap solutions (see Ordering Information). Students should allow their disk to soak in a solution for 1-2 minutes. These should be drained by touching one edge to a sterile paper towel (sterile gauze squares may be used to blot off excess fluid).

(7) soybean plants, pot (inoculated with *Rhizobium*)

Either soybeans or clover can be used for this demonstration. Seeds are simply coated with the *Rhizobium* inoculum by gently shaking them in a plastic bag containing the inoculum. The inoculum that we have found to be particularly good is from Nitragin Company (Milwaukee, WI 53209, phone 414-462-7600) and can be ordered by phone. Carolina Biological Supply Company also provides the inoculum with clover seeds as an inexpensive package (see ordering information). ALLOW EIGHT WEEKS FOR GROWTH.

(8) *Vibrio fischeri* bioluminescent bacteria (broth culture)

Order a slant of *Vibrio fischeria* from Carolina Biological Supply Company or from the American Type Culture Collection (see Ordering Information).

1. To prepare a broth culture for photobacteria, mix the following:

(Bacto) tryptone	5.0 g
(Bacto) yeast extract	2.5 g
ammonium chloride	0.3 g
magnesium sulfate	0.3 g
ferric chloride	0.01 g
calcium carbonate	1.0 g
potassium phosphate (monobasic)	3.0 g
sodium glycerophosphate	23.5 g
sodium chloride	30.0 g

distilled water 100 ml

Adjust to pH 7.0 at 25° C with 1 M NaOH or 1 N HCI.

2. Dispense 100 ml of photobacterium broth into each of several 250 ml flasks and cover with aluminum foil. Autoclave for 15 minutes at 121° C and 20 pounds pressure (large batches may be prepared for large laboratories).
3. To one flask of 100 ml of broth, add 1.5 g of (Bacto) agar, cover, autoclave, and aseptically dispense 5 ml into sterile 16 mm test tubes (be sure to mix well before dispensing agar). Slant immediately. These will be used to maintain fresh cultures.
4. Transfer a loop full of *Vibrio* onto a fresh slant (three days before the laboratory) and incubate at 37° C for 24 hours. Prepare slants for as many laboratory periods as needed.
5. Dispense 5 ml of broth aseptically into several sterile tubes. If you use glass tubes, you will need to sterilize rubber corks to seal the tubes before shaking. Alternatively, use sterile, disposable snap-cap tubes.
6. Inoculate the 5 ml broth tubes, each with a generous loop full of *Vibrio* from the fresh (24 hour) slant.
7. Incubate at 37° C for 24 hours in a shaking waterbath.
8. For larger volumes, inoculate 100 ml of broth with 1 ml of the 5 ml culture.
9. Incubate in a shaking waterbath at 37° C for 24 hours and then dispense 5 ml of the culture into each of several 16 mm sterile tubes (glass or snap-cap) and distribute to students.

Broth culture of *Vibrio* may be stored at –80° C using the following procedure:

1. Dispense 0.25 of glycerol into 1.5 ml Eppendorf (microcentrifuge) tubes, cap, and autoclave.
2. Add 1 ml of fresh (24 hour) *Vibrio* culture to the tube, cap, and vortex until thoroughly mixed.
3. Freeze in the ultra-cold freezer until needed. This eliminates transferring cultures during the rest of the semester.
4. Revive by streaking a loop full of frozen culture onto a fresh slant.

(9) insect Ringer's solution

Dissolve 7.5 g NaCl, 3.5 g KCl, and 0.21 g $CaCl_2$ in enough distilled water to make 1 liter of solution.

(10) termites (larvae)

Collect termites from rotten logs on the floor of the woods. Find areas that are not too wet. Open the log using a hatchet or other suitable tool. Use larvae for this exercise and not winged adults. Termites can also be ordered from Carolina Biological Supply Company (see Ordering Information).

(11) Giemsa stain, dropping bottle

It is easiest to order this stain already prepared (see Ordering Information). If you have powdered Giesma on hand, dissolve 0.5 g Giesma powder in 33 ml of glycerin (this will take 1-2 hours). Add 33 ml of acetone-free absolute methyl alcohol. To use, dilute this stock solution 1:10 with distilled water.

(12) glycerol, dropping bottle

Add 50 ml glycerin (glycerol) to 50 ml distilled water. Place in a dropping bottle for easy use in mounting coverslips.

(13) (14) and (15) (Bacto) agar Petri dishes

Using a 500 ml Erlenmeyer flask, dissolve 2 g of (Bacto) agar into each 100 ml of distilled water. Autoclave for 15 minutes at 15 pounds pressure (121° C). After autoclaving let the agar solution cool, then pour enough agar solution to cover the bottom of each Petri dish.

When you autoclave the agar, also autoclave a bottle of distilled water (100 ml), a pair of forceps, and a few Pasteur pipettes.

Physarum is available in two forms—a culture of the plasmodium growing on agar and as the dried resting stage called a sclerotium. After agar has cooled, place the plasmodium or sclerotium on a small piece of filter paper and place it in the center of the agar dish. Place four oat flakes close to the material and wet the filter paper and oat flakes with the sterile distilled water using a sterile Pasteur pipette. Start the slime mold two days before the laboratory. Hold it at room temperature in the dark. Feed as it moves away from the oat flakes. Exposure to light triggers the formation of fruiting bodies (start cultures one week ahead for fruiting bodies).

You can increase the numbers or save for later by simply drying out the agar with the plasmodium by placing the plate in a fume hood. The agar can then be cut into small pieces to give new starts. Alternatively, the plasmodium can be scraped off the agar and transferred to wet filter paper. Refeed with oat flakes until it spreads out and then simply dry out the filter paper and slime mold by exposing it to the air. Cut up the filter paper to make new cultures.

(16) live water mold

Use a medium size culture dish containing a small amount of dirt. Fill the dish half full of pond water (the dirt and water used for growing water mold should come from a brook or pond). Boil several-radish seeds for 5 minutes. Use two dissecting needles to break the seeds into small pieces and place them on top of the water so that they float. Do not include the seed coats with the seeds. Allow four days at room temperature for growth.

(17) water mold mycelium demonstration

Prepare water mold as in **(16)**. Use this culture to provide the mycelium for a wet mount slide. Put petroleum jelly or nail polish around the coverslip so the slide can be used during the entire period.

(18) plankton sample

A plankton net can be made from a pair of support hosiery. Cut the hose off at the ankle and attach it to a small plastic jar (those in which protozoan cultures are shipped by Carolina Biological Supply Company are ideal) with a rubber band. Usually the top of the stocking is a double thickness. Put a hole in it on the top rim and slide an opened coat hanger through the hole and around the rim. Fold the coat hanger to make a suitable opening to the net and a handle. You may wish to centrifuge the sample at low speed to concentrate the sample.

Ward's Natural Science Establishment has an excellent key for free-living protozoa (4 pages) entitled "Dichotomous Key to Free-living Protozoa." If the use of keys has been taught prior to the beginning of the diversity laboratories, this will reinforce the student's learning. Otherwise, the key is simply useful for identification purposes.

V. PREPARATION SUGGESTIONS

Exercise A

This exercise needs to be started one week ahead of time. Students are encouraged to sample many different types of possible sources of bacteria, including feet of animals, shoes, parts of the body, etc. Agar plates for this experiment should be poured very thin to avoid costly expenses for agar.

Exercise B

If you do not wish to prepare your own antibiotic studies, Carolina Biological kit, # 15-4737 or # 15-4740 can be substituted.

Exercise C

For-soybeans inoculated with *Rhizobium,* we emphasize that plants will take eight weeks to grow and show *Rhizobium* infection.

Extending Your Investigation

This is a fairly simple investigation that allows students to practice their aseptic techniques learned in Laboratory 6. Make sure that everything is kept sterile. Autoclave forceps and use autoclaved paper towels (wrap in foil) or sterile gauze pads to drain disks. Let students select disinfectants. Cleaning solutions and facial scrubs or medicines are best. If liquid soap is to be used, dilute the soap until you have a thin rather than thick liquid.

Exercise D

When students examine *Vibrio* cultures, they must be shaken so tubes must be tightly closed. Use either sterile rubber stoppers to replace looser caps or use screw cap culture tubes. Snap-cap disposable tubes will also work, but be sure to press down on the cap when shaking the tube. Do not fill the tubes completely—leave space for oxygen.

Exercises E and F

A box of microscope slides and a box of cover slips should be provided for every four people (or whatever is logistically simple for your seating arrangement). Slides are used in Exercises C, D, H. and I. Encourage students to rinse slides between wet mounts and reuse them. Provide a dirty slide container for used slides at the end of the laboratory period.

Living materials or preserved materials and prepared slides can be used for Exercises E and F. Substitutions can easily be made to use what is available in your laboratory. A plankton sample from a pond will provide some good examples of many organisms, both plant and animal.

Exercise G

Termites are easily found in the wild. Supply students with a Petri dish and have them collect their own, but warn them to tape the dish around the edges to prevent escape. Termites will live overnight even if the edges of the dish are taped. We collect termites along with a piece of their decayed wood in a Rubbermaid container. The termites will stay in the wood. Place a damp paper towel in the container for moisture.

The flagellates that students will study in this exercise are very numerous and are easily observed.

Exercise H

Experiments with *Physarum* are easy for students to carry out in their dormitory rooms. This exercise can also give students a chance to work with the scientific method and it fosters some discovery learning.

Exercise I

Be sure to set up cultures for water mold at least four to five days ahead of use. Set up several using dirt from different sources.

Exercises J and K

A box of microscope slides and a box of coverslips should be provided for groups of students. Encourage students to rinse slides and reuse them during the exercise. Provide a container for used coverslips.

Exercise L

A plankton sample will usually be rich in algae and protozoans. If a plankton sample is not available, you may wish to substitute a protozoan survey mixture or an algae survey mixture.

VI. ORDERING INFORMATION

bacilli slide—Carolina Biological, # Bao 050; Ward's, # 90W0132
cocci slide—Carolina Biological, # Bao 040; Ward's, # 90W0131
spirilla slide——Carolina Biological, # Bao O30; Ward's, # 90W0133
nutrient agar—Carolina Biological, # 78-5301 (1 lb); Ward's, # 38W1201
antibiotic agar—Fisher, # DF 0163-01- 1
Erlenmeyer flask, 250 ml—Fisher, # 10-040F (Pyrex) or # 10-039F (Kimax); Carolina Biological, # 72-6672 (Corning, student grade); Ward's, # 17W2974
Erlenmeyer flask, 500 ml—Fisher, # 10-040H (Pyrex) or # 10-039H (Kimax); Carolina Biological, # 72-6676 (Corning, student grade); Ward's, # 17W0604
Bunsen burner—Carolina Biological, # 70-6645; Ward's, # 15W0604
alcohol lamp with cap—Carolina Biological, # 70-6604; Ward's, # 17W0211
inoculating loop—Carolina Biological, # 70-3060; Ward's, # 14W0957
Petri dishes, sterile disposable, 100 × 15 mm—Fisher, # 08-757-12; Carolina Biological, #74-1350; Ward's, # 18W7101
Difco antibiotic disks—Carolina Biological, # 80-5080; Ward's, # 38W1602
tryptic soy broth—Carolina Biological, # 78-8440; Ward's, # 88W1817
Escherichia coli (freeze-dried vial)—Carolina Biological, # 15-5065A; Ward's, # 85W0400
Staphylococcus aureus (freeze-dried vial)—Carolina Biological, # 15-5555A; Ward's, # 85W1941
E. coli (prepared slide)—Carolina Biological, # Ba 90; Ward's, # 90W2042
S. aureus, gram stain (prepared slide)—Carolina Biological, # Ba 235; Ward's, # 90W2080
S. epidermidis—Carolina Biological, # 85W-1035 (living culture); # 85W-1940 (freeze dried); Ward's, # 90W2076
paper disks—Carolina Biological, # 38W-1600; Ward's, # 38W1600
forceps—Carolina Biological, # 62-4020 (straight) or 62-4030 (curved); Ward's, # 14W0541
wax pencil—Carolina Biological, # 65-7730 (red), # 65-7732 (blue), # 65-7734 (black); 15W1159
Rhizobium inoculant for soybeans—Nitragin Company, Milwaukee, WI 53209; Carolina Biological, # 15-4720; Ward's, # 20W6051
soybean seeds—Nitragin Company or Carolina Biological, # 17-8200; Ward's, # 86W8089
clover seed—Carolina Biological, # 15-9310; Ward's, # 86W8130
Rhizobium meliloti (prepared slide)—Carolina Biological, # Ba 169M; Ward's, # 90W0546
Vibrio fischeri, slant—Carolina Biological, # 15-5723; Ward's, # 85W0650; American Type

Culture Collection, # 7744[1]
microcentrifuge tubes, Eppendorf (1.5 ml)—Carolina Biological, # 19-9684; Fisher, # 05-406-16
(Bacto) tryptone—Carolina Biological, # 79-4420; Fisher, # DF0123-02-0; Ward's, # 39W3600
(Bacto) yeast extract—Carolina Biological, # 79-4780; Fisher, # DF0127-02-6; Ward's, # 38W5800
ammonium chloride—Fisher, # A661-500; Carolina Biological, # 84-3810; Ward's, # 37W1965
magnesium sulfate— Fisher, # M63-500; Carolina Biological, # 87-3400; Ward's, # 37W2861
ferric chloride— Fisher, # I88- 100; Carolina Biological, # 86-2460; Ward's, # 37W1764
calcium carbonate— Fisher, # C64-500; Carolina Biological, # 85- 1760; Ward's, # 37W0921
potassium phosphate (monobasic) (KH_2PO_4)—Fisher, # P285-500; Carolina Biological, # 88-4250; Ward's, # 37W4841
sodium gylcerophosphate—Fisher, # S314-100
sodium chloride (NaCl)—Fisher, # S271-3; Carolina Biological, # 88-8903; Ward's, # 37W0015
(Bacto) agar[2]—Fisher, # DF0140-02-9; Carolina Biological, # 21-6720 or 21-6721 (500 g); Ward's, # 38W0015
16 × 100 mm culture tubes, glass—Carolina Biological, # 73-1458; Ward's, # 17W1343; Fisher, # 14-923K
caps for 16 × 100 mm culture tubes—Carolina Biological, # 73-1472; Ward's, # 18W7172; Fisher, # 14X127-28B
17 × 100 mm culture tubes, disposable—Carolina Biological, # 21-5078; Fisher, # 14-956-6B
Nostoc (live)—Carolina Biological, # 15- 1845; Ward's, # 86W2150
Cylindrospermum (live)—Carolina Biological, # 15-1755; Ward's, # 86W1930
Oscillatoria (live)—Carolina Biological, # 15- 1865; Ward's, # 86W2300
Anabaena (live)—Carolina Biological, # 15-1710; Ward's, # 86W1800
Gloeocapsa (live)—Carolina Biological, # 15- 1800; Ward's, # 86W2000
Trypanosoma gambiense (prepared slide)—Carolina Biological, # PS-310; Ward's, # 92W4330
Amoeba proteus (live)—Carolina Biological, # 13-1306 (class of 30); Ward's, # 87W0390
Paramecium caudatum (live)—Carolina Biological, # 13-1554 (class of 30); 87W1310
Plasmodium vivax—Carolina Biological, # PS 600; Ward's, # 92W4660
microscope slides, 75 × 25 mm—Fisher, # 12-550A; Carolina Biological, # 63-2000; Ward's, # 14W3500
coverslips, 22 × 22 mm—Fisher, # 12-542B; Carolina Biological, # 63-3095; Ward's, # 14W3251
potassium chloride (KCl)—Fisher, # P217-500; Carolina Biological, # 88-2910; Ward's, #

---

[1] American Type Culture Collection, 12301 Parklawn Drive, Rockville, MD 20852. Your first order must be received on your company letterhead. This order must include a company profile, billing address, shipping address, name, telephone, and fax number. Afterward you may order by phone (800-638-6597; Maryland 301-881-2600) or FAX (301-816-4361). ATCC maintains a Web page at http://www.atc.org.

[2] Bacto is a designation used on Difco Laboratories products available through Fisher Scientific, Inc. and Carolina Biological Supply Company. Difco Laboratories, PO Box 331058, Detroit, MI 48232-7058; FAX 313-462-8517, Voice, 800-521-0851. Difco maintains a Web page at http://difco.com (consult Reference Guide under Technical Support for information on agar).

37W4631

calcium chloride ($CaCl_2$)—Fisher, # C79-500; Carolina Biological, # 85-1810; Ward's, # 37W0920

glycerol—Fisher, # G-33-1; Carolina Biological, # 86-5530; Ward's, # 39W1438

Physarum (live)—Carolina Biological, # 15-6193 or *Sclerotium*—Carolina Biological, # 15-6190; Ward's, # 85W4750

agar (Difco)—Fisher, # DF-0140-02-9; Carolina Biological, # 79-6201; Ward's, # 38W0016

Giemsa stain—Fisher, # G146-10; Carolina Biological, # 86-5171; Ward's, # 38W7932

dropping bottle, 30 ml— Fisher, # 02-980; Carolina Biological, # 71-6525; Ward's, # 17W6011

termites, live—Carolina Biological, # L743; Ward's, # 87W6160

Pasteur pipette, 9"—Fisher, #13-678-20C; Carolina Biological, # 73-6062; Ward's, # 17W1146

rubber bulb—Fisher, # 14-065B

radish seeds—Carolina Biological, # 15-9000; Ward's, # 86W8200

dissecting needles—Carolina Biological, # 62-7200; Ward's, # 14W-650

petroleum jelly, white—Fisher, # P66-1LB; Carolina Biological, # 87-9510; Ward's, # 15W9832

Euglena (live)—*Carolina* Biological, # 13-1768 (class of 30); Ward's, # 87W0100

diatomaceous earth—Carolina Biological, # PB 115; Ward's, # 63W0240

diatoms (prepared slide)—Carolina Biological, # B21; Ward's, # 91W1010

Peridinium (prepared slide)—Carolina Biological, # B83P; Ward's, # 91W1540

Fucus (preserved)—Carolina Biological, # PB 133C; Ward's, # 63W0238

Laminaria (preserved)—Carolina Biological, # PB 136; Ward's, # 63W0335

Ectocarpus (live)—Carolina Biological, # 15-3360; Ward's, # 86W1950

Polysiphonia (live)—Carolina Biological, # 15-3580; Ward's, # 86W2910

coralline algae (preserved)—Carolina Biological, # PB 154; Ward's, # 63W0368

Dasya (preserved)—Carolina Biological, # PB 154B

Chlamydomonas (live)—Carolina Biological, # 15-2030; Ward's, # 86W0102

Spirogyra (live)—Carolina Biological, # 15- 1320; Ward's, # 86W0650

Gonium (live)—Carolina Biological, # 15-2264; Ward's, # 86W1085

Volvox (live)—Carolina Biological, # 15- 2655; Ward's, # 86W0805

Zygnema (live)—Carolina Biological, # 15-2695; Ward's, # 86W0900

Stigeoclonium (live)—Carolina Biological, # 15-2600; Ward's, # 86W0690

Ulva (preserved)—Carolina Biological, # PB 90; Ward's, # 63W0223

Ulothrix (live)—Carolina Biological, # 15-2640; Ward's, # 86W0750

desmids (live)—Carolina Biological, # 15-1260; Ward's, # 86W0270

algae survey mixture—Carolina Biological, # 15-1216; Ward's, # 86W3022

protozoa survey set—Carolina Biological, # 13-1008; Ward's, # 87W1550

Ward's dichotomous key to free-living protists—Ward's, # 32W2205

Notes:

LABORATORY 23 Diversity—Fungi and the Nontracheophytes

I. FOREWORD

This laboratory is designed to introduce students to the divisions within the Kingdom Fungi and to a few of the phyla in the Kingdom Plantae. Students work individually, but the entire laboratory can be set up as a series of stations (two per exercise) and students can do the exercises in any order. This saves on cost of prepared slides, preserved materials, and reduces the number of live specimens required.

II. TIME REQUIREMENTS

PART I—KINGDOM FUNGI
- Exercise A—Phylum Zygomycota (30 minutes)
- Extending Your Investigation: Conditions for Fungal Growth
- Exercise B—Phylum Ascomycota (30 minutes)
- Exercise C—Phylum Basidiomycota (20 minutes)
- Exercise D—Phylum Chytridiomycota (15 minutes)
- Exercise E—Phylum Deuteromycetes (15 minutes)
- Exercise F—Identification of Collected Fungi (15 minutes)
- Exercise G—Diversity Among the Lichens (15 minutes)

PART II—KINGDOM PLANTAE
NONTRACHEOPHYTES
- Exercise H—Nontracheophytes—Mosses, Liverworts, and Hornworts (30 minutes)

III. STUDENT MATERIALS AND EQUIPMENT

	Per Student	Per Pair (2)	Per Group (4)	Per Class (24)
Exercise A				
Rhizopus stolonifer (bread mold, live) **(1)**	1			24
dissecting microscope	1			24
microscope slides, box (75 × 25 mm)			1	6
coverslips, box (22 × 22 mm)			1	6
distilled water (dropping bottle)			1	6
Rhizopus (prepared slide)	1			24
Extending Your Investigation				
bread molds grown by students **(2)**	1			24
compound microscope	1			24
Exercise B				
Sordaria (live) **(3)**			1	6
microscope slides, box (75 × 25 mm)			1	6
coverslips, box (22 × 22 mm)			1	6

	Per Student	Per Pair (2)	Per Group (4)	Per Class (24)
Exercise B—continued				
distilled water, dropping bottle			1	6
yeast culture **(4)**				1
Schizosaccharomyces (prepared slide)	1			24
compound microscope	1			24
Exercise C				
edible mushroom **(5)**	1			24
microscope slides, box (75 × 25 mm)			1	6
coverslips, box (22 × 22 mm)			1	6
distilled water, dropping bottle			1	6
Coprinus (prepared slide)	1			24
basidiomycetes (demonstration) **(6)**				1
compound microscope	1			24
Exercise D				
Allomyces, living culture **(7)**			1	6
Pasteur pipette		1		12
rubber bulb		1		12
coverslips, box (22 × 22 mm)			1	6
distilled water, dropping bottle			1	6
compound microscope	1			24
Exercise E				
Aspergillus (prepared slide)		1		12
Penicillium (prepared slide)		1		12
compound microscope	1			24
Exercise F				
fungus collection—prepared by students **(8)**				1
3 × 5 inch cards	1			24
Exercise G				
selection of crustose, foliose and fruticose lichens **(9)**				1
lichen thallus (prepared slide)	1			24
compound microscope	1			24
Exercise H				
moss plants (live, showing sporophyte and gametophyte) **(10)**			1	6
moss antheridium (prepared slide)		1		12
moss archegonium (prepared slide)		1		12
Polytrichum antheridia (moss, preserved)				1
Polytrichum archegonia (moss, preserved)				1
Marchantia antheridia (liverwort, preserved)				1
Marchantia archegonia (liverwort, preserved)				1
Marchantia (live) **(11)**				1

	Per Student	Per Pair (2)	Per Group (4)	Per Class (24)
Exercise H—continued				
compound microscope	1			24

IV. PREPARATION OF MATERIALS AND SOLUTIONS

(1) live *Rhizopus*

Gather together the following materials:

Rhizopus culture
corn meal agar
dextrose
yeast extract
Petri dishes (100 × 15 mm)

Procedure:

Maintaining a *Rhizopus* culture can be accomplished using corn meal agar plates. Into each 100 ml of distilled water, put 1.7 g of corn meal agar, 0.2 g of dextrose, and 0.1 g of yeast extract. Autoclave for 15 minutes at 15 pounds pressure (121° C). After autoclaving, let the corn meal agar cool and pour in Petri dishes. Using a sterile inoculating loop place some sporantia of two strains (+ and –) of *Rhizopus* (from purchased cultures) onto the corn meal agar plates, keeping the lid over the dish as much as possible to prevent contamination. The asexual spores begin forming around the edge of the mycelium two days after inoculation: zygospores form a day or two later on the line where the two strains meet.

To see zygospores, you will need to part the mycelium. The zygospores will be on the agar surface and may be seen even better by looking at the bottom of the dish by holding it up to the light.

Sporangia can also be placed on bread. Put the bread in a plastic bag to keep it moist.

(2) bread molds

At the end of the preceding laboratory, distribute a piece of bread (without preservatives), a 3 × 5 inch index card, and three plastic bags to each student. Students should design an experiment to test conditions necessary for fungus growth. [You may wish to ask students to take a few minutes to plan their experiments so that you do not dispense more bags or bread than necessary.] Warn students to not breathe mold spores. Keep bags closed.

(3) *Sordaria*

Gather together the following materials:

corn meal agar plates—see preparation in **(1)**
Sordaria (live culture)

Procedure:

After agar plates have jelled, inoculate one or several plates with *Sordaria fimicola.* Obtain a culture of *Sordaria fimicola,* wild type. Cut out a small agar block with mycelia and place this upside down on an agar plate. Place a second block into a new sterile tube of agar to keep the culture going (put this tube in the refrigerator). It will take about 8 days from inoculation to see perithecia. Alternatively, purchase *Sordaria* on a plate if you do not wish to keep a stock culture in the laboratory.

(4) yeast culture

Prepare 200 ml of a 10% molasses or syrup solution. Add 2 g of Fleischmann's dried yeast and 0.5 g peptone to the flask. Incubate in a warm place (25-30° C or 77-86° F) for approximately 12 hr before use and check for budding. Methylene blue (0.l g in 1 liter) or neutral red (0.1 g in 1 liter) can be used as vital stains to make observation easier.

(5) edible mushrooms
Obtain mushrooms from a grocery store.

(6) selected basidiomycetes
These can be collected during wet weather and are best preserved in FAA fixative (formalin-acetic acid):

95% ethanol	50 ml
glacial acetic acid	2 ml
formalin (40% formaldehyde)	10 ml
water	40 ml

(7) *Allomyces arbuscula,* living culture
Order a plate of each of the 1*n* gametophytic and 2*n* sporophytic stages. Using a toothpick, remove a small amount of each on a chunk of agar and place the sample in separate culture dishes of distilled water. Add split or crushed hemp seed (Carolina Biological, # 15-8660).

(8) and (9) fungus and lichen collections
These should be collected by students and preserved dry or in FAA fixative (see **(6)**). The collection should be augmented by your own specimens.

(10) and (11) moss and liverwort collections
Mosses, *Marchantia* (liverworts), and hornworts can be obtained from a woodland area nearby or from a supply house (see Ordering Information).

V. PREPARATION SUGGESTIONS

Exercise A

The experiment on conditions necessary for fungus growth can be used to give students further practice with hypothesis formation and the scientific method. Each student should include a card stating his or her hypothesis, procedure, results, and conclusions. Warn students not to open bags and to avoid breathing mold spores.

Exercise B

Sordaria are simple to grow but the cultures must be set up 8-10 days prior to the laboratory period. Yeast cultures can be best studied using vital stains (methylene blue or neutral red) which will slowly be taken up by the living cells.

Exercise C

Large mushrooms can be obtained from the grocery store. It is also fun to have students do spore prints. Simply place a mushroom cap, gills down, on a piece of white paper and cover with a small glass aquarium dish (or a plastic cup or drinking glass). Wait for one week. As the mushroom dries, spores will be released, reproducing the pattern of the gills on the paper.

Exercise D

Use separate cultures of the 1*n* gametophytic and 2*n* sporophytic forms of *Allomyces arbuscula* for this investigation. In the culture dish containing the 1*n* individuals, gametes will emerge and syngamy should occur within 30 minutes to 1 hour. Examine samples of the water culture using the microscope at 40×. The 2*n* individuals will produce zoospores when mature. If the spread area of the fungus on the agar plate you receive upon ordering is only the size of a quarter, remove the Parafilm from the Petri dish and allow the fungus to grow for 1 week to 10 days until mature before transferring to water. The fungus will appear to be a bright orange. Once again, allow 30 minutes to 1 hour for zoospore production.

Exercise E and F

Most of the demonstration materials can be collected by students or by the instructor. If not available, live or preserved examples of all fungi and lichens can be obtained from Carolina Biological Supply Company. Specimens can be preserved dry or in FAA (see materials).

Exercise G

Depending upon the time of year during which this laboratory is done, it might be difficult to find mosses and liverworts with both sporophyte and gametophyte obvious. A combination of live materials (vegetative forms) and preserved materials (showing archegonia and antheridia) or slides can be used. To obtain live *Marchantia* archegonia and antheridia in season (January to spring), orders must be placed at least four weeks in advance since sexual stages are artificially induced.

VI. ORDERING INFORMATION

Rhizopus (prepared slide)—Carolina Biological, # B224; Ward's, # 91W2110
Rhizopus (live)—Carolina Biological, # 15-6223 (+) and # 15-6224 (–); Ward's, # 85W8900
corn meal agar—Carolina Biological, # 78-2461; Ward's, # 38W0321
yeast extract—Carolina Biological, # 79-4781; Ward's, # 38W5800
dextrose—Fisher, # D16-500; Carolina Biological, # 74-1350 (dish only); Ward's, # 18W7101
Petri dish, 100 × 15 mm—Fisher, # 08-747C (Pyrex) or 08-746C (Kimax); Carolina Biological, # 74-1158; Ward's, # 18W7101
Sordaria fimicola—Carolina Biological, # 15-6290 (tube) or # 15-6291 (plate); Ward's, # 85W5100
Schizosaccharomyces (prepared slide)—Carolina Biological, # B261B
Coprinus (prepared slide)—Carolina Biological, # B270; Ward's, # 91W3210
Allomyces arbuscula, (*n*) plate—Carolina Biological, # 15-5911
Allomyces arbuscula, (2*n*) plate—Carolina Biological, # 15-5912
Pasteur pipette, 9"—Fisher, #13-678-20C; Carolina Biological, # 73-6062; Ward's, # 17W1146
rubber bulb—Fisher, # 14-065B
hemp seed—Carolina Biological, # 15-8660
Aspergillus (prepared slide)—Carolina Biological, # B234; Ward's, # 91W2410
Penicillium (prepared slide)—Carolina Biological, # B252; Ward's, # 91W2471
lichen thallus (prepared slide)— Carolina Biologic al , # B293; Ward's, # 91W3950
Polytrichum (moss, live)—Carolina Biological, # 15-6730; Ward's, # 86W4360
Polytrichum antheridia (preserved)—Carolina Biological, # PB385 or *Mnium* antheridia (preserved)—Carolina Biological, # PB 375A
Polytrichurn archégonia (preserved)—Carolina Biological, # PB386 or *Mnium* archegonia (preserved)—Carolina Biological, # PB376A
moss antheridium (prepared slide)—Carolina Biological, # B361B; Ward's, # 91W4322
moss archegonium (prepared slide)—Carolina Biological, # B363; Ward's, # 91W4326
Marchantia vegetative (live)—Carolina Biological, # 15-6540; Ward's, # 86W4200

Marchantia antheridia (live)—Carolina Biological, # 15-6544 and *Marchantia* archegonia (live)—Carolina Biological, # 15-6546 (available January through spring)
Marchantia antheridia (preserved)—Carolina Biological, # PB320; Ward's, # 63W1030
Marchantia archegonia (preserved)—Carolina Biological, # PB321; Ward's, # 63W1030
Anthoceros (live)—Carolina Biological, # 15-6520

LABORATORY 24 Diversity—The Tracheophytes (Vascular Land Plants)

I. FOREWORD

This laboratory is designed to introduce students to the vascular plants. As much live material as available should be used. The exercises can be set up as stations to cut down on expenses. Students can do the exercises in any order if necessary.

II. TIME REQUIREMENTS

PART I—Tracheophytes (Vascular Plants) Without Seeds
  Exercise A—Examining Seedless Tracheophytes (60 minutes)
PART II—Tracheophytes (Vascular Plants) with Seeds—Gymnosperms and Angiosperms
  Exercise B—Gymnosperms (60 minutes)
  Exercise C—Angiosperms (Phylum Angiospermae) (60 minutes)

III. STUDENT MATERIALS AND EQUIPMENT

	Per Student	Per Pair (2)	Per Group (4)	Per Class (24)
Exercise A				
Psilotum (live or preserved)				1
Psilotum stem (cross section, prepared slide)	1			24
Lycopodium (club moss, live or preserved)				1
Selaginella (live or preserved)				1
Equisetum (live or preserved)				1
fern leaves with sori (live) **(1)**				1
herbarium sheets (several)				1
gooseneck lamp (100 w bulb)			1	6
fern prothallus (live, class of 12)				1
water, dropping bottle			1	6
microscope slides, box (75 × 25 mm)			1	6
coverslips, box (22 × 22 mm)			1	6
compound microscope	1			24
fern prothallus with antheridia and archegonia (prepared slide)	1			24
fern prothallus with attached sporophyte (live, class of 12)				1
Exercise B				
representative gymnosperm branches **(2)**				1
male cones		1		12
water, dropping bottle			1	6
microscope slides, box (75 x 25 mm)			1	6
coverslips, box (22 x 22 mm)			1	6

	Per Student	Per Pair (2)	Per Group (4)	Per Class (24)
Exercise B—continued				
Cycas, megasporophyll (preserved)				1
compound microscope	1			24
male cone (longitudinal section, prepared slide)		1		12
female cones		1		12
female cone (longitudinal section, prepared slide)		1		12
Exercise C				
flower types (assorted for dissection) **(3)**				1
forceps	1			24
dissecting microscope	1			24
10 % sucrose solution, dropping bottle **(4)**			1	6
microscope slides, box (75 × 25 mm)			1	6
coverslips, box (22 × 22 mm)			1	6
compound microscope	1			24

IV. PREPARATION OF MATERIALS AND SOLUTIONS

(1) fern leaves with sori (live)

These can be obtained from a greenhouse. Carolina Biological Supply Company will guarantee several fertile fronds.

(2) representative gymnosperm branches

Obtain locally. If male and female cycads are available, place these on demonstration.

(3) flower types

Obtain from gardens or from florist. Include examples of both monocots and dicots for dissection. Suggestions include:

dicots—morning glory, azalea, water lily, rose, snap dragon, chrysanthemum, dogwood, buttercup, daisy, violet, etc.

monocots—daylily, yucca, orchid, iris, gladiolus, tulip, etc.

(4) 10% sucrose solution

Mass 10 g sucrose. Add water to make 100 ml. As an alternative mix 25 ml of white Karo syrup with 75 ml of distilled water.

V. PREPARATION SUGGESTIONS

Exercise A

Provide each student with a soaked peat pellet (used to germinate seeds). You can put these in a bucket at the beginning of the laboratory so they swell. Students should place their fern prothalli on their peat pellet. Spores from the sori can also be placed on the pellet. Place the pellet on a Petri dish bottom and cover it with a clear plastic cup. Students can take the materials home and can watch the development of the young sporophyte plant and look for growth of new prothalli during the following weeks.

Exercise B

Have students make a dichotomous key to the pines on display. This is usually quite straightforward and will give them good practice.

Exercise C

Since this exercise comes at the end of the laboratory period, you may wish to tell students to select a flower for dissection early in the period. The students should then tap the anthers to see if any pollen can be obtained. If placed on a slide in a sugar solution early in the period, pollen germination can be observed during the period. If *Impatiens is* available in the greenhouse, use their flowers for obtaining pollen—they usually work well for this.

Have students work in a group of four with each student dissecting a different type of flower. After dissecting a flower, each student should demonstrate the structure to the other students in his or her group.

This exercise could be supplemented with a study of flower adaptations for pollination. (See *Instructor's Manual: Answer Guide.)*

VI. ORDERING INFORMATION

Psilotum (live or preserved)——Carolina Biological, # 15-700 (live), # PB396A (preserved); Ward's, # 86W5100

Psilotum stem (cross section, prepared slide)—Carolina Biological, # B396; Ward's, # 91W4514

Lycopodium (live)—Carolina Biological, # 15-6980

Selaginella (live)—Carolina Biological, # 15-7015; Ward's, # 86W5200

Equisetum (live or preserved)—Carolina Biological, # 15-6960 or # PB494 or # PB 496; Ward's, # 86W5300

fern with sori—Carolina Biological, # 15-6910

herbarium mounting paper—Carolina Biological, # 66-3200; Ward's, # 20W4300

dropping bottle, 30 ml— Fisher, # 02-980; Carolina Biological, # 71-6525; Ward's, # 17W6011

microscope slides, 75 × 25 mm—Fisher, # 12-550A; Carolina Biological, # 63-2000; Ward's, # 14W3500

coverslips, 22 × 22 mm—Fisher, # 12-542B; Carolina Biological, # 63-3095; Ward's, # 14W3251

fern prothallia (antheridia and archegonia)—Carolina Biological, # 15-6878

fern prothallia (young sporophytic stage)—Carolina Biological, # 15-6880

fern prothallus with antheridia and archegonia (prepared slide)—Carolina Biological, # B410; Ward's, # 91W4865

pine staminate cone—Carolina Biological, # PB560

pine ovulate cones—Carolina Biological:
- at pollination, # PB561
- first winter, # PB562
- second summer, # PBS63C or # PB563
- second autumn, # PB564

pine winged seed—Carolina Biological, # PB565

pine staminate cone (prepared slide)—Carolina Biological, # B495; Ward's, # 91W6528

pine ovulate cone (prepared slide)—Carolina Biological, # B499; Ward's, # 91W6534

Cycas megasporophylls, with seed (preserved)—Carolina Biological, # PB585
Cycas, microsporophyll (prepared slide)—Ward's, # 91W6014
sucrose (certified)—Fisher, # S5-500; Carolina Biological, #89-2870; Ward's, # 39W3280

LABORATORY 25 Diversity—Porifera, Cnidaria, and Wormlike Invertebrates

I. FOREWORD

This laboratory begins a series of three laboratories (25, 26, and 27) that survey animal diversity. One of the goals of these exercises is to interest students in the overall diversity of animal body forms and adaptations—alternative solutions to recurring selective opportunities. To this end, added specimens should be used freely to augment or expand the basic overview of animals presented in these laboratories.

II. TIME REQUIREMENTS

Exercise A— Phyla Porifera and Placozoa (10 minutes)
Exercise B— Phyla Cnidaria and Ctenophora (30 minutes)
Exercise C—Wormlike Animals: Platyhelminthes, Rhynchocoela, Nematoda, Annelida)
  Part 1 Phylum Platyhelminthes: Flatworms — (30 minutes)
  Part 2— Phylum Rhynchocoela: Ribbon Worms (10 minutes)
  Part 3— Phylum Nematoda (Roundworms) and Other Wormlike Phyla (30 minutes)
Extending Your Investigation: Nematode Diversity (30 minutes)
  Part 4— Phylum Annelida: Segmented Worms (60 minutes)

III. STUDENT MATERIALS AND EQUIPMENT

	Per Student	Per Pair (2)	Per Group (4)	Per Class (24)
Exercise A				
sponges (demonstration specimens)			1	6
Grantia (preserved)	1			24
Grantia (prepared slide)	1			24
Spongilla (live)			1	6
vinegar (dropper bottle)			1	6
bleach (dropper bottle)			1	6
microscope slides, box (75 × 25 mm)	1			24
coverslips, box (22 × 22 mm)	1			24
compound microscope	1			24
dissecting microscope	1			24
Exercise B				
Hydra (live)	1			24
culture dish (1½ inches)	1			24
Daphnia (culture) **(1)**				1
brine shrimp (culture) **(2)**				1
Obelia (whole mount—prepared slide)	1			24
Physalia (Portuguese man-of-war, preserved)				2
Aurelia (plastic mount)			1	6

	Per Student	Per Pair (2)	Per Group (4)	Per Class (24)
Exercise B—continued				
Metridium (preserved)			1	6
corals (demonstration specimens), set				1
Pleurobranchia (comb jelly)			1	6
compound microscope	1			24
dissecting microscope	1			24
Exercise C (Part 1)				
Dugesia (live)	1			24
Petri dish (100 mm)(glass)	1			24
pond water, flask (2000 ml)				1
food for planaria **(3)**	1			24
aluminum foil (12 inch square)	1			24
dropper (plastic or eye dropper)	1			24
carmine powder (bottle)			1	6
Clonorchis (liver fluke, whole mount, prepared slide)			1	6
Taenia (tapeworm, preserved)			1	6
Taenia (whole mount, prepared slide)	1			24
compound microscope	1			24
dissecting microscope	1			24
Exercise C (Part 2)				
Cerebratulus (ribbon worm, preserved)			1	6
Exercise C (Part 3)				
Turbatrix (vinegar eel, live)	1			24
0.2% neutral red, dropping bottle **(4)**			1	6
microscope slides, box (75 × 25 mm)	1			24
coverslips, box (22 × 22 mm) 1			24	
Trichinella (prepared slide)	1			24
compound microscope	1			24
apparatus shown in Figure 25C-7		1		12
Extending Your Investigation				
ring stand	1			24
funnel with rubber tubing extension	1			24
beaker, 250 ml	1			24
cheese cloth	1			24
soil sample	1			24
pinch clamp	1			24
microscope slides, box (75 × 25 mm)	1			24
coverslips, box (22 × 22 mm)	1			24
tap water, bottle	1			24
Exercise C (Part 4)				
Lumbricus (earthworm, preserved)	1			24
dissecting pans	1			24

Exercise C (Part 4)—continued	Per Student	Per Pair (2)	Per Group (4)	Per Class (24)
dissecting pins	20			480
fine scissors or razor blades			1	6
dissecting needles	2			48
Lumbricus (earthworm, cross section, prepared slide)	1			24
Lumbricus (earthworm, live)		1	6	
segmented worms (preserved), collection				1
compound microscope	1			24
dissecting microscope	1			24

IV. PREPARATION OF MATERIALS AND SOLUTIONS

(1) *Daphnia* culture

A culture of *Daphnia* can be used to seed an aquarium for raising more *Daphnia.* Simply put pond water and some aquarium plants in the aquarium. Provide a regular aquarium lamp and some nutrients (small amounts of fish food).

(2) brine shrimp culture

Brine shrimp can be hatched from eggs. Do this 2-3 days before use in laboratory.

Make a 3.5-5% salt water solution (use non-iodized NaCl or lab grade NaCl) and place it in a 1 liter Erlenmeyer flask. Add approximately 1 teaspoonful of eggs to the water. Aerate with an aquarium pump. To collect hatched shrimp, place a lamp to one side and above the flask. The brine shrimp will swim toward the light. Collect the brine shrimp with a turkey baster. Put these into a beaker for students to use in feeding *Hydra.* Unhatched eggs will sink to the bottom.

Make sure containers of eggs are stored in a cool place to remain viable from one year to the next.

(3) food for planaria

Fresh liver (0.5 lb. or 500 g) should be purchased and diced into small cubes using a single-edged razor blade (2-3 mm in each dimension). Dried liver or egg yolk can be substituted but is less effective. If egg yolk is used, dye it with red food color so that students can see the food (red color) become distributed throughout the gastrovascular cavity. Each student should use only one piece. Fed planaria should be returned to a separate culture bottle marked "fed." They can be reused in later labs (after 1-2 days).

(4) 0.2% neutral red

Obtain a mass of 0.2 g of neutral red. Add absolute (100%) ethanol to make 100 ml.

V. PREPARATION SUGGESTIONS

Exercise A

Chalina, a Demospongiae with spongin, and a collection of dry specimens are suggested as demonstration specimens in the section on ordering information. In courses that have been taught through the years, many other demonstration specimens may be available.

Slides of spicules or other materials may be made available as time and resources permit. Chlorox (full strength) or vinegar (acetic acid) can be used to digest the sponge body to expose spicules.

Exercise B

Individual *Hydra* can be transferred to any small dish (Stender, Petri, etc.) for study under the dissecting microscope. Provide a dropper with the culture or cultures. Whatever dish is available for student use should be deep enough to allow the *Hydra* to extend and expand its arms without encountering the surface film. *Daphnia, Artemia* (brine shrimp), or other food could be provided if time permits observation of feeding behavior.

Whole and dissected preserved anemones should be available in the demonstration area. *Metridium is* commonly used.

A variety of corals should be on demonstration. To show the relation of a living polyp to the dried skeleton, a preserved coral might be available for demonstration under a dissecting microscope.

Cladocera, Daphnia, Artemia, or other food should be provided to test feeding behavior. introduce only 2 or 3 *Daphnia* or brine shrimp into the culture dish containing the *Hydra.* Return fed *Hydra* to a collecting dish. Yeast can be used to test for reactions to non-motile food.

Exercise C (Part 1)

Have a supply of pond water available (lake, river, artificial pond water, or tap water allowed to stand for 24 hours to declorinate).

For artificial pond water, make a stock solution as follows

NaCl	0.1 g
KCl	0.004 g
$CaCl_2$	0.006 g

Add distilled water to 1 liter and adjust pH to 6.0-7.0 using 0.1 *N* HCl or 0.1 *N* NaOH.

If you use egg yolk that has been stained with red food coloring, students will be able to see the outlines of the entire gastrovascular cavity as it fills up with red coloring that comes from the egg yolk.

5. Planaria lay down a slime trail as they move around in the dish. Carmine powder is sprinkled onto the surface of the water in the dish (after removing the planarian). Allow time for the carmine to sink to the lower surface of the dish and become attached to the slime trail. Swirl the dish quickly and decant the water. If you place the Petri dish on a white piece of paper, you will be able to determine whether the taxis response of planaria is toward or away from the light (planaria move away from the light). Be sure that students have drawn a line down the middle of the Petri dish—on the bottom—to show which side was dark vs. light.

Exercise C (Part 2)

Most students are unfamiliar with ribbon worms. Make sure that sutdents are acquainted with the evolutionary advances demonstrated by members of this phylum; two digestive openings, closed circulatory system, and protonephridia that are excretory rather than osmoregulatory. The rhynchocoel is not a true coelom—its lining includes endoderm.

Exercise C (Part 3)

Vinegar eels are only one type of roundworm that students can observe. If you do not plan for students to do the *Extending Your Investigation* that follows, you may want to provide students with nematodes from a soil sample. Follow the *Extending Your Investigation* directions to obtain a sample of soil nematodes. Also, *Ascaris* is a common nematode that can be purchased as a preserved specimen. If you obtain live *Ascaris* from a veterinarian or local abattoir, make sure that you have thoroughly washed your hands after handling the material or any of the containers associated with it. The eggs remain viable for a lengthy period of time even outside the *Ascaris* body.

Extending Your Investigation: Nematode Diversity

Use soil from local areas. Rich organic soil usually has the highest number of nematodes. Compare student samples taken from different locations. Students may want to test soil pH. If a local laboratory performs soil tests, you may wish to have some of the samples tested for mineral content. Alternatively, a soil test kit that tests for pH, nitrogen, phosphorous, and potash is available from Carolina Biological Supply Company (see Ordering Information). Students may make hypotheses that link nematode diversity to soil type or specific soil characteristics such as pH or nitrogen content.

Exercise C (Part 4)

Wax-lined dissecting pans and pins should be available. Dissecting needles should also be provided if students do not have their own dissecting kit. Provide scissors with fine, sharp blades for opening the earthworms (a single edged razor blade will also work if students are careful to cut only the outer body wall). Some of the internal organs of the earthworm may be better visualized if the dissected specimen is covered with a thin layer of water.

Various annelids should be available on demonstration. Oligochaetes such as *Lumbricus* lack the segmental appendages (parapodia) typical of polychaetes, are hermaphroditic, and are a less diverse group. Several polychaetes, including *Nereis* (live if possible) should be available on demonstration. Leeches should also be on demonstration. Live leeches are excellent for observing the actions of longitudinal and circular muscle layers.

VI. ORDERING INFORMATION

Grantia (preserved)—Carolina Biological, # P20; Ward's, # 68W0100
Grantia (cross section, prepared slide)—Carolina Biological, # Z500; Ward's, # 92W0550
Spongilla (live)—Carolina Biological, # L54G; Ward's, # 87W2600
microscope slides, box (75 × 25 mm)—Carolina Biological, # 63-1920; Ward's, # 14W3500
coverslips, box (22 × 22 mm)—Carolina Biological, # 63-3075; Ward's, # 14W3251
Chalina (Demospongiae with spongin)—Carolina Biological, # P40; Ward's, # 68W0129
commercial sponge set (dry)—Carolina Biological, # P49; Ward's, # 68W0128
Hydra (live)—Carolina Biological, # L55 (green Hydra, # L60); Ward's, # 87W2020
1½" culture dish—Carolina Biological, # 74-0996
Daphnia (culture kit)—Carolina Biological, # L563; OR
brine shrimp combo—Carolina Biological, # L609E
Obelia (whole mount, prepared slide)—Carolina Biological, # Z690; Ward's, # 92W0730
Physalia (Portuguese man-of-war, preserved)—Carolina Biological, # P115C; Ward's, # 68W0259
Aurelia (plastic mount)—Carolina Biological, # POM825; Ward's, # 55W0350
Metridium (preserved)—Carolina Biological, # P160C; Ward's, # 68W0412
corals—Carolina Biological, # P175A (deerhorn coral); # P180A (brain coral); # 26-1350 (sea fan); # 26-1330 (pipe organ coral); or # 26-1312 (cluster coral); Ward's, # 68W0432
Pleurobranchia (comb jelly, preserved)—Carolina Biological, # P 200F; Ward's, # 68W0442
Dugesia (flatworm, live)—Carolina Biological, # L210; Ward's, # 87W2500
Petri dish, 100 × 15 mm—Carolina Biological, # 74-1158; Ward's, # 17W0700
dropper—Carolina Biological, # 73-6904; Ward's, # 17W0230

carmine—Carolina Biological, # 85-3070; Ward's, # 38W7320
Clonorchis (human liver fluke, whole mount, preserved slide)—Carolina Biological, # PS 1210; Ward's, # 92W4900
Taenia (tapeworm, preserved)—Carolina Biological, # P245C; Ward's, # 68W0551
Taenia (tapeworm, prepared slide)—Carolina Biological, # PS 1810; Ward's, # 92W5400
Cerebratulus (ribbon worm, preserved)—
Turbatrix (vinegar eel, live)—Carolina Biological, # L258; Ward's, # 87W2900
neutral red stain—Carolina Biological, # 87-6830; Ward's, # 38W8385
garden soil test kit—Carolina Biological, # 66-5410; Ward's, # 29W7858
neutral red—Carolina Biological, # K3-87-6830; Ward's, # 38W8385
Trichinella (teased muscle, prepared slide)—Carolina Biological, # Z1040; Ward's, # 92W5766
Trichinella (whole mount, prepared slide)—Carolina Biological, # Z1050; Ward's, # 92W5760
Lumbricus (preserved)— Carolina Biological , # P403C or P405C; Ward's, # 68W2200
dissecting pans, 13 × 9"—Carolina Biological, # 62-9002 (wax); Ward's, # 14W8500
dissecting pins (2")—Carolina Biological, # 62-9122; Ward's, # 14W0200
dissecting needles—Carolina Biological, # 62-7200; Ward's, # 14W0650
fine dissecting scissors—Carolina Biological, # 62- 1820; Ward's, # 14W0600
ring stand, medium—Carolina Biological, # 70-7192; Ward's, # 15W0692
ring, 3" diameter—Carolina Biological, # 70-7273; Ward's, # 15W0707
funnel, 100 mm diameter—Carolina Biological, # 73-4062; Ward's, # 17W4333
beaker, 250 ml— Fisher, # 02-540K (Pyrex) or # 02-539K (Kimax); Carolina Biological, # 72-1223 (Corning, student grade); Ward's, # 17W4040
cheesecloth—Fisher, # 06-665-18 (70 yard bolt); Carolina Biological, # 71-2690 (5 yard package); Ward's, # 15W0015
pinch clamp—Carolina Biological, # 70-7420; Ward's, # 15W0650
Lumbricus (cross section, preserved slide)—Carolina Biological, # Z1250; Ward's, # 92W1840
Lumbricus (live)—Carolina Biological, # L400; Ward's, # 87W4661
worms:
  leech (preserved), Carolina Biological, # P416C; Ward's, # 68W2402
  Nereis (preserved), Carolina Biological, # P412C; Ward's, # 68W2302
  sabellid plume worm (preserved), Carolina Biological, # P431C
  Aphrodite (sea mouse, preserved), Carolina Biological, # P432C; Ward's, # 68W2308
  Arenicola (lug worm, preserved), Carolina Biological, # P434C; Ward's, # 68W2208
  Chaetopterus (preserved), Carolina Biological, #P436C; Ward's, # 68W2230

LABORATORY 26 Diversity—Mollusks, Arthropods, and Echinoderms

I. FOREWORD

This laboratory continues a survey of animal diversity, including Laboratories 21 and 23. In Exercise B. we also introduce students to biochemical methods for studying taxonomy. If live material is available, it can be substituted for preserved specimens. If a marine aquarium is available, point out and discuss feeding in clams and the use of tube feet in locomotion in echinoderms, etc. Supplement with other available specimens as time permits.

Shells and preserved specimens used in this laboratory are not dissected. There will be some wear and tear in their use, but if students are careful, organisms can be reused for a number of laboratories (over several years).

II. TIME REQUIREMENTS

Exercise A—Phylum Mollusca
- Part 1—Chitons (5 minutes)
- Part 2—Bivalves (20 minutes)
- Extending Your Investigation: Filter-feeding (20 minutes)
- Part 3—Gastropods (20 minutes)
- Part 4—Cephalopods (15 minutes)
- Part 5—The Scaphopoda (5 minutes)

Exercise B—Phylum Arthropods
- Part 1—Crustacea: External Morphology of the Crayfish (30 minutes)
- Part 2—Arachnida: External Morphology of a Spider (20 minutes)
- Part 3—Chelicerata: Biochemical Taxonomy of the Horseshoe Crab (20 minutes, optional)
- Part 4—Insecta: External Morphology of the Grasshopper (20 minutes)
- Part 5—Diplopooda and Chilopoda (15 minutes)
- Part 6—Insect Mobility (20 minutes)

Exercise C—Protostomes: Minor Groups (5 minutes)

Exercise D—Phylum Echinodermata (30 minutes)

III. STUDENT MATERIALS AND EQUIPMENT

	Per Student	Per Pair (2)	Per Group (4)	Per Class (24)
Exercise A (Part 1)				
Katharina (chiton, preserved)			1	6
Exercise A (Part 2)				
clam, oyster, or mussel shell (dry)	1			24
venus clam (preserved)			1	6
pelecypod shells (dry)—demonstration				1

	Per Student	Per Pair (2)	Per Group (4)	Per Class (24)
Extending Your Investigation				
freshwater mussels, live **(1)** OR			1	6
clams, live **(1)**			1	6
Exercise A (Part 3)				
snail shell (dry)			1	6
pulmonate snail (live)			1	6
gastropod shells (dry)—demonstration				
Exercise A (Part 4)				
Loligo (squid, preserved)			1	6
cuttlebone or pen (dry)				1
octopus (preserved)				1
Nautilus shell (whole or split)			1	
Exercise A (Part 5)				
scaphopod shells (dry)—demonstration				
Exercise B (Part 1)				
crayfish (preserved)	1			24
dissecting pan with wax	1			24
Exercise B (Part 2)				
spider (preserved)	1			24
cotton pad	1			24
Petri dish (100 mm)	1			24
forceps			1	6
dissecting microscope	1			24
Exercise B (Part 3) (Optional)				
test tubes	4			96
lactic acid assay mixture, tube (3 ml) **(2)**	1			24
Spectronic 20 (mid-range phototube)			1	6
DPIP, tube (1 ml) **(3)**	1			24
Parafilm (small squares)	4			96
spider supernatant **(4)**	1			24
crab supernatant **(5)**	1			24
Limulus supernatant **(6)**	1			24
Exercise B (Part 4)				
Romalia (grasshopper, preserved)	1			24
dissecting microscope	1			24
Exercise B (Part 5)				
Spirobolus (millipede, preserved)		1	6	
Scolopendra (centipede, preserved)		1	6	
Exercise B (Part 6)				
spiracle (whole mount, preserved slide)	1			24

	Per Student	Per Pair (2)	Per Group (4)	Per Class (24)
Exercise B (Part 6—cont.)				
tracheal system (whole mount, preserved slide)	1			24
compound microscope	1			24
insect collection **(7)**				1
Exercise C				
water bears (live, inactive)	1			24
other protostomes (demonstration)				
Exercise D				
sea lily (preserved or dry)				1
starfish (preserved or dry)				1
brittle star (pre served)				1
sea cucumber (preserved)				1
sea urchin (preserved)		1		12
sea urchin shell		1		12
sand dollar shell		1		12
dissecting pan	1			24

IV. PREPARATION OF MATERIALS AND SOLUTIONS

(1) freshwater mussels, live

Live clams or oysters can be obtained from a local grocery store or freshwater mussels can be obtained from Carolina Biological Supply Company (see Ordering Information). To open bivalve mollusks, insert a scalpel into the shell next to the hinge and cut the adductor muscles. Be careful not to insert the scalpel too far because you do not want to damage the visceral mass. Carbon particles can be prepared ahead of time. Black chalk dust, charcoal powder from a charcoal rubbing, or even small fibers can be used to watch filter feeding and the movement of water over the gill surface.

(2) lactic acid assay mixture

Prepare the following three solutions:

a. 0.1 M potassium phosphate, monobasic

Dissolve 2.72 g NaH_2PO_4 in 200 ml of distilled water. Adjust to pH 8.9 by adding 4M NaOH (160 g NaOH/liter), drop by drop.

b. lactic acid, 1:10 dilution

Add 4 ml of lactic acid to 36 ml of distilled water.

c. 0.01 M NAD (nicotinamide adenine dinucleotide)

Add 0.07 g NAD to 8 ml distilled water. Adjust to pH 7.0.

Mix solutions A, B. and C and store in a brown bottle until use.

(3) DPIP

Dissolve the following in 420 ml of distilled water:

DPIP (dichlorophenol indophenol; also called DCIP)	0.06 g
PMS (phenazine methosulphate)	0.02 g

Store in a brown bottle and refrigerate. Make fresh (up to one week ahead).

(4) (5) and (5) spider, crab, and Limulus supernatant

Prepare homogenization buffer for making supernatants as follows:

sucrose	8.56 g
Tris	0.12 g
EDTA (disodium salt)	0.04 g

Add distilled water to 100 ml and adjust to pH 7.0 using concentrated HC1, drop by drop.

For spider supernatant, collect 30-40 large spiders. Freeze to kill and remove abdomens. Determine the mass of the spider tissue and homogenize in 4 times the number of grams of homogenate buffer in ml. Use a glass homogenizer or blender and homogenize for 1 minute—no longer! Centrifuge in a clinical centrifuge, top speed, for 5 minutes. Store the supernatant in the refrigerator.

For crab and *Limulus* supernatant, follow the spider directions, using thorax and walking leg material only. Use the same weights for *Limulus* and crab as used for spider so that all supernatants contain equal concentrations of tissue (usually one walking leg from the crab or *Limulus* will be sufficient to equal the spider tissue).

For the remainder of the exercises in this laboratory, you will be using preserved specimens. Save the preservative from packed specimens to keep them moist. Additional Caro-Safe preservative can be purchased and may be used on both formalin- and alcohol-preserved specimens. The use of this solution reflects the trend away from formaldehyde for specimen storage. However, original preservation of material requires either 10% formaldehyde or 70% alcohol. To mix 10% formalin, add 100 ml of formaldehyde to 900 ml of distilled water and mix—avoid breathing the fumes. To prepare 70% alcohol, add distilled water to 700 ml of 95% ethanol to make the total amount 950 ml.

Cover exposed preserved specimens with moist paper towels (water or 10% glycerine solution) between successive laboratories. Return specimens to their jars and immerse them in preservative overnight.

Ask students to use their fingers or blunt probes in pointing at shells used in this laboratory. Pencils, ball point pens, and dissecting needles should not be used for this purpose.

(7) representative insect collection

Have a variety of insects (dried, pinned) available for inspection. Include apterous forms (silverfish, earwig, flea, louse); forms with one pair of wings (beetles—first pair of wings modified as elytra, flies, mosquitoes); and forms with two pairs of wings (dragonflies, butterflies, hymenopterans—bees, wasps, ants, grasshopper). You might also wish to contrast hemimetabolous forms like the grasshopper, cockroach, etc., with holometabolous forms (complete metamorphosis) such as a fly, moth, etc. If you have a life history demonstration, you could use it with this exercise. If you have live insects (cockroaches, a demonstration honeybee hive, etc.), work them in too!

V. PREPARATION SUGGESTIONS

Exercise A (Parts 1 and 2)

If a marine aquarium is present in the laboratory, small chitons are excellent for cleaning the algae off the sides. Students enjoy being able to see live specimens from all phyla (see Preparation Suggestions, Exercise D.

An assortment of mollusk shells should be kept on hand. These will give students some idea of the diversity of shell types. If live clams or mussels are available, have groups of students observe their anatomy "on the half-shell." Insert a one-piece scalpel into the shell and cut both adductor muscles against one of the shells leaving the gills and visceral mass intact. Place the dissected clam in a finger bowl covered with water (pond water for a freshwater clam, seawater for a marine specimen) and place it on demonstration under a dissecting microscope.

Extending Your Investigation

You may want to have individuals participate in the *Extending Your Investigation* activities with the clams or conduct the activities as a demonstration.

The Carolina Biological video on *The Anatomy of the Freshwater Mussel* (# 49-2364V) is an excellent supplement to this study.

Exercise A (Parts 3 and 4)

A selection of gastropod shells can be used to show differences in shell coiling patterns. Show right-hand and left-hand examples. Have students differentiate between a whelk and a conch shell. Dried whelk egg cases can be found on nearby beaches. Split one of the casings open to show the students the tiny whelks inside.

Display a chambered nautilus shell cut longitudinally. Have students understand why this is an example of a cephalopod and not a gastropod. Cuttlebones can be purchased in most pet stores or pet sections in larger food stores. The "pen" of a squid can be dissected out and dried for demonstration.

Exercise B (Parts 1 and 2)

Dissections can be done as demonstrations. You can supplement with an excellent videotape (*Dissection of the Crayfish,* Carolina Biological, # K3-49-2403V). For spider dissections, have students collect their own spiders (use the freezer to euthanize specimens) or order preserved specimens (see Ordering Information).

Exercise B (Part 3)

This exercise is fairly easy for students to do if the supernatants and lactic assay mixture have been made ahead of time. Spiders can be ordered from Carolina Biological Supply Company or students can collect them for you. Land hermit crabs or fiddler crabs can also be ordered from Carolina biological Supply Company; *Limulus* can be ordered from Wards Biological Supply Company. Freeze these animals to kill them or simply remove a walking leg from the crabs or *Limulus* if they are kept live in aquaria.

The lactic acid mixture (assay mixture) is used to zero the spectrophotometer. Spider and *Limulus* homogenates will decolorize to light blue (absorbance or OD = 0.2-0.6), while crab homogenate will remain medium blue (also 0.25-0.3) with the control a dark blue (absorbance or OD = 0.45-0.5), read at 600 nm.

The horseshoe crab, *Limulus, is* classified with the spiders, Class Arachnida, rather than the crabs, Class Crustacea. Why? This exercise will help students answer this question by looking at one example of biochemical evidence. The enzyme, lactic acid dehydrogenase (LDH) is found in the muscle tissue of all organisms and is used to interconvert lactic acid and pyruvic acid during cellular respiration. Biochemically, LDH may differ slightly from one organism to another—the sequence of amino acids in the protein may differ. The more similar the sequences, the more similar the DNA coding the protein and, thus, the more likely that two organisms share a common origin. (See Laboratory 21.)

As LDH converts pyruvic acid + NADH + H^+ to lactic acid + NAD^+, the electrons carried by NADH + H^+ are accepted by pyruvic acid. For this laboratory, we will study the reverse reaction in which hydrogen electrons are removed from lactic acid as it is converted to pyruvic acid. The electrons are picked up by NAD^+ and then are transferred to EMS

(phenazine metosulphate), and finally to DPIP (dichlorophenol indophenol) indicator in the reaction mixture. Originally blue, the DPIP indicator in the reaction mixture will turn colorless. The decoloration of DPIP can be followed by using the spectrophotometer (similar to our experiment on cellular respiration in lima bean seeds, Laboratory 11, Exercise B) or, qualitatively, without the spectrophotometer by simply observing color changes.
This structure (amino acid sequence) can affect the activity of the enzyme. Thus, differences or similarities in the rate of decoloration and even color (qualitatively) suggest differences or similarities in origin and relatedness.

Exercise B (Part 4)

Supplement the dissection of a grasshopper with a videotape from Carolina Biological Supply Company, #K3-49-2402V.

Exercise B (Part 5)

Collect examples of centipedes and millipedes from the yard or woods. Have students add to the collection—preserve in 70% alcohol. Warn students to be careful when collecting centipedes. Some are fairly toxic and students may react to having these organisms on their skin.

Exercise B (Part 6)

Have an enterprising student capture a live adult cockroach. Pin the body firmly to the wax in a dissecting pan and spread the wings on one side of the body, pinning them to hold them extended. Place the insect under a dissecting microscope focused on the proximal veins of the second (metathoracic) wing illuminated with a bright dissecting light. Have students watch blood flow through the wing. Have them note respiratory movements of the abdomen.

Exercise C

Use preserved specimens of representatives from the minor phyla. Be sure to put "lamp shells" on demonstration for comparison with clams and other bivalves. Use glass depression slides to observe Tardigrades.

Exercise D

If a marine aquarium is available, have students watch the actions of tube feet as a starfish moves. If a specimen attaches to the side of the tank, feed it a small piece of raw shrimp by placing the food between two of its arms. Students will be able to watch the feeding behavior, including the extrusion of the stomach to digest the shrimp.

A 30 gallon aquarium can be set up with an under-gravel filter or a filter-pump combination. Make sure to specify that gravel is for a marine aquarium when you purchase your supplies—this gravel is composed of CaCO3 and contains pieces of shells and corals. Instant Ocean can be used for making salt water. Start with a single fish while setting up the tank. Several additives can be purchased to "seed" the tank. Make sure to purchase a kit for testing water quality—pH and ammonia levels must be checked regularly. You must also test specific gravity. The aquarium will give students an excellent opportunity to study nutrient cycling in an ecosystem—especially the nitrogen cycle. You can order an assortment of marine animals from Gulf Marine Specimen Company. These are shipped overnight by Parcel Post. Be sure not to put too many organisms in a single tank (general rule—"an inch of fish per two gallons").

VI. ORDERING INFORMATION

Caro-Safe (preservative)—Carolina Biological, # 85-3341; Ward's, # 39W1693
Katharina (chiton, preserved)—Carolina Biological, #P460C; Ward's, # 68W7012
venus clam *(Mercenaria,* preserved)—Carolina Biological, # P5 l 3C; Ward's, # 68W7132
clam shell—beach, local seafood store, etc.
clams, live—local grocery store
freshwater mussels—Carolina Biological, # L500; Ward's, # 87W4420
video, *the Anatomy of the Freshwater Mussel*—Carolina Biological, # 49-2364-V
gastropod shell collection—Carolina Biological, # 26-1791
pulmonate snails (live)—Carolina Biological, # L480; Ward's, # 87W4312
Loligo (squid, preserved)—Carolina Biological, # P526D; Ward's, # 68W7462
cuttlebone or pen—pet shop, dissection
octopus (preserved)—Carolina Biological, # P539C; Ward's, # 68W7481
Nautilus shell—Carolina Biological, # 26-1854 (split) or # 26-1856 (whole); Ward's, # 53W2140
mollusk shell collection—Carolina Biological, # 26-1788 or # 26-1790
crayfish (preserved)—Carolina Biological , # P5 90D; Ward's, # 68W2802
dissecting pan, 13 × 9"—Carolina Biological, # 62-9002 (wax); Ward's, # 14W8500
dissecting pins (2")—Carolina Biological, # 62-9122; Ward's, # 14W0200
spider (preserved)—Carolina Biological, # P659; Ward's, # 68W3052
Petri dish, 100 × 15 mm—Fisher, # 08-747C (Pyrex) or 08-746C (Kimax); Carolina Biological, # 74-1158; Ward's, # 17W0700
spiders (live)—Carolina Biological, # L658A
hermit crab (live)—Carolina Biological, # L600; Ward's, # 87W7570
fiddler crab (live)—Carolina Biological, # L607; Ward's, # 87W7574
Limulus (live)—Ward's, # 87W7530
test tubes, 16 × 125 mm—Fisher, # 14-925H; Carolina Biological, # 73-1418; Ward's, # 17W0630
test tube rack—Carolina Biological, # 73-1896
potassium phosphate (monobasic) (KH_2PO_4)—Fisher, # P285-500; Carolina Biological, # 88-4250; Ward's, # 37W4841
lactic acid—Fisher, # A162-500; Sigma, # L 1250; Carolina Biological, # 87-1690; Ward's, # 39W2095
Parafilm—.Fisher, # 13-274-12 (dispenser, # 13-374-18); Carolina Biological, # 21-5600 (dispenser, # 21-5602); Ward's, # 15W1940
Trizma base—Sigma, # T 1503; 37W50935
phenazine methosulphate (PMS)—Sigma, # P9625; Fisher, # 1360155
DPIP—Sigma, # D 1878; Fisher, # S286-5
PMS (phenazine methosulphate)—
EDTA, disodium salt—Sigma, # E4884; Fisher, # S311-100; Carolina Biological, # 21-7430; Ward's, # 39W1255
lactic acid (85% syrup)—Sigma, # L 1250; Fisher, # A159-500; Carolina Biological, # 87-1690; Ward's, # 39W2095
NAD—Sigma, # N 3014; Fisher, # 1361252
sodium hydroxide (NaOH)—Fisher, # S318-100; Carolina Biological, # 88-9470; Ward's, # 37W5560

sodium phosphate (dibasic, heptahydrate) ($Na_2HPO_4 \cdot 7H_20$)—Fisher, # S471-3; Carolina Biological, # 89-1370; Ward's, # 37W5655

sucrose—Fisher, # 55-500; Carolina Biological, #89-2870; Ward's, # 39W3182

Romalia (grasshopper, preserved)—Carolina Biological, # P727C; Ward's, # 68W4052

Spirobolus (millipede, preserved)—Carolina Biological, # P644; Ward's, # 68W3122

Scolopendra (centipede, preserved)—Carolina Biological, #P643; Ward's, # 68W3112

spiracle (whole mount, prepared slide)—Carolina Biological, # Z1860; Ward's, # 92W2384

tracheal system (whole mount, prepared slide)—Carolina Biological, # Z1850; Ward's, # 92W2385

water bears—Carolina Biological, # L 340; Ward's, # 87W3300

Peripatus (onychophoran), plastic mount—Carolina Biological, # POM-1560; Ward's, # 55W2500

bryozoans (moss animals), preserved—Carolina Biological, # P310C; Ward's, # 68W1208

brachiopod (fossil)—Carolina Biological, # GEO6850; Ward's, # 52W6840

fossil crinoid (sea lily)—Carolina Biological, # GEO6684; Ward's, # 50W0930

starfish (preserved)—Carolina Biological, # P331S; Ward's, # 68W7692

brittlestar (preserved)—Carolina Biological, # P351; Ward's, # 68W7672

sea urchin (preserved)—Carolina Biological, # P359C; Ward's, # 68W7702

sea urchin (test)—Carolina Biological, # P363

sand dollar (preserved)—Carolina Biological, # P377; Ward's, # 68W7732

sand dollar (shell)—Carolina Biological, # 26-1552

sea cucumber (preserved)—Carolina Biological, # P380C; Ward's, # 68W7802

LABORATORY 27 Diversity—Phylum Chordata

I. FOREWORD

This laboratory completes our survey of multicellular animals. Specimens are not dissected and should last for many laboratories and over several years if they are handled carefully and stored properly.

Students should be encouraged to pay careful attention to the Overview introducing this Laboratory—the classification scheme and phylogeny presented will be reinforced by comparative study of representative vertebrates in Laboratories 33 through 36.

II. TIME REQUIREMENTS

Exercise A—Hemichordates (20 minutes)
Exercise B—Tunicates (20 minutes)
Exercise C—Sea Lancelets (20 minutes)
Exercise D—Vertebrates—Fishes (30 minutes)
Exercise E—Amphibians and Reptiles (30 minutes)
Exercise F—The Avian and Mammalian Skeletal System (45 minutes)

III. STUDENT MATERIALS AND EQUIPMENT

	Per Student	Per Pair (2)	Per Group (4)	Per Class (24)
Exercise A				
Balanoglossus (acorn worm, plastic mount)			1	6
Balanoglossus (acorn worm, median longitudinal section, prepared slide)		1		12
compound microscope	1			24
dissecting microscope	1			24
Exercise B				
tunicate tadpole larva (whole mount, prepared slide)			1	6
Molgula (sea squirt, preserved) or *Corella* (glass tunicate, preserved)			1	6
compound microscope	1			24
dissecting microscope	1			24
Exercise C				
Branchiostoma (lancelet, preserved) or (whole mount, prepared slide)		1		12
Branchiostoma (cross section, pharyngeal region, prepared slide)		1		12
compound microscope	1			24
dissecting microscope	1			24
Exercise D				
Entosphenus (lamprey, preserved)			1	6
Squalus (shark, preserved)			1	6

	Per Student	Per Pair (2)	Per Group (4)	Per Class (24)
Exercise D—continued				
perch (preserved)			1	6
blunt probe			1	6
hand lens			1	6
Exercise E				
frog (preserved)				1
Necturus (preserved)				1
garter snake (preserved)				1
alligator (display mount)				1
turtle (preserved)				1
frog skeleton (optional)				1
turtle skeleton (optional)				1
frog eggs (cluster of 12)				1
turtle eggs (group of 12)				1
dissecting microscope	1			24
Exercise F				
pigeon skeleton			1	6
cat skeleton			1	6
OR human skeleton				1

IV. PREPARATION OF MATERIALS AND SOLUTIONS

None.

V. PREPARATION SUGGESTIONS

Exercise A

Preserved acorn worms or plastic mounts can be used.

Exercise B

Cut the tunic away from one side of a preserved tunicate to expose internal organs.

Exercise C

If you have a saltwater aquarium in the lab, lancets will do well. Students will be able to observe their undulating swimming pattern.

Exercise D

Preserved specimens can be used. Some are available in glass tubes. Others are available as plastic mounts. Plastic mounts or preserved displays of the cartilaginous skeleton showing the chondrocranium and pharyngeal arches are also available.

Exercise E

While frog and turtle skeletons would be helpful, they are not essential.

Exercise F

If a human skeleton is purchased for this exercise, it is worthwhile to obtain one hanging in a cabinet. Be sure students touch skeletons only with their fingers and only when necessary—warn them not to point with probes, dissecting needles, pencils or pens!

VI. ORDERING INFORMATION

Balanoglossus (acorn worm, plastic mount)—Carolina Biological, # POM 2690
acorn worm (median longitudinal section)—Carolina Biological, # Z 2625
tunicate tadpole larva (whole mount, prepared slide)—Ward's, # 92W8400, or Carolina Biological, # E958
Molgula (sea squirt, preserved)—Carolina Biological, # P1226C; Ward's, # 69W0032
OR *Corella* (glass tunicate, preserved)—Carolina Biological, # P1224C
Branchiostoma (preserved)—Carolina Biological, # P1250C; Ward's, # 69W0052
OR *Branchiostoma* (whole mount, prepared slide)—Carolina Biological, # 2706
Branchiostoma (cross section, pharyngeal region, prepared slide)—Carolina Biological, # Z2720
Entosphenus (lamprey, preserved)—Carolina Biological, # P1263D; Ward's, # 69W1002
Squalus (shark, preserved)—Carolina Biological, # P1305S; Ward's, # 69W1111
perch (teleost, preserved)—Carolina Biological, # P1410AD; Ward's, # 69W1542
frog (preserved)—Carolina Biological, # P1496D; Ward's, # 69W2102
Necturus (preserved)—Carolina Biological, # P1429D ; Ward's, # 69W2412
garter snake (preserved)—Carolina Biological, # P1639AD; Ward's, # 69W3372
alligator (display mount)—Carolina Biological, # 26-3760
turtle (preserved)—Carolina Biological, # P1608D; Ward's, # 69W3302
frog skeleton (optional)—Carolina Biological, # 24-3710; Ward's, # 65W2210
turtle skeleton (optional)—Carolina Biological, # 24-4120; Ward's, # 65W3330
frog eggs (fertilized)—Carolina Biological, # P1530AF; Ward's, # 87W8205
turtle eggs—Carolina Biological, # P1609C
pigeon skeleton—Carolina Biological, # 24-5120; Ward's, # 75W4470
cat skeleton—Carolina Biological, # 24-5835; Ward's, # 65W5205
OR human skeleton—Carolina Biological, # 24-6860; Ward's, # 82W3000

Notes:

LABORATORY 28 Plant Anatomy—Roots, Stems, and Leaves

I. FOREWORD

This laboratory is designed to introduce students to leaf, stem, and root anatomy. Students work individually but slides can be shared among pairs or groups to cut down on cost. Exercise B can also be split into several shorter activities if desired. Exercise C is optional. Students can make their own thin sections of a variety of materials and can use their knowledge to identify monocot and dicot structures as well as cell types. Students find this to be much more fun than using prepared slides.

II. TIME REQUIREMENTS

Exercise A—Plant Tissues (45 minutes)
Exercise B—The Monocot and Dicot Angiosperm Body Plan (120 minutes)
  Part 1—Structure of Dicot and Monocot Root
  Part 2—Structure of Dicot and Monocot Stem
  Extending Your Investigation: Is It a Monocot or a Dicot?
  Part 3—Leaf Structure
Exercise C—The Pine Leaf (15 minutes, optional)

III. STUDENT MATERIALS AND EQUIPMENT

	Per Student	Per Pair (2)	Per Group (4)	Per Class (24)
Exercise A				
lettuce (head)				1
celery (stalk)				1
pear				1
microscope slides, box (75 × 25 mm)			1	6
coverslips, box (22 × 22 mm)			1	6
compound microscope	1			24
Exercise B				
Ranunculus (buttercup, dicot) root (prepared slide)		1		12
Zea mays (corn, monocot) root (prepared slide)	1			24
Salix (willow) branch root (prepared slide)		1		12
Mendicago saliva (alfalfa, dicot) stem (prepared slide)		1		12
Zea mays (monocot) stem (prepared slide)		1		12
Coleus plant (live)				2
single edge razor blade	1			24
nut-and-bolt "microtome" **(1)**	1			24
melted paraffin **(2)**				1
50% ethanol, bottle **(3)**			1	6
toluidine blue O stain, dropping bottle **(4)**			1	6

	Per Student	Per Pair (2)	Per Group (4)	Per Class (24)
Exercise B—continued)				
distilled water, bottle			1	6
glycerine, bottle			1	6
Petri dishes (100 mm)	2			48
Ligustrum (privet, dicot) leaf (prepared slide)			1	12
Zea mays (monocot) leaf (prepared slide)			1	12
compound microscope	1			24
Extending Your Investigation				
use materials from Exercise B (above)				
Exercise C (optional)				
pine leaf, cross section (prepared slide)	1			24

IV. PREPARATION OF MATERIALS AND SOLUTIONS

(1) nut-and-bolt "microtome"

Buy large 2" bolts (approximately ¼" diameter) with fitting nuts in a local hardware store. These may be used from year to year.

(2) paraffin

Melt paraffin used for canning (purchased in a grocery store) in a 500 ml beaker placed in a water bath at about 50° C. Provide an insulated glove or large tongs for handling the paraffin. Warming the "microtome" before pouring will keep the paraffin from solidifying too fast to orient the plant tissues for cutting. Have students pour the paraffin over a covered lab bench.

(3) 50% ethanol

Add 450 ml distilled water to 500 ml 95% ethanol and mix.

(4) toluidine blue O

Mix phosphate buffer as follows:

(A) 4.24 g Na_2HPO_4. Add distilled water to 200 ml.

(B) 4.13 g NaH_2PO_4. Add distilled water to 200 ml.

Mix 51 ml (A) to 49 ml (B), pH 6.8, and add 0.05 g toluidine blue O. Mix 1:1 with distilled water for use.

V. PREPARATION SUGGESTIONS

Exercise A

Make sure that plant tissue sections are thin. Have instructor-prepared slides available for this tissue in the event that student slices are not acceptable. You can make your own preparations. Mount with glycerin and seal with colorless nail polish or paraffin. A tin can of paraffin or candle wax can be used. Melt it until a small pool of liquid wax is present. Bend a dissecting needle so that 1 inch of the needle is at a right angle to the handle. Submerge the needle in the wax and lay the needle along the side of the cover slip.

Exercise B

Students should be encouraged to continually refer to Diagram 28B-1 as they try to summarize the monocot and dicot comparisons of root, stem, and leaf anatomy.

Other monocot and dicot roots and stem slides can be used if on hand, but students should be warned that the diagrams might not be an exact match (but major cell types and regions can still be identified by comparison.) *Smilax* or greenbriar (a monocot) root is often used instead of the *Zea mays* or corn root. Likewise, *Helianthus* or sunflower (a dicot) stem is often used in place of the *Mendicago sativa* or alfafa root.

Toluidine blue O staining of living plant tissues is fun for students and results are usually colorful. Make sure you use toluidine blue O, not just toluidine blue. The former is metachromatic (and will turn a variety of colors but the latter is not). Other plants can be used in addition to *Coleus*. It is effective to have students collect weeds from the roadside and to make toluidine blue stained sections of the stems or roots of the plants to try to discern whether the plants are monocots or dicots. This gives students a chance to apply what they have learned about plant anatomy in a "living unknowns" laboratory.

Toluidine blue O staining is also used in Extending Your Investigation, a part of this exercise. This gives students a chance to apply what they have learned about plant anatomy to study new or unknown plants.

Exercise C (optional)

A cross section of a pine needle can be used to provide an interesting comparison with angiosperm leaves.

VI. ORDERING INFORMATION

microscope slides, 75 × 25 mm—Carolina Biological, #63-200; Ward's, # 14W3500
coverslips, 22 × 22—Carolina Biological, # 63-3015; Ward's, # 14W3521
Ranunculus (buttercup, dicot) root (prepared slide)—Carolina Biological, # B520; Ward's, # 91W8139
Zea mays (corn, monocot) root (prepared slide—Carolina Biological, # 97-8050; Ward's, # 91W7447
Salix branch root (prepared slide)—Triarch, #1312-C; Carolina Biological, # B522; Ward's, # 91W8202
Mendicago sativa (alfafa, dicot) stem (prepared slide)—Carolina Biological, # 97-8220; Ward's, # 91W8010
Zea mays (monocot) stem (prepared slide)—Carolina Biological, # B571; Ward's, # 91W7452
Coleus plant, live—buy locally; Carolina Biological, # 15-7310; Ward's, # 86W6825
single edge razor blades—buy locally; Carolina Biological # 62-6930; Ward's, # 14W4172
paraffin—buy locally; Carolina Biological, # 87-9090; Ward's, # 39W2860
toluidine blue O—Carolina Biological, # 89-6638; Ward's, # 38W8795
dropping bottle, 30 ml—Carolina Biological, # 71-6525; Ward's, # 17W6011
glycerine—Carolina Biological, # 86-5560; Ward's, # 39W1435
Petri dish, 100 × 15 mm—Carolina Biological, # 74-1158; Ward's, # 17W0700
Ligustrum (dicot) leaf (prepared slide)—Carolina Biological, # B598D; Ward's, # 91W7940
Zea mays (monocot) leaf (prepared slide)—Carolina Biological, # B630; Ward's, # 91W7454
pine leaf (cross section, prepared slide)—Carolina Biological, # 97-7219; Ward's, # 91W6501

Notes:

LABORATORY 29 Angiosperm Development—Fruit, Seeds, Meristems, and Secondary Growth

I. FOREWORD

This laboratory is an extension of Laboratory 28 in which we studied vascular plants. Formation of fruits and development of the seedling are emphasized. Both primary growth and secondary growth are considered. The laboratory exercises can easily be done singly and could be combined with Laboratory 31 that examines the effects of hormones on plant development.

II. TIME REQUIREMENTS

Exercise A—Fruits (60 minutes)
Exercise B—Seed Structure
  Part 1—Examining the Dicot Bean Seed (20 minutes)
  Part 2—Examining the Monocot Corn Seed (20 minutes)
Exercise C—Found a Peanut (20 minutes)
Exercise D—Seedling Development
  Part 1—Comparing Germination in Beans, Peas, and Corn (30 minutes)
  Part 2—Observing the Germination and Development of Seeds (20 minute setup, 10 day experiment)
Exercise E—Studying the Stem Tip and Root Tip
  Part 1—Examining the Stem Tip (15 minutes)
  Part 2—Examining the Root Tip (15 minutes)
Exercise F—Secondary Growth of Angiosperms—The Woody Stem (*Tilia*) (20 minutes)
Exercise G—Plant Tissue Culture (45 minutes, preparation); 1-2 weeks, incubation)

III. STUDENT MATERIALS AND EQUIPMENT

	Per Student	Per Pair (2)	Per Group (4)	Per Class (24)
Exercise A				
assorted fruits **(1)**				1
Exercise B (Parts 1 and 2)				
Phaseolus vulgaris (lima bean) seeds **(2)**	1			24
Zea mays (sweet or field corn) seeds **(3)**	1			24
Exercise C				
peanuts, bag	1			24
Exercise D (Parts 1 and 2)				
germinated lima bean seedling **(4)**		1		12
germinated pea seedling **(5)**		1		12
germinated sweet or field corn seedling **(6)**		1		12
Lugol's solution, dropping bottle			1	6

	Per Student	Per Pair (2)	Per Group (4)	Per Class (24)
Exercise D (Parts 1 and 2)—continued				
plastic cup	1			24
blotter paper **(7)**	1			24
soaked seeds (pea, bean, corn) **(8)**	. 2			48
Petri dish half (100 mm)	1			24
Exercise E (Parts 1 and 2)				
Coleus plant (live)			1	6
Coleus stem tip (prepared slide)	1			24
cabbage			1	6
brussel sprout	1			24
germinated radish or rye seed **(9)**	1			24
compound microscope	1			24
Extending Your Investigation				
pea seedlings **(10)**	3			72
glass tubing (4 cm × 5 mm ID)	3			72
Sharpie marker	1			24
ruler (mm markings)	1			24
plastic cup	1			24
towel (moist)	1			24
Exercise F				
Tilia (basswood—prepared slide)	1			24
tree (cross section—prepared slide) **(11)**	1			24
compound microscope	1			24
Exercise G				
Erlenmeyer flask (250 ml)	1			24
70% ethanol (300 ml bottle)			1	6
95 % ethanol (beaker)		1		12
forceps (sterile)	1			24
Bunsen burner or alcohol lamp		1		12
distilled water (sterile in 100 ml glass bottle) **(12)**	1			24
Petri dish (sterile, disposable)	4			96
scalpel (sterile)	1			24
10% bleach or Clorox, bottle (500 ml) **(13)**			1	6
cauliflower (head)				2
dishwashing detergent (Joy) or Tween 20 (bottle)				1
medium A, Petri dish **(14)**	1			24
medium B. Petri dish **(15)**	1			24
Parafilm, strip	2			48
medium A, jar **(14)**	1			24
fluorescent light source				1
timer to control light				1
Styrofoam pots (2 inches) with potting soil	4			96

IV. PREPARATION OF MATERIALS AND SOLUTIONS

(1) fruits

An assortment of simple fruits, cut in half and wrapped in plastic wrap should be provided on a demonstration table. Types of fruits may differ according to locality and time of year. Some suggestions include:

peach (drupe)	squash or pumpkin (pepo)	nectarine (drupe)
cherry (drupe)	coconut (drupe)	plum (drupe)
apple (pome)	cucumber (pepo)	grapefruit (hesperidium)
pear (pome)	banana (berry)	orange (hesperidium)
avocado (drupe)	peanut (legume)	pea (legume)
tomato (berry)	milkweed (follicle)	bean (legume)
grape (berry)	sunflower (achene)	corn (grain)
pepper (berry)	maple (samara)	olive (drupe)
acorn (nut)		

(2) lima bean seeds

Soak *Phaseolus vulgaris* (garden or lima bean) seeds overnight in tap water. If they are soaked longer, place them in a refrigerator to retard mold growth. *Vicia faba,* the broad bean, is also excellent for dissection and can be substituted for lima beans. Due to their size, *Vicia faba* beans need to be soaked longer—at least 24 hours.

(3) corn seeds

Soak *Zea mays* seeds overnight in tap water to make dissection easier. Either sweet or field corn can be used.

(4) (5) and (6) germinated seeds

Seeds of lima bean, pea, and corn need to be germinated approximately 6 days before use. Small cups of soil (with holes punched in the bottom) or wooden flats can be used. Plant seeds approximately ¼ inch below the soil surface. Alternatively, seeds can be germinated by laying them on wet paper towels that are then covered with more wet paper towels and rolled up and covered with aluminum foil. Older plants, 8 to 12 days, should also be planted for comparison. If this lab is prepared during the winter and the room is cold, place seeds in a warm incubator (about room temperature—21 to 22° C).

(7) blotter paper

Blotter paper, purchased from a local stationery or office supply store, works best for this. Cut blotter paper to the proper size to fit closely around the inside of the plastic cup. Alternatively, have students supply straight-sided jars and cut their own. Wet paper towels can also be used if blotter paper is not available. The plastic cup is covered with the top or bottom half of a Petri dish.

(8) soaked seeds

Soak pea, bean, or corn seeds overnight in tap water.

(9) germinated radish or rye seeds

Germinate radish seeds for 2 days by wrapping them in wet paper towels. These can be made into rolls by laying the seeds on wet paper towels, covering them with a second layer of wet paper towels, and then rolling the towels up. Cover the roll with aluminum foil. Alternatively, for small batches, simply fold a wet paper towel around some seeds and place this in a Petri dish. Place a rubber band around the Petri dish.

To germinate rye seeds, simply float them on a Petri dish of tap water. Allow 2 days for germination before use.

Place seeds in a warm incubator (21-22° C) if the room is cold.

(10) pea seeds

Germinate pea seeds by wrapping them in wet paper towels. These can be made into rolls (see **(9)** above). Place in an *open* plastic bag and put in a warm place or in an incubator (21-22° C).

(11) tree cross section

Cross sections of tree trunks can be obtained from pine, cedar, oak, or other hardwood. Be sure to label the display so that a distinction is made between gymnosperms and angiosperms.

(12) glass jar containing 100 ml sterile water

Use wide mouth jars. Large size baby food jars work fine. Pour 100 ml of distilled water into each jar. Cap loosely and autoclave for 15 minutes at 121° C and 15 pounds of pressure. Let cool and then tighten caps.

(13) 10% bleach solution

Add 50 ml bleach (Clorox) to 450 ml of distilled water. One 500 ml bottle will serve a group of four to five students.

(14) medium A (Petri dish)

1. Mark the 1000 ml (1 liter) level on a 2 liter flask (do this by pouring in 1000 ml of water, mark, and pour the water out.
2. Pour 800 ml of distilled water into the 2 liter flask
3. Add one package of Murashige and Skoog Minimal Organics Medium (see Ordering Information) while stirring. Use a small amount of distilled water to wash out the insides of the package and add it to the flask.
4. Add 30 g sucrose and stir until dissolved.
5. Adjust the pH to 5.8 using 1 N NaOH or 1 N HC1.
6. Add 8 g agar while heating gently. Stir continuously until clear (do not boil).
7. Pour 12-15 ml into enough baby food jars for your class—1 per student. Loosely cap and set aside to autoclave.
8. Cover the flask containing the remainder of the medium with aluminum foil.
9. Autoclave jars and flask of medium for 15 minutes at 121° C, 15 pounds pressure.
10. Cool and pour the medium into sterile Petri plates. Tighten caps on the baby food jars.

(15) medium B (Petri dish)

Prepare the following stock solutions of growth regulators:

indole acetic acid (IAA), 1 mg/ml

Dissolve 0.1 g (100 mg) of IAA in 1-2 ml of 1 N NaOH. Add distilled water to bring the final volume to 100 ml.

kinetin, 1 mg/ml

Dissolve 0.1 g (100 mg) of kinetin in 5 ml of 1 N NaOH. Add distilled water to bring the final volume to 100 ml.

1. Follow steps 1 through 4 given in the directions for preparing medium A (13).
2. Add 8 ml of IAA stock solution and 2.5 ml of kinetin stock solution to the flask of medium.
3. Adjust pH to 5.8 using 1 N NaOH or 1 N HCI.
4. Add 8 g of agar while boiling gently. Stir continuously until clear (do not boil).

5. Cover the flask with aluminum foil and autoclave for 15 minutes at 121° C and 15 pounds pressure.
6. Cool and pour into sterile, disposable Petri dishes.

V. PREPARATION SUGGESTIONS

Exercise A

A variety of fruits from the grocery store can be used for this exercise. See Preparation of Materials and Solutions **(1)**.

Exercise B

Vicia faba, if available, is excellent for these dissections.

Exercise C

Peanuts have been chosen since most students are familiar with these fruits. Dissection is easier if fresh peanuts are "roasted" at 275° F for approximately 40 minutes.

Exercise D

If a large aquarium is available, place approximately 6 inches of soil in it. Plant seeds (previously soaked overnight) against the glass so that students can observe the process of germination. These can be planted on different days to give stages of 4 days through 12 days and can be turned in different directions to show that position will not affect the direction of epicotyl and hypocotyl growth.

Exercise E

Onion root tip slides (used for mitosis in Laboratory 9) may be used as a supplement for this exercise.

Extending Your Investigation

Soak pea seeds overnight. Wrap in wet paper towels and place in warm dark area to germinate for 24-48 hours or until roots are 2-3 cm long. Alternatively purchase bean and pea sprouts from your local grocery store. You can use small cocktail straws instead of glass tubing to place roots into, but wet them before sliding roots in. The only drawback is that most straws are opaque and you must remove the pea seeds to make your second set of measurements. This can be done as a demonstration if students do not perform the experiment on their own

Exercise F

A cross section through the trunk of a palm tree provides an excellent comparison with a monocot whose trunk is made primarily of fibers formed during primary rather than secondary growth.

Exercise G[1]

Make sure that all containers and tools for this exercise are sterile (except for general washing flasks or bottles). All steps can be carried out using baby food jars, but sterile plastic Petri dishes are easier to handle. Likewise, medium A and B can be poured into tubes and slanted so that capped tubes can be used as with culturing bacteria. This makes it easier for some students to maintain aspetic conditions. Be sure the work area is clean, wiping it down with alcohol or bleach solution. You may even wish to hook up a UV light to shine on the

[1] Reference: Pietraface, William J. 1988. Plant regeneration. *Amer. Biol Teacher* 50:4.

work surface overnight. For a fluorescent light source, we use an old ceiling fixture, with 4 fluorescent tubes, attached to a wooden frame.

Be sure that Petri dish cultures, both medium A and B cultures, are covered around the edges with strips of Parafilm cut from the roll type dispensers.

If you wish to try tobacco leaf culture, plant tobacco seeds and let them grow in a sunny place until the leaves are fairly large. Wash the tobacco leaves (see Ordering Information) in 20% bleach (Clorox) for 10 minutes and then wash them 3 times in flasks or baby food jars of sterile distilled water. Punch out leaf disks (use areas between the veins) using a sterile # 4 cork borer and aseptically transfer the disks to flasks or large baby food jars containing growth medium.

Prepare growth medium for tobacco leaves as follows:

1. Dissolve one package of Murashige and Skoog salts (see Ordering Information) in 800 ml of distilled water.
2. Add the following

sucrose	30.0 g
inositol	0.1 g
glycine	0.2 g
agar	8.0 g

Then add 1 ml each of solutions a, b, c, d, and e (IAA, kinetin, nicotinic acid, pyridoxine-HCl, thiamine-HCl); see below.

a. Dissolve 0.2 g IAA in 1-2 ml of 1 N NaOH and add distilled water to make 100 ml. Add 1 ml to the medium for a final concentration of 2 mg/1.
b. Dissolve 0.2 g kinetin in 5 ml of 1 N HC1 and add distilled water to male 1000 ml. Add 1 ml to the medium for a final concentration of 0.2 mg/1.
c. Dissolve 0.25 g of nicotinic acid in 500 ml of distilled water. Add 1 ml to the medium for a final concentration of 0.5 mg/1.
d. Dissolve 0.25 g of pyridoxine-HC1 in 500 ml of distilled water. Add 1 ml to the medium for a final concentration of 0.5 mg/1.
e. Dissolve 0.2 g of thymine-HC1 in 500 ml of distilled water. Add 1 ml to the medium for a final concentration of 0.4 mg/l.

3. Make up to 1000 ml with distilled water. Autoclave for 15 minutes at 121° C, 15 pounds pressure, and pour 20-25 ml into each flask or jar. It will take approximately 3-4 weeks for callus to develop and then 4-8 weeks additional time for roots and shoots to develop.

VI. ORDERING INFORMATION

Phaseolus vulgaris (bush lima bean) seed—Carolina Biological, # 15-8330
Zea mays seed—Carolina Biological, # 15-9243
sweet corn—Carolina Biological, # 15-9283; Ward's, # 86W8080
field corn—Carolina Biological, # 15-9243
pea (Little Marvel) seed—Carolina Biological, # 15-9993; Ward's, # 86W8240
bean, lima—Carolina Biological, # 15-8335; Ward's, # 86W8008
Lugol's solution—Carolina Biological, # 87-2793 (100 ml) or # 87-2795 (500 ml); Ward's, # 39W1684
dropping bottle, 30 ml— Fisher, # 02-980; Carolina Biological, # 71-6525; Ward's, # 17W6011

Petri dish, 100 × 15 mm—Fisher, # 08-747C (Pyrex) or 08-746C (Kimax); Carolina Biological, # 74-1158; Ward's, # 17W0700
glass tubing (8 mm OD)—Carolina Biological, # P7-71-1188; Ward's, # 17W0961
radish seed—Carolina Biological, # 15-9000; Ward's, # 86W8200
rye seed—Carolina Biological, # 15-9313; Ward's, # 86W8130
Coleus stem tip (prepared slide)—Carolina Biological, # B563B; Ward's, # 91W7833
Tilia (basswood—three year stem, prepared slide)—Carolina Biological # B591A; Ward's, # 91W8310
Erlenmeyer flask, 250 ml—Fisher, # 10-040F (Pyrex) or # 10-039F (Kimax); Carolina Biological, # 72-6672 (Corning, student grade); Ward's, # 17W2982
forceps—Carolina Biological, # 62-4020 (straight) or 62-4030 (curved); Ward's, # 14W0541
Bunsen burner—Carolina Biological, # 70-6645; Ward's, # 15W0604 OR alcohol lamp with cap—Carolina Biological, # 70-6604; Ward's, # 17W0211
bottle, 4 oz—Carolina Biological, # 71-6204; Ward's, # 17W0003
scalpel—Carolina Biological, # 62-6001; Ward's, # 14W4320
single edge razor blade—bur locally; Carolina Biological, # 62-6930; Ward's, # 14W7142
Murashige and Skoog Minimal Organics Medium (powdered)—Carolina Biological, # 19-5740
sucrose (certified)—Fisher, S5-500; Carolina Biological, # 89-2870; Ward's, # 39W3280
Petri dishes, sterile disposable, 100 × 15 mm—Fisher, # 08-757-12; Carolina Biological, #74-1350; Ward's, # 18W7101
indole acetic acid (IAA)—Carolina Biological, # 20-7625 (100 mg vial) or #20-7626 (1 g); Ward's, # 39W1661
kinetin—Carolina Biological, # 20-7644 (500 mg); Ward's, # 39W0056
agar (TC)—Carolina Biological, # 19-8200
tobacco seeds—Carolina Biological, # 15-9080; Ward's, # 86W8340
Murashige and Skoog salts—Carolina Biological, # 19-5700
inositol—Carolina Biological, # 19-8458
nicotinic acid—Carolina Biological, # 87-7298
pyridoxine-HCl—Carolina Biological, # 88-5582
thiamine-HCl—Carolina Biological, # 89-5810
glycine—Carolina Biological, # 84-3630; Ward's, # 39W1421
Parafilm—Fisher, # 13-274-12 (dispenser, # 13-374-18); Carolina Biological, # 21-5600 (dispenser, # 21-5602); Ward's, # 15W1940
Styrofoam pots—purchase cups locally
potting soil—purchase locally

Notes:

LABORATORY 30 Water Movement and Mineral Nutrition in Plants

I. FOREWORD

This laboratory is designed to introduce students to the ways in which plants take up and lose water. The importance of water and minerals to the plant is also emphasized.

BioBytes *Seedling*, included on the CD-ROM that accompanies *Biology in the Laboratory* is an excellent addition to this laboratory. It offers an opportunity for students to study how plants transpire and distribute photosynthetic products. This program also simulates reactions to environmental conditions such as dispersion (crowding), temperature, light, and moisture.

II. TIME REQUIREMENTS

Exercise A—Observing Stomata (30 minutes)
Exercise B—Guttation (10 minutes)
Exercise C—Transpiration
Part 1—(10 minutes)
Part 2—(75 minutes)
Exercise D—The Pathway of Water Movement Through a Plant (10 minutes)
Exercise E—Plant Mineral Nutrition (30 minutes, demonstration; or 45 minutes set-up , experiment runs for 3-6 weeks as desired)

III. STUDENT MATERIALS AND EQUIPMENT

	Per Student	Per Pair (2)	Per Group (4)	Per Class (24)
Exercise A				
Zebrina (wandering Jew), green plant				1
forceps	1			24
tap water, dropping bottle		1		12
nail polish			1	6
scotch tape (roll)			1	6
microscope slides, box (75 × 25 mm)			1	6
coverslips, box (22 × 22 mm)			1	6
compound microscope	1			24
20% sucrose, dropping bottle **(1)**			1	6
Exercise B				
young barley seedlings, pot **(2)**				1
1000 ml beaker				1
Exercise C (Part 1)				
125 ml Erlenmeyer flask				3
Coleus shoots **(2)**				2
600 ml beaker				3
petroleum jelly, jar				3

	Per Student	Per Pair (2)	Per Group (4)	Per Class (24)
Exercise C (Part 2)				
Wheaton bottle potometer **(4)**		1		12
1 ml in 1/100 pipette		1		12
2-hole rubber stopper (size 8 ½)		1		12
plant mister				3
fan				3
flood light (150 w)				3
plastic bag				3
petroleum jelly, jar				3
Exercise D				
carnation (live)				1
two test tubes, taped together **(5)**				1
food coloring (red, green and blue—l each)				1
Coplin jar				1
celery stalk with leaves				1
50 ml beaker				1
microscope slides, box (75 × 24 mm)			1	6
dissecting microscope		1		24
Exercise E				
aquaria (20 gal) **(6)**				8
Styrofoam float **(6)**				8
OR quart Mason jars (optional) **(7)**			2	12
sunflower plants **(8)**		5		60
cotton (roll)				1
mineral deficiency solutions for aquaria (30 l each) **(9)**				8
OR mineral deficiency solutions for individual quart Mason jars (1 l each) **(10)**			2	12

IV. PREPARATION OF MATERIALS AND

(1) 20% sucrose

Mass 20 g sucrose. Add distilled water to make 100 ml. Dispense in 6 dropping bottles.

(2) barley seedlings

Sprinkle 25-30 barley seedlings on the surface of soil in a 6 inch Styrofoam pot. Cover with soil so that the seeds are 1/4 inch below the soil. Plant seeds approximately one week before use. Invert a 1000 ml beaker over the pot of seeds a few hours before use (water seedlings before placing them under the beaker).

(3) Coleus shoots

Cover the leaves on one *Coleus* shoot with petroleum jelly. Allow leaves on the other shoot to remain uncovered. Place plants in 125 ml flasks filled with water. Place these and a flask of water (without a *Coleus* shoot) under the three 600 ml beakers. Leave for 1-2 hours before use if room is warm or use a flood lamp. Three or four hours is better if possible.

(4) Wheaton jar potometer

Place a pipette (1 ml in 1/100) tip down in one hole of a rubber stopper (size 8). Place a stalk of an appropriate plant (we use red tip, *Photinia,* in the southeast) into the other hole. The stem should be tight enough to fit very snugly. The bark may be stripped off as you insert the stem. If you wish, you can grow *Impatiens* plants up through the hole in the stopper. Cut off the roots for use in the potometer. You may need to try several types of plants prior to the laboratory in order to decide on a local genus that will work well.

(5) test tubes for pathway of water in plants

Tape two test tubes together. Put green food coloring in one and red food coloring in the other. Use concentrated food coloring and fill the tubes up to the top. With a razor blade, split the stem of a carnation lengthwise and place one-half in each of the two different colors of dye.

Use blue or red food coloring for celery.

(6) aquaria and floats

Use 20 gal. aquaria for large classes or 10 gal. aquaria for small classes. Alternatively, you can use quart Mason jars (see (7)). For Styrofoam flats use R-Max Styrofoam insulation, ¾" thick (available from you local building materials supplier). Cut a piece to just fit inside the aquarium. It should not be wedged tightly but should be able to move up and down, floating on the surface as the level of solution changes.

If you use R-Max insulation, it has a pale blue cellophane covering on both sides. Peel the covering off one side and place the insulation white-side up. Use a cork borer to make 8 mm holes. The holes will be conical (larger on the blue side because the blue material pulls some Styrofoam away as the cork borer moves through). Flip the float over so that the larger opening of the holes are now upward and peel off the blue cellophane covering. (If you use a power drill, the holes will be of uniform diameter which is fine—peel off the cellophane from both sides.) Other types of ¾" Styrofoam board can be used if R-Max is not available.

Make 5 rows of 10 holes each to form a grid. Use a waterproof marker to label the grid 1-10 along the length and A, B, C, D, E along the width. (When you assign positions for plants, do it randomly and not all in one row.)

(7) Mason jars (Ball Mason jars, 32 oz)

Drill or punch 3 holes, 2-3 cm apart (each hole approximately 8 mm in diameter) in the lid of the Mason jar. If re-using jars, sterilize them if contamination occurred during previous experiments.

(8) sunflower seeds

Germinate sunflower seeds in wet sterile sand or in a wet paper towel roll (lay seeds on several damp paper towels, cover with additional damp paper towels, and roll up). Cover with aluminum foil. Allow 2-3 weeks for growth. (See Preparation Suggestions for treating seedlings.)

(9) nutrient solutions for aquaria

[If your class is small and you are using a 10 gallon aquarium, you may want to order a Mineral Requirements set that contains packets of chemicals for making 10 liters each of the eight different solutions. Alternatively, a Basic Mineral Requirements set provides enough for 20 liters of three solutions (complete, –N, –Fe, and – N). See Ordering Information.]

If using 20 gallon aquaria, proceed as follows:

To make 30 liters of each solution for a 20 gal. aquarium make 1 liter each of the stock solutions in Table I.

Table I Stock Solutions

Solution	g/liter of Solution
1 M $Ca(NO_3)_2$	236.15
1 M KNO_3	101.11
1 M $MgSO_4$	246.48
1 M KH_2PO_4	136.09
Fe Na EDTA	32.90
1 M $NaNO_3$	84.99
1 M $MgCl_2$	203.31
1 M Na_2SO_4	142.04
1 M NaH_2PO_4	137.99
1 M $CaCl_2$	147.02
1 M KCl	74.56

To make a micronutrient solution, dissolve the indicated amounts of the combination of chemicals given below in 500 ml of distilled water and then add additional distilled water to make one liter.

Table II Micronutrient Solution

Chemical	Amount (g)
boric acd	2.86
manganese chloride, $MnCl_2{\cdot}4H_2O$	1.81
zinc chloride, $ZnCl_2$	0.11
cupric chloride, $CuCl_2\ {\cdot}2H_2O$	0.05
sodium molybdate, $Na_2MoO_4{\cdot}2H_2O$	0.025

Add 10 liters of distilled water to each aquarium and add the amounts listed in Table III for each of the stock solutions prepared according to Tables I and II.

Table III Stocks for Preparing Mineral Nutrient Solutions

Stock Solution	Complete (ml)	–Ca (ml)	–S (ml)	–Mg (ml)	–K (ml)	–N (ml)	–P (ml)	–Fe (ml)
1 M Ca $(NO_3)_2$	150	0	150	150	150	0	150	150
1 M KNO_3	150	150	150	150	0	0	150	150
1 M $MgSO_4$	60	60	0	0	60	60	60	60
1 N KH_2PO_4	30	30	30	30	0	30	0	30
Fe Na EDTA	30	30	30	30	30	30	30	0
micronutrients	30	30	30	30	30	30	30	30
1 M $NaNO_3$	0	150	0	0	150	0	0	0
1 M $MgCl_2$	0	0	60	0	0	0	0	0
1 M Na_2SO_4	0	0	0	60	0	0	0	0
1 M NaH_2PO_4	0	0	0	0	30	0	0	0
1 M $CaCl_2$	0	0	0	0	0	150	30	0
1 M KCl	0	0	0	0	0	5	0	0

To each aquarium add an additional 19,450 ml (19 liters + 450 ml) to make a total of 30 liters.

To make additional 1 liter amounts to replenish solutions due to evaporation, see **(10)**.

(10) nutrient solutions for quart Mason jars (alternate):

To make 1 liter of nutrient solutions for Mason jars, use the following chart, assigning each student group to a particular deficiency.

The student group should fill 2 Mason jars two-thirds full with distilled water. Add all contents for "complete" to one jar. Add all contents for the selected deficiency to the second jar. Add distilled water to the top of the jars.

Stocks for preparing mineral nutrient solutions are given in Tables I and II above.

Stock Solution	Complete (ml)	–Ca (ml)	–S (ml)	–Mg (ml)	–K (ml)	–N (ml)	–P (ml)	–Fe (ml)
1 M Ca $(NO_3)_2$	5	0	5	5	5	0	5	5
1 M KNO_3	5	5	5	5	0	0	5	5
1 M $MgSO_4$	2	2	0	0	2	2	2	2
1 N KH_2PO_4	1	1	1	1	0	1	0	1
Fe Na EDTA	1	1	1	1	1	1	1	0
micronutrients	1	1	1	1	1	1	1	1
1 M $NaNO_3$	0	5	0	0	5	0	0	0
1 M $MgCl_2$	0	0	2	0	0	0	0	0
1 M Na_2SO_4	0	0	0	2	0	0	0	0
1 M NaH_2PO_4	0	0	0	0	1	0	0	0
1 M $CaCl_2$	0	0	0	0	0	5	0	0
1 M KCl	0	0	0	0	0	5	0	0

V. PREPARATION SUGGESTIONS

Exercise A

Other types of plants can be used for this exercise, including *Bryophyllum* or *Kalanchoe* or other types of *Zebrina* (wandering Jew). Some types of privet can also be used. If these plants are not available, the structural features of the epidermis on other types can be studied by using cellulose acetate film.

Exercise B

Barley seeds must be planted at least one week before use. Water them well before covering with a beaker. If you do not water the seeds, you will not observe guttation.

Exercise C

If you wish to have data expressed in the most accurate way, students can strip all leaves from the plant after taking transpiration readings. Trace around all leaves on 20 lb. mimeograph paper. Cut out the leaf tracings and weigh them as an estimate of the leaf surface area of the plant. Mimeograph paper weighs 0.0077 g/cm^2. By dividing the total weight of the paper tracings by 0.0077 and multiplying by 2 (for both surfaces), you will determine the total leaf area. Each student should express average rate for the two conditions tested (room and experimental) as ml water translocated/hour/cm^2 of leaf surface. Make sure that you use a 100 w flood lamp for this experiment.

Note—it is important to remind students to calculate surface area for both surfaces, no matter what method is used.

Exercise D

This is a simple experiment to set up. If you do not have a carnation, any white flower will do—mums or daisies. Celery works well and demonstrates the same principle of flowers are not available.

Exercise E

This experiment can be set up either in aquaria or in quart Mason jars for individual groups. The advantage of hydroponics in aquaria is that all plants are subjected to the same conditions and it is easy to see differences when you view large numbers of plants together. The Styrofoam float retards evaporation of the solution in the aquarium and sinks as the solution is used up by the plants. This is an added advantage since roots do not dry out as the solution level goes down (a problem with Mason jars). Have enough solution of each type on hand (store it in milk bottles) for students to add solution to Mason jars or for the instructor to add to aquaria when necessary.

To place sunflowers in the float, carefully rinse roots with distilled water and place the plant in the hole with roots reaching out into the hydroponics solution. Secure *gently* with a piece of cotton. Make sure the cotton does not touch the solution. Cover the outside of the portion of the aquarium that contains solution with black paper to prevent light from reaching the roots. Place the aquaria beneath grow lights (or fluorescent lights) in fixtures (4 per "shop" fixture). A stand for lights can be made with wood or PVC pipe. Chains can be used to suspend the fixture about half a meter (20 inches or so) above the aquaria. Alternatively, use a Plant Light Bank System (available from Carolina Biological Supply Company, # 15-8998).

If using Mason jars, drill or punch 3 holes, 2-3 cm apart (each hole approximately 8 mm in diameter) in the lid of the Mason jar. To insert plants, gently rinse the roots in distilled water and wrap cotton around the stem (the cotton should not be wrapped so tightly that there is no room for stem expansion). Insert each seedling into a hole in the Mason jar lid. Make sure the cotton does not hang down and touch the solution. Cover the glass part of the jar with aluminum foil and label the jar. If a greenhouse is available, put the plants in the greenhouse. Alternatively, if lab bench space is available, use grow lights (or just fluorescent lights) suspended in fixtures (4 per "shop" fixture) above the jars. A stand for lights can be made with wood or PVC pipe. Chains can be used to suspend the fixture. Alternatively, use a Plant Light Bank System (available from Carolina Biological Supply Company, # 15-8998).

If students are filling their own Mason jars, provide the bottles of nutrient solution, each with a pipette (labeled and held in a test tube attached to the matching bottle). Avoiding mixing of solutions is important. If you desire to use only two plants per jar, plug the third hole with cotton. Make sure students label their jars! Provide cotton and aluminum foil during the following weeks. Students will need to uncover jars to observe roots and will need to unwrap the cotton in order to pull plants up far enough to make measurements of height.

Note: if plants are grown for only 3 weeks prior to placing them in nutrient solutions, they will still have cotyledons attached. Differences become apparent as plants use the minerals in the solution rather than what is present in the cotyledons (approximately week 5-6 of growth). However, you cannot wait until plants are 5 weeks old to put them in floats or jars because the root systems will be too large, so you must begin this experiment ahead of time when plants are small (2-3 weeks after germination). Data collection can begin at week 5 of growth (or earlier if you begin to see differences). Students should be encouraged to observe plants each week, noting observations on color, height, bushiness, etc., for each treatment.

Class data can be analyzed using the statistics package available on the CD-ROM or the Chi square worksheet with Laboratory I in this *Preparator's Manual*. Students can compare each treatment to the complete medium or one treatment to another. Students should also be able to graph data on chlorophyll content, biomass, height, and leaf spread.

BioBytes *Seedling* is an excellent addition to these exercise. It explores how plants use their photosynthetic products as well as transpiration activity.

VI. ORDERING INFORMATION

Zebrina (wandering jew)—Ward's, # 86W7490
forceps—Carolina Biological, # 62-4020 (straight) or 62-4030 (curved); Ward's, # 14W0541
dropping bottle, 30 ml—Carolina Biological, # 71-6525; Ward's, # 17W6011
sucrose—Carolina Biological, # 89-2860; Ward's, # 39W3182
barley seeds—Carolina Biological, # 15-9220
Erlenmeyer flask, 250 ml—Fisher, # 10-040F (Pyrex) or # 10-039F (Kimax); Carolina Biological, # 72-6672 (Corning, student grade); Ward's, # 17W292
Coleus plant, live—buy locally; Carolina Biological, # 15-7310; Ward's, # 86W6825
beaker, 600 ml— Fisher, # 02-540M (Pyrex) or # 02-539M (Kimax); Carolina Biological, # 72-1225 (Corning, student grade)
petroleum jelly, jar—Carolina Biological, # 87-9510; Ward's, # 15W9832
Wheaton bottle, collecting (16 oz) 500 ml Wheaton # 214209:
Van Waters and Rogers Scientific, # 16175-160
American Scientific, #B7545-250
pipette, 1 ml—Fisher, # 13-665F; Carolina Biological, # 73-6270; Ward's, # 17W4852
two hole rubber stopper (size 8), Fisher # 14-140K, or Carolina Biological, # 71-2472; Ward's, # 15W8518
plant mister—purchase locally; Carolina Biological, # 66-5565; Ward's, # 18W2260
carnation—purchase locally
test tubes, 16 × 125 mm—Fisher, # 14-925H; Carolina Biological, # 73-1418; Ward's, # 17W1403
Coplin jar—Carolina Biological, #74-2160; Ward's, # 17W1200
beaker, 50 ml—Fisher, # 02-540G (Pyrex) or # 02-539G (Kimax); Carolina Biological, # 72-1220 (Corning, student grade); Ward's, # 17W4010
sunflower plants, seeds—Carolina Biological, # 15-9063; Ward's, # 86W8322
aquaria. 20 gal.—local pet store; Carolina Biological, # 67-1234; Ward's, # 21W5273
Styrofoam, sheet—purchase locally
boric acid (H_3BO_3)—Fisher, # A74-500, or Carolina Biological, # 84-8440; Ward's, # 37W0865
manganese chloride ($MnCl_2 \cdot 4H_2O$)—Fisher, # M87-100, or Carolina Biological, # 87-3860
zinc chloride ($ZnCl_2$)—Fisher, # Z33-500, or Carolina Biological, # 89-9518; Ward's, # 37W6345
sodium molybdate ($Na_2MoO_4 \cdot 2H_2O$)—Fisher, # S336-500, or Carolina Biological, # 88-9810
cupric chloride ($CuCl_2 \cdot 2H_20$)—Fisher, # C455-500, or Carolina Biological, # 85-6440; Ward's, # 37W2221
ethylenediamine tetraacetic acid (EDTA)—Fisher, #E-478-500, or Carolina Biological, # 86-1790; Ward's, # 39W6125
calcium nitrate ($Ca(NO_3)_2$)—Fisher, # C108-3, or Carolina Biological, # 85-2208; Ward's, #

37W0950

potassium nitrate (KNO_3)—Fisher, #P263-500, or Carolina Biological, # 88-3940; Ward's, # 37W4825

magnesium sulfate ($MgSO_4 \cdot 7H_20$)—Fisher, # M63-500; Carolina Biological, # 87-3376, Ward's, # 37W2861

potassium phosphate (monobasic) (KH_2PO_4)—Fisher, # P285-500; Carolina Biological, # 88-4250; Ward's, # 37W4841

sodium nitrate ($NaNO_3$)—Fisher, # S342-3, Carolina Biological, # 88-9900; Ward's, # 37W5630

magnesium chloride ($MgCl_2$)—Fisher, # M33-500, Carolina Biological, # 87-3240; Ward's, # 37W2830

sodium sulfate (Na_2SO_4)—Fisher, # S420-3, Carolina Biological, # 89-1710; Ward's, # 37W3145

sodium phosphate (monobasic) (NaH_2PO_4)—Fisher, # S369-1; Carolina Biological, # 89-1350; Ward's, # 37W5655

calcium chloride ($CaCl_2$)—Fisher, # C79-500; Carolina Biological, # 85-1800; Ward's, # 37W0936

potassium chloride (KCl)—Fisher, # P217-500; Carolina Biological, # 88.2900; Ward's, # 37W4634

plant light bank system—Carolina Biological, # 15-8998

Mineral Requirements Set—Carolina Biological, # 20-7922

Basic Mineral Requirements Set—Carolina Biological, # 20-7920

LABORATORY 31 Plant Responses to Stimuli

I. FOREWORD

This laboratory is designed to introduce students to the effects of various plant hormones on development and growth. Several of the experiments in Exercise A can be set up by students at the end of the previous week's laboratory and the results can be considered during the laboratory itself. Other experiments, included in Exercise B. can be set up during the laboratory period and the results can be analyzed during the following week.

II. TIME REQUIREMENTS

Exercise A—Auxins (30 minutes, set-up; 30 minutes interpretation, second week)
  Part 1—Bud Inhibition and Apical Dominance
  Part 2—Leaf Abscission
Exercise B—Gibberellins (30 minutes, set up; 30 minutes interpretation, second week)
Exercise C—Tropisms (60 minutes)
  Part 1—Gravitropism (Geotropism)
  Part 2—Phototropism
Exercise D—Light-Induced Germination (40 minutes)
  Part 1—The Role of Phytochrome
  Part 2—The Forms of Phytochrome
  Part 3—Germination and Gibberellins (Optional)
Exercise E—Photoperiodism (20 minutes)

III. STUDENT MATERIALS AND EQUIPMENT

	Per Student	Per Pair (2)	Per Group (4)	Per Class (24)
Exercise A (Part 1)				
Coleus, pot			1	6
forceps or razor blade			1	6
lanolin paste, jar			1	6
lanolin paste, jar, containing 5000 ppm IAA **(1)**			1	6
plant markers, tags			1	6
toothpicks			2	12
metric ruler (30 cm)			1	6
Exercise A (Part 2)				
Coleus plant			1	6
single-edged razor blade			1	6
reinforcement rings			3	18
lanolin paste, jar—as in part 1			1	6
lanolin paste, jar, containing 5000 ppm IAA **(1)**			1	6
Exercise B				
dwarf pea plants, pot **(2)**			2	12

	Per Student	Per Pair (2)	Per Group (4)	Per Class (24)
Exercise B—continued				
metric ruler (30 cm)			1	6
GA_3 solution, dropping bottle **(3)**			1	6
control solution, dropping bottle **(4)**			1	6
plant markers, tags			2	12
Exercise C (Part 1A)				
chicken gizzard plant **(5)**				2
Exercise C (Part 1B)				
corn seeds **(6)**	3			72
Petri dish (100 mm)	1			24
paper towels, stack	1			24
masking tape, roll			1	6
black pencil	1			24
forceps	1			24
Exercise C (Part 1C)				
bush beans (germinated) **(7)**	4			96
Petri dish (100 mm)	1			24
single edged razor blade	1			24
paper towels, stack	1			24
wheat seedlings, pot **(9)**				1
Exercise C (Part 2)				
germinating wheat seedlings, pot **(8)**				1
tomato plant				1
150 w flood lamp				1
Exercise D (Part 1)				
Great Lakes lettuce seeds (germinated in continuous darkness)				50
Grand Rapids lettuce seeds (germinated in continuous darkness)				50
Great Lakes lettuce seeds (germinated in light)				50
Grand Rapids lettuce seeds (germinated in light)				50
Exercise D (Part 2)				
Grand Rapids lettuce seeds (germinated in 30 minutes of red light) **(9)**				50
Grand Rapids lettuce seeds (germinated in 30 minutes of far-red light) **(9)**				50
Grand Rapids lettuce seeds (germinated in 30 minutes of blue light) **(9)**				50
Grand Rapids lettuce seeds (germinated in 30 minutes of green light) **(9)**				50

	Per Student	Per Pair (2)	Per Group (4)	Per Class (24)
Exercise D (Part 3) (optional)				
Petri dishes containing blotter paper soaked in water	2			48
Petri dishes containing blotter paper soaked in gibberellic acid solution **(10)**	2			48
aluminum foil (10-inch square)	2			48
Exercise E				
petunia or morning glory plants **(11)**				2
spinach or radish plants **(11)**				2
kidney bean plants **(11)**				2

IV. PREPARATION OF MATERIALS AND SOLUTIONS

(1) lanolin paste containing 5000 ppm IAA

Order prepared.

(2) dwarf peas

Plant peas about 12 days before lab. Place in moist soil, four to a pot.

(3) GAS solution

Gather together the following materials:

triethanolamine
Tween-20
gibberellic acid

Procedure:

Part A: Dissolve 0.1 g gibberellic acid in 4 ml of triethanolamine and dilute up to 100 ml with distilled water.

Part B: Mix 10 ml of Part A solution with 90 ml of distilled water. Add 2 drops of Tween-20 as a wetting agent.

(4) control solution

Mix 2 drops of Tween-20 into 100 ml of distilled water.

(5) chicken gizzard plants (*Iresine* sp.)

Place plants in the dark (a dark cabinet will do). Allow one to remain vertical and place the other on its side. You may also wish to place a plant on a ring stand upside down (cover the soil withr gauze so it will not fall out). Do this 2-3 days prior to the laboratory.

(6) corn seeds

Soak overnight in tap water.

(7) germinated bush beans

Soak bush bean seeds overnight in tap water. Lay beans flat on several layers of moist paper towels. Cover seeds with another layer of paper towels. Roll the towels up and wrap in aluminum foil. Allow two days for germination. If done during the winter season and the room is cold, place the seed roll into a warm oven or incubator (21-23° C).

(8) pot of wheat seedlings

Sprinkle 25-30 seeds in a pot. Cover with a sprinkling of soil so that seeds are about ¼ inch below the surface. Plant seeds approximately one week before use. Put light on one side after they have germinated so that the seedlings bend toward the light.

(9) action spectrum of light-stimulated seed germination

Place filter paper in each of 8 Petri dishes and dampen with 8 ml of distilled water. In a darkened room, scatter lettuce 25 seeds (soaked overnight) in each dish. Place all dishes in the dark for 12 hours for imbibition to occur. After 12 hours remove the dishes to be light treated, put them in shoe boxes with the appropriate color filter, and place them under the appropriate light source for at least 30 minutes. Seeds receiving fluorescent light should be placed 30 cm from a 40 watt fluorescent tube. Place the seeds receiving incandescent light 50 cm from a 150 w incandescent lamp. After exposure, return the dishes to the dark cabinet. Keep the dark control in the dark at all times.

Use Great Lakes (light insensitive) and Grand Rapids (light sensitive) seeds in the following dishes:

Dish 1 — Dark Control—Great Lakes variety
Dish 2 — Dark Control—Grand Rapids variety
(wrap both dishes in aluminum foil and keep in the dark)
Dish 3 — Light Control—Great Lakes variety
Dish 4 — Light Control—Grand Rapids variety
(expose both dishes in fluorescent light, cover with clear polyethylene)

Use only Grand Rapids variety (light sensitive) in the following dishes:

Dish 5 — Blue Light (390-590 nm)—expose to fluorescent light, 3 layers of dark blue cellophane
Dish 6 — Green Light (480-630 nm)—expose to fluorescent light, 3 layers of dark green cellophane
Dish 7 — Red Light (580-630 nm)—expose to fluorescent light, 2 layers of dark red cellophane
Dish 8 — Far-red Light (670 nm)—expose to incandescent light, 3 layers of dark blue and 2 layers of dark red cellophane

Observe the seeds in each treatment 24 hours after exposure and determine the % germination in each dish. Which light treatment was most effective? How do the two varieties differ in their response to light?

(10) gibberellic acid solution

Prepare a solution of gibberellic acid as follows:

Dissolve 100 mg of gibberellic acid in 1000 ml of distilled water (premeasured vials of 100 mg are available from Carolina Biological Supply Company). If you have GA_3 solution left over from Exercise B. see **(4)**, you may use the left over solution for this exercise.

(11) petunia, morning glory, spinach, or radish plants

Expose one plant of each type to 8 hours of light and 16 hours of dark (16L:8D) and the other to 16 hours of light and 8 hours of dark (16L:8D). This is best done in a growth chamber under controlled temperature (approximately 22° C). Start plants from seeds. It will take approximately 35 days (6-8 weeks) for flowering. Spinach and radishes are long day plants. Petunias and morning glories are short day plants. Kidney beans are day neutral. Do about 6-8 weeks before use.

V. PREPARATION SUGGESTIONS

Exercises A and B

The hormone experiments in Exercises A and B take 2 weeks to complete. Their set up is simple and can be done during the previous week. Results can then be analyzed during this laboratory period.

Exercise C

Tropism experiments with seeds can be set up during this laboratory and analyzed at the beginning of the following week's laboratory period. Phototropism experiments should be set up two weeks prior to this laboratory.

Exercise D

This experiment must be set up two days before the laboratory. Shoeboxes with colored cellophane are made by cutting a window in the top of the box (almost the same size as the box lid) and gluing the appropriate "filters" in place. An optional exercise demonstrates how gibberellic acid substitutes for light and can induce germination in light sensitive seeds.

Exercise E

This can best be done in a growth chamber. Plants should be started from seeds (or from very small plants) and will need approximately 6-8 weeks on the correct light regime. Follow preparation directions **(12)** for choice of plants.

VI. ORDERING INFORMATION

Coleus, pot—obtain locally; Carolina Biological Supply, # 15-7310; Ward's, # 86W6825
single edge razor blade, Carolina Biological, # 626930; Ward's, # 14W4172
forceps—Carolina Biological, # 62-4020 (straight) or 62-4030 (curved); Ward's, # 14W0541
indole-3-acetic acid paste (5000 ppm IAA)—Carolina Biological, # 20-7602; Ward's, # 39W0031
control lanolin paste—Carolina Biological, # 20-7850; Ward's, # 39W2081
triethanolamine (reagent)—Carolina Biological, # 89-6850
Tween 20—Fisher, # CS279-500; Carolina Biological, # 20-7861
gibberellic acid—Carolina Biological, # 20-7565; Ward's, # 39W1445
chicken gizzard plant (kalanchoe), *Dryophyllum*—purchase locally; Carolina Biological, # 15-7400; Ward's, # 86W6910
Petri dish, 100 × 15 mm—Fisher, # 08-747C (Pyrex) or 08-746C (Kimax); Carolina Biological, # 74-1158; Ward's, # 17W 0700
single edge razor blade—buy locally; Carolina Biological, # 626930; Ward's, # 14W4172
wheat seeds—Carolina Biological, # 15-9394; Ward's, # 86W8360
dark red, green, and blue cellophane—Fisher, Life Science catalog # S52570
sunflower seeds—Carolina Biological, # 15-9063; Ward's, # 86W8322
dwarf pea seeds (Little Marvel)—Carolina Biological, # 15-8883; Ward's, # 86W8251
corn seeds—Carolina Biological, # 15-9243; Ward's, # 86W8080
bush bean seeds—Carolina Biological, # 15-8330; Ward's, # 86W8006
light insensitive (Great Lakes) lettuce seeds—Ward's, # 86W8155
light sensitive (Grand Rapids) lettuce seeds—Ward's, # 86W8150
gibberellic acid—Carolina Biological, # 20-7565; Ward's, # 39W1445
wheat seeds—Carolina Biological, # 15-9395; Ward's, # 86W8361
petunia seeds—local seed store; Carolina Biological, # 15-8942
morning glory seeds—Carolina Biological, # 15-8780; Ward's, # 86W8165
spinach seeds—local seed store

radish seeds—Carolina Biological, # 15-9000; Ward's, # 86W8200
kidney bean seeds—Carolina Biological, # 15-8420; Ward's, # 86W8006

LABORATORY 32 Animal Tissues

I. FOREWORD

This laboratory is designed to introduce students to different types of animal tissues. Some comparisons are made with living tissues to add interest to the laboratory. All exercises can be enhanced by using color transparencies of slides of the tissue types being studied. This facilitates discussion and identification of structures which students will need to locate on different slides. If a microscope attached to a video camera is available, slides can also be shown directly to the class as a whole.

II. TIME REQUIREMENTS

Exercise A—Epithelial Tissue (45 minutes)
- Part 1—Squamous Epithelium
- Part 2—Columnar Epithelium
- Part 3—Cuboidal Epithelium

Exercise B—Connective Tissue (60 minutes)
- Part 1—Cartilage
- Part 2—Bone
- Part 3—Loose and Dense Connective Tissue
- Part 4—Adipose (Fat) Tissue
- Part 5—Blood

Exercise C—Muscle Tissue (30 minutes)
- Part 1—Skeletal Muscle
- Part 2—Cardiac Muscle
- Part 3—Smooth Muscle

Exercise D—Nervous Tissue (15 minutes)

III. STUDENT MATERIALS AND EQUIPMENT

	Per Student	Per Pair (2)	Per Group (4)	Per Class (24)
Exercise A				
frog (live)				7
frog (pithed)				1
forceps				1
frog Ringer's solution, bottle **(1)**				1
frog skin (prepared slide) (optional)		1		12
stratified squamous epithelium (dog esophagus, prepared slide)		1		12
simple columnar epithelium (human kidney, prepared slide)		1		12
simple cuboidal epithelium (human thyroid, prepared slide)		1		12
microscope slides, box (75 × 25 mm)			1	6

	Per Student	Per Pair (2)	Per Group (4)	Per Class (24)
Exercise A (continued)				
coverslips, box (22 × 22 mm)			1	6
compound microscope	1			24
board (8 × 8 inches)			1	6
string, ball			1	6
cork particles, container **(2)**			1	6
methylene blue, dropping bottle **(3)**			1	6
board covered with cheese cloth				1
scalpel or single edge razor blade				1
Exercise B				
hyaline cartilage (human trachea, prepared slide)			1	12
elastic cartilage (human fetus, prepared slide)		1		12
decalcified bone (human, prepared slide)		1		12
areolar tissue (rat or cat, prepared slide)		1		12
tendon (monkey, prepared slide)		1		12
adipose tissue (cat, osmium, prepared slide)		1		12
adipose tissue (fat dissolved, prepared slide)		1		12
human blood smear (Wright stain, prepared slide)		1		12
blood film (frog, Giesma stain, prepared slide)		1		12
microscope slides, box (75 x 25 mm)			1	6
coverslips, box (22 x 22 mm)			1	6
compound microscope	1			24
Wright's stain, dropping bottle **(4)**			1	6
buffer for Wright's stain **(5)**			1	6
frog Ringer's solution, bottle **(1)**			1	6
single-edge razor blades				1
frog (pithed—same as in A)				1
Exercise C				
skeletal muscle (dog, cat, rat or rabbit, prepared slide)		1		12
cardiac muscle (dog, cow, or sheep, prepared slide)		1		12
smooth muscle (cat or frog, prepared slide)		1		12
frog (pithed—same as in A)				1
dissecting needles				2
methylene blue, dropping bottle **(3)**			1	6
frog Ringer's solution, dropping bottle **(1)**			1	6
microscope slides, box (75 × 25 mm)			1	6
coverslips, box (22 × 22 mm)			1	6
compound microscope	1			24
Exercise D				
neuron (cow, prepared slide)		1		12
compound microscope	1			24

IV. PREPARATION OF MATERIALS AND SOLUTIONS

(1) frog Ringer's solution

Weigh 0.42 g KCl, 9.0 g $NaCl$, 0.24 g $CaCl_2$, and 0.20 g $NaHCO_3$. Make up to 1 liter with distilled water.

(2) cork particles

Use a kitchen grater to break a cork into fine particles.

(3) methylene blue

Order prepared solution or weigh 0.01 g methylene blue and add absolute alcohol (100%) to make 100 ml.

(4) Wright's stain

Order prepared stain.

(5) buffer for Wright's stain

Weigh 1.63 g KH_2PO_4 and 3.2 g K_2HPO_4. Add distilled water to make 1 liter.

V. PREPARATION SUGGESTIONS

To pith a frog—

> Note: the use of live vertebrate animals must be approved by your institution's animal care and use committee (IACUC). Be sure to submit a protocol for this laboratory in advance so that you may proceed with this experiment.

Place the frog on crushed ice in a dish pan at least an hour before it is to be pithed—this induces cold narcosis.

If right handed, grasp the chilled frog firmly in your left hand, anchoring the body with your thumb and holding the tip of the snout firmly between your second and third fingers (reverse this if your are left handed). Bend the head down slightly to emphasize the posterior margin of the cranium. Draw an imaginary line across the back of the head connecting the anterior edges of the tympanic membranes on each side of the head. Insert a dissecting needle into the spinal column precisely in the middle of this line—muscle tremors will indicate that you have severed the spinal cord. Keep the tip of the dissecting needle in the neural canal and point it backwards. Gently push it to the posterior, keeping it in the neural canal (if you have to force it, it is in the wrong place). Withdraw the needle, keeping the tip in the neural canal and push it forward into the brain and rotate it within the cranium to destroy brain tissue. Withdraw the needle—the frog should be flaccid. You may wish to do this out of sight of the class.

If appropriate you may wish to discuss the use of live vertebrates in teaching and research, indicating that there are certain experiences and experiments that can only be done with live animals. However, the ethical use of live animals requires procedures which minimize pain and suffering (such as chilling or anesthetizing the frog before the central nervous system is destroyed in this exercise).

Exercise A

If frogs have been kept in an aquarium for a few days, some of the epidermis is usually shed or becomes loosely attached to the body. This can be removed easily. If frogs are in short supply, make a single slide at the beginning of the laboratory and place it on demonstration.

Exercise B

Examples of different types of bones are helpful in this exercise. A human femur (longitudinally sectioned—Carolina Biological, # 24-7811) or a cross section of a femur (Carolina Biological, # 24-7820), is excellent for studying compact and spongy bone tissue. Bird bones can be used for comparisons. If funds are scarce, road killed-mammals and birds provide a source of bones that you can section. Chicken bones can be demineralized by soaking them in vinegar for 1-2 weeks (or until soft). They can then be cut longitudinally or in cross section more easily.

Make sure that bone slides are decalcified bone. Be sure to order adipose tissue slides both stained with osmium and with fat dissolved so students will have a complete understanding of fat cell structure.

In order to avoid exposing students and instructors to human bodily fluids, prepared slides of human blood are used-in this exercise. You may wish to prepare fresh blood smears and stain them with Wright's stain under appropriate conditions (see Ordering Information).

To prepare a blood smear from the frog, place a drop of blood plus a drop of frog Ringer's solution on the end of a glass slide (the slide should be on a paper towel). Pull a second glass slide, held at a 45° angle, down the length of the flat slide (alternatively, you can push the slide). Let the slide dry and then stain as follows.

To stain a blood smear, place the slide on several layers of paper towels to absorb spills. Cover the dried blood film completely with Wright's stain for 1-3 min. This fixes the blood film. Next add Wright's buffer solution, drop by drop, until a metallic greenish scum forms on the surface of the stain. Continue until the stain is diluted in half and then let the slide stand for approximately 2 min (lengthen to 3 min if staining is light). Tip the slide and allow stain and buffer to drain onto the paper towels. Place the slide in a Coplin jar of distilled water for 2-3 minutes and air dry. Inspect without a coverslip. Use oil immersion for closer study (focus under high power and rotate the objective away from the slide, apply immersion oil to stained spear directly and rotate the objective into the drop and focus). Basophilic granules should be bright blue, eosinophilic granules should be red, and neutrophil granules should be lilac or purple. Rotate the oil immersion objective away from the slide and clean the objective tip. Blot the slide with lens paper (do not wipe) and store it if you wish to keep good smears for future use. Be sure that the students are not confused by the nucleated red blood cells found in the frog.

Exercise C (Part 3)

Note that only the first instruction deals with smooth muscle using a fixed and stained microscope slide. Students could attempt to retrieve smooth muscle from the intestine but the small size of the smooth muscle cells and their lack of striations make it unlikely that they will be able to visualize smooth muscle tissues adequately. If you have a pithed frog, you might open the intestinal cavity and have students study movements of the smooth muscle tissue in the small intestine (open the pericardium and also study the contractions of cardiac muscle). A section of small intestine could be removed and placed in a Petri dish with frog Ringer's solution to study spontaneous movements. If the heart is removed and placed in a Petri dish of Ringer's solution, will it continue to beat? (Are the contractions of the vertebrate heart myogenic or neurogenic in origin?)

Note that thigh muscle is striated tissue. Instructions 2-4 to should be included with Exercise C (Part 1). Make sure that the muscle tissue removed from the frog's thigh is an extremely thin strip. You will need to tease it into thin strands and use methylene blue stain to see the muscle cell structure.

VI. ORDERING INFORMATION

frog (live)—Carolina Biological, # L1502 or L 1505 (large); Ward's, # 87W8200
forceps—Carolina Biological, # 62-4020 (straight) or 62-4030 (curved); Ward's, # 14W0541
epithelial cells (frog, whole mount)—Carolina Biological, # K3-H120; Ward's, # 92W3643
stratified squamous epithelium (dog esophagus, prepared slide)—Carolina Biological, # H230; Ward's, # 93W3036
simple columnar epithelium (human kidney, prepared slide)—Carolina Biological, # H6008; Ward's, # 93W5238
simple cuboidal epithelium (human thyroid, prepared slide)—Carolina Biological, # H6006; Ward's, # 93W6636
microscope slides, 75 × 25 mm—Fisher, # 12-550A; Carolina Biological, # 63-2000; Ward's, # 14W3555
coverslips, 22 × 22 mm—Fisher, # 12-542B; Carolina Biological, # 63-3095; Ward's, # 14W3555
methylene blue—Carolina Biological, # 87-5684 (powder), or # 87-5911(liquid); Ward's, # 38W8323
cheesecloth—Fisher, # 06-665-18 (70 yard bolt); Carolina Biological, # 71-2690 (5 yard package); Ward's, # 15W0015
scalpel—Carolina Biological, # 62-6001; Ward's, # 14W4320
single edge razor blades—buy locally; Carolina Biological # 62-6930; Ward's, # 14W4172
hyaline cartilage (human trachea, prepared slide)—Carolina Biological, # H6002; Ward's, # 93W6003
elastic cartilage (human fetus, prepared slide)—Carolina Biological, # H6026; Ward's, # 93W6130
decalcified bone (human, prepared slide)—Carolina Biological, # H6320
areolar tissue (rat or cat, prepared slide)—Carolina Biological, # H570; Ward's, # 93W3224
tendon (monkey, prepared slide)—Carolina Biological, # K3-H9064
adipose tissue, osmium (cat, prepared slide)—Carolina Biological, # H610; Ward's, # 93W3232
adipose tissue, fat dissolved (prepared slide)—Carolina Biological, # H600; Ward's, # 93W4338
human blood smear (Wright's stain, prepared slide)—Carolina Biological, # H1155; Ward's, # 92W6540
blood film (frog, Giesma stain, prepared slide)—Carolina Biological, # H1060; Ward's, # 93W3640
Wright's stain solution—Carolina Biological, # 89-8533; Ward's, # 38W9940
buffer solution for Wright's stain—Carolina Biological, # 84-9751
dropping bottle, 30 ml—Carolina Biological, # 71-6525; Ward's, # 17W6011
skeletal muscle (dog, cat, rat, or rabbit, prepared slide)—Carolina Biological, #H1310; Ward's, # 93W3541
cardiac muscle (dog, cow, or sheep, prepared slide)—Carolina Biological, # H1350; Ward's, # 93W3527
smooth muscle (cat or frog, prepared slide)—Carolina Biological, # H1260; Ward's, # 93W3520

dissecting needles—Carolina Biological, # 62-7200; Ward's, # 14W0650
neuron (cow, prepared slide)—Carolina Biological, # H1660; Ward's, # 93W3617
potassium chloride (KCl)—Carolina Biological, # 88-2910; Ward's, # 37W4631
sodium chloride (NaCl)—Fisher, # S271-500; Carolina Biological 88-8901; Ward's, # 37W5487
calcium chloride ($CaCl_2$)—Carolina Biological, # 85-1810; Ward's, # 37W0929
sodium bicarbonate ($NaHCO_3$)—Carolina Biological, # 88-8403; Ward's, # 37W5450
potassium phosphate (monobasic) (KH_2PO_4)—Fisher, # P285-500; Carolina Biological, # 88-4250; Ward's, # 37W4845
potassium phosphate (dibasic) (K_2HP0_4)—Fisher, # P288-500; Carolina Biological, # 88-4290; Ward's, # 37W4841

LABORATORIES 33-36 The Anatomy of Representative Vertebrates

I. FOREWORD

Laboratory 33 introduces the study of anatomy with a selection of terms describing locations and directions and discussions of techniques of dissection and the organ systems of vertebrates.

The remainder of Laboratory 33 begins a sequence of exercises continued in Laboratories 34, 35, and 36, in which four representative vertebrates are examined in an attempt to convey information about the phylogenetic development and evolution of vertebrates while introducing the basic organ systems of vertebrate animals. Structures are compared in most of the organ systems between the shark, frog, turtle, and rat—a chondrichthyian fish, an amphibian, a reptile, and a mammal. It is suggested that these animals be dissected by groups of four students with adequate time allotted for students to share their learning so that all can gain a feel for the comparative differences in organs and organ systems between these representatives—in these exercises students will learn by working together. If less time is available for these dissections, the turtle, or turtle and frog, can be omitted with the shark and rat illustrating relatively different ends of the evolutionary spectrum. If these laboratory dissections are not done, the introductory material presented with each section may be of interest as a supplement to the brief discussion of vertebrate evolution in introductory text books.

A practical examination, required drawings, or a graded dissection may be useful in evaluating student learning in these exercises. Encourage students to add notes and drawings to their manual as they do these laboratories.

II. TIME REQUIREMENTS

Laboratory 33—Introduction to the Study of Anatomy, and the External Anatomy and Integument of Representative Vertebrates (2.5-3 hours)
- Part 1—Introducing Dissection: Anatomical Locations, Techniques of Dissection, and Anatomical Locations
- Part 2—Introduction to the Representative Vertebrates
  - Exercise A—Life History and External Anatomy of Four Representative Vertebrates
  - Exercise B—Integument—A Dynamic Interface

Laboratory 34—The Anatomy of Representative Vertebrates: Behavioral Systems (2.5-3 hours)
- Exercise A—Sensors (Affectors)
  - Part 1—The Lateral-Line System
  - Part 2—The Inner Ear
  - Part 3—Other Sensors of the Head
- Exercise B—Control Systems
  - Part 1—The Nervous System
  - Part 2—The Endocrine System
- Exercise C—Effectors: Muscles and Bones
  - Part 1—Muscles
  - Part 2—Bones
  - Part 3—Other Effectors

Laboratory 35—The Anatomy of Representative Vertebrates: Digestive and Respiratory Systems (2.5-3 hours)
- Exercise A—Coelom and Mesenteries
- Exercise B—The Digestive System
- Exercise C—The Respiratory System

Laboratory 36—The Anatomy of Representative Vertebrates: Circulatory and Urogenital Systems (2.5-3 hours)
- Exercise A—The Circulatory System
  - Part 1—Arteries
  - Part 2—Veins
  - Part 3—Heart
- Exercise B—The Urogenital System

III. STUDENT MATERIALS AND EQUIPMENT

	Per Student	Per Pair (2)	Per Group (4)	Per Class (24)
Laboratories 33, 34, 35, and 36				
materials contained in the manual				
latex gloves, disposable (one pair/laboratory) **(1)**	4			96
shark (preserved) **(2)**			1	6
frog (preserved) **(2)**			1	6
turtle (preserved) **(2)**			1	6
rat (preserved) **(2)**			1	6
dissecting trays (large for sharks)			1	6
dissecting pans (small for other vertebrates)			3	18
dissecting pins	10			240
dissecting kits **(3)**	1			24
Laboratory 33, Part 2, Exercise B				
compound microscope			1	6
microscope slides, box (75 × 25 mm)			1	6
coverslips, box (22 × 22 mm)			1	6
shark skin (whole mount, prepared slide)			1	6
OR prepared shark skin (pieces) **(4)**			1	6
fish skin (cross section, prepared slide)			1	6
frog skin (cross section, prepared slide)			1	6
turtle skeleton				1
human scalp skin (cross section, prepared slide)			1	6
Laboratory 34, Exercise A				
shark chondrocranium (cleared and plastic embedded)			1	6
finger bowl			1	6
Laboratory 34, Exercise B				
brain models (set) **(5)**				1
Laboratory 34, Exercise C				
demonstration preserved shark **(6)**				1

	Per Student	Per Pair (2)	Per Group (4)	Per Class (24)
Laboratory 34, Exercise C—continued				
shark chondrocranium, plastic mount **(7)**				1
shark skull, liquid preserved **(7)**				1
shark skeleton, liquid preserved **(7)**			1	6
demonstration skeletons of other vertebrates **(8)**				
Laboratory 35				
amphibian tadpoles (live) **(9)**		1	6	
cat or rat skull sectioned to show oral and pharyngeal cavities **(10)**				1
Laboratory 36, Exercise A				
artery and vein (cross section, prepared slide)(optional)			1	6
heart models (set) **(11)**				1
sheep or beef heart **(11)**				
pregnant shark (optional) **(12)**				1
pregnant rat (optional) **(12)**				1

IV. PREPARATION OF MATERIALS AND SOLUTIONS

(1) latex gloves

Disposable latex gloves should be available in small, medium, and large sizes. Students with sensitivity to preservatives should be encouraged to wear gloves while handling specimens.

(2) preserved vertebrates

There are a variety of choices of size, preservative, and injection, with differences in cost. We recommend the largest size specimens, preserved with formalin if it is possible to obtain them, with a single latex injection (arteries). You may also wish to order 1 or 2 doubly-injected specimens for demonstration purposes. You should be sure that you receive at least one representative of each sex for each laboratory section (and might like to order one pregnant shark and rat—see **(12)** below.) Individually packaged specimens are easier to store than those that come in buckets. Additional comments are given below.

(3) dissecting kit

It is desirable to have students buy their own dissecting kits. They should have a good pair of surgical scissors, a probe and seeker, and two dissecting needles. A one-piece scalpel or cartilage knife should be used for dissecting cartilage and bone. Be sure that the scalpel is not used to section tissues—encourage students to use fingers and probes to separate organs from surrounding tissues without cutting. Learning to dissect is a skill students should obtain from these laboratories—have instructors spend some time showing each student how to expose a difficult structure or complete a dissection so that all aspects of the organ can be seen.

(4) shark skin

If desired, small pieces (about 2×2 mm) of skin can be removed from one of the preserved sharks and placed in a 10% solution of glycerin overnight. This clears the skin so those whole mounts can be made for study.

(5) brain models
If brain models are available, they will be useful to illustrate phylogenetic trends among the vertebrates. Other demonstrations could include intact and hemisected sheep's brains.

(6) demonstration shark
Section the tail region of one shark and remove the skin from the surface on one side for two inches behind the cut to show muscles as drawn in Figure 34C-2.

(7) and (8) demonstration skeletons
Each group should have a complete shark skeleton available for study. One chondrocranium (the cartilaginous brain case without the visceral skeleton) and one complete skull will help students see the structures of the head more clearly. These preserved skeletons and skulls should not be removed from their display bottles—they are fragile and can be destroyed easily. If a chondrocranium embedded in plastic is available, it can be substituted for the preserved chondrocranium (without the visceral skeleton).

If skeletons of the frog, turtle, and rat (or cat or human) are available, allow students to study them to supplement their study of the shark.

(9) live tadpoles
Amphibian tadpoles are generally available in the field or from biological supply houses in the spring. They can be kept in an aquarium in the laboratory and development can be followed over a period of weeks. Note that your use of live vertebrates in the laboratory will require the approval of your institution's animal care and use committee.

(10) skull showing oral and pharyngeal cavities
Purchase a demonstration of the cat skull sectioned to show the oral and pharyngeal cavities or prepare a median longitudinal section from one of the preserved rat specimens for study.

(11) heart models or demonstration dissection
If available, heart models can be used to study the evolution of the chambers and valves in relation to the modification of aortic arches and the separation of pulmonary and systemic circulations. A dissected sheep or beef heart may be useful as a demonstration.

(12) gravid viviparous representatives
A pregnant shark and a rat (or other mammal) can be used to contrast maternal-fetal interchange and adaptations for viviparity in these diverse groups.

V. PREPARATION SUGGESTIONS

All Laboratories

Preserved animals lack the colors, textures, and smells of fresh tissue. If surplus live rats are available in your animal room, you might consider using them in place of preserved animals for the initial dissections, including muscles which are much easier to decipher in unpreserved animals. Be sure that rats are humanely killed by animal room personnel before class (use carbon dioxide euthanasia or other methods approved by your institutional animal care and use committee). Have enough preserved rats available for common use by all sections for Laboratory 33 and then introduce the fresh rats in Laboratory 34 for skinning and muscle dissection. At the end of the laboratory, have students open the body cavity as described in Laboratory 34, tag their specimen, and immerse them in 10% formaldehyde for subsequent study (these animals will not have the arterial system injected but the arteries should be easy to find without this treatment). If time permits, students might examine the viscera before preserving the rats, noting the texture and color of various internal organs.

If preserved animals are purchased in bulk packages, the buckets and shipping preservative can be retained and used to store specimens between laboratories. If possible, use one bucket for each section and mark it with the section number and meeting time. Give each student a heavy tag on which they can write their name and section number using a water-resistant marker. Tie the tag to an appendage such as the pelvic fin or hind leg.

If specimens come individually packaged, have students cut the end of the bag carefully and keep the bag to store specimens between laboratories (provide twist ties or rubber bands to secure the bags). They can tag their specimen or write their name and section number on the bag using an indelible marker. Be sure that specimens remain moist—they can be wrapped in wet paper towels or towels moistened with fixative if necessary (spray bottles with 10% glycerine may be helpful to keep everything wet). Bagged specimens can be stored on shelves or in boxes between laboratories.

When preserved animals are used for the first time and when the abdominal cavity is first opened, they should be rinsed thoroughly in a large sink. Specimens shipped in buckets can be soaked overnight to help remove some of the odor. Ask students to wear old clothes to laboratory—stains may be impossible to remove from good clothing. Insist that students wear glasses rather than contact lenses. It might be a good idea to have several pairs of safety glasses available in case any students do come with contact lenses.

As dissection proceeds, bits and pieces of tissue will be discarded. Maintain a trash can with a plastic liner for animal parts and remains ONLY (no towels or gloves). Be sure that students clean out sinks in which dissecting pans are washed, placing all residue in the designated trash can. Empty this can daily or tie and place the bag aside for pickup as required by your institution's hazardous materials rules.

Provide added demonstration materials as available. Various skeletal elements may be displayed for examination. A fresh cow or pig heart, brain, kidney, or lung may be ordered from your butcher and used to demonstrate various internal structures (and their natural color) in the fresh organ. If other materials are available from the comparative anatomy laboratory in your school, feel free to supplement appropriate exercises.

VI. ORDERING INFORMATION

latex gloves—Carolina Biological, # 70-6345, 70-6346, and 70-6347 (small, medium, and large); Ward's, # 15W1073

shark (preserved)—Carolina Biological, # P 1301D (P 1301PD, pregnant); Ward's, # 69W1121

frog (preserved)—Carolina Biological, # P 1476D; Ward's, # 69W2315

turtle (preserved)—Carolina Biological, # P 1601C; Ward's, # 69W3311

rat (preserved)—Carolina Biological, # P 1876D (P 1876AS, pregnant); Ward's, # 69W5425

dissecting pans, 13 × 9" (aluminum)—Carolina Biological, # K3-62-9000 (without wax), # K3-62-9002 (with wax); Ward's, # 14W8510

dissecting trays, 16 × 30" (polyethylene)—Carolina Biological, # K3-62-9068; Ward's, # 18W3660

dissecting pins (2 inch), box—Carolina Biological, # K3-62-9122; Ward's, # 14W0200

dissecting kits—see Carolina Biological catalog (include # 62-2225 scissors, # 62-6164 scalpel, 2 # 62-7200 dissecting needles, and # 62-7400 probe and seeker if ordering individual items); Ward's, # 14W0988

microscope slides, 75 × 25 mm—Fisher, # 12-550A; Carolina Biological, # 63-2000; Ward's, # 14W2501
coverslips, 22 × 22 mm—Fisher, # 12-542B; Carolina Biological, # 63-3095; Ward's, # 14W3213
shark skin (dogfish, placoid scales, w.m.)—Carolina Biological, # Z2905; Ward's, # 92W3230
fish skin (cross section)—Carolina Biological, # H 2050; Ward's, # 92W3146
frog skin (cross section)—Carolina Biological, # H 2060 (whole mount, # Z 4195 to show pigment cells); Ward's, # 92W3658
turtle skeleton—Carolina Biological, # 24-4120; Ward's, # 65W3330
human skin—Carolina Biological, # H 7475 (scalp), # H 7455 (l.s. of hair shafts); Ward's, # 93W7018
brain models—Carolina Biological, # 56-4650
sheep, half brain—Carolina Biological, # K3-P2105S; Ward's, # 69W7131
shark chrondrocranium—Carolina Biological, # 24-31-30; Ward's, # 65W1111
shark chrondrocranium with gill arches—Carolina Biological, # 24-3140 or 24-3150
shark chondrocranium, stained to show contours with semicircular canals injected (plastic mount)—Carolina Biological, # POM 2755
shark skeleton—Carolina Biological, # 24-3110 or 24-3120; Ward's, # 65W1100
bullfrog tadpoles (large, live)—Carolina Biological, # L 1492E; Ward's, # 87W8212
bisected cat skull—Carolina Biological, # 24-5930; Ward's, # 65W5235
artery and vein (cross section)—Carolina Biological, # K3-H1750; Ward's, # 93W4042
preserved heart—Carolina Biological, # K3-P141S (sheep, in pericardium) or # K3-P2300S (cow, with base of aorta attached); Ward's, # 69W7201
heart models—Carolina Biological, # 56-4655

LABORATORY 37 The Basics of Animal Form: Skin, Bones, and Muscles

I. FOREWARD

This laboratory is designed to introduce students to the support and effector systems of the body. It is the first in a series of four laboratories that explore the anatomy and physiology of vertebrate systems. Human anatomy and physiology are emphasized throughout the series. In this laboratory, students will study the origins and diversity of epithelial coverings and will explore how muscles and bones work together to allow movement. Muscle contraction is studied at both anatomical and biochemical levels.

II. TIME REQUIREMENTS

Exercise A—The Language of the Body (20 minutes—at home)
Exercise B— Vertebrate Coverings: Skin, Scales, Feathers, and Hair (60 minutes)
Exercise C Bones and Joints
  Part 1— Bones (20 minutes)
  Part 2— Joints (20 minutes)
Exercise D Muscles and Bones Working Together
  Part 1— Isotonic and Isometric Muscle Contractions (30 minutes)
  Part 2— Muscle Fatigue (45 minutes)
Extending Your Investigation: Interactions of Muscles and Bones (30 minutes)
Exercise E— The Biochemistry of Muscle Contraction (45 minutes)

III. STUDENT MATERIALS AND EQUIPMENT

	Per Student	Per Pair (2)	Per Group (4)	Per Class (24)
Exercise A				
materials contained in manual				
Exercise B				
shark skin, placoid scales (whole mount, prepared slide)		1		12
fish skin (cross section, prepared slide)		1		12
frog skin (cross section, prepared slide)		1		12
snake skin (cross section, prepared slide)		1		12
turtle skeleton				1
bird feather, contour **(1)**		1		12
bird feather, down **(1)**		1		12
human skin, scalp (cross section, prepared slide)		1		12
microscope slides (75 × 25 mm)	1			24
coverslips (22 × 22 mm)	1			24
glycerol (dropping bottle)			1	6
ethanol (100%), 125 ml beaker (optional) **(1)**			1	6
xylene, 125 ml beaker (optional) **(1)**			1	6
balsam, Canada, dropping bottle (optional) **(1)**			1	6

	Per Student	Per Pair (2)	Per Group (4)	Per Class (24)
Exercise B—continued				
compound microscope	1			24
Exercise C (Part 1)				
mammalian bone (fresh or dry, cut longitudinally, or human bone, plastic mount)				1
demineralized bones **(2)**				1
Exercise C (Part 2)				
materials contained in manual				
Exercise D (Part 1)				
tape measure		1		12
tape (roll)			1	6
Exercise D (Part 2)				
rubber ball (tennis ball size)		1		12
Extending Your Investigation				
chicken wings (fresh, uncooked)	1			24
diagram of chicken wing **(3)**	1			24
forceps	1			24
probe	1			24
scissors	1			24
Exercise E				
glycerinated rabbit psoas muscle **(4)**				1
ATP solution, containing Ca^{2+} (5 ml) **(5)**				1
$MgC1_2$ + KC1 solution (5 ml) **(5)**				1
ATP + $MgCl_2$ + KC1 solution (5 ml) **(5)**				1
ATP—muscle kit				1
microscope slides (75 × 25 mm)	1			24
coverslips (22 × 22 mm)	1			24
glycerol,50% (dropping bottle) **(6)**			1	6
dissecting needles	2			48
millimeter ruler (clear plastic)	1			24
lens paper, pieces			1	6
compound microscope	1			24

IV. PREPARATION OF MATERIALS AND SOLUTIONS

(1) bird feathers

Students can prepare their own feather slides for study. Temporary mounts can be prepared using glycerin. To make permanent slides, cut a 3-4 mm long section from the vane or one side of each feather that includes the central shaft (down feathers can be mounted whole). If a flight feather (wing or tail feather) is available, pull the barbs apart in at least one place along the length of the section before mounting the feather. Rinse the piece of feather in absolute (100%) ethanol and then in xylene. Place on a microscope slide and cover with a drop of

Canadian balsam in xylene. Moisten a coverslip with xylene and hold it at a 45° angle to the slide. Gently move it until one edge touches the balsam. Gently lower the coverslip, using care to avoid bubbles under the feather and coverslip. Keep slides flat until the mounting medium sets. These slides can be used in subsequent years. The same technique can be used to mount hairs or vibrissae from mammals if desired.

(2) demineralized bones

Long bones from a chicken or other small bones from squirrels or other types of "road kills" (be sure to wear gloves when removing bones) can be demineralized by soaking the bones in vinegar for 3-10 days, depending on the size of the bone. With minerals removed, the bones become fairly rubbery and can be cut longitudinally with ease.

Alternatively, ask the butcher in your local grocery to use a bone saw to cut a larger beef bone longitudinally. After students have examined the bone, clean off any remaining tissue (marrow can be rinsed out with hot water). Place the bone on a windowsill to dry. Direct sunlight will also bleach the bone. Keep dried bones for use in following years.

(3) diagram of chicken wing

Copy the diagram at the end of this guide for class distribution.

(4) glycerinated rabbit psoas muscle

This material is perishable. Have shipped close to the date of use and store in the freezer at –20° to –10° C. Keep at this temperature. The muscles can be used for up to 6 months.

(5) ATP, KC1, and $MgCl_2$ solutions

The ATP solution containing Ca^{2+} must be used within 7 days after arrival or activity will be lost. Store in the refrigerator at 4° to 10° C. Refrigerate all three solutions: ATP; $MgCl_2$ + KC1; and ATP + $MgCl_2$ + KC1. Solutions must be kept on ice during the experiment.

(6) glycerol, 50%

Add 50 ml glycerol to 50 ml distilled water, mix, and pour into a dropping bottle.

V. PREPARATION SUGGESTIONS

Exercise A

Have students complete exercise A before coming to class. Give a brief oral quiz using a cat skeleton and a member of your class for comparison.

Exercise B

Supplement this exercise with additional specimens showing modifications of epidermal coverings—hoofs, nails, claws, porcupine quills, horns. Compare horns to antlers that are formed by bone formed under a covering skin which is shed after growth of the antlers (the "velvet"). Slides of dermal fish scales (ganoid, cycloid, and placoid—Carolina Biological, # Z3115) or snake skins (epidermal scales) can also be used (ask students to bring in shed snake skins for study). Examine a turtle shell if available.

Have students look at hairs of different colors and textures.

Also examine feathers of different colors. Study slides under both reflected light (dissecting microscope) and transmitted light (compound microscope). If the color is due to melanin, it will appear dark brown to black in transmitted light (but viewed with reflected light, feather structure and other pigments may modify the brown color of melanin to provide green, blue, purple, and iridescent colors). If the color is due to a true pigment (as in the case of brown, yellow, red, or orange feathers), the feather will appear to be the same color in both transmitted and reflected light. Visit your local pet store and collect feathers molted from parakeets, canaries, or other captive birds.

Have students examine and explain the interlocking mechanisms revealed by the flight feather slide. Why does a bird "preen"?

Exercise C (Part 1)

Use a bone mounted in plastic for comparison of longitudinal and cross-section pieces. Have the butcher cut beef long bones longitudinally or demineralize bones (see Preparation of Materials and Solutions). Visit your state's medical school and request specimens of dried bones. Supplement this exercise with bones that have been broken and repaired (also from local medical schools) and with X-rays of bones with fractures, cancer, etc. from local doctors or medical schools.

If students did not look at the histology of bones during Laboratory 32 (Animal Tissues), you might want to have students refer to Exercise B and examine a cross section of decalcified bone (Carolina Biological, # 740).

Exercise C (Part 2)

Supplement this exercise with X-rays of joints obtained from a local hospital. Show both healthy joints and joints with arthritis. Discuss what it means to have a "torn" ligament—something that seems to be common among young, athletic students.

Exercise D (Parts 1 and 2)

This investigation is designed to get students to think about exercising. Many students talk about isometrics but have no idea what is meant. Make sure that the ball used for Part 2 is a spongy rubber bulb so that the hand muscles actually squeeze the ball.

Extending Your Investigation

Simple, but messy. Have plenty of paper towels and soap on hand—the chicken is greasy. As students examine the chicken wing and try to locate the different muscles (see diagram included at the end of this section of the *Preparator's Guide*), have them compare these muscles to those present in the human arm. Discuss what muscles and tendons are involved in moving the human arm and decide what the counterpart is in the chicken wing. Note the comparison of the upper arm, lower arm and hand to the upper and lower parts of the wing plus the "hand," composed of fused digits. Primary feathers attach to the hand while the secondaries attach to the posterior bone of the forelimb (ulna).

If available, have students study a skeleton of a bird. This is also a very good time to review homologies—look at the wing of a bat or flipper of a whale for comparison. Cover diagrams or reference materials in plastic for protection.

Exercise E

Glycerinated muscles should be ordered in advance of the laboratory and should be stored in the freezer. All solutions must be kept cold during the experiment. The ATP solution will lose its effectiveness after 7 days.

Instructors should try this exercise before introducing it to your class. Students must have some idea of what the muscle looks like and some notion of what they might see as the muscle contracts. They cannot read directions and descriptions at the same time they are looking through the microscope! Clean microscope objectives and oculars are essential as well as good lighting. You may wish to supplement this exercise with a video demonstration (Carolina Biological Supply Company, # 49-8349-V: *What is a Muscle ?).*

VI. ORDERING INFORMATION

shark skin, placoid scales, whole mount (prepared slide)—Carolina Biological, # Z2905; Ward's, # 92W3130
fish skin, cross section showing origin of scales (prepared slide)—Carolina Biological, # K3-H2050; Ward's, # 92W3146
frog skin, (prepared slides)—Carolina Biological, # H2060 (cross section); # Z4195 (whole mount to show pigment cells); Ward's, # 92W3658
snake skin, cross section (prepared slide)—Carolina Biological, # H2065; Ward's, # 92W3770
turtle skeleton—Carolina Biological, # 14-4120; Ward's, # 65W3330
bird feather (prepared slides)—Carolina Biological, # Z4975 (contour); # Z4980 (down); # Z4985 (feather types—contour, down, filoplume); Ward's, # 92W3807
human skin (prepared slides)—Carolina Biological, # H7475 (cross section); # H7455 (scalp, longitudinal section of hair shafts); Ward's, # 93W7001
microscope slides, 75 × 25 mm—Fisher, # 12-550A; Carolina Biological, # 63-2000; Ward's, # 14W3500
coverslips, 22 × 22 mm—Fisher, # 12-542B; Carolina Biological, # 63-3095; Ward's, # 14W3251
glycerol—Carolina Biological, # 86-5560; Fisher, # G-33-500; Ward's, # 39W1435
wide-mouthed glass balsam bottle, 45 ml—Carolina Biological, # 71-6690; Ward's, # 17W5405
balsam, Canada, in xylene—Carolina Biological, # 84-6558 (30 g); # 84-8560 (100 g); Ward's, # 37W9505
ethanol, absolute (100%)(denatured) —Carolina Biological, # A-962-4; Fisher, # A-407-500
xylene—Carolina Biological, # 89-8743; Fisher, # X3S-4; Ward's, # 39W4516
human bone (Plastomount)—Carolina Biological, # POM 9100
tape measurers—local hardware store
rubber ball—local store
chicken wings—local grocery store
glycerinated rabbit psaos muscle—Carolina Biological, # 20-3520; Ward's, # 38W0025
ATP solution—Carolina Biological, # 20-3530
$MgCl_2$ + KC1 solution—Carolina Biological, # 20-3530
ATP + $MgCl_2$ + KC1 solution—Carolina Biological, # 20-3530
ATP muscle kit—Carolina Biological, # 20-3525; Ward's, # 36W5417
video—*what is a Muscle*—Carolina Biological, # 49-8349-V

Notes:

MUSCLES OF THE CHICKEN WING

Figure 1. Muscles of the right wing of the pigeon. Ventral view. The overlying pectoralis muscle has been removed on the right side.

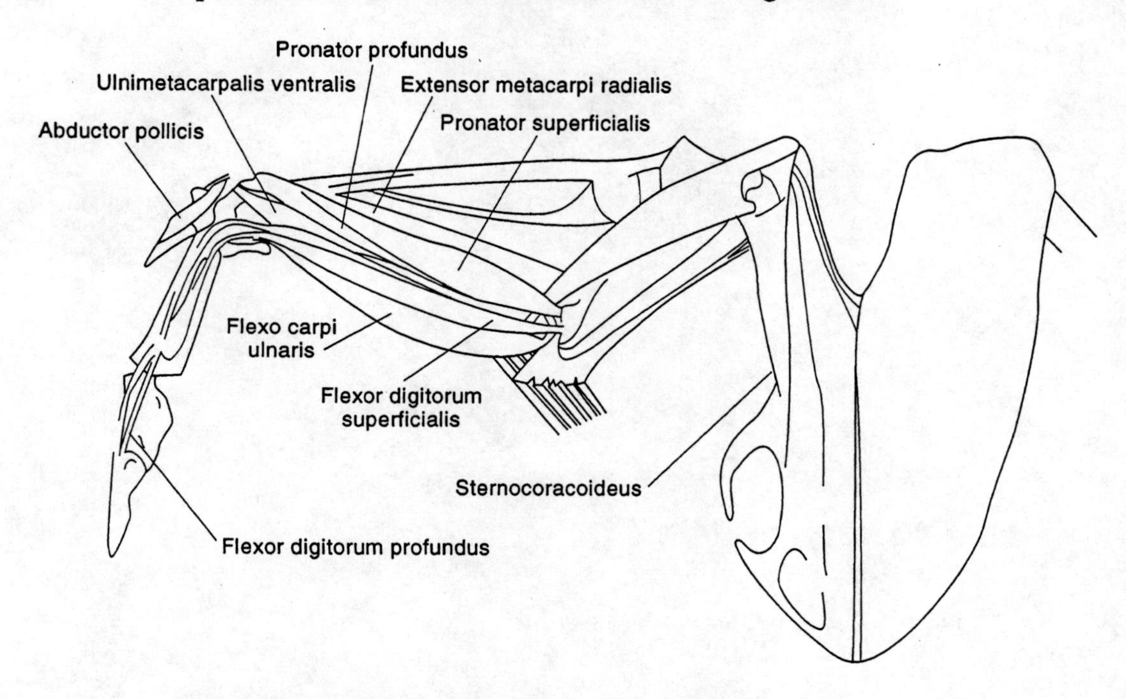

Figure 2. Muscles of the right wing of the pigeon, dorsal view.

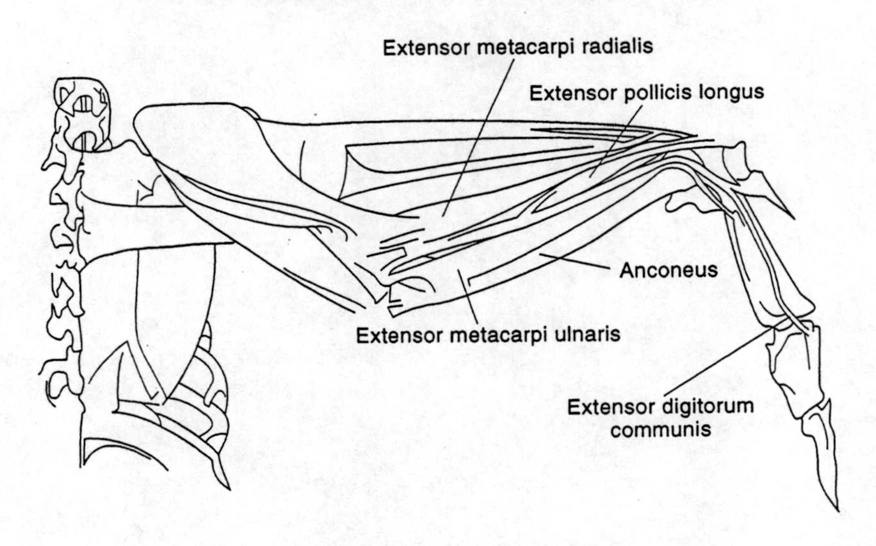

From Pettingill, O. S. Jr. 1970. *Ornithology in Laboratory and Field.* 4 th. ed. Burgess Publishing Company, Minneapolis. p. 73

LABORATORY 38 The Physiology of Circulation

I. FOREWORD

This laboratory is designed to give students an opportunity to learn about the structure and function of the circulatory system. The structural features of the blood vessels and pumps that circulate this fluid through the body are studied with prepared slides and preserved materials. Heart rate and pulse rate are measured. Additionally, the effects of temperature and hormones on heart rate are measured.

II. TIME REQUIREMENTS

Exercise A—Microscopic Examination of Human Blood Cells (30 minutes)
Exercise B—Pumps—The Vertebrate Heart (45 minutes)
Exercise C—The Structure of Arteries and Veins (20 minutes)
Exercise D—Movement of Blood in a Goldfish Tail (10 minutes)
Exercise E—Determining Blood Pressure (30 minutes)
Exercise F—Measuring Pulse Rate and Blood Pressure (30 minutes)
  Part 1—Relation of the Heartbeat to Circulation
  Part 2—Variability of the Heart Rate and Blood Pressure
  Part 3—The Effect of Exercise on Heart Rate
  Part 4—The Effect of Smoking on Heart Rate *(Optional)*
Exercise G—Factors Influencing Heart Rate in the Water Flea, *Daphnia* (45 minutes)
  Part 1—The Effect of Temperature Change
  Part 2—The Effect of Chemicals

III. STUDENT MATERIALS AND EQUIPMENT

	Per Student	Per Pair (2)	Per Group (4)	Per Class (24)
Exercise A				
human blood (prepared slide)	1			24
human blood, sickle cell anemia in crisis (prepared slide)				1
human blood, mononucleosis (prepared slide)				1
frog blood (prepared slide)	1			24
compound microscope	1			24
Exercise B				
sheep heart (1)		1		12
dissecting pan		1		12
scalpel	1			24
probe	1			24
scissors	1			24
Exercise C				
artery, cat (cross section, prepared slide)		1		12
vein, cat (cross section, prepared slide)		1		12

	Per Student	Per Pair (2)	Per Group (4)	Per Class (24)
Exercise C—continued				
atherosclerosis (human, prepared slide)				1
compound microscope	1			24
Exercise D				
goldfish, live				2
cotton (box)				1
microscope slides, box (75 × 25 mm)				1
coverslips, box (22 × 22 mm)				1
compound microscope				1
Exercise E				
sphygmomanometer				4
stethoscope				4
Exercise F				
stethoscope		1		12
Exercise G				
Daphnia magna (live culture) **(2)**				1
Pasteur pipette	1			24
petroleum jelly, jar			1	6
finger bowl or Petri dish	1			24
ice bucket with ice			1	6
thermometer	1			24
glass file			1	6
depression slide	1			24
acetylcholine, dropping bottle **(3)**			1	6
epinephrine (adrenaline), dropping bottle **(4)**			1	6
dissecting microscope	1			24

IV. PREPARATION OF MATERIALS AND SOLUTIONS

(1) sheep heart

Purchase preserved (see Ordering Information) or obtain from local abattoir. Use gloves when handling fresh material.

(2) *Daphnia magna*, live culture

Order only *Daphnia magna*. Other species of *Daphnia* are so small that they cannot be trapped as easily in the Pasteur pipettes and their heart rate is more difficult to observe.

(3) acetylcholine, dropping bottle

Order prepared (see Ordering Information). Keep refrigerated.

(4) epinephrine (adrenaline), dropping bottle

Order prepared (see Ordering Information). Keep refrigerated.

V. PREPARATION SUGGESTIONS

Exercise A

Because it is not advisable to use human bodily fluids in a general biology laboratory, prepared slides have been substituted.

Exercise B

Sheep hearts can be purchased from Carolina Biological Supply or Ward's Natural History. These are preserved hearts. Fresh hearts can be obtained from the local abattoir. If fresh material is used, make sure that students wear gloves. Have a flexible knitting needle and some string on hand. It is fairly easy to follow the blood vessels as they go to and from the heart (even before you dissect the heart) by probing with the needle.

Exercise C

If you dissect a fresh heart, have students observe the difference in the thickness of the walls of the aorta and vena cava. Relate this to the prepared slides used for this exercise.

Exercise D

You may need approval of your institution's animal use and care committee for this experiment.

The goldfish can be wrapped in cotton and taped directly to the microscope stage or it can be laid out in a Petri dish, wrapped in cotton and taped to the dish. In this case, only one glass slide or coverslip will be necessary to observe the tail. If the fish does not move its tail very much, no covering will be necessary.

Exercises E and F

The stethoscopes, which come with the sphygmomanometers if purchased in kit form, can supplement the number of stethoscopes needed in Exercises E and F. To save on funds, have half of the class do Exercise G while the other half does Exercises E and F. If you have a smoker in the room, do you see any differences in heart rate or pulse?

Exercise G (Parts 1 and 2)

Use only *Daphnia magna* for these experiments. Once a *Daphnia* is trapped in the capillary tube, the heart beat will be easy to follow. Use medium from the shipping culture or pond water as needed to suspend the *Daphnia.* Softened clay can be used to seal the end of the tube as an alternative to petroleum jelly. This method of restraining *Daphnia* is preferred for Q_{10} analysis because it is easy to change temperatures. You can, however, use the same method as that used for Part 2. Simply place the slide on top of a closed Petri dish of hot water or ice. Heat or cool the slide to the desired temperature and quickly transfer it to the microscope for observation.

VI. ORDERING INFORMATION

human blood smear, Wright's stain (prepared slide)—Carolina Biological, # HI 155; Ward's, # 93W6540

sickle cell anemia in crisis (prepared slide)—Carolina Biological, # PH1015; Ward's, # 93W8120

infectious mononucleosis (prepared slide)—Carolina Biological, # PH 1085; Ward's, # 93W9052

frog blood smear (prepared slide)—Carolina Biological, # H1060; Ward's, # 92W3640

sheep heart—Carolina Biological, # P2140C; Ward's, # 69W7201; Ward's, # 69W6201
dissecting pan—Carolina Biological, # K3-62-9000 (without wax) or # K3-62-9002 (with wax); Ward's, # 14W8510
artery, cat (cross section, prepared slide)—Carolina Biological, # H1700; Ward's, # 93W4022
vein, cat (cross section, prepared slide)—Carolina Biological, # H1745; Ward's, # 93W4050
atherosclerosis, human (prepared slide)—Carolina Biological, # PH810
goldfish, live (small)—local pet store
microscope slides, 75 × 25 mm—Fisher, # 12-550A; Carolina Biological, # 63-2000; Ward's, # 14W3500
coverslips, 22 × 22 mm—Fisher, # 12-542B; Carolina Biological, # 63-3095; Ward's, # 14W3251
sphygmomanometer (blood pressure kit)—Carolina Biological, # 69-1030; Ward's, # 14W5023
stethoscope—Carolina Biological, # 69-1644; Ward's, # 15W8818
Daphnia magna—Wards, # 87W5210 (pure culture)
petroleum jelly, jar—Carolina Biological, # 87-9510; Ward's, # 39W2765
Pasteur pipette, 9"—Fisher, #13-678-20C; Carolina Biological, # 73-6062; Ward's, # 17W116
rubber bulb—Fisher, # 14-065B
acetylcholine (1: 10,000)—Carolina Biological, # 84-1611; Ward's, # 38W2021
epinephrine (adrenaline, 1: 10,000)—Carolina Biological, # 842091; Ward's, # 38W2052
dissecting kit (with probe)—Carolina Biological, # 62-1136; Ward's, # 14W0776
scalpel—Carolina Biological, # 62-6031; Ward's, # 14W0875
scissors—Carolina Biological, # 62-2505; Ward's, # 14W0980
probe—Carolina Biological, # 62-7400; Ward's, # 15W0950
latex gloves, disposable—Carolina Biological, # 70-6340, 70-6341, and 70-6342 (small, medium, and large)

LABORATORY 39 Gas Exchange and Respiratory Systems

I. FOREWORD

This laboratory is designed to explore the structures and functions of vertebrate respiratory systems, blood pigments as carriers of respiratory gases (oxygen and carbon dioxide) and includes exercises on lung capacity and the effects of exercise on heart and respiratory rates.

II. TIME REQUIREMENTS

Exercise A—The Vertebrate Respiratory System
 Part 1—How a Fish Breathes—Gills as a Respiratory Surface (30 minutes)
 Part 2—How a Mammal Breathes—Lungs as a Respiratory Surface (30 minutes)
Exercise B—Respiratory Pigments (40 minutes)
Exercise C—Lung Capacity (20 minutes)
Exercise D—How Does Smoking Affect Lung Capacity? (20 minutes)
Exercise E—The Effect of Exercise on Heart and Respiratory Rates (30 minutes)

III. STUDENT MATERIALS AND EQUIPMENT

	Per Student	Per Pair (2)	Per Group (4)	Per Class (24)
Exercise A (Part 1)				
perch, preserved **(1)**		1		12
gill filaments (prepared slide)		1		12
scissors	1			24
probe	1			24
dissecting pan		1		12
Exercise A (Part 2)				
sheep pluck **(2)**			1	6
lung tissue, cat (cross section, prepared slide)		1		12
trachea, cat (cross section, prepared slide) **(3)**		1		12
dissecting pan			1	6
compound microscope	1			24
Exercise B				
cow's blood, tube (1 ml) **(4)**	1			24
5 ml pipette (to pipette blood)				1
conical centrifuge tube	1			24
0.9% saline, bottle (100 ml) **(5)**			1	6
clinical centrifuge				1
Pasteur pipette	1			24
rubber bulb	1			24
test tube (10 × 75 mm), corked (#1)	1			24
activated yeast solution, bottle (50 ml) **(6)**			1	6

	Per Student	Per Pair (2)	Per Group (4)	Per Class (24)
Exercise B—continued				
Petri dish	1			24
Exercise C				
balloon	1			24
ruler	1			24
Exercise D				
same materials as exercise C				
Exercise E				
stethoscope		1		12

IV. PREPARATION OF MATERIALS AND SOLUTIONS

(1) perch
Preserved specimens can be used. Any bony fish (teleost) is acceptable. If you have an enterprising student who likes to fish, fresh material can also be used. Fish heads, free from the grocery, can be frozen and stored until use.

(2) sheep pluck
Preserved specimens can be used or fresh specimens can be obtained from the local abattoir. Use gloves to handle fresh specimens and wash hands afterwards. *Ascaris* may be present.

(3) trachea, cat (cross section)
Two slides are necessary to observe both lung epithelium surrounding alveoli and the cartilage rings of the trachea (see Ordering Information).

(4) cow's blood
Obtain cow's blood from the local abattoir—ask for heparanized blood so that it will not clot. Centrifuge it at a very low setting (3-4) in a clinical centrifuge and remove the top layer of serum proteins. Have instructor pipette 1 ml of blood into each student's centrifuge tube.

(5) 0.9% saline
Weigh out 9 g NaC1 and add distilled water to make 1000 ml of solution. Bottles of 100 ml will be sufficient for groups of four students. Attach a test tube to the bottle to hold a 5 ml pipette. Label pipette "NaCl."

(6) activated yeast solution
Add one packet of dried yeast to 100 ml of warm 0.9% saline (do not use distilled water).

V. PREPARATION SUGGESTIONS

Exercise A (Part 1)

This exercise can be done by each student or you can place a fresh or preserved fish on demonstration. Fish heads can be obtained free from the local grocery store and can be frozen until you are ready for this laboratory. If this is done, each student can have his or her own material and learning will be improved beyond what occurs with demonstration material. Unpreserved (fresh or frozen) material is more pliable and easier to use if it is available.

Exercise A (Part 2)

The sheep pluck can be obtained as a preserved specimen (see Ordering Information) or can be obtained from the local abattoir. Lung inflation is more obvious in fresh material than in fixed material and students may gain a better appreciation for the delicacy of lung structure using fresh organs. If you use fresh material, make sure students wear gloves.

You do not necessarily need an air compressor to inflate the lung preparation—a simple foot pump like those used to inflate rafts or swimming floats will work fine. If you have an air supply in the laboratory, use it.

Exercise B

For this experiment, obtain blood from the abattoir. Make sure that cows are free of brucilosis. Use dry yeast in packets obtained in the grocery store. When adding yeast suspension to the blood, be sure that you can still see the reddish color (approximately 5 parts of blood to 1 part of yeast). Add less yeast and more blood if the mixture becomes opaque or colorless. The more yeast you add, the faster the reaction will take place. If you use larger tubes than those recommended (e.g., 10 ml), increase the amount of blood you start with to 2-3 ml and wash with 10 ml saline. After the final rinse, suspend cells in 8 ml of saline and yeast suspension. Don't forget to balance centrifuge tubes in the centrifuge. Directions call for 1 tube per person (or pair). This should be balanced with another student's tube.

Exercise C

Make sure that students follow directions carefully. Most students want to completely empty their lungs and forcibly exhale. Have one student hold the ruler with an end on the table and the rest of the ruler sticking upward. Use a piece of paper or cardboard on top of the balloon to be sure that he or she is measuring the place where the diameter is maximum (this is very much like measuring someone's height by placing a book on the top of the head).

Exercise D

Combine this investigation with Exercise C. Students will find that smokers have a reduced lung capacity. You might also have students investigate whether there are differences in the lung capacities of males and females.

Exercise E

Stethoscopes are not really necessary—respiration rates can be determined by observation alone

VI. ORDERING INFORMATION

perch, preserved—Carolina Biological, # P1410A; Ward's, # 69W1542
fish gills, filaments (prepared slide)—Carolina Biological, # Z3116; Ward's, # 92W3148
dissecting kit (with probe)—Carolina Biological, # 62-1136
scalpel—Carolina Biological, # 62-6031
scissors—Carolina Biological, # 62-2505
probe—Carolina Biological, # 62-7400
latex gloves, disposable—Carolina Biological, # 70-6340, 70-6341, and 70-6342 (small, medium, and large)
dissecting pans, 13 × 9" (aluminum)—Carolina Biological, # 62-9000 (without wax), # 62-9002 (with wax); Ward's, # 14W8500
sheep pluck—Carolina Biological, # P 2150C; Ward's, # 69W7311
lung tissue, cat (cross section, prepared slide)—Carolina Biological, # H2460; Ward's, # 93W4883

trachea, cat (cross section, prepared slide)—Carolina Biological, # H2430 and H2435; Ward's, # 93W4871
pipette, 1 ml—Fisher, # 13-665F; Carolina Biological, # 73-6270
pipette, 5 ml—Fisher, #13-665K; Carolina Biological, # 73-6276
conical centrifuge tubes—Carolina Biological, # 73-2014; Ward's, # 17W1330
Pasteur pipettes—Fisher, # 13-678-20C; Carolina Biological, # 73-6062; Ward's, # 17W1146
rubber bulbs—Fisher, # 14-065B
test tubes—Fisher, # 14-957A (10 × 75 mm) or # 07-781D (6 × 50 mm); Ward's, # 17W0600
yeast—purchase locally
corks (size #l)—Fisher, # 07-781D; Ward's, # 15W8361
Petri dish, 100 × 15 mm—Fisher, # 08-747C (Pyrex) or 08-746C (Kimax); Carolina Biological, # 74-1158; Ward's, # 17W0700
stethoscope—Carolina Biological, # 69-1638 or 69-1659; Fisher, # 14-409-110; Ward's, # 15W8840

LABORATORY 40 The Digestive, Excretory, and Reproductive Systems

I. FOREWORD

This laboratory is designed to examine the anatomy and physiology of the digestive, excretory and reproductive systems. The chemistry of digestion is explored using amylase, pepsin, and lipase. The kidney is studied at the macroscopic and microscopic levels with reference to filtration functions. Reproductive structures are introduced as a part of the urogenital system..

II. TIME REQUIREMENTS

Exercise A—Examining the Digestive System
  Part 1—Microscopic Anatomy of the Digestive System (45 minutes)
  Part 2—The Role of Peristalsis (15 minutes)
Exercise B—The Chemistry of Digestion
  Part 1—Carbohydrates (15 minutes)
Extending Your Investigation: Where Is It Digested? Enzymes and pH (15 minutes)
  Part 2—Proteins (20 minutes and overnight incubation)
  Part 3—Fats (20 minutes and overnight incubation)
Exercise C—The Mammalian Kidney
  Part 1—Anatomy of the Mammalian Kidney (20 minutes)
  Part 2—Microscopic Anatomy of the Kidney (10 minutes)
  Part 3—The Urogenital System (5 minutes)
Exercise D—Gamete Formation (30 minutes)

III. STUDENT MATERIALS AND EQUIPMENT

	Per Student	Per Pair (2)	Per Group (4)	Per Class(24)
Exercise A (Part 1)				
salivary glands, human (prepared slide)		1		12
pancreas, human (prepared slide)		1		12
liver, human (prepared slide)		1		12
small intestine (duodenum), human (prepared slide)		1		12
compound microscope	1			24
Exercise A (Part 2)				
glass or paper cup	1			24
stethoscope		1		12
Exercise B (Parts 1, 2, and 3)				
test tubes	6			144
starch solution (1%), bottle **(1)**			1	6
amylase solution (2%), bottle **(2)**			1	6
Parafilm				
porcelain spotting plate		1		12

	Per Student	Per Pair (2)	Per Group (4)	Per Class(24)
Exercise B (Parts 1, 2, and 3)—continued				
Lugol's solution (I_2KI), dropping bottle **(3)**			1	6
maltase solution (1%), bottle **(4)**			1	6
TesTape			1	6
pepsin solution (5%) **(5)**			1	6
albumin solution **(6)**			1	6
2 N HC1 **(7)**			1	6
photographic film **(9)**		2		24
Petri dish		2		24
biuret reagent **(9)**			1	6
ninhydrin reagent **(10)**			1	6
oil-emulsion agar plates **(11)**		2		24
copper sulfate ($CuSO_4$) solution **(12)**			1	6
lipase solution **(13)**			1	6
pipettes, 5 ml	4			96
Pasteur pipettes	6			144
water bath (37° C)				1
incubator (30° C)				1
Extending Your Investigation				
same materials as used for Exercise B				
Exercise C (Part 1)				
sheep or beef kidney **(15)**		1		12
scalpel		1		12
scissors		1		12
probe		1		12
dissecting pan		1		12
Exercise C (Part 2)				
kidney tissue, monkey (prepared slide)	1			24
Exercise C (Part 3)				
written laboratory materials				
Exercise D				
rat testis (prepared slide)		1		12
rat or human spermatazoa (prepared slide)		1		12
cat ovary (prepared slide)				1

IV. PREPARATION OF MATERIALS AND SOLUTIONS

(1) starch solution (1%)

Bring 900 ml of distilled water to a boil. Mix 10 g of soluble potato starch in 100 ml distilled water and slowly pour this mixture into the 900 ml of boiling water. Heat the entire solution to a second boil, mix, and remove from the heat. Allow the solution to cool (cover with aluminum foil while cooling). Some starch may settle out of solution. Prepare fresh for each day.

(2) amylase solution (2%)

Dissolve 2 g of alpha amylase (fungal; breaks glucose into maltose) in enough distilled water to make 100 ml solution. Test this solution by adding 2 drops of amylase to 2 drops of starch. Then add 2 drops of Lugol's solution. If the solution immediately turns yellowish-gold, dilute amylase 1:10 (10 ml amylase to 90 ml distilled water). Repeat the test. Amylase strength varies (in units of activity) when purchased. You should obtain a brownish-red color with this test. As the time of incubation increases, a gold color will develop, indicating that starch has been catabolized to maltose.

(3) Lugol's solution (I_2KI)

Dissolve 10 g potassium iodide in 100 ml distilled water and add 5 g of iodine. Store in a dark bottle.

(4) maltose solution

Dissolve 10 g of maltose in enough distilled water to make 1 liter of solution.

(5) pepsin (5%)

Dissolve 50 g of pepsin in enough distilled water to make 1 liter of solution.

(6) albumin (1%)

Dissolve 1 g powdered albumin in enough distilled water to make 100 ml of solution.

(7) 2N HCl

Purchase already prepared (see Ordering Information).

(8) photographic film

Remove black and white film from case, expose to light, and develop (the film will be opaque black). Alternatively, use undeveloped film (translucent). Make sure you can identify the emulsion side on the developed film.

(9) biuret reagent

Order prepared solution. Approximately 500 ml per class of 24.

(10) ninhydrin

Dissolve 0.2 g of ninhydrin in 100 ml of 95% ethanol. Avoid breathing poisonous fumes.

(11) oil emulsion agar plates—

To 5 ml olive oil, add a pinch of bile salts (sodium taurocholate) and shake vigorously. Allow the mixture to stand for 10-15 minutes while you prepare agar.

Bring 100 ml of 2% agar (2 g Bacto-agar dissolved in distilled water to make 100 ml) to a boil.

Prepare 100 ml of a 5% soluble starch solution (5 g starch dissolved in distilled water to make 100 ml).

Combine the agar and starch. Shake the oil and combine it with the agar/starch mixture. Stir the mixture well and pour into Petri dishes, forming a thin layer.

(12) copper sulfate ($CuSO_4$)

Prepare 100 ml saturated $CuSO_4$. Add $CuSO_4$ to 100 ml distilled water until no more will dissolve (approximately 65 g).

(13) lipase (5%)

Dissolve 50 g lipase in enough distilled water to make 1 liter of solution.

(14) beef or sheep kidney

Order preserved material (see Ordering Information) or obtain from local abattoir. Use gloves to handle fresh material.

V. PREPARATION SUGGESTIONS

Exercise A (Part 1)

Students should examine prepared slides of organs responsible for production of enzymes and secretions used in the process of digestion, as well as a cross section of the intestine. Discuss the role of microvilli as part of the intestinal lining.

Exercise A (Part 2)

You will need a good stethoscope for this exercise. Make sure to use disposable paper or plastic cups for drinking and that students do not share reuse them.

Exercise B (Part 1)

Salivary amylase (α amylase) is responsible for the breakdown of starch into a mixture of dextrins and maltose. It works on $\alpha(1,4)$ linkages between glucose units of amylose, but not the $\alpha(1,6)$ linkages of amylopectin. Recall that starch is composed of both amylose and amylopectin. Further digestion of dextrins and maltose residues occurs in the intestine. The enzyme maltase is responsible for breaking down maltose to make glucose molecules.

Make sure that you test your a amylase solution before you start (see Preparation of Materials and Solutions (2)). You should see a progressive change in color from black (α amylase + starch gives black rather than blue color when it reacts with Lugol's solution) to brownish-red to gold. This indicates a breakdown of starch into dextrins and finally into maltose. When you add 1% maltose, you will find that α amylase does not break this down further. The TesTape will NOT indicate that glucose is present. You may wish to add 2 ml of 5% maltase to the solution and retest with TesTape to verify the presence of glucose after maltase breaks down maltose.

Extending Your Investigation

This investigation asks students to think about where amylase and pepsin work. Students should realize that amylase does not work in the stomach but pepsin does. They should make appropriate hypotheses. Why isn't starch digested to dextrins in the stomach? Why couldn't pepsin work in the mouth? Students will be able to investigate the pH requirements for these enzymes. They can design experiments and collect qualitative results.

Exercise B (Part 2)

To test for the action of pepsin, you will use the biuret test and the ninhydrin test, both described in Laboratory 5, Organic Molecules. Pepsin works in the stomach and has an optimum pH of 1.5 to 2.2. It is important that both your albumin and pepsin solutions should be at pH 7 before beginning this part of the experiment. The addition of 2 N HC1 to tube P_1 will lower the pH so that digestion will occur. Pepsin, however, does not reduce proteins to amino acids. Further hydrolysis by trypsin, chymotrypsin, and carboxypeptidase in the intestine continues to break down proteins. Tests with biuret reagent will indicate the continual presence of proteins (remember, pepsin is itself a protein). Ninhydrin will turn purple in the presence of amino acids. More amino acids should be present in tubes containing pepsin at a low pH.

This experiment can also be done by treating a piece of exposed photographic film with pepsin. Tape a piece of film to the bottom of a disposable Petri dish. Make sure that the dull side (emulsion side) is upward. Place a drop of each of the 4 solutions on the film. Wait overnight and wash the film, rubbing with your finger. A clear spot should result from the contents of tube 1. Note: you can substitute this procedure for the albumin procedure instead of doing both. In this case, you do not need the albumin in the 4 tubes—you just need pepsin, HCl, or water.

Exercise B (Part 3)

Make sure that you do not heat dishes to more than 30° C because the oil will run out of the agar. It is best to just leave this experiment incubating overnight. After treatment of your plates with copper sulfate ($CuSO_4$), you will need to pour off the $CuSO_4$ and rinse with water. Do this gently and be careful not to let the agar slip out of the plate (agar containing oil is very soft).

You might want to talk about saponification and the process of soap-making as students complete this exercise. The process of breaking the ester bonds between glycerol and the fatty acids produces free fatty acids. This can be done by using heat and a strong alkali such as NaOH. The fatty acids will react with Na^+ to form soaps. The soaps aggregate of form micelles with their hydrophobic fatty acid parts on the inside of the cluster of molecules and their hydrophilic Na^+ ends on the outside. Cu^{2+} interacts with fatty acids in the same way.

Exercise C (Parts 1 and 2)

You can use a fresh or preserved kidney for this exercise. Make sure that students wear gloves if fresh material is used. Students should be able to identify anatomically the region of the kidney to be studied histologically and then locate them on the prepared slide. Have students trace the pathway for the flow of excretory wastes through the kidney.

Exercise C (Part 3)

Make sure that students understand the relationship between the excretory system and the reproductive system—why the two are called the "urogenital system." Explain how the ducts of the anterior kidney are "taken over'" by the male gonad and the old opisthonephric duct becomes the vas deferens while a new duct, the ureter, forms to drain the kidney of organisms with metanephric kidneys (reptiles, birds, and mammals).

Exercise D

You might like to add a demonstration slide of a rat (or human) ovary to Part 2. You might explain why the second "o" in oögenesis, oögonia, etc., *should* have the diacritical mark (even though editors like to omit it)—it means that the second vowel must be pronounced (o'-o'-genesis rather than ou-genesis)!

VI. ORDERING INFORMATION

salivary glands, sublingual gland, human (prepared slide)—Carolina Biological, # H8845; Ward's, # 93W4483
pancreas, human (prepared slide)—Carolina Biological, # H8122
liver, human (prepared slide)—Carolina Biological, # H8145; Ward's, # 93W4570
small intestine, duodenum, human (prepared slide)—Carolina Biological, # H8000 or H8010; Ward's, # 93W4540
disposable paper or plastic cups—local supplier
stethoscope—Carolina Biological, # 69-1634; Ward's, # 15W8840
test tubes, 16 × 125 mm—Fisher, # 14-925H; Carolina Biological, # 73-1418; Ward's, # 17W1403
test tube rack—Carolina Biological, # 73-1896; Ward's, # 18W4213

starch, soluble potato—Carolina Biological, # 89-2530; Ward's, # 39W3275
amylase (alpha amylase, fungal) Carolina Biological, # 20-2350; Ward's, # 39W0058
Parafilm—Fisher, # 13-274-12 (dispenser, # 13-374-18); Carolina Biological, # 21-5600 (dispenser, # 21-5602); Ward's, # 15W1940
Lugol's solution (I_2KI)—Carolina Biological, # 87-2793 (100 ml) or # 87-2795 (500 ml); Ward's, # 39W1685
spotting plates—Carolina Biological, # 70-0600; Fisher, # 13-745; Ward's, # 18W2009
maltase—Sigma Chemical Company, # M3145
TesTape—Carolina Biological, # 89-3840; Ward's, # 14W4107
pepsin—Carolina Biological, # 87-9378; 39W2865
albumin—Carolina Biological, # 84-2250; Ward's, # 39W1097
2N HC1—Fisher, # SA431-500
Petri dish, 100 × 15 mm—Fisher, # 08-747C (Pyrex) or 08-746C (Kimax); Carolina Biological, # 74-1158; Ward's, # 17W0700
sodium taurocholate (bile salts)—Ward's, # 38W2179
(Bacto)agar[1]—Fisher, # DF0140-02-9; Carolina Biological, # 21-6720 or 21-6721 (500 g); Ward's, # 38W0015
soluble starch, potato—Carolina Biological, # 89-2532 (500 g); Ward's, # 39W3276
biuret reagent—Carolina Biological, # 84-8211 (120 ml); Ward's, # 37W0790
ninhydrin (powder)—Carolina Biological, # 87-7460; Ward's, # 37W3389
lipase—Sigma Chemical Company, # L-1754; Ward's, # 39W2090
copper sulfate (CuSO4)—Carolina Biological, # 85-6550; Fisher, # C 495-500; Ward's, # 37W2258
pipette, 5 ml—Fisher, #13-665K; Carolina Biological, # 73-6276; Ward's, # 17W1307
Pasteur pipette, 9"—Fisher, #13-678-20C; Carolina Biological, # 73-6062; Ward's, # 17W1145
rubber bulb—Fisher, # 14-065B
sheep or beef kidney—local grocery or abattoir
dissecting pans, 13 × 9" (aluminum)—Carolina Biological, # 62-9000 (without wax), # 62-9002 (with wax); Ward's, # 14W8500
kidney tissue, monkey (prepared slide)—Carolina Biological, # H0699 (H&E), H0601 (Mallory); Ward's, # 93W8701
testis (rat, cross section, prepared slide)—Carolina Biological, # H4159; Ward's, # 93W5441
cat (ovary, prepared slide)—Carolina Biological, # H3785; Ward's, # 93W5532
spermatozoa (human, prepared slide)—Carolina Biological, # 8790; Ward's, # 93W7241

[1] Bacto is a designation used on Difco Laboratories products available through Fisher Scientific, Inc., Carolina Biological Supply Company, and Ward's. (See Suppliers.)

LABORATORY 41 Control—The Nervous System

I. FOREWORD

This laboratory is designed to introduce selected aspects the structure and function of neurons, affectors (sensors), and the mammalian control center—the brain.

II. TIME REQUIREMENTS

PART I—Neurons
- Exercise A—Examining Nerve Cells (Neurons) (10 minutes)

PART II—Sensory Receptors
- Exercise B—Chemoreception: The Sense of Taste (30 minutes)
- Exercise C—Chemoreception: Individual Differences in Taste (20 minutes)
- Exercise D—Chemoreception: Smell Discrimination and Its Influence on Taste (20 minutes)
- Exercise E—Photoreception: Vision—Structure of the Eye (30 minutes)
- Exercise F—Photoreception: How We See (15 minutes)
  - Part 1—Peripheral Vision and Color Vision
  - Part 2—The Blind Spot
- Exercise G—Mechanoreception: The Role of Sensory Receptors in Touch (15 minutes)
- Exercise H—Thermoreception: Discriminating Temperature (15 minutes)
- Exercise I—Proprioception: The Role of Proprioceptors in Determining Position (10 minutes)
- Exercise J—Mechanoreceptors of the Ear (10 minutes)

PART III—The Brain
- Exercise K—The Structure of the Mammalian Brain (30 minutes)

III. STUDENT MATERIALS AND EQUIPMENT

	Per Student	Per Pair (2)	Per Group (4)	Per Class (24)
Exercise A				
neuron, cow (prepared slide)				1
compound microscope				1
Exercise B				
unknown solution A, bottle **(1)**			1	6
unknown solution B. bottle **(2)**			1	6
unknown solution C, bottle **(3)**			1	6
unknown solution D, bottle **(4)**			1	6
unknown solution E, bottle **(5)**			1	6
magnesium sulfate solution (10%), bottle **(6)**			**1**	**6**
cotton swabs	3			72
paper cups	1			24
Exercise C				
PTC taste paper (small pieces)	1			24
thiourea taste paper (small pieces)	1			24

	Per Student	Per Pair (2)	Per Group (4)	Per Class (24)
Exercise C—continued				
sodium benzoate taste paper (small pieces)	1			24
control taste paper (small pieces)	1			24
Exercise D				
Life Savers® pieces in Petri dish (1 piece of each flavor: strawberry, tangerine, butterscotch, peppermint)	1			24
Exercise E				
sheep eye (preserved)	1			24
dissecting pan (with wax)	1			24
single edge razor blade	1			24
disposable gloves (pairs)	1			24
Exercise F (Parts 1 and 2)				
colored objects (balls, boxes, cardboard cut-outs)				12
Exercise G				
metric ruler (30 cm)		1		12
protractor (compass)		1		12
Exercise H				
600 ml beaker			3	18
Exercise I				
8 inch piece of string	1			24
Exercise J				
rotating stool				4
Exercise K				
sheep brain (preserved)		1		12
dissecting pan				
disposable gloves (pairs)				

IV. PREPARATION OF MATERIALS AND SOLUTIONS

(1) solution A—10% NaC1
Obtain a mass of 100 g NaCl. Add distilled water to 1 liter and mix.

(2) solution B—water
Use 1 liter of distilled (or tap) water.

(3) solution C—10% sucrose
Obtain a mass of 100 g sucrose. Add distilled water to 1 liter and mix.

(4) solution D—quinine
Dissolve 1 capsule of quinine into each 1 liter of "almost" boiling distilled water.

(5) solution —5% acetic acid
Add 50 ml of glacial acetic acid slowly to 90 ml of distilled water.

(6) magnesium sulfate solution

Add 58 g anhydrous magnesium sulfate to 900 ml of distilled water and bring to 1000 ml using distilled water (0.4 M $MgSO_4$) or add 10 g of $MgSO_4 \cdot 7H_2O$ to 90 ml of distilled water and dilute to 100 ml (10%).

V. PREPARATION SUGGESTIONS

Exercise B

Give students five small paper cups (bathroom cup size) and have them label the cups A-E. Then, pour a small amount of the corresponding solution into each cup. Discard all cups when through. This might appear wasteful, but it is more sanitary since students may forget and dip a used cotton swab into one of the bottles of solution.

Exercise C

Cut tasting strips into squares to economize.

Exercise D

Break LifeSavers into thirds (otherwise you will go broke on LifeSavers).

Exercise F

Use single color objects with distinctly different shapes.

Exercise K

Sheep brains can be used from year to year. We find that there is less wear and tear on the brains if we remove them from their containers and place them on dissecting trays, ready to use by students. Cover the brains with wet paper towels so they do not dry out (put them back with preservative overnight).

VI. ORDERING INFORMATION

neuron, cow (prepared slide—Carolina Biological, # H1660; Ward's, # 93W3617
sodium chloride (NaCl)—Fisher, # S271-500; Carolina Biological 88-8901; Ward's, # 37W5487
sucrose—Fisher, # 55-500; Carolina Biological, #89-2870; Ward's, # 39W3182
acetic acid, glacial—Carolina Biological, # 84-1293; Ward's, # 39W0125
magnesium sulfate—Carolina Biological, # 87-3370; Ward's, # 14W4105
taste paper:
- PTC—Carolina Biological, # 17-4010; Ward's, # 14W4105
- sodium benzoate—Carolina Biological, # 17-4020; Ward's, # 37W9968
- thiourea—Carolina Biological, # 17-4030; Ward's, # 37W9962
- control—Carolina Biological, # 17-4000; Ward's, # 14W4116

magnesium sulfate—Carolina Biological, # 87-3400 (anhydrous) or # 87-3370 (crystal); Ward's, # 37W2861
sheep eye (preserved)—Carolina Biological, # P2130D; Ward's, # 69W7192
dissecting pans, 13 × 9" (aluminum)—Carolina Biological, # 62-9000 (without wax), # 62-9002 (with wax); Ward's, # 14W8500
single edge razor blades—buy locally; Carolina Biological # 62-6930; Ward's, # 14W4172
latex gloves, disposable—Carolina Biological, # 70-6340, 70-6341, and 70-6342 (small, medium, and large)

beaker, 600 ml— Fisher, # 02-540M (Pyrex) or # 02-539M (Kimax); Carolina Biological, # 72-1225 (Corning, student grade); Ward's, # 17W4060
sheep brain (preserved)—Carolina Biological, # P21SOF; Ward's, # 69W7101
quinine sulfate capsules—available at local pharmacies; Carolina Biological, # 88-6050; Ward's, # 39W3099

LABORATORY 42 Behavior

I. FOREWORD

This laboratory is designed to introduce students to the behavior of animals. Students work in pairs as they study simple responses to external stimuli, a behavior chain involved in courtship, and learning. Exercise C requires that students return on subsequent days to complete the training of their larvae. Alternatively, subsequent sections may continue this exercise through the week, sharing their results at the beginning of the following week.

II. TIME REQUIREMENTS

Exercise A—Reactions of Isopods to Light and Humidity (40 minutes)
Exercise B—Courtship Behavior in Fruit Flies (60 minutes)
Exercise C—Social Behavior in Crickets (120 minutes)
Part 1—Aggressive Behavior and Social Dominance (60 minutes)
Part 2—Aggressive Behavior, Dominance, and Courtship Among Male and Female Crickets (60 minutes)
Exercise D—Learning in the Mealworm (120 minutes, continued over several days)

III. STUDENT MATERIALS AND EQUIPMENT

	Per Student	Per Pair (2)	Per Group (4)	Per Class (24)
Exercise A				
paper lids (opaque)		2		24
plastic lids (clear)		1		12
aluminum dissecting pan		1		12
pill bugs (isopods) **(1)**		10		120
Exercise B				
Drosophila anesthetized				
vial (5 male flies) **(2)**		1		12
vial (5 virgin female flies) **(2)**		1		12
mating chamber		1		12
fly morgue			1	6
Exercise C (Part 1)				
plastic shoe box			1	6
male crickets, marked **(3)**			4	24
matchbox **(4)**			1	6
Exercise C (Part 2)				
plastic shoe box			1	6
male crickets, marked **(3)**			4	24
female crickets, marked **(3)**			2	12

	Per Student	Per Pair (2)	Per Group (4)	Per Class (24)
Exercise D				
black box		1		12
T-maze		1		12
mealworm larvae **(5)**		1		12
model paint (set with several colors)				1
small paint brushes (one for each color paint)				1
flashlight			1	12

IV. PREPARATION OF MATERIALS AND SOLUTIONS

(1) isopods

Pill bugs (sow bugs or "rolly pollies") are easily collected in many regions of the country by examining the undersides of rocks, bricks, or boards lying on the soil in damp areas. Place them in a plastic refrigerator container with some moist moss or leaf litter until needed. These crustaceans are generally available year round in many parts of the country. They may also be purchased (see Ordering Information).

(2) male and female *Drosophila*

Use wild type flies—mating behavior is aberrant in some mutant phenotypes. Separate the sexes and prepare vials of flies for students before the laboratory. To collect virgin females, empty culture vials of all adults early in the morning (place them in the morgue) and remove females that emerge before they are 8-10 hours old (they will not mate in this time interval). Do not use flies that have been anesthetized within the past 24 hours. Complete directions for culturing and handling fruit flies are given in the Carolina *Drosophila* Manual.

(3) crickets, marked

Order crickets from a cricket farm. Do not purchase them from a bait store because you will not know age and they will have already established dominance patterns. Suggested vendor—Fluker's Farms, 1333 Plantation Road, Port Allen, LA 70767 (504-343-7035), http://www.flukerfarms.com/. (See Ordering Information for other sources.)

Sex crickets when they arrive and separate males into one aquarium and females into another. Females can be identified by the long, needle-like ovipositor extending from their abdomens.

Place cardboard-type egg cartons upside down in the bottom of the aquarium, giving enough room for the crickets to move around. Fill a Petri dish with aquarium gravel and add water. Place this in the tank. Do not use a container of open water. Keep the cage as dry as possible. Place small wedges of potato on a Petri dish as food. Change food and water about every 4 days. Crickets have a life span of two weeks.

Approximately 3-4 days before using crickets for experiments, you must isolate both males and females. Pick the crickets up carefully using small containers. We use either chili containers (Wendy's) or urine specimen cups with small holes poked in the top. Do not use regular paper cups—the crickets will eat through. Place a small wedge of potato in each cup for food and moisture. The day before use, obtain tempera paint and a small brush. Place a dot of colored paint (red, yellow, blue, or green) on the back of each cricket. Place a corresponding dot on the cup. Only the males need to be color marked. You will need 4 male crickets (each of a different color) for each group of students. You will need 2 females for each group of students.

Have enough crickets isolated so that each set is used for only one laboratory. Have students place "used" crickets back into the aquarium

(4) matchbox
Use the outer sleeve from small, safety match boxes.

(3) mealworms (*Tenebrio*)
Mealworms can be obtained in most pet stores or they can be purchased from Carolina Biological Supply Company or Ward's (see Ordering Information), or obtained from a local culture.

To develop a culture, use a flat, opaque plastic or metal storage container. Place a 1" layer of uncooked Cream of Wheat or bran meal and four layers of paper towels in the pan and add larvae (and adults if available). The towels should be sprinkled lightly with water every 4-5 days and replaced as necessary. Add cereal when the level decreases (every several months). A slice of apple or other fruit also adds moisture to the colony that can be maintained indefinitely. Start your colony well in advance of use and maintain it from year to year.

Individual larvae should be marked with dots of model paint in different color combinations for individual identification. Keep mealworms "in training" in a small separate culture dish with only a small amount of food so they can be easily located.

V. PREPARATION SUGGESTIONS

Exercise A

Collect opaque lids from pint ice cream containers, Wendy's chili cartons, or other food containers for use in this exercise. Cut 2 "doors" in opposite sides of the lip. The lids should be approximately the same diameter. Plastic Petri dishes can be used as clear shelters. Cut "doors" in the sides opposite one another (use a heated wire cheese-slicer to make cuts). Clean, empty dissecting pans without wax or vinyl or disposable aluminum trays can be used to hold the experiments—they should not be larger than necessary so that the isopods are constrained to some extent and cannot wander off.

Exercise B

For safety, a non-explosive anesthetic should be used in anesthetizing *Drosophila*—do not use diethyl ether. Use FlyNap or carbon dioxide. A CO_2 generator can be made using a 125 ml sidearm flask. Connect a piece of rubber tubing to the sidearm and place a Pasteur pipette on its end (this can be slipped between the sponge rubber stopper and side of the vial holding the flies). Place 2 Alka-Seltzer tablets in 50 ml of water and close the mouth of the flask with a rubber cork. Place the Pasteur pipette into the bottle of flies, sliding it between the foam cork and side of the container. When you anesthetize flies, dump them into a clean vial without medium. Do not anesthetize flies in vials containing soft medium or the flies will get stuck in the medium.

A fly morgue should be available for each table or group. Use a wide mouth jar with oil or a detergent solution. Be sure that used flies are not released to roam freely in the laboratory.

A *Drosophila* mating chamber can be fashioned from a variety of materials—a small Petri dish would be ideal, but it is difficult to transfer alert, unanesthetized flies into such a shallow container. A 1.25" diameter culture vial can also be used—heat a small wire in a flame and melt several small holes in the bottom of the plastic vial (be sure that they are not large enough to allow flies to escape). Insert an applicator stick into a plastic foam plug and glue it in place. After the flies are added to the chamber, the plug can be pushed down the tube to restrict the flies to the bottom inch or so of the chamber for observation.

Other chambers are described in the following references:

Larson, J. R. and D. M. Meyer. 1984. Animal behavior experiments using arthropods. In *Tested Studies for Laboratory Teaching.* C. L. Harris (ed.). Kendall/Hunt Publishing Company, Dubuque, Iowa. p. 127.

Marler, P. 1975. Courtship behavior of *Drosophila.* In *Animal Behavior in Laboratory and Field.* 2nd ed. E. P. Price and A. W. Stokes (eds.). W. H. Freeman and Company, San Francisco. p. 67 (see p. 73).

Exercise C (Parts 1 and 2)

Use a plastic or translucent Rubbermaid shoebox as a container for your experiment. Some students may wish to make a grid the size of the bottom of the box to help map territories. In this case, you must be able to see through the box. The grid can be labeled A through K or L along the top and 1 to 5 or 6 down the side.

Crickets prefer low light, so keep the room light dimmed. Crickets are also sensitive to vibrations—do not move the plastic container during the experiment and do not bump the table it is sitting on. Give students small paper cups to handle crickets.

Students will have fun with this lab. When one group establishes its most dominant cricket, you may wish to have crickets from different groups "face off." If you allow students to do this, use other males for Part 2, Courtship.

Exercise D

You will need to construct the training apparatus used in this exercise. Use a box measuring about 8 × 10 × ½" to hold the T-maze. This can be made from 0.25" plywood or cardboard (small, flat gift boxes are adequate). It should be sprayed with flat black paint on the inside surfaces and allowed to dry thoroughly to dissipate all paint odor.

A floorless T-maze should be constructed with an alley (the upright member of the "T") measuring 5" (13 cm) by 1/8" (3 mm) and the two arms measuring 4" (10 cm) by about 0.5" (1.3 cm). The maze should be about 0.5-0.75" high (1.3-2 cm). The mealworm larva should not be able to turn around in the alley but should be able to do so in the arms of the "T." The maze should be made of plywood and should be spray-painted flat black.

Be sure that the room lights are kept dim and offer no directional information that can affect behavioral tendencies during the training period.

VI. ORDERING INFORMATION

dissecting pans, 13 × 9" (aluminum)—Carolina Biological, # 62-9000 (without wax), # 62-9002 (with wax); Ward's, # 14W7000

pill or sow bugs—Carolina Biological, # L 624; Ward's, # 87W5520

Drosophila (wild type)—Carolina Biological, # 17-2100 (or # 17-2781); Ward's, # 87W6550

Carolina *Drosophila* Manual—Carolina Biological, # 45-2620

FlyNap/anesthetizer kit—Carolina Biological, # 17-3015

Drosophila culture vials—Carolina Biological, # 17-3085; Ward's, # 18W4957

Drosophila vial plugs—Carolina Biological, # 17-3086; Ward's, # 18W4960

crickets—Fluker's Cricket Farm, 1333 Plantation Road, Port Allen, LA 70767 (504-343-7035);

Carolina Biological, # L715; Ward's, # 87W6100[1]

mealworm larvae, *Tenebrio—Carolina* Biological, # L 891 (or # L 893 for starting a new culture); Ward's, # 87W6250

plastic shoe boxes—purchase locally

[1] Other potential vendors for crickets include Nature's Way, PO Box 7268, Hamilton, OH 45013-7268, 800-318-2611 (http:/www.herp.com/nature/nature/html) and Armstrong's Cricket Farm, PO Box 125, West Monroe, LA 71294, 800-345-8778 (http:/www.best.com/watr/cricket/cricket.html).

Notes:

LABORATORY 43 Communities and Ecosystems

I. FOREWORD

This laboratory examines several properties of interacting organisms and their environments, beginning with a repetition of Gause's classic competition experiments in *Paramecium* (this exercise also illustrates the nature of population growth in monocultures). Invertebrate organisms in leaf litter communities are sampled to show the diversity of organisms in different samples taken from areas with different environmental conditions. Finally, some features of climate that affect biomes are examined.

II. TIME REQUIREMENTS

Exercise A—Observing Competition Between Species Sharing Resources (60 minutes)
Exercise B—Measuring the Diversity of a Community (60 minutes)
Exercise C—Using Climate Data as an Index to Vegetation (30 minutes)

III. STUDENT MATERIALS AND EQUIPMENT

	Per Student	Per Pair (2)	Per Group (4)	Per Class (24)
Exercise A				
Paramecium caudatum, P. aurelia, and *P. caudatum-P. aurelia* cultures **(1)**				9
microscope slides, box (75 × 25 mm)			1	6
0.01 ml (10 μl) micropipette	1			24
compound microscope	1			24
Exercise B				
organisms from leaf litter **(2)**				1
figure of representative organisms (included)	1			24
dissecting microscope	1			24
small Petri dishes		1		12
Exercise C				
weather data (included)	1			24
world map				1
graph paper	1			24

IV. PREPARATION OF MATERIALS AND SOLUTIONS

(1) *Paramecium* cultures

Gather together the following materials:

Paramecium caudatum (live culture)
Paramecium aurelia (live culture)
2000 ml Erlenmeyer flask
9 sterile Erlenmeyer flasks, 250 ml
10 sterile cotton plugs (cheesecloth covered)

2 2 ml sterile pipettes
timothy hay

Procedure:

Order 1 culture (for a class of 30 students) each of *Paramecium caudatum* and *P. aurelia* for delivery 3 weeks before the laboratory, a second set of cultures (one of each species) for delivery 2 weeks before the laboratory, and a third set of cultures (one of each species) for delivery 1 week before the laboratory. These sets of cultures will be sufficient for multiple laboratory sections, but should be ordered fresh to start each population flask.

On the day before each culture set is to arrive (i.e., 3 weeks and a day before the lab, 2 weeks and a day before the lab, and 1 week and a day before the lab), prepare a hay solution as follows: add 1000 ml of pond water and a handful of timothy hay to a 2000 ml Erlenmeyer flask and bring the mixture to a boil. Boil slowly (avoid boiling over) for 10 minutes and plug with a sterile, cheesecloth wrapped, cotton plug. Remove from the burner or hot plate and cool at room temperature overnight. Pour 100 ml of supernatant (filter through 4 layers of sterile cheesecloth if necessary) into each of three 250 ml sterile Erlenmeyer flasks and stopper each with a sterile, cheesecloth wrapped, cotton plug.

Label one flask "A." Stir a *Paramecium aurelia* culture with a sterile 2 ml pipette to suspend the organisms evenly in the stock jar (be careful to dislodge any organisms adhering to the sides). Use the pipette to transfer 4 ml of medium (containing about 400 organisms) to the flask and replace the sterile plug. Label a second flask "C." Aseptically, transfer 2 ml of *P. aurelia* culture to this flask (containing about 200 organisms) and stopper. Stir the *P. caudatum* culture with a second sterile 2 ml pipette and add 1.6 ml of medium (also containing about 200 organisms) to the same flask and stopper. Label a third flask "B." Aseptically, transfer 3.2 ml (about 400 organisms) to this flask and replace the sterile stopper. Add the age (3 weeks, 2 weeks, or 1 week) to the label on each flask and hold them at room temperature out of direct sunlight until the laboratory period. Note that this process needs to be repeated three times to supply a total of 9 cultures for each laboratory: 6 control flasks (3 of each species) and 3 experimental flasks. These cultures can be used by several successive laboratories over 2 days or so, if cultures are not spilled or badly contaminated.

There is the potential for large sampling errors in this experiment, but barring problems from contamination, population growth and competition should both be demonstrated. Remember to use aseptic technique when preparing the cultures and when sampling from the cultures if you will be using them for more than one laboratory period.

Alternatives:

Other methods exist for the preparation of the culture medium used in this experiment. One involves the use of protozoan pellets (Carolina Biological Supply Company) and sterile pond water. A second method involves boiling grains of wheat (50 g) in 500 ml of water to make a stock medium which is then diluted using 10 ml stock to 100 ml of sterile pond water. Try to minimize contamination in whatever method is used—we want students to see how the two species of paramecia interact, not bacteria or other microorganisms!

(2) leaf litter

Gather together the following materials:

litter samples
plastic bags
shelf for extraction apparatus

100 w light bulbs
plastic funnels (150 mm top diameter, 28 mm stem diameter)
250 ml Erlenmeyer flasks
70% ethanol

Procedure:

Prepare an extraction rack for collecting soil invertebrates. A wooden rack should be built with three shelves to hold the apparatus: the lowest shelf will hold 250 ml Erlenmeyer flasks directly below the small opening of funnels protruding through holes drilled in the middle shelf. The top shelf holds 100 w light bulbs used to provide light and gentle heat to drive organisms from the soil. Construct the rack to hold 3-10 "collectors" depending on the numbers of samples needed. The tip of each funnel should be slightly below the top of each flask so organisms extracted fall into the flask. Light bulbs should be about 6" above the tops of each of the funnels. Alternatively, set up a comparable apparatus using ring stands and clamps.

Several days before this laboratory, collect samples of leaf litter taken from shaded areas that are relatively moist. Place them in plastic bags. [Samples should include decomposing leaf litter down into the upper soil layer at least 1 cm deep. Each sample should be approximately 500 ml in volume.] Try to collect samples from contrasting habitats—deciduous woodland, coniferous forest, riparian habitat, etc. Label your samples. Students could make these collections if you wish to expose them to different habitats and the sampling process.

Place a small piece of screen wire in the bottom of each funnel to keep nonliving portions of your sample from falling into the flasks below. Use a different funnel for each sample. After the funnels with samples are in place, carefully place a 250 ml Erlenmeyer funnel with 50 ml of 70% ethanol below each. [Mix 700 ml of 95% ethanol with 250 ml of distilled water to make 70% alcohol.] Turn on the bulb over each funnel (be sure that nothing flammable can come in contact with a bulb and that the wiring is safe). Leave the apparatus overnight (longer if the soil is especially wet).

Carefully remove each flask (do not bump the funnels—the dry litter will contaminate samples easily). Stopper and label each flask for use in the laboratory.

Provide small dishes for pairs of students to share samples while they study them under the dissecting microscope. Samples can be used in several laboratories if students are careful to return all organisms to the proper flask at the end of the laboratory.

If time permits, some students might find it especially interesting to examine fresh samples and see some of the organisms live. Carefully remove intact samples of leaf litter about 2-3 cm deep and wrap in plastic and aluminum foil to keep them moist. Allow students to work on these samples with dissecting needles under high magnification on a dissecting microscope.

V. PREPARATION SUGGESTIONS

Exercise A

See notes in section IV, Preparation of Materials and Solutions **(1)**.

Exercise B

The Shannon-Weaver index for small samples has a large variance, so caution students not to attach importance to minimal differences in indices. On the whole we would expect mature forests to be more diverse than early succession stages and deciduous woodlands should be more diverse than coniferous areas—particularly monoculture plantations. Riparian habitats should be intermediate, depending on the sites sampled.

A simple key to microarthropods and drawings of representative soil invertebrates are included at the end of this exercise in the *Preparator's Guide*.

Exercise C

Weather data for several cities are tabulated at the end of this exercise in the *Preparator's Guide*. Make copies for students to use. This part of the laboratory could be done at home if there is not time in the laboratory period to complete the exercise. Use a textbook map of biomes to display in class or refer students to the map if the exercise is to be done at home.

VI. ORDERING INFORMATION

Paramecium caudatum (live culture)—Carolina Biological , # L2A; Ward's, # 87W1310
Paramecium aurelia—Carolina Biological, # L2B; Ward's, # 87W1300
Erlenmeyer flask, 250 ml—Fisher, # 10-040F (Pyrex) or # 10-039F (Kimax); Carolina Biological, # 72-6672 (Corning, student grade)
Erlenmeyer flask, 2000 ml—Fisher, # 10-040M (Pyrex) or # 10-039M (Kimax); Carolina Biological, # 72-6702 (Pyrex)
plugging cotton, bacteriological—Carolina Biological, # 71-2600; Ward's, # 15W3828
cheesecloth—Fisher, # 06-665-18 (70 yard bolt); Carolina Biological, # 71-2690 (5 yard package); Ward's, # 15W0015
pipette, 5 ml—Fisher, #13-665K; Carolina Biological, # 73-6276; Ward's, # 17W4854
protozoan pellets—Carolina Biological, # L50P
sterile spring water—Carolina Biological, # L5 1X (1 liter)
100 ml microcapillary pipette—Fisher, # 21-4517
plastic funnel, polypropylene (OD 150 mm top, 28 mm bottom)—Fisher, # 10-348D or # 73-4016 (6" polypropylene); Ward's, # 18W1425

CLIMATOLOGICAL DATA FROM DIFFERENT BIOMES[1]

Sta	Var	Jan	Feb	Mar	Apr	May	Jun	Jul	Aug	Sep	Oct	Nov	Dec
1	T	0.3	0.6	2.4	6.8	17.2	19.9	22.7	22.8	14.9	11.8	7.9	0.5
	PT	8.5	11.8	11.7	5.8	4.7	11.6	11.1	19.2	12.9	8.8	4.5	7.4
2	T	26.8	25.4	24.6	19.4	13.3	11.8	11.6	11.4	13.9	19.3	23.3	25.8
	PT	0.5	3.3	4.0	0.8	2.8	1.4	0.3	3.1	1.4	0.0	2.3	2.3
3	T	5.4	5.4	13.6	18.6	19.4	25.2	26.5	26.3	21.6	17.0	9.7	4.3
	PT	2.0	0.5	6.3	14.5	10.8	5.8	3.1	3.6	8.0	7.4	1.0	2.6
4	T	8.4	11.3	12.7	12.3	16.4	19.5	22.4	23.5	22.4	17.3	13.3	11.8
	PT	3.2	2.3	5.4	8.7	2.3	4.7	2.1	0.6	3.8	8.4	0.0	2.0
5	T	24.4	29.0	32.5	30.0	28.2	26.3	24.9	24.9	25.3	26.0	24.5	23.8
	PT	0.0	0.0	0.0	4.1	13.6	11.1	23.6	45.0	18.5	2.6	0.0	0.0
6	T	25.9	26.6	27.1	27.1	27.5	27.2	26.9	27.4	26.8	26.8	25.9	25.7
	PT	29.0	9.3	18.8	14.6	17.5	19.7	16.7	13.0	18.5	14.3	35.7	67.0
7	T	–31.9	–33.0	–32.7	–20.5	–12.6	–0.7	1.0	–0.4	–1.3	–9.8	–19.1	–30.0
	PT	0.8	2.0	0.6	1.7	1.2	2.0	2.6	0.7	1.7	1.8	1.8	0.4
8	T	–24.8	–18.7	–21.4	–6.8	4.2	8.1	12.5	11.0	4.1	–2.1	–11.7	–19.9
	PT	4.0	3.7	4.8	1.9	7.0	16.4	17.2	21.6	11.4	8.0	4.7	3.2
9	T	26.2	26.6	26.4	26.6	27.2	25.1	24.8	24.4	25.8	26.0	26.0	26.4
	PT	0.8	27.5	15.0	14.2	29.0	8.7	30.8	15.3	4.2	6.5	1.1	6.5

Sta—station, see below for location
Var—variable: T = mean monthly temperature (° C); PT = mean monthly precipitation (cm)
Jan, Feb,...Dec—months

Data sites include:

1. Pittsburgh, PA, USA; latitude 40° 30' N, longitude 80° 13' W (**temperate deciduous forest**)
2. Kalgoorlie, Australia; latitude 30° 46' S, longitude 121° 27' E (**desert**)
3. Oklahoma City, OH, USA; latitude 35° 24' N, longitude 97° 36' W (**temperate grassland**)
4. Barcelona, Spain; latitude 41° 24' N, longitude 2° 9' E (**Mediterranean, chapparal**)
5. Moundau, Chad; latitude 8° 37 N, longitude 16° 4' E (**savannah**)
6. Singapore; latitude 1° 22' N, longitude 103° 55 E (**tropical forest**)
7. Cape Cheljuskin, USSR; latitude 77° 43' N, longitude 104° 17' E (**tundra**)
8. Nitchequon, Canada; latitude 53° 12' N, longitude 70° 54' W (**taiga**)
9. Aracaju, Brazil; latitude 10° 55' S, longitude 37° 3' (**monsoon forest**)

[1] Gale Research. 1981. Weather of US Cities. Vol. 2. Gale Research, Detroit; National Climatic Center. 1978 and 1979. NOAA, Asheville.

Key to Soil Microarthropods[1]

1a Six legs, distinct head, segmented thorax and abdomen .... **Class Insecta**
1b More than six legs **5**

2a Antennae, eyes, and cerci absent, 9-12 abdominal segments................ **Order Protura**
2b Antennae present, cerci absent or present **3**

3a Antennae with 10-12 segments; 2-3 cerci on tip of abdomen as filaments or pincers .... **Order Thysanura**
3b Antennae present, cerci absent **4**

4a Antennae with 4-8 segments, 6 abdominal segments, forked appendages ("spring") for leaping (usually on fourth abdominal segment) .... **Order Collembola**
4b Antennae with 5 segments, body cylindrical, claws absent on legs .... **Order Thysanoptera**

5a Head and thorax fused together, 8 legs (**Class Arachnida**) **6**
5b More than 8 legs on body **8**

6a Abdomen distinctly segmented, first pair of leglike appendages (palpi) contain large pincerlike claws, 4 pairs of true legs .... **Pseudoscorpions**
6b Abdomen not segmented **7**

7a Body smooth, without hairs or projections .... ticks, **Order Acarina**
7b Hair on body and legs .... mites, **Order Acarina**

8a One pair of legs on each abdominal segment................. centipedes, **Order Chilopoda**
8b Two pairs of legs on each segment .... millipedes, **Order Diplopoda**

[1] See Borrer, D. J. and R. E. White. 1970. *A Field Guide to the Insects of America North of Mexico.* Houghton Mifflin Company, Boston. for a pictorial key to the orders of insects.

Soil Invertebrates

PHYLUM ARTHROPODA
Class Insecta

Order Protura
(proturans)

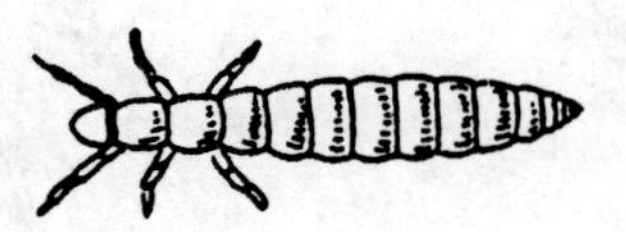

Order Thysanura
(silverfishes)

Order Collembola
(springtails)

Order Thysanoptera
(thrips)

Order Coleoptera
(beetles)

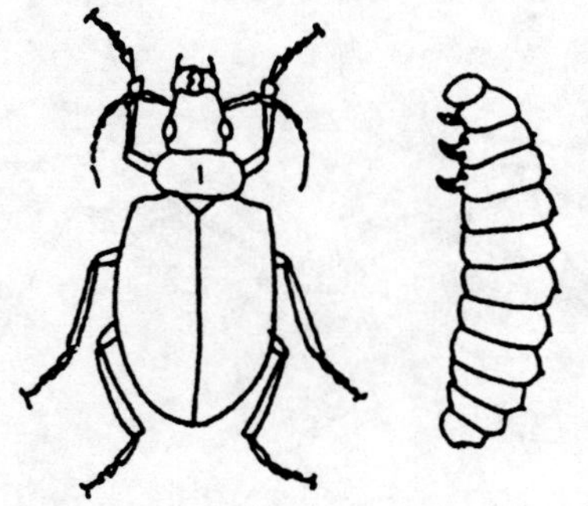

Order Orthoptera
(roaches)

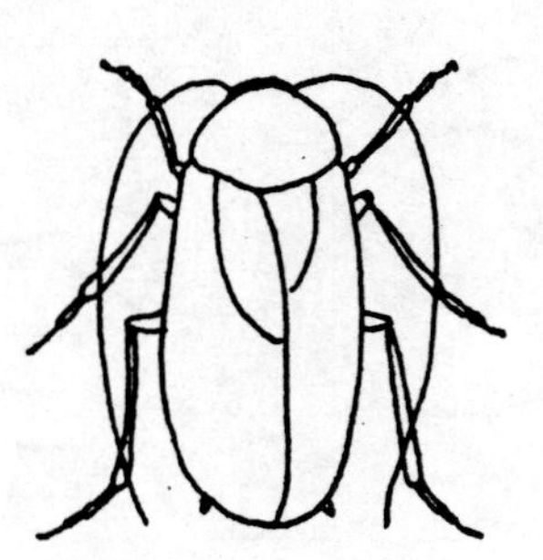

Order Hymenoptera
(ants)

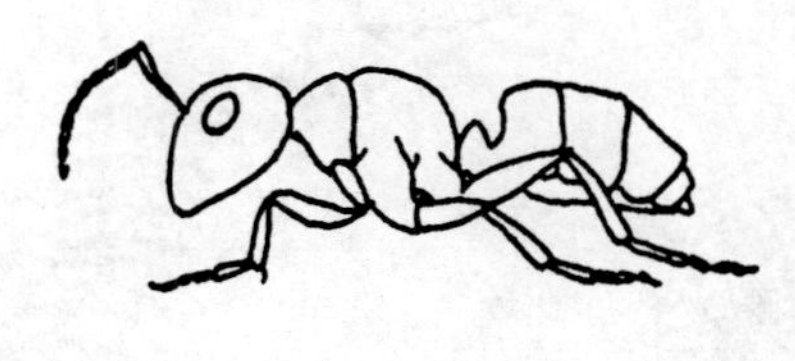

Order Siphonaptera
(fleas)

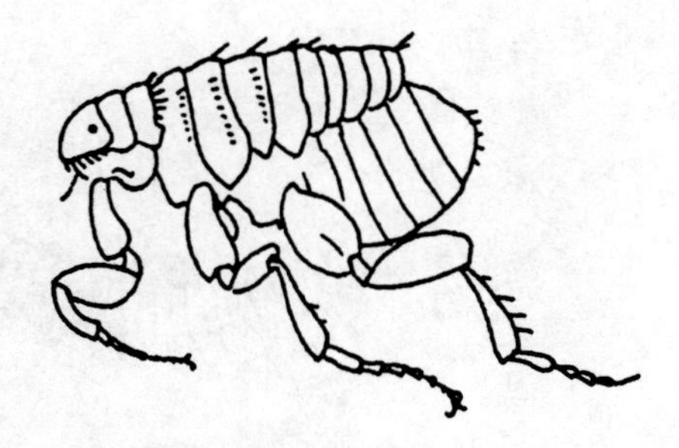

Order Hemiptera
(true bugs)

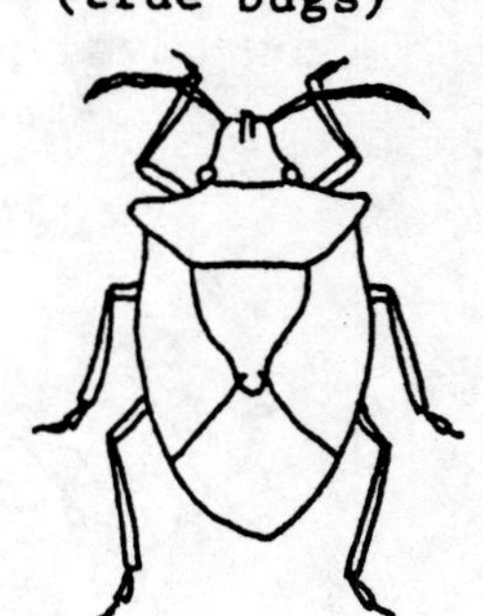

Order Diptera
(flies)

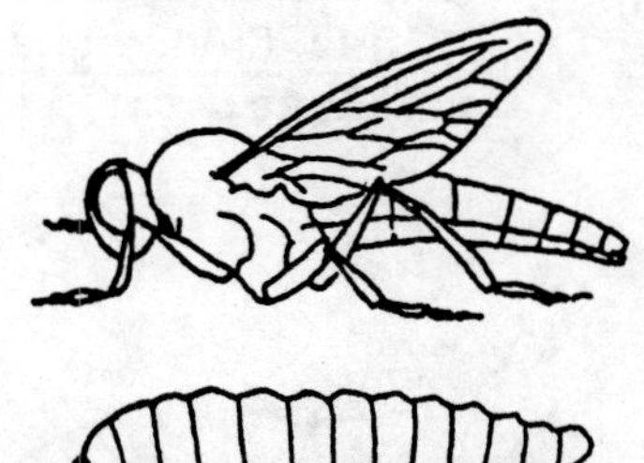

Order Psocoptera
(bark lice)

Order Isoptera
(termites)

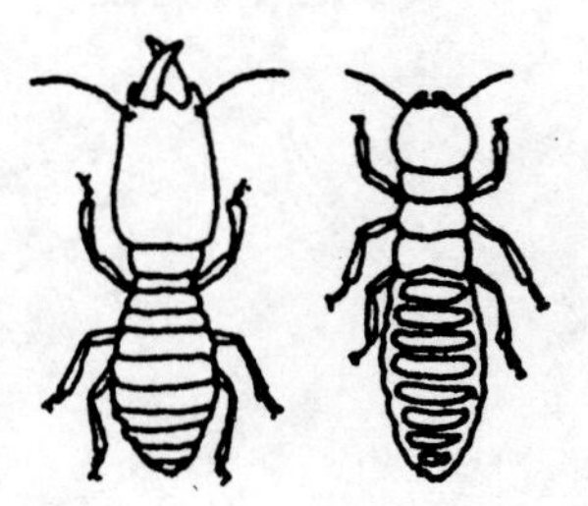

PHYLUM ARTHROPODA
Class Arachnida

Order Araneae
(spiders)

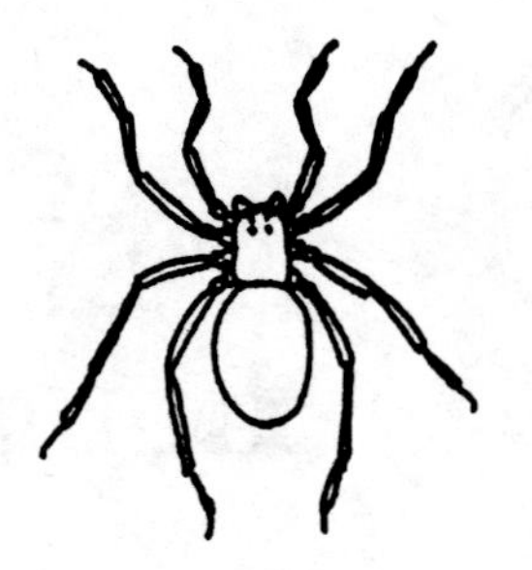

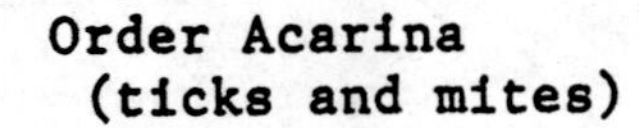

Order Acarina
(ticks and mites)

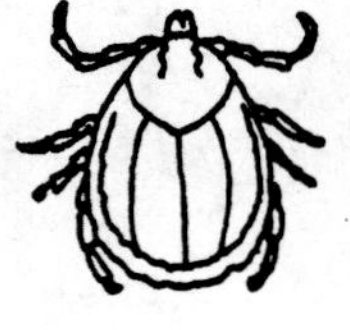

Order Pseudoscorpiones
(pseudoscorpions)

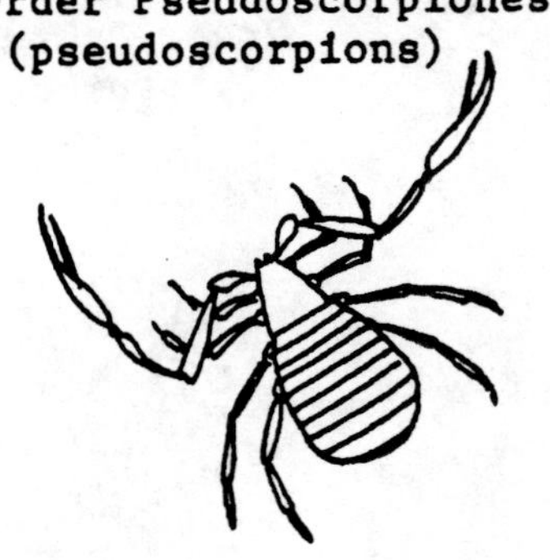

Order Opiliones
(daddy long-legs, harvestmen)

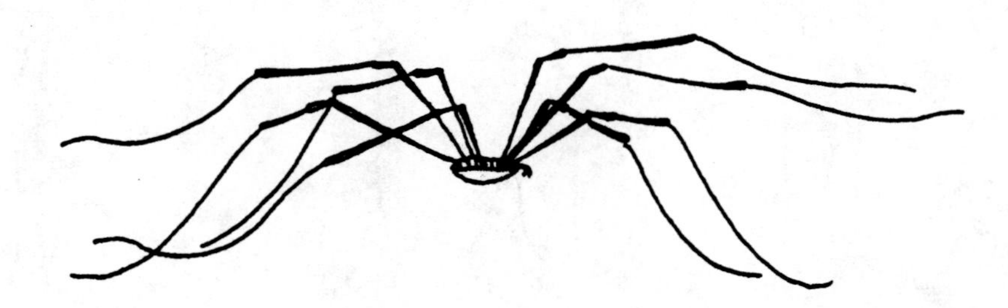

Class Crustacea
Order Isopoda
(sow bugs, pill bugs)

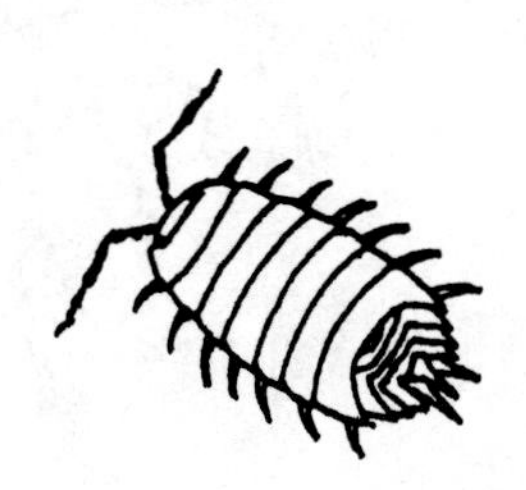

Class Diplopoda
(millipedes)

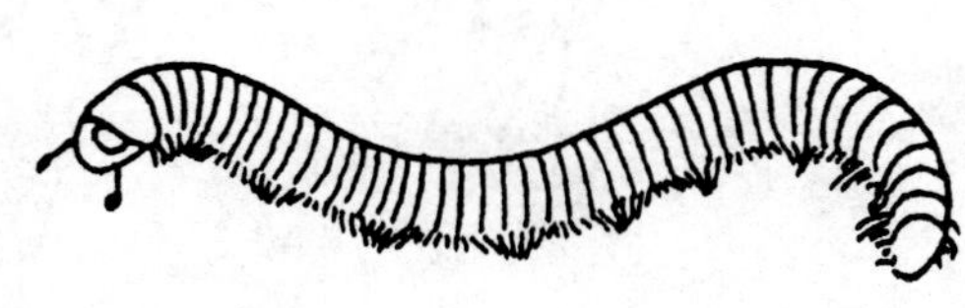

Class Chilopoda
(centipedes)

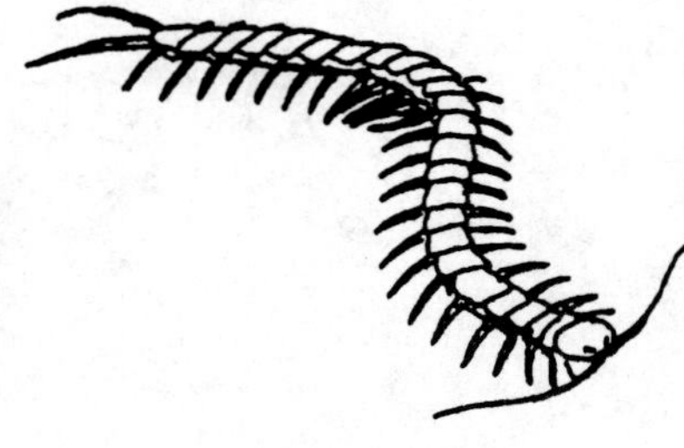

PHYLUM NEMATODA
(roundworms)

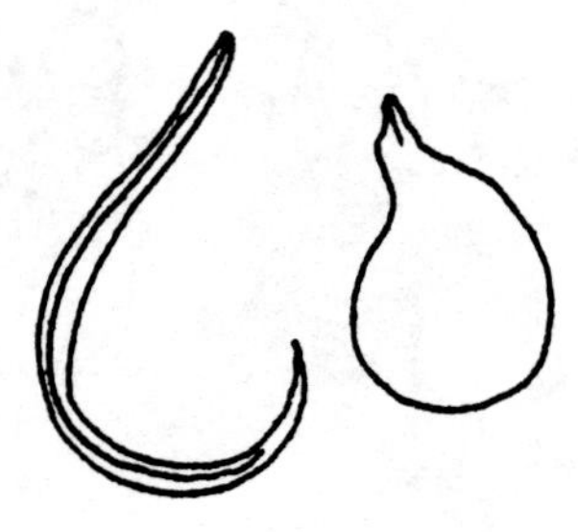

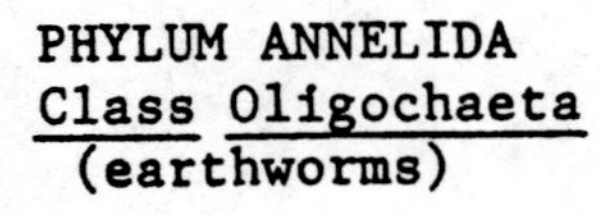

PHYLUM ANNELIDA
Class Oligochaeta
(earthworms)

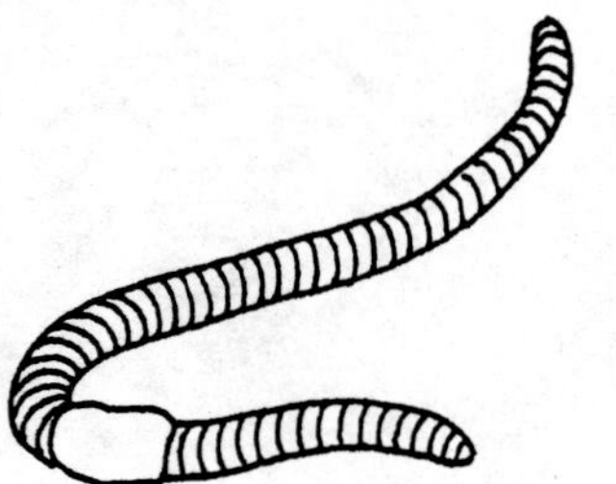

PHYLUM MOLLUSCA
Class Gastropoda
(slugs, snails)

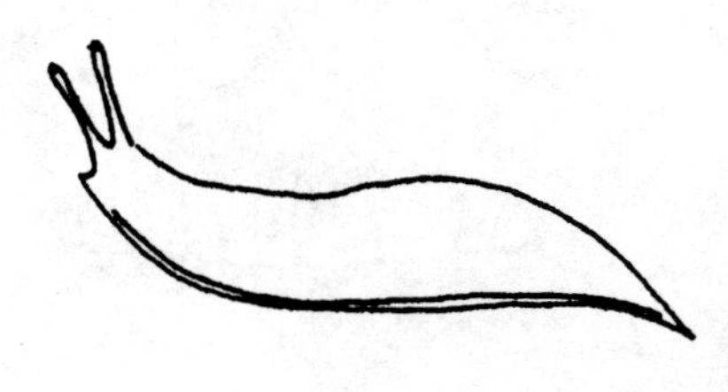

JOHN NORTON

LABORATORY 44 Predator-Prey Relations

I. FOREWARD

These exercises use simulations to illustrate predation, functional responses by prey, and competition.

II. TIME REQUIREMENTS

Exercise A—Predation (60 minutes)
Exercise B—Functional Responses by Predators (60 minutes)
Exercise C—Competition (40 minutes)

III. STUDENT MATERIALS AND EQUIPMENT

	Per Student	Per Pair (2)	Per Group (4)	Per Class (24)
Exercise A				
macaroni (box)		1		12
dish or box lid		1		12
string or tape to establish grids **(1)**				
tape measure (10 m)				1
flags **(2)**	1			24
Exercise B				
materials from Exercise A				
bags		1		12
Exercise C				
materials from Exercise A				

IV. PREPARATION OF MATERIALS AND SOLUTIONS

(1) grids

Use masking tape or string to delimit 10 × 10 or 20 × 20 foot (3 × 3 or 6 × 6 m) grids.

(2) flags

Use small triangles of red material supported by 1 foot long lengths of heavy gauge wire. Use flags to mark grids. This is especially important for multiple groups/laboratory section and because you must move to a clean new grid space for each exercise

V. PREPARATION SUGGESTIONS

Exercises A, B, and C

See directions in the exercise. These exercises are to be carried out outdoors. Macaroni is used because it is biodegradable. Use similar materials if you wish to substitute. Mark grids with flags and make sure that, as new grids are established, they do not overlap onto old spaces where residual food may be present.

VI. ORDERING INFORMATION

string or masking tape—purchase locally
measuring tape (10 m)—purchase locally
wire and cloth to make flags—purchase locally

LABORATORY 45 Productivity in an Aquatic Ecosystem

I. FOREWORD

This laboratory is designed to introduce students to some basic techniques in aquatic ecology. Students will learn how to use the Winkler method to determine the amount of dissolved oxygen in water samples and how this quantity can be used to measure primary productivity. Exercise A is designed to allow students to practice the Winkler titration. Exercise B introduces productivity in aquatic ecosystems using light and dark bottles. Exercise C gives students a chance to simulate measuring the primary productivity of a column of water in a lake or marine environment. Exercise D explores thermal stratification in temperate fresh water lakes and its effects on dissolved oxygen patterns during the year.

II. TIME REQUIREMENTS

Exercise A—Measuring Dissolved Oxygen: Effects of Temperature and Salinity (45 minutes)
Exercise B—Measuring Production in an Aquatic Ecosystem (60 minutes, set up; 24 hours, incubation; 60 minutes, analysis)
Exercise C—Productivity of Phytoplankton in a Water Column (as in Exercise B)
Exercise D—Thermal Stratification and Dissolved Oxygen Patterns in Lakes (30 minutes)

III. STUDENT MATERIALS AND EQUIPMENT

	Per Student	Per Pair (2)	Per Group (4)	Per Class (24)
Exercise A				
freshwater (distilled), O°, 20°, and 40° C (bucket, 2 gal) **(1)**				1
saltwater, 0°, 20°, and 40° C (optional) (bucket, 2 gal.) **(2)**			1	
LaMotte Dissolved Oxygen test kit			1	6
aluminum foil (20 × 20 cm)			1	6
45 ml BOD bottles			2	12
Exercise B				
water sample (2 liters), eutrophic **(3)**			4	24
water sample (2 liters), oligotrophic **(3)**			4	24
water sample (2 liters), polluted **(3)**			4	24
LaMotte Dissolved Oxygen test kit			1	6
aluminum foil (20 × 20 cm)			1	6
45 ml BOD bottles			2	12
grow light or fluorescent light fixture **(4)**			3	18

	Per Student	Per Pair (2)	Per Group (4)	Per Class (24)
Exercise C				
phytoplankton sample (3 liters) **(5)**				**1**
LaMotte Dissolved Oxygen kit			1	6
BOD bottles, 45 ml			5	30
aluminum foil (20 × 20 cm)			1	6
plastic screen for wrapping bottles (9 × 12 inches)			8	48
nitrogen enrichment solution, bottle (500 ml) **(6)**				1
phosphorus enrichment solution, bottle (500 ml) **(7)**				1
grow lights or fluorescent light fixture **(4)**			1	6
Exercise D				
Spectronic 20 (510 nm)			1	6
spectrophotometer tubes			5	30
electric fan			1	6
meter stick			1	6
wax pencil (or Sharpie)			1	6
beaker, 250 ml (waste)			1	6
Pasteur pipettes			4	24
rubber bulbs			4	24
container, translucent (approximately 13 × 24 cm)			2	12
sodium chloride, table salt (24 g/group)			1	6
small funnel			1	6
hypolimnion solution (200 ml) **(9)**			2	12

IV. PREPARATION OF MATERIALS AND SOLUTIONS

(1) water, 0°, 20°, and 40°

Collect water in advance. Let stand overnight, if possible, in a water bath at 40° and in the refrigerator. If you take water directly from the tap, you will have additional air in the water and results will differ among groups. Place water in a bucket, a 2 liter beaker, or a cut-off gallon milk bottle.

(2) saltwater, 0°, 20°, and 40° (optional)

Use a 3.5% NaCl solution. Dissolve 35 g of NaCl in enough distilled water to make 1000 ml of solution.

(3) eutrophic, oligotrophic, and polluted water samples

One day prior to the laboratory, prepare 4 liters of Alga-Gro solution (1 tube Alga-Gro/liter spring water) and 150 ml tryptic soy broth (according to manufacturer's directions). For:

oligotrophic—mix 125 ml of normal strength Algo-Gro and mix it with 375 ml of distilled water

eutrophic—use 500 ml normal strength Alga-Gro

polluted—use 495 ml normal strength Alga-Gro supplemented with 5 ml tryptic soy broth.

Inoculate all media with equal amounts of *Chlorella*. If purchased in tubes, resuspend the *Chlorella* and divide evenly.

(4) grow lights or fluorescent light

A fixture that holds 4 lights (4 feet each) can be attached to a wooden stand so that the lights are suspended 12 inches from the counter top.

(5) water sample

Obtain a 3 liter sample from a nearby lake. If sources are questionable or not very productive, you may wish to use a culture of algae. We have found *Chlorella* (see Ordering Information) to work best. Fill 10 test tubes, each with 9 ml of the medium and add 1 ml of *Chlorella* from the original tube. Let incubate on a windowsill or beneath a grow light until it is greenish in color. Prepare enough additional medium for approximately 1500 ml per group of students. To each 3 liters, add 1 tube of *Chlorella* culture. Place the culture under a grow light, fluorescent light (close) or in a sunny (but not too hot) window for 2 days or until the culture appears greenish. Use a magnetic stir bar to mix the culture. If not available, swirl the culture at intervals of several hours during the next two days.

If you use Alga-Gro, however, you will not be able to test for the effects of nitrogen (N) or phosphorus (P) since Alga-Gro provides this enrichment. *Chlorella* can be grown in spring water (available from your grocery store) but it will take approximately 4-5 days for a culture to be ready for use.

(6) plastic screen

Obtain a roll (or several yards) of flexible plastic screen wire for screen doors. Cut pieces of 9 × 12 inches.

(7) nitrogen enrichment solution

Stock solution:

KNO_3	30.33 g
NH_4Cl	10.605 g

Dissolve in 100 ml distilled water.

Working solution:

Mix 10 ml of stock solution with 490 ml of distilled water. One ml of this solution is used to enrich the sample in a 300 ml BOD bottle by 0.04 μM/ml.

(8) phosphorus enrichment solution

Stock solution:

Dissolve 1.065 g Na_2HPO_4 in 100 ml of distilled water.

Working solution:

Mix 10 ml of stock solution with 490 ml of distilled water. One ml of this solution is used to enrich the sample in a 300 ml BOD bottle by 0.005 μM/ml.

(9) hypolimnion solution

Dissolve 4 g of sodium chloride (table salt) in 196 ml tap water (2% solution). Add several drops of red food coloring until the solution turns bright red.

V. PREPARATION SUGGESTIONS

Exercise A

Although you can measure dissolved oxygen using reagents that have been prepared in the laboratory, the LaMotte Dissolved Oxygen Kit is simple and requires smaller water samples. For under $40, you can perform 50 dissolved oxygen tests. A $18 refill kit allows 200 additional tests. The kit comes in a plastic carrying case and can be used in the field for other experiments or coursework. Additional collection bottles and microburet titrators can be ordered from LaMotte so that several groups of students can work simultaneously from a single kit. Directions for using the kit are also found in the lid of the carrying case for fast reference. Students find this easy to use and follow the steps without problems.

Make sure that students submerge their sample bottle completely when filling it—otherwise, excess oxygen will bubble into the bottle. If you wish to use a siphon to fill sample bottles from a single source, make sure that the source is above the level of the collecting bottles to be filled. Overfill the bottles to exclude air. Place bottles in a bucket to avoid puddles on the floor!

Make sure water for Exercise A has been sitting overnight. To get cold water as close to 0° C as possible, place it in a freezer for a short while prior to the laboratory. As an option to this exercise, substitute salt water for fresh water (see Preparation of Materials and Solutions **(2)**. You may also simply want to compare a saltwater sample to a freshwater sample at room temperature.

Note: when using the LaMotte kit titrators with 45 ml sample bottles, the syringes are 1 ml syringes graduated in 0.1 ml amounts. A reading of 0.46 is equal to 4.6 ppm. If the numbers on your microburet (titrator) read 0-9 the read ppm directly; a reading of 4.6 is 4.6 ppm. **A mistake appears in the legend for Figure 45A-3. This should read… "marks 3, 4, 5., etc. as 3.0, 4.0, 5.0 parts per million (ppm)."**

Exercise B

Be sure that the BOD bottles are exposed to fluorescent or plant growth lights that will not heat up the bottles. Keep them at an 12" distance from the light source. (You may wish to check for heating effects by placing a flask of water under the light and determining its temperature over a 24 hour interval before proceeding with this experiment.) If heating becomes a problem and you cannot keep bottles in a lighted environmental chamber, immerse all bottles in dishpans of water to serve as a heat shield.

If you use *Chlorella* culture for your water sample, do not let the *Chlorella* become too dense—the experiment works best if the *Chlorella* are in the log phase of growth. You can complete this experiment using a three-hour incubation or you can leave the bottles overnight if necessary. Longer incubations may result in respiration using up all O_2 produced, especially in polluted and eutrophic bottles.

BE SURE TO SHAKE SAMPLES AT THE BEGINNING OF THE EXPERIMENT TO INTRODUCE OXYGEN BEFORE FILLING SAMPLE BOTTLES. This is especially important for the polluted sample which may already be low in oxygen. Since samples contain bacterial growth, fill bottles by pouring. Place bottles in a Petri dish so you can overfill them and dispose of the excess.

Exercise C

Use LaMotte kits as in Exercises A and B for this exercise. To cut down on the number of BOD bottles, you may wish to assign each group to only a light, dark or "screened" bottle. The effects of nitrogen and phosphorus enrichment can be studied by separate groups (two groups for data reliability). Recall that if AlgaGro is used for *Chlorella* cultures, the use of N and P is not necessary since AlgaGro introduces these enrichments (see Preparation of Materials and Solutions **(5)**).

You can use lake samples for this exercise if you have a nearby source. Be sure that bottles are exposed to adequate light and are not heated (see Exercise B above).

Exercise D

This exercise simulates lake thermal stratification and overturn with a model system. A colored 2% salt solution represents a dense hypolimnion that is placed under a layer of tapwater in a container. The salt solution will form a stable, red layer underneath the clear tapwater. Then, using an electric fan placed at various distances from the container, students experiment with how much "wind" is necessary to mix the "epilimnion" and "hypolimnion." A spectrophotometer is used to monitor the changing color of the surface and bottom solutions as mixing of the two layers proceeds.

Students should operate in groups of four. Each group will need at least 1 m of table space and two containers. Each model "lake" is a plastic shoebox or other container containing a layer of clear tapwater overlying a layer of colored, salty water. The container also must be correctly placed with respect to the fan.

For each container, prepare the "hypolimnion" solution (see **(9)** in Preparation of Materials and Solutions). At each group's station, place one container in front of the fan, either 0, 0.5, or 1 m away from it. This container should be raised so its upper edge is at the same level as the center of the fan. The other container is placed somewhat behind the fan where the fan will not influence it.

Place 2000 ml of tapwater in each container. Let the water become still (about 1 minute). Carefully add the 200 ml of colored salt solution. One method is to slowly pour it into a funnel propped up in the corner of each container. The funnel will convey the solution to the bottom of the container with a minimum of mixing. Once the heavy, colored solution settles down into a stable layer under the tap water, students can turn on the fan and start sampling according to the laboratory directions.

VI. ORDERING INFORMATION

LaMotte Water Test Kits (dissolved oxygen)—LaMotte[1], Model EDO, Code # 7414; Carolina Biological, # 65-2865 (kit), # 65-2866 (refills); Ward's # 21W0054 (kit), # 21 W 0050 (refills)

BOD (Biological Oxygen Demand)/water sampling bottle, 60 ml—LaMotte, Code # 0688-DO; Carolina Biological, # 74-6980; Ward's, # 17W0570

microburets for dissolved oxygen test kit—LaMotte, Code # 0377

grow light or fluorescent light fixture with fluorescent tubes—buy locally

sodium chloride (NaCl)—use table salt; Fisher # 5271-500; Carolina Biological, # 88-8880; Ward's, # 37W5480

Alga-Gro concentrated medium—Carolina Biological, # 15-3750; Ward's, # 99W8350

Chlorella culture, tube—Carolina Biological, # 15-3750; Ward's, # 86W0126

tryptic soy broth—Fisher, # DF0370-01-1; Carolina Biological, # 78-8440 (dehydrated) or # 77-6840 (bottle); Ward's, # 88W0817

electric fan—purchase locally

meter stick—Carolina Biological, # 70-2620; Ward's, # 15W4065

wax pencil—Carolina Biological, # 65-7730 (red), # 65-7732 (blue), # 65-7734 (black); Ward's, # 15W1159

beaker, 250 ml— Fisher, # 02-540K (Pyrex) or # 02-539K (Kimax); Carolina Biological, # 72-1223 (Corning, student grade); Ward's, # 17W4040

[1] LaMotte Chemical, PO Box 329, Chestertown, MD 21620; 800-344-3100 or 301-778-3100 (Maryland)

Spectronic 20 (Milton Roy, formerly B&L)—Fisher, # 07-143-1; Carolina Biological, # 65-3300; Ward's, # 14W5551
wide range phototube and filter—Fisher, # 07-144-10; Carolina Biological, # 65-3312; Ward's, # 14W5553
Spectronic 20 tubes—Fisher, # 14-385-9008; Carolina Biological, # 65-3310; Ward's, # 14W5554 OR
Spectronic 20 tubes, Pyrex (100 × 13 mm)—Fisher, # 14-957C; Carolina Biological, # 73-1408
Spectronic 20 tubes, Kimax (100 × 13 mm)—Fisher, # 14-923D; Carolina Biological, # 73-1408A
Pasteur pipette, 9"—Fisher, #13-678-20C; Carolina Biological, # 73-6062; Ward's, # 17W1146
rubber bulb—Fisher, # 14-065B
plastic screen—local hardware store
potassium nitrate (KNO_3)—Fisher, # P263-500, or Carolina Biological, # 88-3908; Ward's, # 37W4824
ammonium chloride (NH_4Cl)—Fisher, # A661-500, or Carolina Biological, # 84-3810; Ward's, # 37W1965
sodium phosphate, dibasic (Na_2HPO_4)—Fisher, # S374-100, or Carolina Biological, # 89-1370; Ward's, # 37W5655
red food coloring—obtain locally
plastic boxes—obtain locally
funnel, 65 mm—Carolina Biological, # 73-4284

BioBytes

BioBytes INTRODUCTION

Using the BioBytes Simulations to Extend the Learning Experience

The BioBytes computer simulations will allow you to give students learning experiences that would never be possible in the "real world." Using BioBytes, students can watch days of plant growth in a few minutes, observe the collapse of homeostasis in a human who is running on top of Mt. Everest, track fertility cycles in human females and study their variations, or watch the fluctuations in the frequency of an allele over thousands of generations.

The use of BioBytes simulations as learning tools require planning, decision-making, and data interpretation. To prepare students to make he best use of this experience, they should read the appropriate student materials and plan their activities in advance of the laboratory activities. This information will be found on their computer's hard drives after the accompanying CD-ROM has been properly installed. Thus, student information for *Alien* is found in the file "manual_a," *Cycle* in "manual_c," *Dueling Alleles* in "manual_d," and *Seedling* in "manual_s." Copies of this student material are also provided following the description of each simulation in the *Preparator's Guide* and can be duplicated for distribution to those lacking appropriate computer resources.

BioBytes programs are also most successful when the instructor takes an active role while students are using the programs. Having an instructor present to explain objectives, help with questions, and generally offer encouragement will make the experience much more rewarding for the students.

A Note About Simulations

The BioBytes simulations are computer models of biological systems, and, as such, are far simpler than the real systems portrayed. This is not necessarily a defect. Models are usually somewhat simplified to suit them to their purpose. For example, a road map depicts roads as lines and towns as dots, omits detail that doesn't pertain to highway travel (buildings and even mountains), and emphasizes small features (such as tolls and access roads) that do pertain to the transportation system. To say that the road map is not an accurate depiction of the Earth's surface misses the point that these distortions and simplifications make the map more useful for its purpose helping the traveler.

Starting The BioBytes Simulations

Directions for installation of the BioBytes simulations are given in the *Preparator's Guide* before Laboratory I (pages CD-ROM-1 –4). To start BioBytes, open (double-click) on the BioBytes folder and the BioBytes icon (Macintosh), open (double-click) on the BioBytes icon in the Windows 3.1 program group named "biobytes," or select Start...BioBytes in Windows 95. Select the appropriate simulation from the BioBytes menu and enjoy!

Notes:

BioBytes *Alien*

Purpose

The purpose of *Alien* is to demonstrate the relationship between several of the major variables in heart-lung physiology.

The *Alien* Simulation—An Overview

Alien is a simulation of cardiopulmonary physiology that can be used in two ways: it can serve as a data-generator to perform three relatively directed experiments (Exercises A, B, or C), or to perform a more demanding and less confining experiment (Exercise D).

Exercise A—Demonstration Experiments

Part 1—Exercise at Sea Level

Part 2—Exercise at High Altitude

Part 3—Increasing Atmospheric CO_2

This exercise offers step-by step directions for exploring the effects of exercise, the effects of exercise at high altitude, and effects of increasing CO_2 in the atmosphere. The student is told exactly what values to put into the simulation and what data to record.

Exercise B—The Effect of Altitude on Running Stamina

Exercise C—Running Distance and Running Speed

Exercises B and C explore the effects of altitude and running speed on running stamina. These exercises require more active participation because the student must determine the altitudes and running speeds to use to accomplish the experimental objectives.

Exercise D—Answering the Questions

Finally, Exercise D is both demanding and gives the student a high degree of freedom. The scenario is that NASA has returned a humanoid alien to earth. The student, a world-famous physiologist, has been hired by NASA to answer 16 questions about the alien's physiology. This exercise has no step-by-step directions. The student must determine the best order in which to answer the questions and which experiments to use.

Alien has the flexibility to accomplish these varied objectives. It simulates either a human or five fictitious extraterrestrials from varied home planets. It outputs key physiological variables in either digital or graphical format, and allows the student to change the speed of a treadmill, the partial pressure of oxygen or carbon dioxide, to inject acids, bases, tranquilizers and stimulants, to administer electric shock, and to perform more specialized experiments such as determination of lung volume and cardiac output.

As either the humans or the aliens, experience changes in levels of stress, they exhibit realistic physiological changes such as changes in heart rate, blood pressure, and metabolic rate. If the individuals' various homeostatic limits are exceeded, they may drop from fatigue, faint, experience cardiac arrest, and finally die, if not saved in time by CPR. As students observe the changes in various physiological variables, they learn about the relationships that exist among them, for example, that fatigue correlates with a drop in blood pH, and that a rise in arterial CO_2 causes an increase in ventilation rate.

Alien's Simplifications

Alien leaves out many details. For example, one of the major ways that blood pressure is regulated over the long term is through the kidneys. If there is a loss of blood, the kidneys will secrete less urine in order to conserve fluids and maintain blood pressure; if the volume of blood rises, the kidneys will secrete more urine to bring blood pressure down to normal again. *Alien* does not allow the subject to ingest of lose fluids, so this role of the kidneys is omitted. To take another example, *Alien* models fatigue solely as an accumulation of lactic acid due to anaerobic metabolism. A consequence of this is subject to running so slowly (about 2 m/sec for the human) that lactic acid is not formed, he will never fatigue. This cannot happen in reality because after 4 or 5 hours, even a slow runner will deplete the glycogen in his muscles. *Alien* is only meant to simulate events in the first few minutes of exercise, and so omits the role of glycogen.

To summarize, Alien will demonstrate many of the major features of cardiopulmonary physiology but it is not, by any means, a mathematical reconstruction of an organism's total physiology.

Instructional Objectives

Alien can help to develop students cognitive abilities at several levels:

- As students work through Exercises A-D in "Manual_a" on the CD-ROM (which should also be on the hard drive if installation was done properly), they will develop a working knowledge of cardiopulmonary physiology, including heart rate, blood pressure, ventilation rate, cardiac output, and stroke volume, and the partial pressures of blood gases. The text and questions in the exercises help students understand the dynamic relationships among these variables.
- The exercises require students to interpret data by using calculations and graphs. By working with the data, students develop an understanding of how patterns of numbers reveal information about physiological processes, such as the maintenance of homeostasis.
- The student can observe the cardiopulmonary system's ability to maintain homeostasis during rest, during moderate physiological stress, and during stress that exceeds the body's ability to maintain homeostasis. Exercise A, for example, emphasizes how the circulatory and pulmonary systems work together to maintain homeostasis when stressed by exercise, both at sea level and at high altitude, and by increases in the concentration of atmospheric carbon dioxide.
- *Alien* activities help students to develop synthetic thinking and problem-solving skills, by far the most important objectives of these simulations. This is the emphasis of almost every experiment in *Alien*.
- One additional, and important, outcome of the *Alien* exercises is an affective one: a sense of accomplishment. At the beginning, students are told that they must do calculations, plan experiments, and even respond to a medical emergency such as a cardiac arrest. Many may have doubts about their ability in these areas, and success (especially the revival of a cardiac-arrest subject) is thrilling.

Advice on Classroom Use of *Alien*

Alien allows the student to experiment with a physiological system. It does not teach physiology from the group up. Because of this:

a) Students must have some background in cardiopulmonary physiology before they begin the program. They should be familiar with the basic operation of the circulatory and respiratory systems, the concept of partial pressure of gases, and pH. Other physiological terminology such as cardiac output and stroke volume will be taught by the program. Reading pages 3-5 in "Manual_a" on the CD-ROM will give them the background they need.

b) The written directions should be followed just as they would be followed for a "wet lab" experiment. The program is not hard to operate, but deciding what to do requires thought. Especially for beginning students, the highly directed Exercise A is useful for familiarization with both the program and some dramatic physiological responses (including cardiac arrest).

c) The instructor's role is very important. The program is complex and students will face many choices. Even after the choices have been made, there are many responses to examine. Therefore, an instructor who can provide advice and encouragement will be valuable. A good general strategy is to pause the simulation, focus student attention on the important responses, and to ask leading questions. For example, if exercise is being simulated, the instructor should ask, "You have the subject running in a low-oxygen atmosphere. Is venous oxygen going to be the same as when you had the subject running at sea level? It will go lower? Why? I thought homeostasis could maintain internal conditions *regardless* of what the external environment was."

d) Focus on changes in variables rather than on the magnitude of the variables. For example, it may not mean much to a student that the venous partial pressure of oxygen during exercise is dipping below 20 mm Hg, but this change would be more meaningful if the student realized that the resting value is 40 mm Hg. This is why the directions often ask the student to record initial conditions.

Physiological Variables Displayed by *Alien*

The *Alien* display contains a right portion that could show either text or a graphics of several variables. The left portion of the screen usually shows control buttons for the current experiment (in this case, changing the partial pressure of oxygen). The bottom of the screen contains tabs that allow the student to select different experiments.

Figures 2 and 3 show close-ups of the displays. In the graphical display, a human subject has been exercised for about a minute and then allowed to recover from exercise over the next two minutes.

The variables used in *Alien* are:

- ❑ **Time (minutes).** *Cumulative* time since the beginning of the session, updated every 0.025 simulation minutes. The program runs faster than real clock time (1.6 simulation minutes/minute).
- ❑ **Ventilation (l/min)**(liters/minute). This is the rate at which air moves in and out of the lungs. Ventilation is the product of tidal volume (the amount of air inhaled per breath, 0.5 liters in humans) and the breathing rate (about 12 breaths/minute in resting humans).

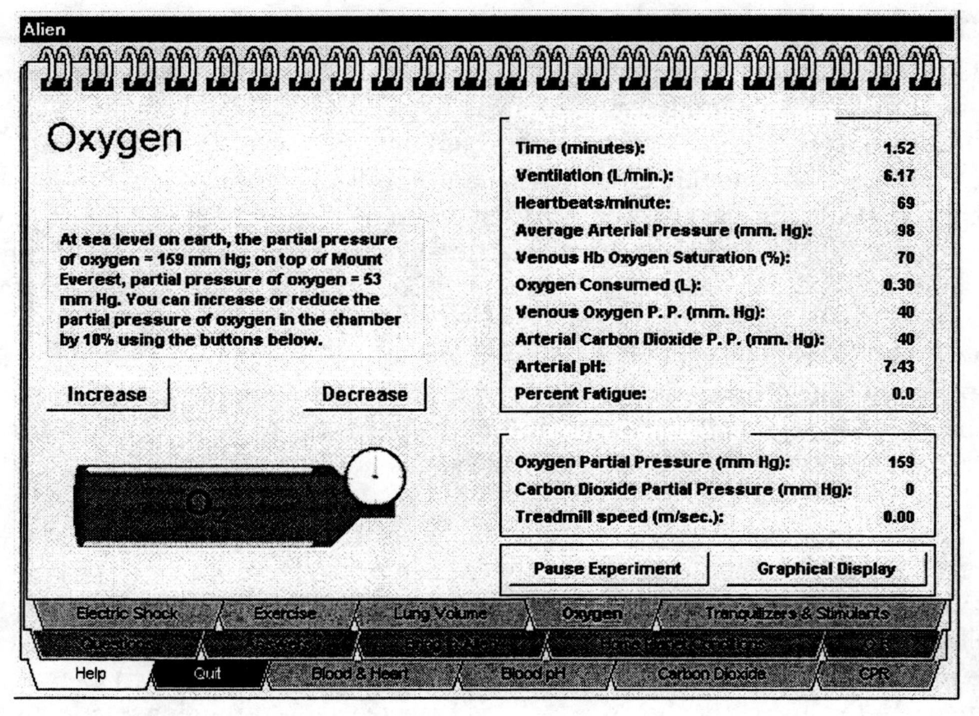

Figure 1 *Alien digital display.*

Physiological Status: Human	
Time (minutes):	1.52
Ventilation (L/min.):	6.17
Heartbeats/minute:	69
Average Arterial Pressure (mm. Hg):	98
Venous Hb Oxygen Saturation (%):	70
Oxygen Consumed (L):	0.30
Venous Oxygen P. P. (mm. Hg):	40
Arterial Carbon Dioxide P. P. (mm. Hg):	40
Arterial pH:	7.43
Percent Fatigue:	0.0

Figure 2 *Physiological Status—Data Display.*

- ❑ **Heartbeats/minute**. Heart rate is the same as the pulse rate, with a resting value of about 70 beats/ minute in the young human male adult modeled in *Alien*.
- ❑ **Average arterial pressure (mm Hg)** (millimeters of mercury). Blood pressure is expressed as the height of a column of mercury that the blood pressure would support. Blood pressure in a resting human rises to a peak of about 120 mm Hg when the heart is contracting and falls to 80 mm Hg when the heart is relaxed. *Alien* averages this into a single value of 100 mm Hg for the human.
- ❑ **Venous Hb oxygen saturation (%)**. This variable indicates the percentage of venous hemoglobin that is saturated with oxygen. Students may be surprised to learn that venous hemoglobin remains up to 70% saturated with oxygen in a resting human. The oxygen

remaining in venous hemoglobin furnishes a useful reserve for muscles during strenuous exercise, when venous hemoglobin may be only 20% saturated.

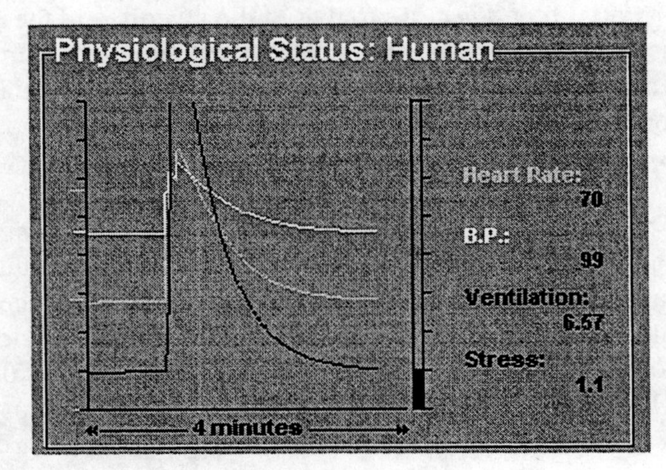

Figure 3 *Physiological Status—Graphic Display.*

- **Oxygen consumed (liters).** *Cumulative* oxygen consumption since the beginning of the session. Students doing Exercise D will be asked to compute the oxygen consumption rate under *both* resting and exercising conditions. To do this, they must divide the *change* in oxygen consumed by the *change* in time. It is a common student mistake to simply divide the total oxygen consumed by the total time.
- **Venous oxygen P. P. (mm Hg).** Indicates partial pressure of oxygen in the venous blood. The total pressure of a mixture of gases is the sum of the pressures of the individual gases in the mixture. Oxygen makes up 21% of the volume of dry air. The pressure resulting from O_2 concentration alone is the oxygen partial pressure (P_{O2}). At sea level, all the gases that make up the air (oxygen, nitrogen, argon, carbon dioxide, and others) exert enough total pressure to raise a column of mercury 760 mm. The partial pressure exerted by O_2 is 21% of 760 mm, sufficient pressure to raise a column of mercury 159 mm.

  Partial pressure is a measure of oxygen concentration in the air, in the blood, or in the tissues. P_{O2} is 159 mm Hg in air but only about 100 mm Hg in the alveoli of the lungs and in the blood leaving the alveolar capillaries on its way to the heart, and from there to the body's tissues (at this point hemoglobin is 100% saturated with oxygen). In the tissues and in the venous capillaries that carry blood away from these tissues, P_{O2} falls to 40 mm Hg, at which point the blood is approximately 70% saturated with O_2. During extremely strenuous exercise, the tissue value of P_{O2} may drop to 20 mm Hg and unconsciousness will occur if it drops to 10 mm Hg.
- **Arterial carbon dioxide P. P (mm Hg).** Indicates the partial pressure of carbon dioxide in the arterial blood. The concentration of CO_2 in the arterial blood is the primary stimulus for controlling ventilation rate. In humans, the resting value of P_{CO2} is 40 mm Hg. A decrease in arterial P_{CO2} to 30 mm Hg can stop ventilation entirely.

  Even small increases in P_{CO2} above the resting value will result in hyperventilation, a major function of which is to expel excess CO_2. Note that CO_2 makes up only 0.03% of dry air and thus has an atmospheric partial pressure (P_{CO2}) of only 0.228 mm Hg (i.e., 0.03% = 0.0003; 760 mm Hg × 0.0003 = 0.228 mm Hg). Thus the *Alien* program rounds normal atmospheric P_{CO2} to 0 mm Hg.

It may seem surprising that even after passing from the body's tissues to the alveoli for oxygenation, blood P_{CO2} remains high, about 40 mm Hg. The carbon dioxide partial pressure in the arterial blood is close to that in the alveoli, and the alveoli are CO_2–enriched because CO_2 is constantly bing discharged into them from the blood. *Alien* displays *arterial* rather than venous P_{CO2} because the arterial partial pressure is monitored by the respiratory centers in the medulla of the brain in order to control ventilation rate. If the lungs are not dispelling CO_2 rapidly enough, the arterial P_{CO2} will rise to reflect this fact.

- **Arterial pH**. The normal pH of human arterial blood is 7.4. Changes in the pH of the blood have many important physiological effects. For example, during exercise, higher levels of cellular respiration result in increased levels of CO_2 in the tissues. Under normal conditions, some of the CO_2 is carried in the blood as carbonic acid. When cellular respiration becomes anaerobic in fatiguing muscles, lactic acid is produced. In both cases, higher acidity (lower pH) of the arterial blood causes the respiratory center in the medulla of the brain to signal for an increase in ventilation, which helps to clear excess CO_2 from the system.

  When blood returning from the body's tissues reaches the lungs, ventilation blows off CO_2 and raises the pH in both the lung capillaries and in the pulmonary vessels through which blood leaves the lungs for the heart. However, if CO_2 is present in concentrations that exceed the lungs' ability to expel the excess, measurements of arterial pH can give an indication that homeostatic mechanisms are having difficulty compensating for these concentrations of CO_2.
- **Percent fatigue**. In this program, the index of fatigue is arbitrary: 100% fatigue causes collapse from exhaustion. In this simulation, the only factor taken into account as a cause of fatigue is an accumulation of lactic acid in the muscles, due to anaerobic respiration. Fatigue increases only during exercise, and diminishes during periods of rest, as increased availability of oxygen allows lactic acid to be metabolized to pyruvate.
- **Stress index** (graphics display only). This is an arbitrary index indicating overall physiological excitement. In this simulation, 1.0 is the resting value in humans, but in some aliens the resting value is higher. A stress index of 45 results in cardiac arrest in humans.

Program Controls

The program controls appear as tabs (buttons) on the bottom of the screen. When a tab is pressed, that "page" in the *Alien* notebook comes to the front and tabs may change position. Of the sixteen tabs, nine are experiments, and the rest are other options such as answering the questions or quitting the program.

Carbon Dioxide

The student may increase or decrease the concentration (partial pressure) of atmospheric carbon dioxide. As blood levels of CO_2 rise, humans experience an increase in lung ventilation rate of up to 10 times the resting rate (by comparison, lack of oxygen can cause only a 1.6 times increase in ventilation rate). However, very high blood levels of CO_2 depress the respiratory center and cause coma. A rapid demonstration of homeostasis and its failure can be done with this experiment (see Exercise A, Part 3).

Oxygen

The student may increase or decrease the atmospheric concentration (partial pressure) of oxygen. *Decreases* in atmospheric oxygen cause several effects: a small increase in ventilation directly due to lack of oxygen, which lowers venous PO_2, a long-term ventilation increase due to lactic acid production by anaerobic respiration, which lowers blood pH, and increased stress, resulting in a higher metabolic rate and higher CO_2 production. The major increase in ventilation is due to this latter effect. *Increases* in atmospheric oxygen partial pressure improve oxygenation of the blood but also have a paradoxical effect: the greater density of the air slows ventilation and this causes an *increase* in CO_2 in the tissues.

Blood pH

The student may inject acids and bases into the blood. Ventilation is significantly controlled by blood pH: respiratory centers in the medulla of the brain are stimulated by decreasing pH, the result of the formation of carbonic acid when P_{CO2} levels rise. Thus acid injections cause an immediate increase in ventilation and a reduction in blood carbon dioxide. With CO_2 reduced, ventilation returns to a normal rate, but the CO_2 concentration remains regulated below normal. *Warning:* A dose of an acid or a base that is harmless to a human might be fatal to an alien.

Electric Shock

A small electric shock causes an immediate increase in heart rate, blood pressure, and stress. The ventilation rate increases after a delay (and as the other rates are declining) because a stress-induced increase in metabolism raises arterial CO_2. Shocking the subject is the quickest way to cause cardiac arrest while maintaining adequate blood oxygenation. Good blood oxygenation at the time of cardiac arrest allows CPR to be successful if it is begun rapidly. The human requires about 12 electric shocks delivered in rapid succession to cause cardiac arrest; the aliens require either more or fewer shocks, but no alien will suffer cardiac arrest due to just one shock.

Tranquilizers and Stimulants

The student may inject various doses of either of these drugs, which take effect only after a delay. Overdoses of either are fatal, and the student will not be able to trust any answers obtained while the subject is under their influence, so the use of these drugs should be carefully considered and infrequent.

Exercise

Exercise is one of the most useful experiments and is used to obtain exercising heart rate, blood pressure, and so on. It is also the basis of several of the experiments in the student exercises. Exercise can have several effects:

Falling—If the treadmill is set at too rapid a pace (more than 10 m/see for a human), even a rested subject will fall immediately. This results in a temporary increase in excitement, but no injury.

Fatigue—Fatigue is modeled as a buildup of lactic acid. The rate of lactic acid synthesis is proportional to the square of the running velocity, and the rate of removal of lactic acid is proportional to venous oxygen concentration. This means that fatigue is more rapid at a faster running speed and at a lower venous oxygen concentration. Lower venous oxygen concentration slows the rate of lactic acid removal from the blood. Thus recovery is slower at a lower oxygen concentration.

Blood acidification—When blood is acidified by the accumulation of lactic acid, ventilation is stimulated, which, in turn keeps CO_2 at a lower level than during the rested state. This also explains why blood CO_2 is elevated during the early stages of exercise, but actually sinks below normal levels if exercise is prolonged and lactic acid accumulates.

CPR

As variables such as shock or excitement cause levels of blood pH or blood oxygen to approach lethal limits, unconsciousness occurs and is followed by cardiac arrest and brain damage. Brain damage occurs most rapidly when venous oxygen is low. CPR allows the student to try to revive the subject by administering heart massage and artificial respiration. When brain damage reaches the lethal limit, the program notifies the student, "We have a flat EEG, Doctor. The patient is dead."

If cardiac arrest occurs, the program will automatically go to the CPR screen.

1. Start heart massage and restore a resting level of ventilation (6 liters/min in humans)
2. Leave CPR and remove any stressful conditions (such as low oxygen or elevated carbon dioxide).
3. The student will be reminded periodically that CPR is still in progress. If the patient lives, the student will be told, "We have a heartbeat!" At that point, take the patient off CPR. It could be damaging to continue CPR once the heartbeat returns because when the patient is on the ventilator, ventilation is controlled by the machine and cannot respond to the normal cues such as arterial CO_2.

If venous oxygen is not too low when cardiac arrest occurs, the chances of restarting the heart are good. But following revival, the subject will be suffering from cardiac insufficiency, a condition in which blood pressure and heart rate will be elevated, venous oxygen will be lower than normal, and the subject will have little stamina and will be prone to another cardiac arrest. If a second arrest occurs, there is a chance that the subject can be revived by rapid CPR, but he will be unable to survive without supplemental oxygen. A third arrest is always fatal. Exercise A, Part 3 provides an opportunity to practice CPR.

Blood and Heart

This experiment presents a data screen that is used to answer the questions in Exercise D. Accessing this screen has no effect on the patient.

Lung Volume

This experiment also accesses a data screen used in Exercise D, and has no effect on the patient.

Other Options

Bring in Alien. This button allows the student to bring in a fresh, rested subject (either alien or human). Therefore, the student can avoid waiting for a subject to recover from excitement, unconsciousness, or fatigue.

Home Planet Conditions. This button shows the home planet conditions for the alien and may give the student clues about alien physiology. The home planets and physiologies of each of the aliens are discussed at the end of this section of the instructor's manual. A sample home planet screen is shown in Figure 4.

Questions. This option allows the student who is doing Exercise D to see a list of the questions.

Answers. This option allows the student who is doing Exercise D to answer the questions. The student is scored on accuracy. Answers may only be given once, although bringing in a new alien clears all the answers and restores the score to zero.

Help. This option accesses some short text screens that explain the basics of the program and Exercises A, B, C, and D.

Quit. This allows students to leave the program and return to the BioBytes title screen.

Characteristic	Earth	Delta Aurigae 1
Diameter (km):	12,576	56790
Surface gravity (g):	1.0	2.6
Surface atmospheric pressure (mm Hg):	760	85440
Major atmospheric constituent:	Nitrogen	Carbon Dioxide
Oxygen Partial Pressure:	159	55
Carbon Dioxide Partial Pressure:	0	79800

Proceed...

Figure 4 *Sample Home Planet Screen.*

EXERCISE A Demonstration Experiments

Exercise A is a highly guided introduction to *Alien.* It offers step-by-step directions for exploring the effects of exercise, the effects of exercise at high altitude, and effects of increasing CO2 in the atmosphere. The student is told exactly what values to put into the simulation and what data to record.

PART 1 Exercise at Sea Level

Vigorous exercise is one of the greatest stresses the body must withstand, and the body's ability to compensate for it is an example of the effectiveness of homeostasis. At rest, the rate of oxygen consumption is about 0.2 liters/ minute; during exertion, consumption may rise to 20 times this amount. It might be expected that such massive increases in oxygen demand might cause oxygen shortages in the tissues within seconds. Yet the partial pressure of oxygen (P_{O2}) in venous blood only drops from 40 mm Hg to about 20 to 30 mm Hg. Carbon dioxide partial pressure (P_{CO2}) in the arterial blood rises only slightly, from 40 to 45 mm Hg, and with prolonged exercise, arterial P_{CO2} returns to the resting value or even lower. Maintaining blood gases within a range consistent with life is the result of homeostatic compensation by the cardiovascular system and the lungs.

As an example of this, suppose a subject exercises by running at 8 m/see for 30 sec. Immediately afterwards, the subject's heart will be beating at almost 200 beats per minute and blood pressure will exceed 200 mm Hg. The blood is literally pounding through the arteries—in only 30 seconds, cardiac output increases from 5.2 liters/min to 21.5 liters/min. Just as important, the lung ventilation rate has gone from only 6 liters/min to over 90 liters/min.

To realize the significance of these observed changes, it is important that students first understand how rapid ventilation keeps CO_2 from accumulating in the lungs and maintains the highest possible concentration of oxygen in the alveoli. At the same time, rapid blood flow purges CO_2 from the body tissues and supplies oxygen almost as fast as it is consumed. Thus tissue oxygen declines only slightly. Without increased ventilation, the air in the lungs would become oxygen-depleted and loaded with CO_2, and even the most rapid exchange between this stale air and the tissues would do no good. Without circulation, the deep tissues would suffocate even though the lungs were filled with fresh air.

As illustrated in the example just given, the major changes in circulation and ventilation that occur during exercise in the *Alien* simulation can be explained as follows. The excitement and body motions caused by exercise are the main reasons that heart rate and blood pressure increase. An increased concentration of CO_2 in the arterial blood reaching the brain is the primary stimulus for increased ventilation. CO_2 produced by cellular respiration in exercising tissue combines with water to form carbonic acid, some of which dissociates to form bicarbonate ions and hydrogen ions. The hydrogen ions diffuse out of the blood into the cerebrospinal fluid where they excite chemoreceptors, which synapse with respiratory centers in the medulla. This stimulates respiratory muscles and increases ventilation. (Lactic acid from anaerobic respiration also contributes slightly to a drop in arterial pH.)

Note: It is important for students to understand why it is the arterial blood that is monitored by the brain: when the current rate of ventilation is inadequate to dispel excess CO_2 from the venous blood, some of the excess CO_2 will "spill over" to the arterial blood, thus providing the signal that a higher rate of ventilation is required.

In the exercising subject, venous P_{O2} declines because oxygen demand slightly exceeds the rate of supply to the tissues. However, in conditions of declining venous P_{O2}, hemoglobin gives up its oxygen more readily, thus ensuring that oxygen becomes more available as the need increases. A rise in temperature, commonly experienced during exercise, also decreases hemoglobin's affinity for oxygen; active tissues are generally warmer than less active ones, so, in this case, too, more oxygen is delivered to the harder working tissues.

Likewise, declining pH brought about by increased arterial P_{CO2} facilitates the release of oxygen from hemoglobin.

Oxygen consumption and CO_2 production rise by a factor of more than seven before the human collapses from exhaustion. After collapse, the production of CO_2 immediately starts to decline and so does ventilation. Arterial P_{CO2} is restored to the resting value of 40 mm Hg almost immediately. But paradoxically, venous oxygen concentration continues to decline, bottoming out at 19 mm Hg about one game minute after collapse, and then recovering slowly to 40 mm Hg. This occurs (a) because ventilation slows down more rapidly than oxygen consumption rate in the tissues, leading to a temporary dip in venous oxygen, and (b) because the oxygen debt of exercise is being repaid during recovery. Lactic acid in the muscles consumes oxygen as it is resynthesized to glucose or glycogen in the liver.

PART 2 Exercise at High Altitude

When the partial pressure of atmospheric oxygen reaches 50 mm Hg, the subject drops from exhaustion after about 0.5 game minutes, rapidly goes unconscious, and then experiences a cardiac arrest. After 0.5 more game minutes, he dies. The reason is lack of oxygen. With less than one third of the normal oxygen in the air, no matter how fast the lungs ventilate and the blood flows, the system cannot deliver enough oxygen to the tissues. Venous oxygen (P_{O2}) dips below the lethal level (5 mm Hg). After cardiac arrest, with no blood flow and no ventilation, venous P_{O2} drops to 0 mm Hg, and brain death occurs. The students should watch venous oxygen most closely throughout this exercise.

An interesting observation about arterial P_{CO2} is that even when cardiac arrest is impending, CO_2 concentration is controlled spectacularly well, barely budging from the normal resting value of 40 mm Hg. The reason is that ventilation and the elimination of CO_2 is easy in the thin atmosphere. But after cardiac arrest and the cessation of ventilation, CO_2 rises rapidly, reaching 57 mm Hg just before brain death.

Even without exercising, the subject would die in a few minutes in 50 mm Hg oxygen, his demise hastened by the conscious knowledge of oxygen starvation, which increases panic and the rate of oxygen consumption. On the other hand, an anesthetized subject will survive at 50 mm Hg oxygen because oxygen demand is kept low.

Suggestions for Helping Students. Advise students to use the pause button frequently and make sure they understand the significance of the changes in venous oxygen and arterial carbon dioxide. **Note:** When cardiac arrest occurs, the program will advise using CPR. However, the patient cannot be saved because there is no oxygen in his venous blood, so tell students to ignore the CPR instruction. CPR will be used in Part 3.

PART 3 Increasing Atmospheric CO_2

This exercise is another exploration of the cardiopulmonary system's ability to maintain homeostasis. Students should be reminded that homeostasis is the body's capacity to maintain a limited range of preferred internal conditions in the midst of a much broader range of environmental conditions. To maintain these preferred conditions, the body's homeostatic mechanisms must often maintain a difference between internal and external conditions.

The more the external conditions vary from the preferred internal conditions, the harder the body's systems must work. As external conditions begin to exceed the capacity of homeostatic mechanisms to maintain the necessary internal conditions, internal conditions begin to leave the preferred range. For example, an air conditioner designed to keep a room at 70° F on a 90° day may be able to keep the room down to only 100° on a day when the outside temperature reaches 115°. In a somewhat similar way, organisms have developed mechanisms that work most efficiently within the range of environmental conditions they are likely to encounter. Physiological systems work much less well or may fail altogether under conditions that are extreme, relative to what is normal in the organism's environment.

In the example of atmospheric P_{CO2} and arterial P_{CO2} emphasize that if every 10 mm Hg rise in levels of atmospheric CO_2 resulted in a 10 mm Hg rise in levels of arterial CO_2, there would be no homeostasis. But what actually happens is that as levels of atmospheric CO_2 rise from 0 mm Hg to 30 mm Hg, human arterial CO_2 rises only from 40 mm Hg to 43 mm Hg and ventilation increases from the normal 6 l/min to 20 l/min. As atmospheric CO_2 continues to rise to 60 mm Hg (and becomes more extremely different from normal conditions), ventilation approaches the maximum value and stabilizes at 109 l/min. The concentration of arterial CO_2 reaches 61 mm Hg. Since a human cannot increase ventilation beyond this point, ventilation has reached the limit of its capacity to rid the body of excess CO_2 Thus the homeostatic system has reached the limit of its ability to compensate for further rises in the level of atmospheric CO_2. At this point, the toxic effects of high levels of CO_2 result in unconsciousness.

As atmospheric levels of CO_2 rise toward 120 mm Hg, arterial CO_2 rises very rapidly to 121 mm Hg, and the ventilation rate declines due to the increasing toxic effects of CO_2 on the brain's respiratory center. Homeostasis fails completely and cardiac arrest occurs.

The ability of the body to compensate for increasing levels of atmospheric CO_2 can be determined from the size of the increase in arterial CO_2 relative to atmospheric CO_2. Make sure that students observe that as levels of atmospheric CO_2 rise from 0 mm Hg to 30 mm Hg, there is only a 3 mm Hg rise in arterial P_{CO2}. With an additional rise of 30 mm Hg in atmospheric CO_2, arterial P_{CO2} experiences a rise of 18 mm Hg. With the jump to 120 mm Hg of atmospheric CO_2, arterial P_{CO2} rises to 121 mm Hg, which means that the rise in arterial P_{CO2} matches the rise in atmospheric P_{CO2}. Both have risen 60 mm Hg, a clear indication that homeostasis is no longer operating.

Provided that atmospheric CO_2 is restored to normal, performance of the CPR routine will easily revive the patient because the hyperventilation that preceded cardiac arrest raised the P_{O2} of the tissues. However, the revived subject will suffer from cardiac insufficiency: heart rate and blood pressure will be slightly high and the oxygenation of venous blood will be poor. The subject will have poor stamina and his next heart attack can be caused by only minor stress.

Suggestions for Helping Students. The increase in ventilation rate is obvious, especially if you remind students that the resting rate of ventilation is about 6 l/min. This will put the new rates in perspective.

EXERCISE B The Effect of Altitude on Running Stamina

Exercise B, pages 13-15 in the "Manual_a" file on the CD-ROM, asks students to determine the relationship between partial pressure of oxygen in the atmosphere and the distance that can be run before exhaustion. It is more demanding than Exercise A because students must choose the oxygen partial pressures to try. Their objective will be to discern the shape of the curve shown in the representative results in Figure 5. Table 1 can be used to relate these oxygen partial pressures to altitude. Table 1 also appears on page 13 of the student materials.

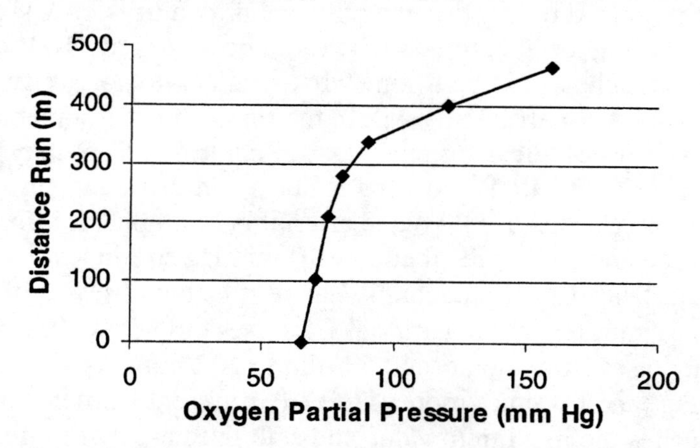

Figure 5 *A graph of running distance versus oxygen partial pressure. At an atmosphere* P_{O2} *of 300 mm Hg (rather than the normal 159 mm Hg), the human can run 650 m before exhaustion.*

Two effects of low atmospheric P_{O2} reduce the distance a human can run at high altitudes. First, at 5 m/see, lactic acid is continuously synthesized in the muscles. Lactic acid diffuses into the blood and travels to the liver and is resynthesized into pyruvic acid, and eventually back into glucose, if oxygen is abundant. But if low atmospheric P_{O2} reduces blood levels of oxygen to a degree that drastically slows the rate of lactic acid conversion, then lactic acid accumulates in the muscles and causes fatigue. Secondly, low atmospheric P_{O2} so reduces resting venous P_{O2} that very little exertion pushes venous P_{O2} below 10 mm Hg and the subject passes out from lack of oxygen (rather than from exhaustion).

The graph in Figure 5 shows a representative set of results. Students should collect enough points to discern the shape of this curve.

Table 1 Oxygen Partial Pressures (mm Hg) at Some Sample Altitudes

Location	Altitude	P_{O2}
New York, NY	10 m	159
Table Rock, SC	955 m	143
Mt. Mitchell, NC	2,030 m	126
Telluride, CO	2,710 m	116
Pike's Peak, CO	4,340 m	94
Mt. McKinley, AK	6,200 m	65
Himalayan peaks	7,000 m	65
Mt. Everest	8,848 m	50
jet airliners	10,000 m	42

Suggestions for Helping Students. This exercise requires that the students bring the subject to the partial pressure of oxygen characteristic of the tested altitude and then leave him or her there for three game minutes. Impatience prods students to start the treadmill too early. If the subject is not given time to equilibriate at each new P_{O2}, exercise data will reflect, in part, the oxygen stored in hemoglobin while the subject was at the previous altitude.

This exercise and the next one, Exercise C, ask the students to choose which partial pressure of oxygen they will test. To do this intelligently, students must have enough understanding of how the system will respond to make simple predictions, such as, "As oxygen decreases, running stamina will decrease." Therefore, it is important to start this experiment by asking students to make a hypothesis about what will happen during the course of their experiment. Once they have formulated a hypothesis, they should try to predict the partial pressures of atmospheric oxygen that will be most useful in testing their hypothesis. For example, it would be reasonable to select values of P_{O2} at intervals above and below the normal sea level concentration of 159 mm Hg.

As students graph their results, as required on p. 14 of the "Manual_a" file on the CD-ROM, they should understand that more data are required in those areas where the curve is undergoing significant change. For example, Figure 5 shows that the major drop in running stamina occurs below a P_{O2} of 90 mm Hg, so more points should be collected in that region than in the relatively uninteresting region at higher oxygen partial pressures.

EXERCISE C Running Distance and Running Speed

In Exercise C, pages 15-17 in the "Manual_a" file on the CD-ROM, students graph the relationship between running speed and the distance a human can run before exhaustion. As explained in the exercise, fast-paced running events rely largely on anaerobic respiration, which provides quick energy. Anaerobic pathways have relatively few steps and don't require oxygen transport from the blood. The tradeoff is that these pathways are only 5% as efficient as aerobic pathways in terms of ATP yield per glucose molecule. This results in a waste of the glycogen stored in the muscles. Anaerobic respiration also produces lactic acid. Lactic acid can be converted to pyruvic acid in the liver, and then to glucose and to glycogen, but this conversion requires oxygen and is so slow that lactic acid can be kept at a low level only if the rate of production is also slow.

Sprint track events (such as a 100-m dash) are mainly anaerobic, and can utterly exhaust a runner in just a few seconds. A mile run (4 minutes) is about 50% anaerobic. A marathon of approximately 2.5 hours is only 2% anaerobic. The marathon runner burns glycogen aerobically, which keeps lactic acid accumulation at a low level.

The aerobic/anaerobic dichotomy also accounts for common observations about track events:

- ❑ Warming up helps performance by raising the metabolism, thus preparing the aerobic system to function efficiently.
- ❑ Eating carbohydrates before a marathon builds up muscle glycogen.
- ❑ "Hitting the wall," a sudden feeling of fatigue 15 miles into a marathon, occurs as glycogen is exhausted and the aerobic system starts burning proteins.
- ❑ Saving energy for a final spurt near the finish makes sense. The spurt will cause the production of lactic acid, but if the runner is near enough to the end of the event, performance will not be drastically impeded.

A typical graph of distance versus speed data is shown in Figure 6.

Exercise C asks what would happen if the subject were running in a low-oxygen atmosphere. The whole curve would be lower, and the highest speed at which indefinite running becomes possible would be slower. For example, at 159 mm Hg of oxygen, the human can run 288 m at 8 m/sec and 619 m at 4 m/sec. (Keep in mind that *Alien* models an average 20-year-old male, not an athlete.) At 90 mm Hg of oxygen, the running distance is only 192 m at 8 m/see and 420 m at 4 m/sec. This effect occurs because lactic acid begins to accumulate at a lower level of exercise due to the lower availability of oxygen.

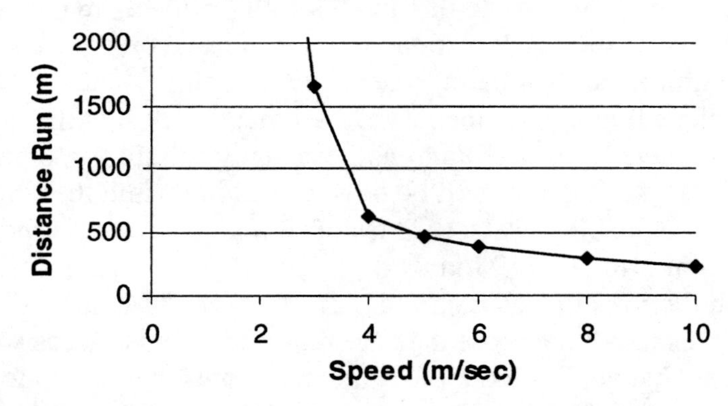

Figure 6 *A typical graph of running distance versus running speed.*

Note: The instructor should point out that the *Alien* simulation is unrealistic in that the only factor taken into account in fatigue is the buildup of lactic acid resulting from anaerobic respiration. In fact, depletion of muscle glycogen will eventually halt even aerobic exercise.

Suggestions for Helping Students. As the previous example shows, running distance increases greatly at very low running speeds. This is the totally aerobic "slow jog" that can be maintained for hours. Therefore it is important that the students collect data for their fastest running speeds first.

See the Exercise B discussion guide for hints about helping students formulate a prediction and use it to design their experiment. The students should be asked to predict how running distance varies with running speed and then to decide which speeds they predict would be the best ones to test.

In the curve presented in Figure 6, only a few points are needed above 5 m/see, and more are needed at the slow running speeds because that is where the slope of the curve changes significantly in its shape and direction. An important (and difficult) point to obtain would be the one that indicates the lowest running speed (about 2.25 m/ see) at which fatigue eventually occurs.

EXERCISE D Answering the Questions

In Exercise D (pages 18-19 in the "Manual_a" file on the CD-ROM), the student must answer 16 questions about the human or alien. This is the most demanding of the *Alien* experiments. First, some questions require the student to convert between units or to compute rates by noting the starting and finishing values of both a variable (such as oxygen consumed) and time. Second, students must read over the questions and decide the most efficient order in which to answer them.

The questions in Exercise D require both resting and "exercising" data. Some can be answered by simple observations of values given on the screen and others require computation. To ensure accuracy, follow computation directions on the screen. "Exercising" data is defined as data obtained after 30 seconds of running at half the maximum rate. Therefore, for the human, take exercising data only after a rested subject has been running at 5 m/see for 0.5 minutes (the maximum human running speed is 10 m/sec). Pause the simulation by pressing the Pause button after precisely 30 seconds of running.

Note that the "resting" and exercising questions alternate, so the questions cannot be answered in numerical order. This arrangement requires students to evaluate the entire task and then decide which parts to tackle first. Instruct students to read all the questions first, answer the resting ones before the subject has been disturbed or damaged, and then exercise the subject and take all the exercising data at once. Question 14 (maximum running speed) must be answered before any of the exercising questions because it allows determination of 50% of the maximum running speed. Warning students that the most *inefficient* way to proceed is to try to answer the questions in numerical order is often an effective way to get them to read all the questions before they start. If the students need to go back and obtain a resting value after the subject has been exhausted, there is no need to wait for his recovery—pressing the "Bring in Alien" tab will allow them to bring in a fresh human or alien.

A discussion of each question follows. (If your students need extra help, you may want to photocopy and hand out the following material.) A table of answers is included in Table 2.

1. **Resting oxygen consumption rate**. During a period when the subject is at rest and not recovering from excitement or fatigue, divide the change in liters of oxygen consumed by the change in time (minutes). Note that both beginning and end times and oxygen consumption must be taken. Simply dividing the total oxygen consumed by the total time results in the average oxygen consumption rate since the beginning of the session, which is not the data desired.
2. **Exercising oxygen consumption rate**. Same as question 1, but after 30 seconds of running at 50% of the maximum rate. Question 14 must be answered first to determine the running speed to be used. This is one of the most difficult questions because the subject must first be brought to the exercising state, and *then* the oxygen consumption rate must be measured over an additional time period. A good way to proceed is to

exercise the subject for 0.5 minutes, pause the simulation with the Pause button, and record the time and oxygen consumed. Then let the subject continue to run for perhaps 0.2 minutes more, pause again, and again record time and oxygen consumed. The *change* in oxygen consumed divided by the *change* in time is the exercising oxygen consumption rate.

3. **Resting heart rate**. Read off the screen during a resting period.
4. **Exercising heart rate**. After 0.5 minutes of running at half the maximum rate (see question 14), pause the simulation and read off the screen.
5. **Resting arterial blood pressure**. Read off the screen during a resting period.
6. **Exercising arterial blood pressure**. After 0.5 minutes of running at half the maximum rate (see question 14), pause the simulation, click on the "Blood & Heart" experiment, and read off the screen.
7. **Total blood volume**. Press the Blood & Heart button and follow the directions. Blood volume is computed via the "dye dilution method": a known quantity of dye is injected into the subject; the more dilute its final concentration in the blood, the greater the blood volume. For example, if 5 mg of dye is injected and the final concentration of dye in the blood is 5 mg/l, this implies that there is only 1 liter of blood. On the other hand, if the final concentration is 1 mg/l, this means that the blood volume is five times as great, or 5 liters. When 5 mg of dye is considered to be evenly dissolved in the whole bloodstream, the formula for determining blood volume is simple:[1]

   blood volume (liters) = 5 mg dye/dye concentration in blood (mg/l)

8. **Resting cardiac output**. Press the Pause button and read off the screen. Cardiac output is the amount of blood pumped out by the left ventricle into the aorta in one minute. In *Alien*, the cardiac output is monitored by a flow meter in one of the subject's systemic arteries. The data are reported in liters per minute.
9. **Exercising cardiac output**. Just after 30 seconds of running is completed, press the Blood & Heart button and read off the screen (see question 8).
10. **Resting cardiac stroke volume**. Press the Blood & Heart button. Stroke volume, the volume of blood output by one heartbeat, is computed by dividing cardiac output (liters/min) by heartbeats/min, to yield liters pumped per heartbeat. However, warn your students that to answer the stroke volume question, they must convert their answer to milliliters/heartbeat.
11. **Exercising stroke volume**. After 0.5 minutes of running at half the maximum rate (see question 14), pause the simulation, click on the Blood & Heart experiment, and compute. This answer will be in liters; the question requires an answer in milliliters.
12. **Beat frequency or stroke volume for cardiac output**. Divide exercising cardiac output by resting cardiac output. Then divide exercising heart rate by resting heart rate. If the second ratio is more than half of the first ratio, then most of the increase in cardiac output is caused by beat frequency. If not, stroke volume is causing most of the increase.
13. **Lung volume**. Press the Lung Volume button and follow the directions. Inflated lung volume cannot be determined merely by asking the subject to inflate his lungs and then exhale to the maximum extent possible. There is a residual volume of air in the lungs and in the air passages (about 1200 ml of the 5500 ml of total lung volume in humans), a

---

[1] A complication not considered in this simulation is that dye is dissolved only in the plasma portion of the blood; dye is not taken up by the cellular portion of the blood—the hematocrit.

reserve necessary to keep the alveoli inflated and which cannot be expelled. However, total lung volume *can* be determined by the dilution method.

Like the blood volume experiment above, this method determines an unknown volume by measuring the dilution of a known quantity of a marker substance in an unknown volume. After a normal exhalation, the subject is allowed to take one normal inhalation of a mixture of air that contains 1% helium. The expired air of the next exhalation is collected, and its helium concentration is measured.

The change in helium concentration in the expired air can be used to find the total lung volume. For example, let's say that the expiration that follows the inspiration of the 1% helium mixture contains only 0.1% helium. This means that the "normal" air in the lungs has diluted the helium to $1/10^{th}$ of its original concentration. If the tidal volume (the volume of a normal inhalation or exhalation) is 500 ml, then lung volume can be computed by this formula:

inflated lung volume = tidal volume (500 ml)/% He in the first exhalation,
or, in this case,
inflated lung volume = 500 ml/0.1 = 5000 ml

To understand this method, consider an analogy. Say we were trying to determine the volume of an irregular container filled with pure alcohol. We could withdraw 100 ml of alcohol, replace it with 100 ml of water, and then withdraw 100 ml of the alcohol/ water mixture. If we analyze the mixture and find that it is 95% alcohol, this means that the 100 ml of water replaced 5% of all the alcohol in the container. That is,

$$100 \text{ ml} = 0.05 \times \text{container volume, and}$$
$$\text{container volume} = 100 \text{ ml}/0.05 = 2000 \text{ ml}$$

14. **Maximum running speed**. Start this question with an unexercised subject. Increase the treadmill speed until the message, "He's fallen! He can't run that fast!" is displayed. (If the message "Subject has dropped from exhaustion!" appears, then the subject has fallen due to fatigue, not inability to run at the treadmill speed—this is *not* the data point you need.) To avoid fatigue, increase the treadmill speed rapidly so that maximum speed is attained before exhaustion overtakes the subject. Percent fatigue can be monitored to make sure that the subject was not at 100% fatigue when falling occurred.
15. **Distance run at 50% of maximum speed**. After answering question 14, bring in a fresh subject ("Bring in Alien" button). Note the starting time. Rapidly raise the treadmill speed to 50% of the maximum determined in question 14 and note the starting time. When the subject drops from exhaustion, note the finishing time. Find the difference between the starting time and finishing time; multiply this number × running speed × 60 (60 is the conversion factor that compensates for the fact that running speed is measured in meters/second, but time is measured in minutes).
16. **Can live in a terrestrial atmosphere**. For the human, of course the answer is "yes." To make this determination for an alien, cautiously bring the atmosphere towards 159 mm oxygen and 0 mm carbon dioxide. If the alien faints or shows unusual stress at any point, quickly bring the atmosphere back to his home planet conditions and conclude that the answer is "no."

To answer the questions most efficiently, wait one game minute after the subject has been delivered and pause the simulation with the Pause button. Then answer 1, 3, 5, 7, 8,10, and 13.

To use the Blood & Heart or Lung Volume buttons (questions 7, 8, 10, and 13), the simulation must be running (not paused). Then answer 14. Bring in a fresh subject and start the subject running at 50% of the maximum rate. After 30 seconds of this, pause the simulation and answer 2, 4, 6, 9, 11, and 12. Start the simulation again and wait until the subject drops from exhaustion, thus answering 15. Then bring in another fresh subject and answer 16.

When students are finished answering the questions, the answer sheet (page 20) should be given out as a handout. This page will allow students not only to see what the correct answers are, but also to compare the physiological conditions of humans and aliens. The fact that a resting alien consumes over 4 liters of oxygen per minute may not be remarkable until the student sees that the human's resting oxygen consumption is only $1/20^{th}$ as much.

Additional Investigations

Alien can also be used as a data generator for original student investigations. For example, students could study the effect of oxygen *enriched* atmospheres on running stamina, the influence of oxygen partial pressures on recovery time from fatigue, the effect of oxygen partial pressure on maximum sustainable running or walking speed, the relation between heart rate and metabolic rate, and the ways in which different aliens are adapted to their home planet conditions. These investigations are quite challenging because students must design the experiment in addition to performing it. Student-designed investigations like these could be evaluated by student reports.

Descriptions of the Aliens

The Aliens and their planets of origin (all outside our solar system) are all fictitious. While there is much evidence of planets (mostly Jupiter-sized or much larger) around neighboring stars, there is no evidence of life outside our solar system.

Planets are designated by their star and their distance from it. By this system, Earth would be called Sol 3 because it is the third planet out from the star called Sol (Latin for "sun"). Vega 1 is the first planet out from Vega, a 0.1 magnitude star in the constellation Lyra (The Harp). Regulus 4 circles around a 1.3 magnitude star in Leo (The Lion). If a star is too obscure to have an individual name, it is named for its constellation and its brightness in that constellation. Thus the fourth brightest star in the constellation Auriga (The Charioteer) is Delta Aurigae (delta is the fourth letter of the Greek alphabet) and the nineteenth brightest star in the constellation Cetus (The Whale) is Tau Ceti (tau is the nineteenth letter). By this system, Vega 1 could also be called Alpha Lyrae 1 because Vega is the brightest (alpha) star in Lyra. Some data about the planets and the aliens follow:

Vega 1

Vega 1 is a large planet with powerful gravity, a thick atmosphere, and more than 15 times the oxygen partial pressure of Earth. This oxygen-rich home and a lifetime of training under 4.6 g of gravity make this alien a "superman" athlete on Earth, with a very high metabolic rate, and small lungs.

Delta Aurigae 1

This is the planet most unlike Earth: a vast, murky world with a crushing CO_2 atmosphere thousands of kilometers thick. The atmosphere includes only a trace of oxygen—0.06%. Despite this, the alien has a high metabolic rate and compensates for the lack of oxygen and the near liquid consistency of the atmosphere by having a high performance circulatory system and huge lungs.

Tau Ceti 2

Although smaller than Earth's moon, Tau Ceti 2 still retains a light atmosphere with a partial pressure of oxygen of only 20 mm Hg. This alien has a very low metabolic rate, a slow and feeble circulatory system, a great tolerance for low-oxygen environments, and very little stamina on Earth.

Tau Ceti 3

Tau Ceti 3 is an earthlike planet. The alien is superficially similar to a human, but has a remarkably large, slow heart. Judging from his lack of athletic ability, this alien seems adapted to a sedentary lifestyle.

Regulus 4

Regulus 4 is also earthlike, but its atmosphere is mostly oxygen. This alien has the highest metabolic rate and the smallest lungs of any of the aliens. His system, however, has little reserve capacity. This makes him prone to cardiac arrest, and brain death will rapidly follow.

Table 1 Answers to the 16 Questions in Exercise D

Question	Human	Vega	Aurigae	Ceti 2	Ceti 3	Regulus
1. Resting oxygen consumption rate (l/min)	0.19	1.39	0.95	0.02	0.286	4.15
2. Exercising oxygen consumption rate (l/min)	1.07	3.33	1.80	0.033	0.367	4.85
3. Resting heart rate (beats/min)	69	170	36	12	23	110
4. Exercising heart rate (beats/min)	153	451	51	17	27	147
5. Resting mean arterial pressure (mm Hg)	98	131	524	12	89	238
6. Exercising mean arterial pressure (mm Hg)	132	208	540	16	108	255
7. Total blood volume (liters)	5.6	7.1	12.5	2.18	16.7	12.8
8. Resting cardiac output (l/min)	5.15	3.2	22.46	0.22	31.05	6.65
9. Exercising cardiac output (l/min)	15.36	17.56	32.54	0.97	66.15	8.41
10. Resting cardiac stroke volume (l/min)	74.6	18.82	624	18.3	1350	60.4
11. Exercising cardiac stroke volume (l/min)	100	38.9	638	5.71	2450	57.2
12. Exercising cardiac output increased by faster beat or by larger stroke volume	beat	beat	beat	stroke	stroke	beat
13. Lung volume (liters)	5.55	1.29	32	10.0	9.2	0.303
14. Maximum speed at which a resting subject can run (m/sec)	10	16	8	4	2	5
15. Running distance (m) at 50% of maximum rate	465	1056	456	108	310	233
16. Ability to live in earth's atmosphere	yes	no	no	yes	yes	no

Notes:

Alien: A Simulation of Cardiopulmonary Physiology

Overview

One of the basic properties of life is homeostasis: Despite changing external conditions and threatened internal changes (due to processes such as consumption of oxygen and production of CO_2), living things can maintain their internal conditions within narrow limits that are suitable for life. As an example, consider how the body responds to exercise. During extremely strenuous exercise, the rate of oxygen consumption and CO_2 production may increase by a factor of twenty times over the resting rates. It might appear that the tissues would die in less than a minute as they run out of oxygen. However, the cardiopulmonary system can maintain homeostasis even with this severe challenge. Ventilation may increase almost twenty-fold, and blood flow to the muscles may increase about twenty-five times. Oxygen concentration in the muscles does go down, but only slightly. Then, when exercise stops, the blood flow and ventilation gradually return to their resting values. Despite drastic fluctuation in oxygen demand and CO_2 production, the system was able to keep blood pH and oxygen and CO_2 content of the blood within normal limits.

Alien will allow you go explore the powerful homeostatic system of cardiopulmonary physiology. Also, because you will be using a computer simulation, you will be able to observe the limits of physiological systems and the collapse of homeostasis when a system is pushed beyond its capacities.

The *Alien* Exercises

Alien can be used either as a game (Exercise D) or as a series of simulated experiments (Exercise A, B, and C). Because the experiments will familiarize you with the program, you should do them first.

In Exercise A, you will follow directions in this manual to observe the effect of exercise at sea level, exercise at high altitude, and increasing CO_2 in the atmosphere on human physiology.

In Exercises B and C, you will observe the effect of altitude on running stamina and the effect of running speed on the maximum distance run before exhaustion by a human. In these two experiments, unlike in Exercise A, you must plan the experiments yourself.

Finally, Exercise D is the *Alien* game. In the game, you assume the role of a cardiopulmonary physiologist hired by NASA to study an extraterrestrial who has been brought back to earth. The program can simulate the responses of five fictional aliens with physiologies well adapted to their home planets, and thus quite different from human physiology. You can subject the alien (or a human) to exercise, changes in the oxygen or carbon dioxide concentrations in the atmosphere, injections of drugs, and electric shock. Depending on how far the subjects are pushed beyond their limits, they will drop from fatigue, faint, have a heart attack, and finally die (if not saved in time by cardiopulmonary resuscitation—CPR).

If you can avoid killing your patient, your objective in the *Alien* game is to experimentally determine the answers to 16 physiological questions. The program will grade you on the accuracy of your answers. Student teams within a laboratory section may compete for the highest score.

Physiological Variables Displayed by *Alien*

You cannot use *Alien* very well unless you understand the variables it displays. These are listed below.

- **Time (minutes)** This is cumulative time since the beginning of the session. *Alien* simulation time runs slightly more rapidly than real clock time, at about 1.6 simulation minutes/minute.
- **Ventilation (l/min)** (liters per minute) This is the rate at which air moves in and out of the lungs. This is not the same thing as the oxygen consumption rate.
- **Heartbeats/minute** Heart rate is the same as the pulse rate.
- **Average arterial pressure (mm Hg)** (millimeters of mercury) Blood pressure is expressed as the height of a column of mercury that the blood pressure would support. Blood pressure in a resting human rises to a peak of about 120 mm Hg when the heart is contracting and falls to 80 mm Hg when the heart is relaxed. *Alien* averages these into a single value of 100 mm Hg for the human.
- **Venous Hb oxygen saturation (%)** indicates the percentage of venous hemoglobin that is saturated with oxygen. Oxygen saturation of venous hemoglobin is about 70% for a resting human and may drop to 20% during strenuous exercise.

  a. *What information can be gained by measuring the oxygen saturation of venous hemoglobin that could not be gained by measuring the oxygen saturation of arterial hemoglobin?*______________________________

  ____

  ____

- **Oxygen consumed (l)** (liters) This is also a value that is cumulative from the beginning of the session. At one point you will be asked to find the rate of oxygen consumption. You must compute the change in oxygen consumed divided by the change in time. To do this, be sure to note both your beginning and ending times and the volume of oxygen consumed at the beginning and at the end of the measurement period.
- **Venous oxygen p. p. (mm Hg)** indicates **partial pressure of oxygen in the venous blood** The total pressure of a mixture of gases is the sum of the pressures of the individual gases in the mixture. Oxygen makes up 21% of the volume of dry air. The pressure resulting from O_2 concentration alone is the oxygen partial pressure (P_{O_2}). At sea level, all the gases that make up the air (oxygen, nitrogen, argon, carbon dioxide, and others) exert enough total pressure to raise a column of mercury 760 mm. The partial pressure exerted by O_2 is 21% of 760 mm sufficient pressure to raise a column of mercury 159 mm.

  Partial pressure is a measure of oxygen concentration in the air, in the blood, or in the tissues. P_{O_2} is 159 mm Hg in air and about 100 mm Hg in the alveoli of the lungs and in the blood leaving the alveolar capillaries (at this point hemoglobin is 100% saturated with oxygen). In the tissues and in the venous capillaries that carry blood away from these tissues, P_{O_2} falls to 40 mm Hg, at which point the blood is approximately 70% saturated. During extremely strenuous exercise, the tissue value of P_{O_2} may drop to 20 mm Hg, and unconsciousness will occur if it drops to 10 mm Hg.

- **Arterial carbon dioxide p. p. (mm Hg)** indicates **partial pressure of carbon dioxide in the arterial blood** In humans, even small increases in P_{CO_2} above the resting value of 40 mm Hg will result in hyperventilation, a major function of which is to expel excess CO_2. Note that CO_2 makes up only 0.03% of dry air.

   a. *Calculate the partial pressure of CO_2 in the air (in mm Hg). (Find 0.03% of 760 mm Hg, the total air pressure.)* ________ *At rest, the P_{CO_2} of venous blood entering the alveoli of the lungs is about 45 mm Hg. The P_{CO_2} of arterial blood leaving the alveoli is about 40 mm Hg.*

   b. *Explain why carbon dioxide concentration in terms of venous P_{CO2} would be expected to be much higher than atmospheric P_{CO_2}.* ____

   ____

   It may seem surprising that even after passing through the alveoli, P_{CO_2} remains high. Keep in mind that venous blood discharges CO_2 into the lungs and picks up O_2. Therefore, alveolar air is enriched in CO_2 and depleted in O_2. Arterial CO_2 levels reflect alveolar gas composition. P_{O_2} is 159 mm Hg in the atmosphere, but only 100 mm Hg in the alveoli (and in the blood leaving the alveolar capillaries). P_{CO_2} of the blood entering the alveoli at 45 mm Hg is reduced only to 40 mm Hg upon exiting from the lungs.

   In this simulation, we measure arterial (rather than venous) P_{CO_2} because an excess CO_2 concentration in the venous blood will continue to be reflected in the arterial blood. Also, levels of CO_2 in the arterial blood are monitored by respiratory centers in the medulla of the brain, which exert significant control over the ventilation rate; this rate will be increased if the current ventilation is not dispelling CO_2 rapidly enough.

   Arterial pH The normal pH of human arterial blood is 7.4. Changes in the pH of the blood have many important physiological effects. For example, during exercise, higher levels of cellular respiration result in increased levels of CO_2 in the tissues, some of which is carried in the blood as carbonic acid. Anaerobic respiration by fatiguing muscles produces lactic acid. Higher acidity (lower pH) of the arterial blood causes the respiratory center in the medulla of the brain to signal for an increase in ventilation, which helps to clear excess CO_2 from the system. If CO_2 is present in concentrations that exceed the lungs' ability to expel it, measurements of arterial pH can give an indication that homeostatic mechanisms are having difficulty compensating for these concentrations of CO_2.

- **Percent fatigue** In this simulation, the only factor taken into account as a cause of fatigue is an accumulation of lactic acid in the muscles, due to anaerobic respiration. In this program, the index of fatigue is arbitrary: 100% fatigue causes collapse from exhaustion. Fatigue increases only during exercise, and diminishes during periods of rest, as increased availability of oxygen allows lactic acid to be metabolized to pyruvate.

- **Stress index** (graphics display only) This is an arbitrary index indicating overall physiological excitement. In this simulation, a stress index of 45 results in cardiac arrest in humans.

To help you understand some of the relationships among the variables, Table AA1 outlines some of the cardiopulmonary changes that occur during exercise.

The *Alien* Digital and Graphical Displays

The *digital* display shows the all variables above as continuously updated numbers.

The *graphical display* is accessed by pressing the **Graphical Display** button. Pressing this button a second time returns you to the digital display. The graphics display shows a plot heart rate, blood pressure, ventilation rate, and stress for about 4 minutes into the past. Tic marks on the y-axis of this graph are multiples of the resting values. Values greater than eight times the resting value are off-scale.

Exercise A **Demonstration Experiments**

In the following experiments, you will observe the physiological effects of exercise at sea level (Part 1), of exercise at high altitude (Part 2), and of increasing atmospheric concentrations of CO_2 (Part 3). These experiments rapidly illustrate some of the more dramatic and important features of *Alien*.

Objectives

- ❑ Determine the changes in metabolic rate, ventilation rate, heart rate, and blood pressure that accompany exercise at sea level.
- ❑ Explain how the body compensates for the stress of exercise.
- ❑ Describe how the conditions in a low-oxygen atmosphere disrupt the exercise homeostasis maintained under normal oxygen conditions.
- ❑ Describe how the body compensates for increasing concentrations of carbon dioxide in the air and how this homeostasis collapses when carbon dioxide concentration exceeds the limits of compensation.

PART 1. EXERCISE AT SEA LEVEL

Vigorous exercise is one of the greatest stresses the body must withstand, and the body's ability to compensate for it is an example of the effectiveness of homeostasis. At rest, the rate of oxygen consumption is about 0.2 liters/min; during exertion, consumption may rise to 20 times this amount. It might be expected that such massive increases in oxygen demand might cause oxygen shortages in the tissues within seconds. Yet the partial pressure of oxygen (P_{O_2}) in venous blood only drops from 40 mm Hg to about 20 to 30 mm Hg. Carbon dioxide partial pressure (P_{CO_2}) in the arterial blood rises only slightly, from 40 to 45 mm Hg, and with prolonged exercise, P_{CO_2} returns to the resting value or even lower. Maintaining blood gases within a range consistent with life is the result of compensation by the cardiovascular system and the lungs. Receptors in the brain and in the carotid artery respond to increased concentrations of CO_2 in the blood by causing a rapid increase in ventilation. Rapid blood flow purges CO_2 from body tissues and supplies oxygen. Then this same rapid blood flow takes venous blood to the lungs where it can take on additional oxygen and discharge its CO_2.

a. Why does venous oxygen concentration (in terms of P_{O_2}*) decrease during exercise?*

____

____

When exercise is so strenuous as to result in extreme fatigue and, eventually, collapse, the physiological changes—in heart rate, ventilation, partial pressures of oxygen and carbon dioxide, and so on—tend to return to normal values.

Table AA1 details the effects of exercise on the *Alien* variables.

Table AA1 The Effects of Exercise on the Variables Used in *Alien*

Variable	Change	Reason
Oxygen Consumption	Increases	ATP is rapidly consumed and ADP becomes more abundant. This increases the rate of most metabolic reactions.
Ventilation rate	Increases	Increased metabolism produces more CO_2, which stimulates the respiratory center in the medulla. Also, the motion of the body directly excites the respiratory center.
Heart rate	Increases	The excitement of exercise increases sympathetic stimulation of the heart, which causes a more rapid rate of firing of the sino-atrial node. Also, lack of oxygen in exercising muscles increases blood flow through them, which increases the rate at which blood is returned to the heart.
Arterial pressure	Increases	Lactic acid and CO_2 plus nervous impulses to contracting pressure muscles excite the "vasomotor" center in the medulla. The center constricts arterioles in nonexercising parts of the body, raising blood pressure and diverting blood to the contracting muscles.
Venous oxygen	Decreases	Exercising muscles consume more oxygen. But reduction in tissue oxygen and acidification caused by increased CO_2 cause hemoglobin to give up oxygen more readily to needy tissues.
Arterial CO_2	Increases then decreases	Exercising muscles generate more CO_2. This and the greater acidity caused by both carbonic acid and lactic acid (from anaerobically respiring muscles) strongly stimulate the respiratory center in the medulla, which stimulates ventilation, and CO_2 is removed more rapidly from the lungs.
Arterial pH	Decreases	First, CO_2 combines with water in plasma to form carbonic acid. Later, lactic acid formed by anaerobic respiration in the muscles causes an additional lowering of blood pH.
Fatigue	Increases	If muscles cannot be supplied with oxygen rapidly enough, anaerobic respiration creates lactic acid. In addition, the muscles' reserves of glycogen become depleted.

Procedure

1. Start the *Alien* program (instructions for installing and running *Alien* are found elsewhere).

2. Indicate that you want to do the experiments and that you do not need instructions.
3. Indicate that you wish to study a human in the **Bring in Alien** screen.
4. When the digital display screen is displayed, wait a few seconds and then press the **Pause Experiment** button to pause the simulation. Fill in the first column of Table AA2 with the resting values of heart rate, ventilation rate, arterial P_{CO_2} and venous P_{O_2}.
5. Start the simulation again by pressing the **Resume Experiment** button. Then select the **Exercise** tab and quickly raise the treadmill speed to 8 m/sec (a 3.4 minute mile). Follow the directions that appear on the screen.
6. Watch the **Percent fatigue** variable; when it reaches 100% the subject will drop from exhaustion. Using both the digital and the graphics displays, watch the trends in venous oxygen as the subject approaches exhaustion. Just before the subject collapses, pause the simulation (**Pause Experiment** button) and record data in the second column of Table AA2.

Table AA2 Human Response to Exercise at Sea Level (Oxygen = 159 mm Hg)

	Resting	Exhaustion	0.5 minutes after collapse*	1.0 minutes after collapse†
Ventilation				
Heartbeats/minute				
Venous P_{O_2}				
Arterial P_{CO_2}				
Arterial pH				

Time of collapse:________
*Time of collapse + 0.5 minutes:________
†Time of collapse + 1.0 minutes:________

7. Continue the simulation. When the subject collapses, pause the simulation, write the simulation time at the bottom of Table AA2, compute what the simulation time will be at 0.5 and 1.0 game minutes after collapse, and record these at the bottom of Table AA2. This will help you keep track of the time. Resume the simulation, pause at 0.5 and 1.0 game minutes after collapse, and record your data in the third and fourth columns of Table AA2.

   b. *Why does ventilation rate increase as the subject exercises?*________________________

   ____

   c. *How do the large increases in heart rate (and blood pressure) during vigorous exercise help to maintain homeostasis?*________________________

   ____

   d. *What evidence do you see that blood-gas homeostasis is being maintained despite the stress of exercise?*________________________

   ____

e. *Of the variables listed in Table AA2, which is the first to return to normal resting values after exercise? Why do you think this variable is the first to normalize?*

f. *Why do all the variables in Table AA2 gradually return to their resting values after exercise stops?*

PART 2. EXERCISE AT HIGH ALTITUDE

Human homeostatic mechanisms operate less efficiently at high altitudes. For example, at the top of Mt. Everest, about 30,000 ft (or 9,000 m) above sea level, there is only one third of the amount of oxygen in the air that would be available at sea level. No matter how fast the lungs ventilate and the blood flows, the system cannot deliver enough oxygen to the tissues. Venous oxygen (P_{O_2}) dips below the lethal level (5 mm Hg) and cardiac arrest occurs. Body tissues continue to respire, but with no blood flow and no ventilation, venous P_{O_2} drops even faster and soon there is no oxygen in the venous blood. Meanwhile, arterial CO_2 rises rapidly, and brain death occurs within a simulation minute (simulation minutes are slightly shorter than actual minutes).

Procedure

1. Press the **New Human** button to remove the fatigued human subject and bring in another one. Pause the simulation after it has been running a few seconds and record the resting values in the first column of Table AA3. Press the **Resume Experiment** button to continue.

Table AA3 Human Response to Exercise at 9,000 m (Oxygen = 50 mm Hg)

	Resting	Unconsciousness	Venous P_{O_2}	
			5 mm Hg	0 mm Hg
Ventilation				
Heartbeats/minute				
Venous P_{O_2}				
Arterial P_{CO_2}.				
Arterial pH				

2. Select the **Oxygen** tab and lower the oxygen from 159 mm Hg to about 50 mm Hg—about the same oxygen content that would be present at the top of Mt. Everest.
3. Select the **Exercise** tab and rapidly raise the treadmill speed to 8 m/sec. If you do not do this rapidly, the subject may faint from lack of oxygen before you start significant exercise.
4. This time watch venous oxygen, which will go much lower than in Exercise A, Part 1. When the subject becomes unconscious, immediately pause the simulation and record the data in the second column of Table AA3. Unpause the program and continue watching venous oxygen.
5. When venous oxygen reaches 5 mm Hg, record the data in the third column of Table AA3. Finally, when venous oxygen reaches 0 mm Hg, record the data in the fourth column.
6. Continue to note the change in arterial CO_2 until the subject dies.

   a. *Why did exercise cause a failure of homeostasis in this case?* ____________________

   ____

   ____

   ____

   b. *How did the physiological responses in this experiment differ from those that occurred in Part 1?* ____________________

   ____

   ____

   c. *Why did arterial CO_2 remain at fairly normal values until cardiac arrest, but then rapidly increase?* ____________________

   ____

   ____

   ____

PART 3. INCREASING ATMOSPHERIC CO_2

Increasing atmospheric carbon dioxide provides another demonstration of how the cardiovascular system and the lungs maintain homeostasis. As moderate amounts of CO_2 are added to the atmosphere, the subject can easily compensate by increasing his ventilation rate. Large increases in atmospheric CO_2 cause only minor increases in arterial CO_2. But eventually the subject is breathing as fast as is possible, and no further compensation can occur—the limits of homeostasis have been reached. After this point, further increases in atmospheric CO_2 cause equal increases in arterial CO_2, and soon the respiratory center in the brain becomes depressed and death follows.

Procedure

1. Bring in another fresh human subject. Pause the simulation (press the **Pause Experiment** button) after it has been running a few seconds and record the data in the first column of Table AA4.
2. Resume the simulation by pressing the **Resume Experiment** button. Select the **Carbon Dioxide** tab and raise the CO_2 to 30 mm Hg. Note how arterial CO_2 and ventilation rise briefly, but then rapidly settle down to a new equilibrium. When equilibrium is reached (little change occurring), pause the simulation, and record data in the second column of Table AA4.

Table AA4 Human Response to Increasing Partial Pressures of CO_2

	0 mm Hg	30 mm Hg	60 mm Hg	120 mm Hg
Ventilation				
Arterial P_{CO_2}				
Arterial pH				

3. Raise the CO_2 to 60 mm Hg. Again note the surge in arterial CO_2 and ventilation. At equilibrium, pause and record data in the third column of Table AA4.
4. Finally, change the atmospheric CO_2 to 120 mm Hg. Note the trends in arterial CO_2 and ventilation. Cardiac arrest will soon occur. When it does, pause the simulation and record data in the fourth column of Table AA4. Don't worry about CPR immediately. The subject has abundant venous oxygen and is not in danger of rapid death.
5. Use CPR to save your subject
   a. Unpause and bring the atmospheric CO_2 back to 0 mm Hg.
   b. Select the **CPR** tab.
   c. Press the **Initiate Heart Massage** button to start heart massage.
   d. Press **Increase Ventilation** and set the respirator at a 15 liters/min ventilation rate.
   e. When the subject's heart starts beating again, press **Quit CPR** to end CPR. As you do, note the surge of ventilation, the decline of CO_2, and the restoration of normal conditions.
6. Answer the following questions:
   a. *The success of homeostasis can be measured by how large a difference the system can maintain between the concentration of atmospheric CO_2 and the concentration of arterial CO_2. That is, without homeostasis, arterial and atmospheric CO_2 would always be equal; if homeostasis operated perfectly and had no limitations, then arterial CO_2 would always remain constant despite all changes in atmospheric CO_2. What evidence do you have that the human system can maintain homeostasis at 30 mm Hg of atmospheric CO_2?*

   ____

   ____

   ____

   b. *How is homeostasis maintained at 30 mm Hg of CO_2?*

   ____

   ____

   ____

   c. *Is homeostasis less effective at 60 mm Hg of CO_2 than at 30 mm Hg?*________ *Why?*

   ____

   ____

   ____

   d. *Why does homeostasis collapse at 120 mm Hg of CO_2?*

   ____

Exercise B The Effect of Altitude on Running Stamina

As altitude increases, oxygen continues to make up 21% of the atmosphere, but its partial pressure declines. The following table shows some representative altitudes and oxygen partial pressures:

Table AB5 Oxygen Partial Pressures (mm Hg) at Some Sample Altitudes

Location	mm Hg
New York, NY (10 m)	159
Table Rock, SC (955 m)	143
Mt. Mitchell, NC (2,030 m)	126
Telluride, CO (2,710 m)	116
Pike's Peak, CO (4,340 m)	94
Mt. McKinley, AK (6,200 m)	73
Himalayan peaks (7,000 m)	65
Mt. Everest (8,848 m)	50
Jet airliners (10,000 m)	42

Because of this decline, the distance a human can run before fatigue is heavily influenced by altitude. Without sufficient oxygen, efficient aerobic respiration cannot occur. Anaerobic respiration produces lactic acid that accumulates in the muscle tissue, lowers the pH of blood in these tissues, and reduces the capacity of the muscle fibers to contract, producing sensations of muscle soreness, stiffness, and fatigue. Anaerobic respiration produces ATP, but inefficiently, and soon muscle glycogen has been exhausted. Without ATP, the muscle must stop contracting. Even if the person rests, low oxygen continues to hamper muscle activity, since recovery from fatigue (repayment of the "oxygen debt") and conversion of lactic acid to pyruvic acid to enable further production of ATP cannot occur if oxygen is not available.

Objectives

- Determine the relationship between partial pressure of oxygen in the atmosphere and the distance that can be run before exhaustion, and explain the results.

Procedure

1. Return to the **O.R.** screen. Press the **New Human** button to bring in another human volunteer.
2. You will choose the oxygen partial pressures you wish to use, but first, begin with 159 mm Hg (normal P_{O_2} at sea level), which will serve as a standard for comparison. The partial pressure of oxygen should be 159 mm Hg when the human subject "arrives" on the screen.
3. Select the **Exercise** tab and increase the treadmill speed to 5 m/see (a marathon runner's pace). Note the starting time on the screen (not on a clock or a watch) as the speed reaches 5 m/sec. Record this value at the bottom of Table AB6.
4. When **Percent fatigue** reaches 100% the subject will drop from exhaustion. While waiting for this to happen, watch changes in the venous partial pressure of oxygen, the arterial partial pressure of CO_2, and arterial pH.

5. When the subject drops from exhaustion, note the finishing time and record it at the bottom of Table AB6. You may want to pause the simulation by pressing the **Pause Experiment** button. In the first column of Table AB6, record the venous P_{O_2}, arterial P_{CO_2} and arterial pH at the moment of collapse.

Table AB6 Effect of Decreasing Partial Pressure of Oxygen on Human Running Performance

	P_{O_2} (159)	P_{O_2}( )	P_{O_2}( )	P_{O_2}( )	P_{O_2}( )
Venous P_{O_2}					
Arterial P_{CO_2}					
Arterial pH					
Distance run*					

Starting time ________
Finishing time ________
*Distance run = speed (m/sec) × 60 × (finishing time - starting time)

6. Calculate and record the distance as the distance run at sea level.
7. From the **O.R.** screen, press the **New Human** button to bring in a new human volunteer.
8. Select the next altitude. Use your judgment to select four or five altitudes you believe will represent the full range of human performance, from unimpaired to total inability to exercise. You may use an oxygen partial pressure not listed in Table AB5.
9. Let the subject equilibrate at the new altitude for at least 3 game minutes. In Table AB6, write the partial pressure of oxygen in the parentheses ( ) at the top of the appropriate column. Repeat steps 2 through 7 at this altitude. Record any episodes of unconsciousness or cardiac arrest. If the subject passes out before the 3-minute waiting period is up, the distance run is zero.
10. Repeat the directions in step 9 until you feel you have collected enough data to describe the relationship between partial pressure of oxygen and running stamina. Mark appropriate intervals on the vertical Y-axis, then graph your results in Figure AB1.

Figure AB1 *Graph of the relationship between partial pressure of oxygen and running stamina.*

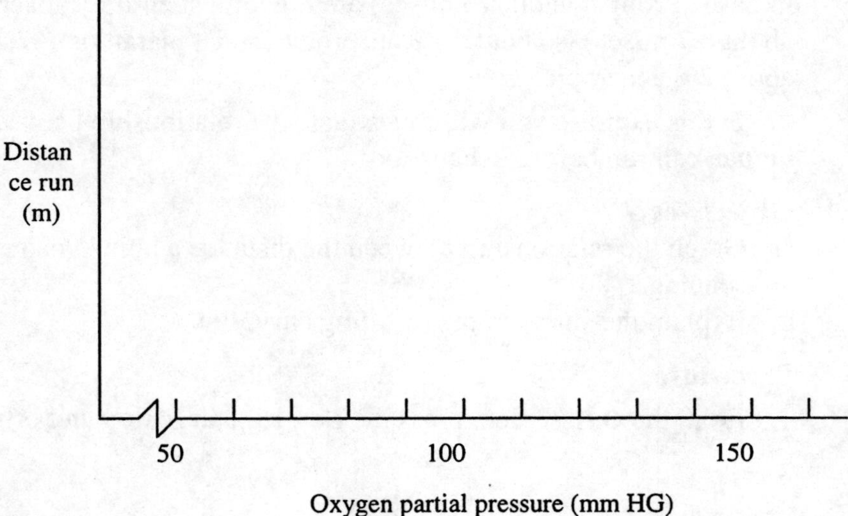

11. Answer the following questions.

a. *Why did the directions advise you to let the human remain at an "altitude" for a few minutes before testing running stamina?*____________________

____

____

b. *Using your graph, predict the oxygen partial pressure (and altitude) at which a human can remain conscious but running is impossible.*____________________

c. *At low partial pressures of oxygen, "exercise" usually ends not in fatigue but in unconsciousness. Why?*____________________

____

d. *In real (not simulated) humans, do you think that unlimited running stamina could be achieved by increasing the partial pressure of oxygen? Why or why not?* (Hint: *Why does lack of oxygen limit running endurance? If oxygen were present in higher concentration, what would then limit running performance?*

____

____

____

Exercise C **Running Distance and Running Speed**

The slower a human runs, the farther the runner can go before becoming exhausted. This is a familiar fact to us, but why is it true? If sprinting 100m at 10 m/sec utterly exhausts an athlete, why isn't the same athlete exhausted by covering the same 100m by walking at 1 m/sec? The answer is that the sprinter is using anaerobic respiration and the walker using aerobic respiration. As we have seen, anaerobic respiration yields energy quickly, but uses glycogen in muscles very inefficiently and produces toxic lactic acid. Aerobic respiration is efficient and produces only CO_2 and water as waste products, but it is slow, in part because there are many more steps in the pathway of aerobic glycolysis, in part because some time is required for oxygen to be delivered from the lungs to the blood to the cells.

Thus, the slower the pace, the larger the role of aerobic respiration, allowing the athlete to consume muscle glycogen with maximum "fuel efficiency." Sprint track events such as the 100-m dash (100m at about 10 m/sec) are run almost entirely anaerobically, a mile run (1,609m at about 6.7 m/sec) is about 50% anaerobic, and a marathon (42,194m at about 5.2 m/sec) is only about 2% anaerobic.

In this exercise you will investigate the relationship between running speed and the distance a human can run before exhaustion.

Objectives

- Graph the relationship between the distance a human can run before fatigue and the speed of running.
- Explain the shape of the resulting curve.

Procedure

1. From the **O.R.** screen, press the **New Human** button and bring in another human subject.

2. It is most efficient to proceed from the fastest running speeds down to the slowest (because eventually you will reach a speed that is so low that the human will be able to keep running indefinitely). The fastest a human subject can run is 10 m/sec. Therefore, select the **Exercise** tab and quickly raise the treadmill speed to 10 m/sec. Note the time on the screen when the treadmill reaches this speed. Record this value as the starting time in Table AC7.
3. When **Percent fatigue** reaches 100%, the subject will drop from exhaustion. While waiting for this to happen, watch changes in the venous partial pressure of oxygen, the arterial pressure of CO_2, and arterial pH.
4. When the subject drops from exhaustion, note the finishing time. You may want to pause the simulation by pressing the **Pause Experiment** button.. At the moment of collapse, write the partial pressure of venous oxygen, partial pressure of arterial CO_2, and arterial pH in the 10 m/sec column of In Table AC7.
5. In Table AC7, record the distance run at 10 m/sec.

Table AC7 Distance Run before Exhaustion at Various Running Speeds

	10 m/sec	m/sec	m/sec	m/sec	m/sec
Starting time					
Finishing time					
Distance run*					
Venous P_{O_2}					
Arterial P_{CO_2}					
Arterial pH					

*Distance run = speed × 60 × (finishing time - starting time)

6. Press the **New Human** button to bring in another human volunteer.
7. You must now choose at least four other speeds. Remember that at some low speed (approximating a jog), the subject will be able to keep running indefinitely. Your lowest speed should be just slightly faster than this, enough to cause fatigue in about 8 to 10 simulation minutes. Your four speeds should be evenly spaced between this low speed and 10 m/sec.
8. Select the next running speed. It should be slower than 10 m/see but still a rapid running speed that you feel an average, college-age human could not maintain for a long period. Record the chosen speed in the first blank space to the right of 10 m/sec in Table AC7.
9. Select the **Exercise** tab to raise the treadmill to this new speed, note the starting time, and repeat steps 3 through 5 for the new speed.
10. Repeat step 9 for the three other speeds.
11. Mark appropriate intervals on the vertical Y-axis, and then graph your results in Figure AC2.

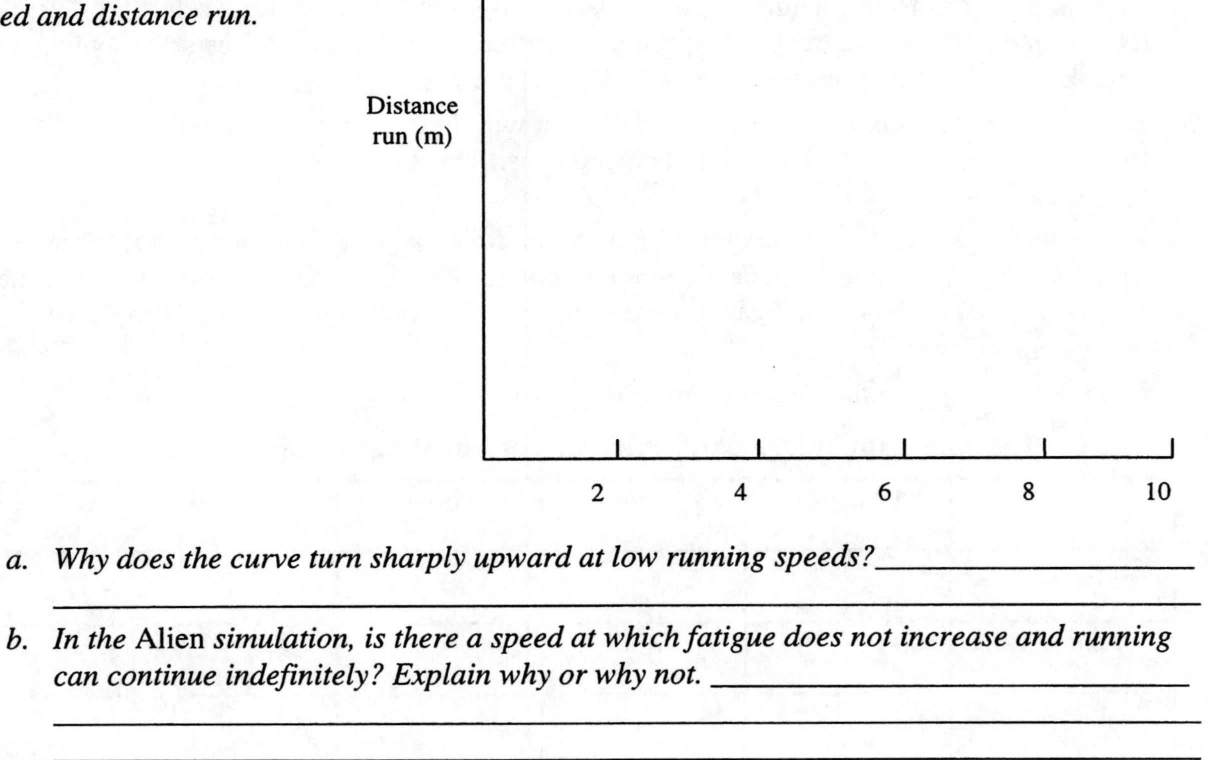

Figure AC2 *Graph of the relationship between running speed and distance run.*

a. *Why does the curve turn sharply upward at low running speeds?*____________________

b. *In the* Alien *simulation, is there a speed at which fatigue does not increase and running can continue indefinitely? Explain why or why not.* ____________________

c. *In the* Alien *simulation, the only factor taken into account as a cause of fatigue is lactic acid accumulation due to anaerobic respiration in the muscles. How is this an oversimplification? What other factors help to explain why an individual eventually becomes exhausted even when jogging very slowly for a long distance?* ________________

d. *If a human were running in a low-oxygen atmosphere (say at 90 mm Hg of oxygen instead of 169), how would your distance versus running speed curve be different?* ______

Exercise D Answering the Questions

The *Alien* game poses 16 questions (shown below) about physiology that allow you to test your skill at problem-solving, gathering data, and accurately computing results. You may choose either a human or a fictitious extraterrestrial subject. These extraterrestrials are somewhat similar to humans in that they use oxygen, expel carbon dioxide, and have lungs and hearts, but they are adapted to the conditions of their home planets, which are quite different from those of Earth. For example, Vega 1 is a larger planet than Earth, with a stronger gravitational force, a denser atmosphere, and more than 15 times the oxygen partial pressure of Earth. Delta Aurigae 1 is a vast, dark world with a crushing, nearly liquid carbon dioxide atmosphere thousands of kilometers thick, and only a trace of oxygen. In contrast, Tau Ceti 2 is smaller than Earth's moon, but retains a light atmosphere. The aliens' physiologies reflect adaptations to these conditions.

Table AD8 Questions to be answered in Exercise D.

What is the subject's:	Human	Alien
1. resting* oxygen consumption rate (liters/min)?		
2. exercising† oxygen consumption rate (liters/min)		
3. resting heart rate (beats/min)?		
4. exercising heart rate (beats/min)?		
5. resting arterial blood pressure (mm Hg)?		
6. exercising arterial blood pressure (mm Hg)?		
7. total blood volume (liters)?		
8. resting cardiac output (liters/min)?		
9. exercising cardiac output (liters/min)?		
10. resting cardiac stroke volume (mL/beat)?		
11. exercising stroke volume (mL/beat)?		
12. In exercise, is increased cardiac output caused more by increased heart rate or larger stroke volume?		
13. lung volume (liters)?		
14. maximum running speed (m/sec)?		
15. maximum distance run at 50% of maximum speed?		
16. Can the subject live in Earth's atmosphere?		

**Resting* indicates that the subject is unstressed, unfatigued, awake, and alert.
†*Exercising* indicates that the subject has been running for 0.5 minutes at 50% of the maximum rate.

Objectives

- ❑ Set up procedures and experiments for answering physiological questions about the cardiopulmonary system.

Procedure

1. Select the **Bring in Alien** tab.
2. You will then need to choose to experiment with a human or with one of the five aliens. Follow your instructor's directions.
3. If you use an alien, note the first screen showing the home planet conditions of the alien. As you experiment with your alien, study its adaptations to its conditions.
4. Answer the questions in any order you choose, but please note that since the "resting" and "exercising" questions alternate, answering questions in numerical order is probably the most time-consuming way to proceed. Instead, read through all questions and first answer the ones requiring resting data. Then answer the exercising questions. Remember that the "exercising" state in *Alien* means after 0.5 minutes of running at 50% of the maximum rate.

5. Before answering any exercising question, you must find the maximum running speed for your alien. The subject may collapse either from exhaustion or from being unable to run faster. You must find the speed just below the one that causes the subject to collapse with the message, "He's fallen! He can't go that fast!" If you exceed the maximum speed, bring in another subject and increase the speed to just under the preceding value. The message, "The subject has collapsed from exhaustion!" does not mean that you have found the maximum speed. Try another subject.
6. Some questions require specific experiments. For example, the questions about total blood volume, cardiac output, and stroke volume can be answered by selecting the **Blood & Heart** tab. The question about total lung volume is answered by selecting the **Lung Volume** tab. Be careful to use the correct units when recording answers.
7. If your subject becomes fatigued or has a cardiac arrest, do not wait for recovery. Select the **Bring in Alien** tab and bring in a fresh subject.
8. Write your answers opposite the questions above. When you want to evaluate your accuracy, select the **Answers** tab and give your answers. If you wish to pass by a question and not answer it, press the **Next** button as directed on the screen. You may return to the experiments after answering any of the questions, but once you have given an answer, you will not get another chance at that question as long as you keep the current subject.
9. When you have answered all the questions, you will receive a percentage grade on your accuracy. Report this grade to your instructor.

Name ______________________ ***Alien* Worksheet**

Section ______________________

To answer the questions below, read pages 1-5 in this manual. Complete this worksheet and bring it to the class period in which the *Alien* simulation will be used.

1. What is homeostasis?

2. How does the cardiopulmonary system help to maintain homeostasis?

3. What is meant by the partial pressure of an atmospheric gas? How is partial pressure measured?

   a) Is a partial pressure of oxygen of 40 mm Hg higher or lower than the partial pressure of oxygen in the atmosphere?_________

   b) Is a partial pressure of carbon dioxide of 40 mm Hg higher or lower than the partial pressure of carbon dioxide in the atmosphere?_________

4. Why does the partial pressure of oxygen in the body's tissues decrease during exercise?

5. Why does blood pH decrease (become more acid) during exercise?

6. Briefly describe some of the effects of exercise on the following: ventilation rate, heart rate, oxygen in the tissues, and carbon dioxide in the blood.

BioBytes *Cycle*

Instructional Objectives

The purpose of *Cycle* is to demonstrate the relationship between several of the more important variables of the reproductive system.

Cycle can be used to attain three increasingly ambitious goals:

a. To teach the basic facts about the fertile period in the menstrual cycle and when it occurs;
b. To teach the physiology of the menstrual cycle;
c. To perform original, simulated experiments.

THE BASICS

Cycle's main messages are that:

- ❑ Fertilization is possible only within a limited time before and after ovulation;
- ❑ The time of ovulation varies between cycles but occurs in a range of days in the middle of the cycle;
- ❑ There are clear hormonal indications that ovulation is about to occur.

If time is short and the instructional goals are limited to these facts, playing the game (Exercise A in "Manual_c" on the CD-ROM) will probably be sufficient. Student success at the game will measure the extent to which the objectives have been achieved.

PHYSIOLOGY OF THE MENSTRUAL CYCLE

Cycle can be used to teach the physiology of the menstrual cycle. This objective will require broadening the discussion from a focus on indicators of ovulation to include a close examination of the graphs of follicle diameter, hormones, and uterine lining. The best way to prepare students for such a discussion is to assign pages 2-8 in the *Cycle* manual ("Manual_c") on the CD-ROM, and to have them complete the CYCLE worksheet at the end of this exercise. Students should also make use of the "Review of the Menstrual Cycle" feature offered at the beginning of the simulation.

To gauge students' understanding of the menstrual cycle, the instructor might show a graph of follicle-diameter, separately number the growth phase, the collapse at ovulation, the constant luteal phase, and the degenerating luteal phase and ask the student to state the major events that are taking place in each of these regions of the curve. The same type of question could be asked for estrogen, progesterone, or LH. For a more difficult variation, ask the student to sketch one of these curves above and explain why it takes the shape that it does. The students could be asked to choose which hormone they would prefer to watch in order to make the most accurate predictions of ovulation, and to justify their choice. Other examples: Contrast the roles of estrogen and progesterone in maintaining pregnancy, and describe the effect of birth control pills on the menstrual cycle.

PERFORMING EXPERIMENTS

The most common *Cycle* experiment is Exercise B, "The Effect of Birth Control Pills" (pages 11-12 in "Manual_c"). Students can manipulate levels of estrogen, progesterone, FSH, and LH, and observe the effects of these changes in the menstrual cycle and on fertility.

Other experiments, from simple demonstrations to student-planned explorations, are described in the "Additional Experiments" section, pages C-12-13. A lab report would be an effective method for evaluating the instructional success of the experiment.

An Overview of *Cycle*

Cycle illustrates some of the basic hormonal features of the human menstrual cycle, female fertility, and pregnancy. It can be used both as a competitive game (Exercise A, pages 8-10 of the student exercises in "Manual_c") and as a data generator (pages 11-12).

In the game, students use data on hormones, follicle size, and other indicators to advise a woman when she can become pregnant during her menstrual cycle. The more the user advises the woman that pregnancy cannot occur, the more points are awarded. However, if intercourse occurs too close to ovulation and a pregnancy results, the user suffers a large point penalty. In the game mode, the user can choose from four levels of difficulty:

Level 1 The cycles are always 29 days long and data are provided for hormone levels, basal body temperature, and the development of the follicle. Predicting ovulation on level 1 is very easy.

Level 2 Provides the same data as level 1, but the cycles are variable in length and some are anovulatory. But by using hormonal patterns and the size of Instructor's Guide for the CYCLE Computer Simulations the follicle, the observant student can still operate nearly error-free.

Level 2 is the most useful in helping the student to understand the physiological and hormonal relationships of the menstrual cycle.

Level 3 More realistically, the cycles vary in length, and only data on basal body temperature and menstrual flow are available to the user.

Level 4 The cycles vary in length and the user has access only to data on menstrual flow.

When *Cycle* is used as an experimental data generator, the user can change the rate of production of any hormone to multiples of the normal value. It is possible to simulate the effects of several birth control methods, and explore the effects of various hormones on each other, on follicle development, on the uterine lining, on pregnancy, and so on.

Advice on the Classroom Use of *Cycle*

Students should read pages 2-8 in the *Cycle* student manual before using the program. You might also want to have them complete the worksheet on page 1 of the student manual.

Beginning with an instructor demonstration is often effective. A projector or large-screen monitor can help with this.

a. Go through a cycle on level 1, with the pregnancy prediction remaining on **Yes**. Switch between the text and graphic displays and point out the indicators that ovulation is about to occur.
b. Then try another level 1 cycle with the pregnancy prediction on **No**. Points will start to

accumulate. Be sure to switch the prediction back to **Yes** well before ovulation! Then switch it back to **No** once the ovum is infertile.

c. Then, try a level 2 cycle. Point out that, although the day of the cycle can no longer be used as an ovulation predictor, the follicle size and estrogen are still excellent sources of information.

d. Finally, allow the simulated woman to get pregnant and follow the pregnancy through to its conclusion. Point out the enormous point loss that pregnancy causes. If the zygote dies, point out that 33% of the zygotes die before they are even detected. A slightly delayed period is the only evidence of their brief existence. Another 17% of embryos die during the first trimester of known pregnancies.

Once the demonstration is completed, the best way to use *Cycle* is to assign 2-4 students per computer and let the program introduce itself. Students will be quick to understand the point of the game and ways to detect ovulation.

If more than six students must share a machine, an alternative format can be used. Begin with the instructor demonstration described above. If there are two or more computers, divide the class into that number of groups, and run a competition among the groups. Use the pause tab ti interrupt the program and allow each person in each group a turn at entering a pregnancy prediction. Choosing students randomly will keep attention from wandering for those who think that that it is not their turn.

Even if only one computer is available, the strategy just described can be employed; simply have the whole class attempt to maximize points. Or to move the game along faster, the instructor can remain in control of the keyboard while the class determines the pregnancy prediction. This can be done by a show of hands. If the class seems reluctant to participate, pause the game, explain where in the cycle the game is, and directly ask if pregnancy can occur now, why or why not. Then ask for a show of hands.

Of course, it is always possible to use one or two computers to provide enrichment material for small groups. Students may also pursue projects independently.

Cycle's Review of the Menstrual Cycle

As the first *Cycle* screen after the title screen appears, students will have a chance to go through a review of the menstrual cycle. This review summarizes the cycle's function of producing an ovum and preparing the uterus for pregnancy. Using a combination of text and graphics, it recounts the changes in follicle diameter, estrogen, progesterone, LH, FSH, the uterine lining, basal body temperature, and menstrual flow as the cycle progresses. For example, the plot of follicle diameter is shown in Figure 1 on page C-4.

The first plot shows the 'normal" course of follicle diameter in a series of histograms. The part of the plot being discussed on the screen (in this case, the growth of the follicle during the follicular phase) is highlighted with solid blocks. The part of the cycle during which the woman can become pregnant is shaded.

Discussion of each variable follows. These variables are reviewed in the student exercises in "Manual_c" on the CD-ROM.

Follicle Diameter. The follicle diameter screens review how the follicle grows from the size of a pinhead to about an inch in diameter, ovulates, survives for about 8 days after ovulation as the corpus luteum, and then starts to degenerate. It is easy to predict the time of ovulation from the follicle data because the sudden decrease in the diameter of the follicle is an obvious indicator that ovulation has occurred, and the further decline in follicle size as the corpus luteum degenerates shows that the end of the cycle is near. The students can see that follicle diameter tells them all they need to know to predict ovulation successfully.

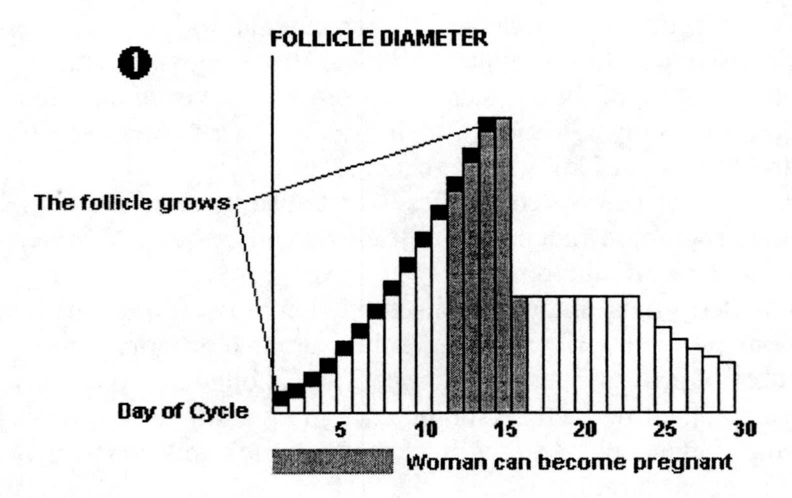

Figure 1 *The first plot of follicle diameter in the menstrual cycle review.*

Changes in Estrogen. Estrogen concentrations (Figure 2) are very helpful in predicting ovulation.

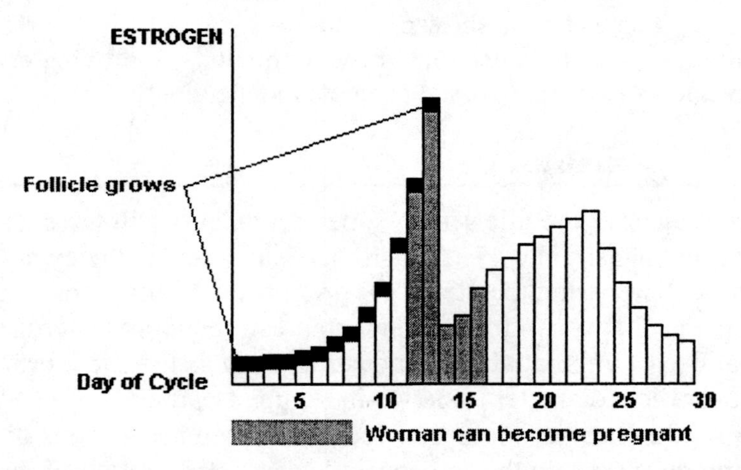

Figure 2 *The first plot of estrogen in the menstrual cycle review screens.*

Estrogen concentrations rise during most of the follicular phase to a peak of about 680 pg/ml, but then drop sharply about two days before ovulation. (This occurs mainly because several follicles may have been developing at once, but at this point the development and estrogen secretions of the smaller follicles are suppressed by the dominant one.) This indicator is particularly useful because it provides sufficient advance warning to avoid intercourse at a time when sperm might be able to survive in the oviducts until ovulation occurs. After ovulation, the rise and fall of estrogen concentration reflects the development and degeneration of the corpus luteum.

In humans, there are three major types of estrogens—estradiol, estrone, and estriol. Of these, the most potent and important is estradiol, and many discussions of menstrual cycle control do not even concern themselves with the other two hormones. CYCLE, too, plots only the concentration of estradiol, but because this term will be unfamiliar to many students, estradiol is indicated by the general name, "estrogen."

Students should take note of the units used to measure hormone concentrations in this simulation, either nanograms per milliliter (10^{-9} g/ml) of plasma or picagrams per milliliter (10^{-12} g/ml) of plasma.[1] A concentration of 1 pg/ml would be achieved by taking a weight of hormone equal to that of a penny (2.46 g) and dissolving it in a circular lake 10 m (33 ft) deep and 400 m (0.25 mi) in diameter.

LH and HCG. Because a spectacular "LH surge" (to about 790 ng/ml, see Figure 41-5 in the student exercises) precedes ovulation and may even cause it, students may conclude that luteinizing hormone provides ample advance warning of ovulation, but this is not true. LH changes little in concentration except during the surge, which occurs so close to ovulation (less than 24 hours before it) that allowing intercourse until the LH peak will probably allow residual sperm to reach the egg.

The digital and graphic displays for LH present the sum of LH and HCG (human chorionic gonadotropin). HCG is an LH-like hormone secreted by the fetal tissues; it keeps the corpus luteum from degenerating during the first three months of pregnancy. If pregnancy does not occur, changes in LH & HCG are caused by changes in LH only. If pregnancy does occur, large increases in LH & HCG are brought about by HCG only.

The profile for follicle-stimulating hormone (FSH) is similar to that of LH, but it is not given a graphic presentation.

Progesterone. Progesterone level is very low throughout the follicular phase, and rises to a sharp maximum in the middle of the luteal phase (see Figure 41-6 in the student exercises), but its profile gives little warning of ovulation.

The Uterine Lining and Menstrual Flow. Neither the thickness of the uterine lining (maximum thickness, about 7 mm) nor menstrual flow (the day with the heaviest flow produces about 22 ml) can be used to predict ovulation. When estrogen and progesterone are high, the uterine lining builds up slowly. When those hormones decline sharply, the lining persists for a few days and then degenerates, and menstrual flow occurs.

The Fertile Period of the Cycle. Even if the time of ovulation is known, there is still an element of chance in determining whether or not fertilization can take place. Sperm can live in the female reproductive tract for up to four days, although they lose fertility continuously during this period. The egg begins to lose fertility just hours after ovlulation, and by 24 hours the egg is infertile. The probability of fertilization in *Cycle* varies as shown in Table 1.

[1] *Note:* Hormone concentrations are usually measured in the plasma portion of the blood.

Table 1 The Probability that a Single Intercourse Will Result in Fertilization

Hours before ovulation:	100	0%
	96	12%
	72	20%
	48	36%
	36	50%
	24	69%
	12	80%
At ovulation:		100%
Hours after ovulation:	12	50%
	24	6%
	30	0%

Variations in Cycle Length. Cycle length varies widely between different women. The following data were reported by Chaizze et al., 1968:

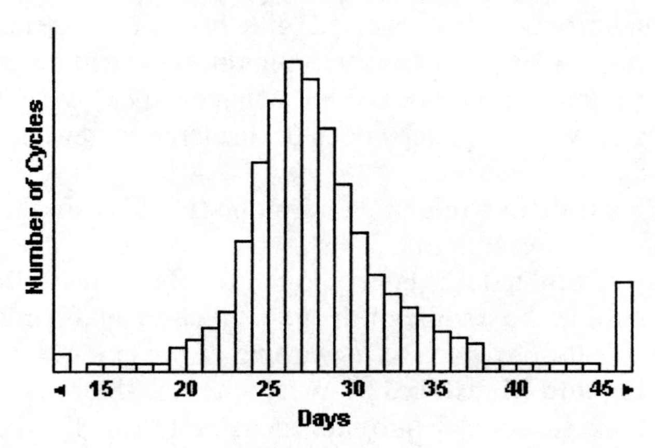

Figure 3 *Distribution of cycle lengths in 2316 women of all ages*[2]

While an individual woman would probably not experience such a range of variability, *Cycle* simulates years of moderate-to-high individual variability into a short time period to show students the diversity of cycles that are possible. In *Cycle,* on difficulty levels 2, 3, and 4, ovulation occurs on days 13, 14, 15, or 16 in 60% of all cycles. However, ovulation might occur at any time from day 6 to day 24. In addition, 16% of the time the follicle grows so slowly that it collapses and becomes "atretic" rather than ovulating.

[2] Chiazze, L., Jr., F. T. Brayer, J. J. Macisco, Jr., M. P. Parker and B. J. Duffy. 1968. The length and variability of the human menstrual cycle. *JAMA* 203: 377-380.

Running the *Cycle* Program

After a graphic of the uterus and ovaries, the first *Cycle* screen appears.

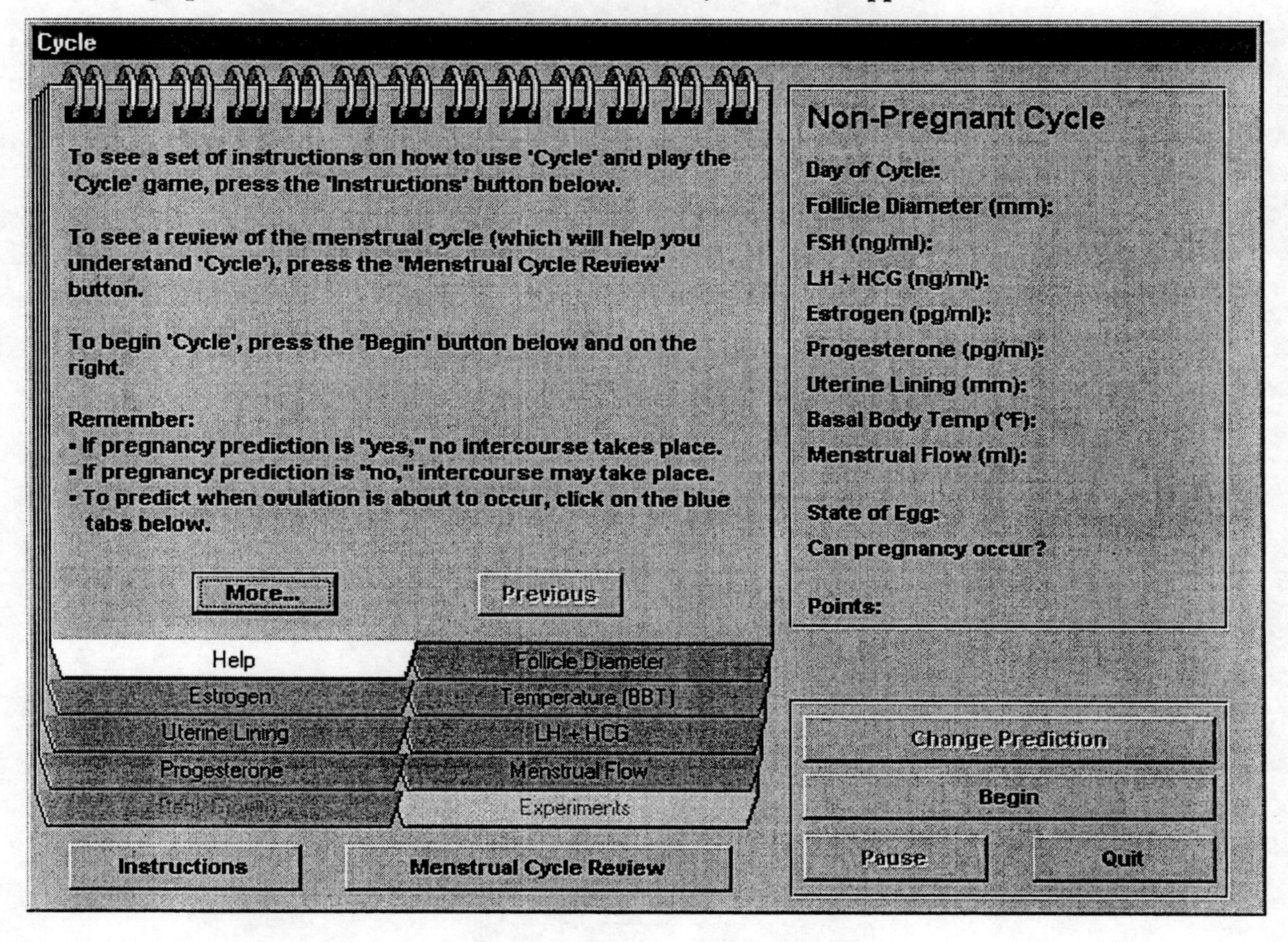

Figure 4 *The first* Cycle *screen.*

The left portion contains help information plus a series of tabs and buttons on the lower left. The right portion contains the digital display plus a series of control buttons on the lower right:

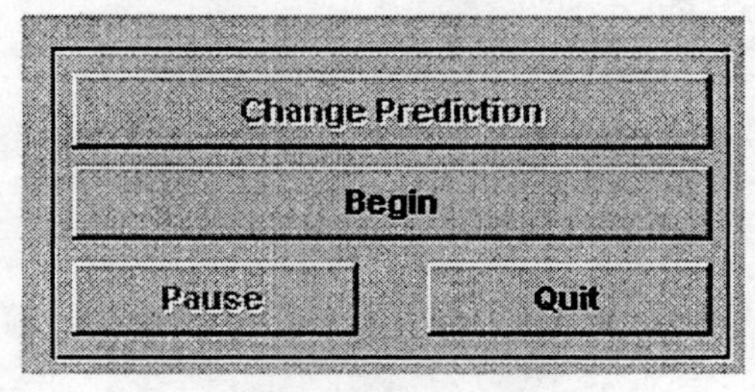

Figure 5 *Some important* Cycle *control buttons.*

To begin the simulation, the user clicks on the "Begin" button and then must choose the level of difficulty:

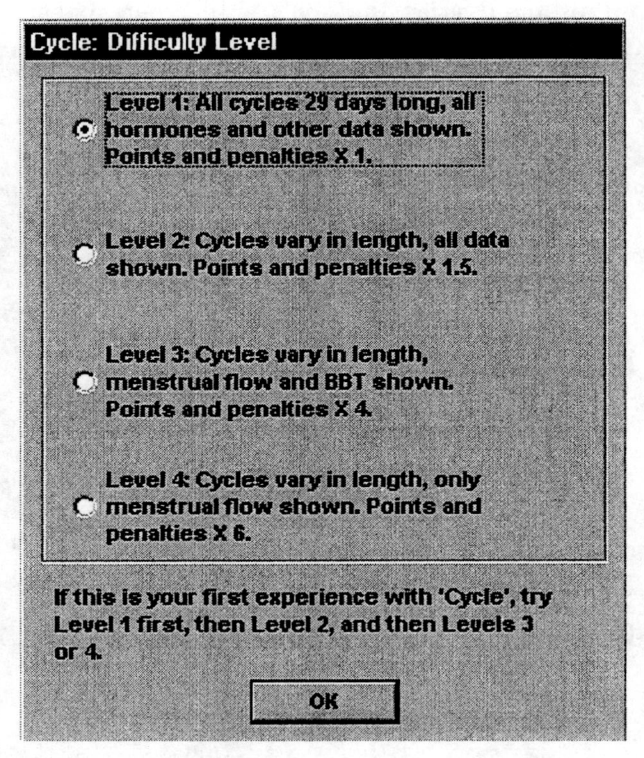

Figure 6 *Choosing the level of difficulty*

Students should begin on level 1. Once the level of difficulty is chosen, the simulation will begin. Now the user may view a display of all the *Cycle* variables. This display appears in the right portion of the screen in Figure 4:

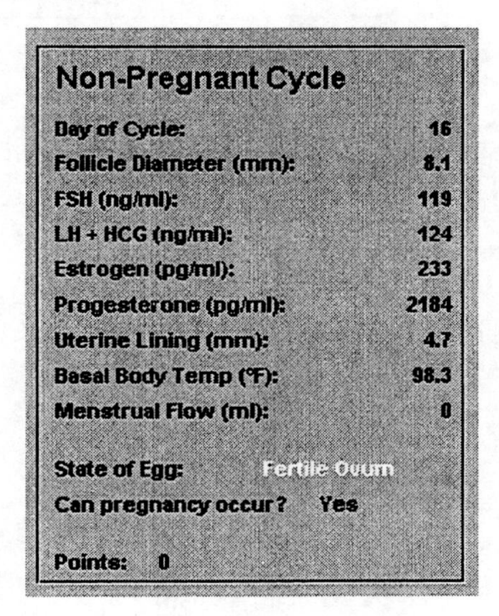

Figure 7 *The* Cycle *display.*

However, most users will prefer to access one of the graphic displays by pressing one of the tabs in the lower left portion of Figure 4.

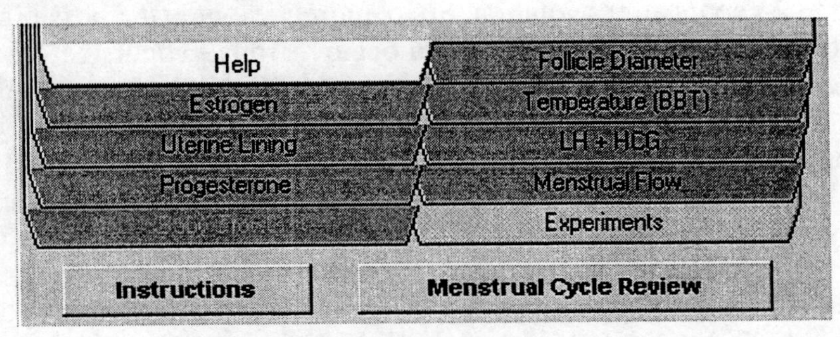

Figure 8 *The display control buttons on the* Cycle *screen. On difficulty level 4, the only buttons that would remain active would be "Help" and "Menstrual Flow."*

Clicking on one of the tabs will show a graph of the hormone or other quantity on the tab. As an example, clicking on the estrogen tab will show the following graph on the left portion of the screen.

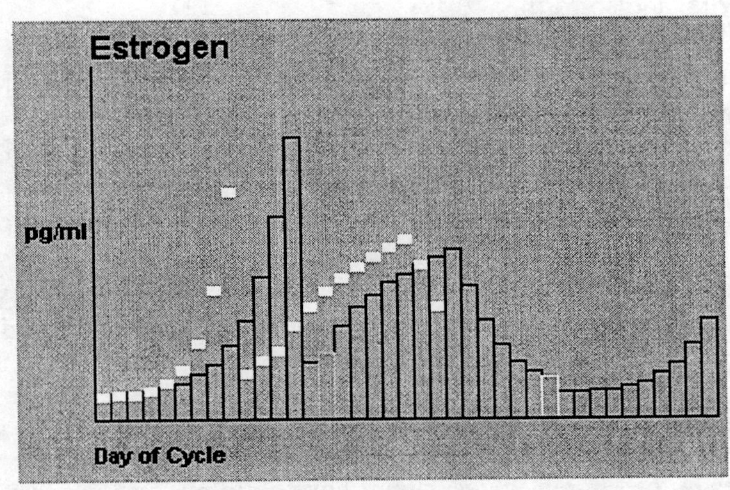

Figure 9 *A mid-cycle graph of estrogen in a cycle in which the follicle grew faster than normal and ovulated early. The "ideal" estrogen values for a 29-day cycle are shown as a histogram, but the values being observed in this cycle are shown as the bright blocks.*

The final choice that the user must make is to change the pregnancy prediction. In Figure 7, the pregnancy prediction is at the default value of **Yes**. This means that no intercourse is occurring and the user is receiving no points. However, by clicking on the "Change Prediction" button shown in Figure 5, students can toggle the prediction from **Yes** to **No** and **No** to **Yes**. If the student feels that pregnancy cannot occur, the choice should read **No**. However, if the student allows intercourse too close to ovulation, a pregnancy will start.

If the cycle ends with no pregnancy, the student may review all the graphs, even if the cycle was run on difficulty levels 3 or 4. Then "Begin" starts the next cycle.

Fertilization and Pregnancy

If the student gives a **No** pregnancy prediction during a time when fertilization can occur, the program moves to a simulation of fertilization and pregnancy. As explained earlier, sperm deposited in the female reproductive tract can remain viable for up to 96 hours. Also, eggs may be fertilized up to 24 hours after ovulation. So to entirely exclude the possibility of pregnancy, it is necessary to predict, "**Yes, pregnancy *can* occur**," from 96 hours before to 24 hours after it. When fertilization occurs, the message **FERTILIZATION!** flashes on the screen, and the student's point total immediately begins to be decremented by 4 points per day (in difficulty level 1). Five days later, the zygote makes contact with the uterine lining, and the message **IMPLANTATION!** appears on the screen. For the next eight days, the student's score is decremented by 8 points per day (in level 1). There is a 33% chance that the fetus will die during this period. If it lives to the eighth day after implantation, the student's score in level 1 is further decremented by 400 points and the "pregnancy screen" appears. Fetuses that survive eight days have a 17% change of dying before the end of the twelfth week, with miscarriages more frequent earlier in pregnancy. Thus only 50% of all fertilized eggs result in a birth.

This display is updated every week. The hormone levels change continuously, and the very high concentrations of estrogen and progesterone (many times those present in the nonpregnant cycle) are worth noting. The points do not change throughout pregnancy. **Fetal Development** is periodically updated with bulletins like, **Fingers and toes are forming, Mother can feel fetus moving in uterus**, and so forth.

At the conclusion of the pregnancy, the user is told the outcome **(Baby was full-term and healthy, Baby was undersized, but survived**, or **Baby was undersized and died)**.

At this point the user can review the course of hormone levels over the last cycle and the beginning of pregnancy. This is very useful for instructor demonstration of the difference between the normal end of a cycle and the changes that occur in pregnancy. It is especially illuminating to point out the high levels attained by estrogen and progesterone (due to the healthy corpus luteum); LH (really LH plus human chorionic gonadotropin) is far above normal also. "Early Pregnancy Test" kits sold in supermarkets and drug stores test for the presence of HCG with a monoclonal antibody preparation. If HCG is detected, then the test is positive.

Playing *Cycle* to Maximize the Chances of Pregnancy

About 10% of couples in the United States cannot have children, and another 10% have low fertility. The most common reason is low sperm counts. Part of the solution for the less severe cases is to time intercourse to match the times of high female fertility. To use *Cycle* in this way, play to maximize *negative* points. That is, students should strive to allow intercourse only at the moment of ovulation or a few hours before it.

Classroom Management of the *Cycle* Game

Students should spend most of their time playing the *Cycle* game on level 2 because this level will allow them to observe a diversity of cycles and to gather information that will allow them to play safely on levels 3 and 4. The more competitive students will want to move to levels 3 and 4 immediately, but emphasize the importance of observing the hormone patterns on level 2. Students find the *Cycle* game very engaging and there will be much hilarity when a previously successful team (or an instructor!) gets (virtually) "pregnant." If most students can play on levels 3 an 4, the game has achieved its objective and it is time to move on to Exercise B (the *Cycle* experiments) or to a wrap-up discussion.

The *Cycle* Experiments

Users conduct experiments with *Cycle* by changing the production rates of hormones. Clicking on the "Experiments" tab gives the user the following screen:

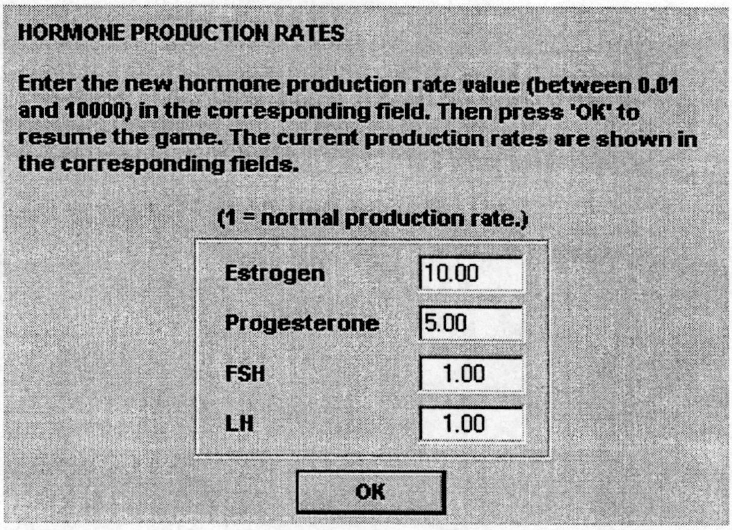

HORMONE PRODUCTION RATES

Enter the new hormone production rate value (between 0.01 and 10000) in the corresponding field. Then press 'OK' to resume the game. The current production rates are shown in the corresponding fields.

(1 = normal production rate.)

Estrogen	10.00
Progesterone	5.00
FSH	1.00
LH	1.00

OK

Figure 10 *The Experiment screen. The student has just changed the production of estrogen to 10 times the normal rate and progesterone to 5 times the normal rate. FSH and LH are still being secreted at the normal rate.*

The Experiment screen can be used to simulate several common types of birth control.

COMBINATION BIRTH CONTROL PILLS

A typical "combination" oral contraceptive package will contain 28 pills. Of these, 21 will contain active hormones, and 7 will be hormonally inert. A typical formulation for the active pills might be 0.5 mg of a synthetic progestin such as norgestrel or norethindrone, plus about a tenth as much synthetic estrogen (such as ethinyl estradiol). The woman takes the active pills for the first 21 days of her cycle, and takes the inactive pills for the remaining 7 days. The high concentrations of estrogen and progestin during the first 21 days have several effects. The most important of these are inhibiting the release of FSH and LH, thus preventing follicle development and ovulation, and changing the character of mucus in the cervix so that sperm cannot penetrate into the uterus. During the last 7 days, the withdrawal of these hormones causes the uterine lining to be shed, simulating a normal menstruation.

To simulate the effects of combination birth control pills, select the experiment menu on day 1 and increase the secretion rate of estrogen to 20 times the normal value and of progesterone to 50 times the normal value. At the beginning of the cycle, this will bring the plasma concentrations of estrogen and progesterone to approximately the levels that would be attained if an additional 0.5 mg of progesterone and 0.05 mg of estrogen were added to the bloodstream; the supplemented estrogen concentration would be 1,818 pg/ml. These concentrations are about 50x the initial concentration of naturally produced progesterone (300 pg/ml) and 20x the natural estrogen concentration (73 pg/ml). Set both the secretion rates back to 1 on day 22. The result of this hormonal manipulation is a slowly growing follicle that never ovulates and becomes atretic on day 24, followed by an abnormal menstrual flow.

NORPLANT

Norplant is a long-term contraceptive. Six straws the size of cardboard matches are implanted in the skin of the woman's upper arm. The straws slowly release a synthetic progesterone called levonorgestrel into the blood, and can prevent pregnancy for up to five years by preventing ovulation, thickening the cervical mucus, and preventing implantation of the zygote.

Cycle cannot exactly simulate the effects of Norplant, but if the Experiment screen is used to set progesterone secretion to four times the normal level, ovulation will be suppressed. If progesterone is set to just two times the normal level, ovulation will still occur but the zygote will be unable to implant properly, and will die shortly after implantation.

RU-86

The French "abortion pill," RU-486, consists of two components. Mifepristone blocks progesterone receptors and either prevents implantation or interferes with the zygote's nutrition if it has already implanted. Then, 36 to 48 hours later, a dose of prostaglandin called misoprostol stimulates uterine contractions and causes expulsion of the dead zygote or fetus. RU-486 can either be used as a post-coital contraceptive (before implantation) or as an abortion-inducing drug (from 5 to 9 weeks after conception).

To stimulate the post-coital action of RU-486, allow conception to occur and then, before the zygote implants, reduce progesterone secretion to 0.25 times the normal rate. The zygote will be unable to implant, and will be lost with premature menstrual bleeding.

Additional Experiments

Cycle can also be used as a data generator for the following student investigations. A user can explore the relationship among the various hormones and follicle maturation. For example, ovulation depends on follicle growth, controlled by FSH, and follicle maturation controlled by LH. If the user increases FSH to 10× normal, the follicle will grow rapidly, but a small LH surge will be induced too soon (about day 7), before the follicle is mature enough to ovulate. Thus the follicle will not ovulate and will become atretic. On the other hand, if both FSH and LH are increased to 10× the level of the beginning of the cycle, both growth and maturation of the follicle will proceed in tandem and ovulation can be induced as early as day 7. The user can experiment with different ratios of FSH and LH to see which one causes the earliest ovulation. The class can hold a competition for inducing the earliest ovulation.

Or to take another example, both LH and FSH are necessary for successful follicular development. But after ovulation, FSH can be set to zero without ill effect. On the other hand, if LH is set to zero after ovulation, the corpus luteum collapses and the cycle comes rapidly to an end. Reductions of LH after ovulation (to about 10% of normal) will not end the cycle immediately, but will cause a decline in the secretory functions of the corpus luteum, including the production of progesterone. This, in turn, will cause reduced development of the uterine lining and may prevent implantation of a zygote.

A more exploratory type of question relates to the role that each of the four hormones have on successful reproduction. This can be investigated by reducing each of the report. hormones, one at a time, to 20% of its normal rate at the beginning of the cycle.

- ❑ FSH produced at 20% of the normal rate retards follicle growth and estrogen production, prevents the LH surge, and leads to a long cycle ending in an atretic follicle.
- ❑ Cutting estrogen production to 20% of normal results in higher FSH and *faster* follicle development, but the estrogen concentrations are so low that the LH surge never occurs and the follicle goes atretic.
- ❑ Reduced LH production results in a rapidly growing but chronically immature follicle because of the failure of luteinization. It never ovulates and produces reduced amounts of progesterone.
- ❑ Finally, reduced progesterone allows ovulation and even implantation, but soon premature labor expels the fetus, and it dies.
- ❑ Finally, the effect of combination birth control pills has already been described. The user could experiment with lower concentrations than 20x estrogen and 50x progesterone to see how small a dose of hormones can successfully prevent ovulation. This is a practical issue because oral contraceptives have some deleterious effects, mainly on the cardiovascular system, so finding the smallest doses that will reliably inhibit ovulation is desirable.

Each of these investigations could result in a student report.

Notes:

Cycle: A Simulation of the Menstrual Cycle and Human Fertility

An Overview of *Cycle*

Cycle is a computer simulation that illustrates some of the basic features of the human menstrual cycle and female fertility.

Cycle can be used both as a competitive game and as a data generator on which experiments can be based. In the game, you use data on hormones, follicle size, and other indicators to advise a woman as to the time during her menstrual cycle when sexual intercourse can result in pregnancy. You receive points for giving correct information that will enable the woman to avoid becoming pregnant. If you indicate that pregnancy can occur, the woman avoids intercourse. If you predict that pregnancy cannot occur at a particular time, intercourse may take place. You receive points for correctly predicting when it is "safe" to have intercourse; you receive no points for indicating that intercourse should be avoided. Therefore, the less you warn that pregnancy is possible, the more points you will receive. However, if you predict that the woman can engage in intercourse at a time that is too close to ovulation, a pregnancy will occur and you will suffer a large point penalty. So the highest scores will be compiled by the user who can predict the time of ovulation as precisely as possible.

Knowing the time of ovulation is not simple. There are four levels of difficulty in *Cycle*.

- **Level 1** The cycles are always 29 days long and you will be given data on all hormones, basal body temperature, and the development of the follicle. Predicting ovulation on level is very easy. Points (and penalties) are awarded at the "base" rate.
- **Level 2** Provides the same data as level 1, but the cycles are variable in length and, in some, there is no release of an ovum. But by using hormonal patterns and the size of the follicle, the observant student can still operate nearly error-free. Points and penalties are awarded at one and a half times the base rate.
- **Level 3** More realistically, the cycles vary in length, and only data on basal body temperature and menstrual flow will be available to you. Points and penalties are awarded at four times the base rate.
- **Level 4** The cycles vary in length and you will have access only to data on menstrual flow. This is nearly useless as a predictor of ovulation, so disaster will befall the score of any user who is not extremely cautious. Points and penalties are given at six times the base rate.

In this laboratory exercise you will use *Cycle* as a game (Exercise A) and as a demonstration of the effect of birth control pills (Exercise B). However, before proceeding, you will find it helpful to review some of the features of the menstrual cycle by examining Figures C-1 to C-10 and by reading pages 2 through 7.

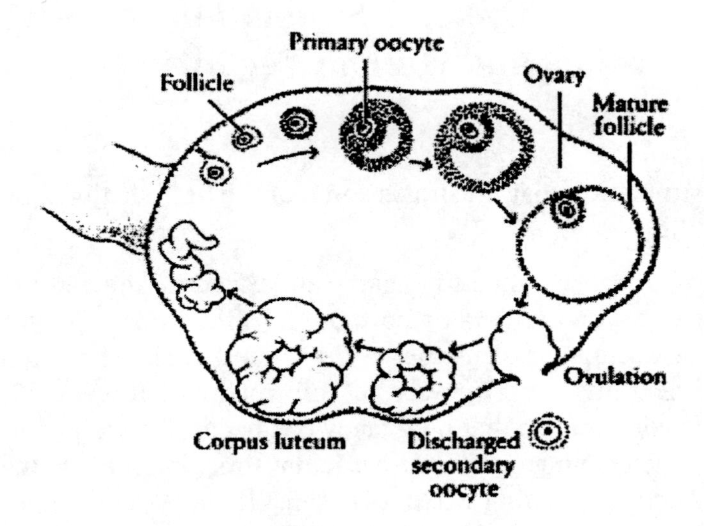

Figure C-1 *The development of follicle within an ovary. After a secondary oöcyte is discharged from a follicle at ovulation, the remaining cells of the follicle give rise to the corpus luteum, which secretes estrogens and progesterone.*

The Menstrual Cycle and Its Fertile Period

The menstrual cycle is the recurring hormonal oscillation that usually causes a woman to produce one ovum per month from her ovaries. The two main purposes of the cycle are to produce the hormonal environment that will allow the ovum to develop, and to prepare the lining of the uterus (the endometrium) to nurture the zygote if fertilization occurs. If fertilization does not occur, then the endometrium breaks down and approximately 40 ml of fluid are shed from the uterus over a five-day period. Although "menstrual flow" is the most obvious part of the menstrual cycle, it is only a "side effect" of the cycle's function: to prepare and release an egg, and to ready the uterus to receive a zygote.

A woman can become pregnant only during a brief period during her cycle. The egg, or ovum, has good fertility for only a few hours after ovulation (its release form the follicle), and by 24 hours after its release, it is incapable of being fertilized. In calculating the fertile period, however, we must also consider the fact that sperm deposited in the female reproductive tract before ovulation could fertilize the egg as it is released. Sperm can live for up to four days in the oviducts, although their fertility only remains high for about 24 hours. This means that intercourse from four days before ovulation to one day after it might cause a pregnancy.

However, restricting sexual abstinence only to this five-day period cannot guarantee that pregnancy will not occur because cycles vary in length. The period from the beginning of menstrual flow to ovulation (the release of an egg from a follicle of the ovary) is especially variable due to "fast" or "slow" development of follicles and varying levels of stress, which can change the secretion rates of follicle-stimulating hormone (FSH) and luteinizing hormone (LH). The part of the cycle between ovulation and the beginning of menstrual flow is controlled by the degeneration of the corpus luteum—a progesterone- and estrogen- secreting structure that develops from the ovarian follicle just after it has ovulated. The corpus luteum has a genetically programmed life-span of about 14 days. The "typical" menstrual cycle is 28 days long, but cycle lengths varying from less than 18 days to more than 45 days have been reported.

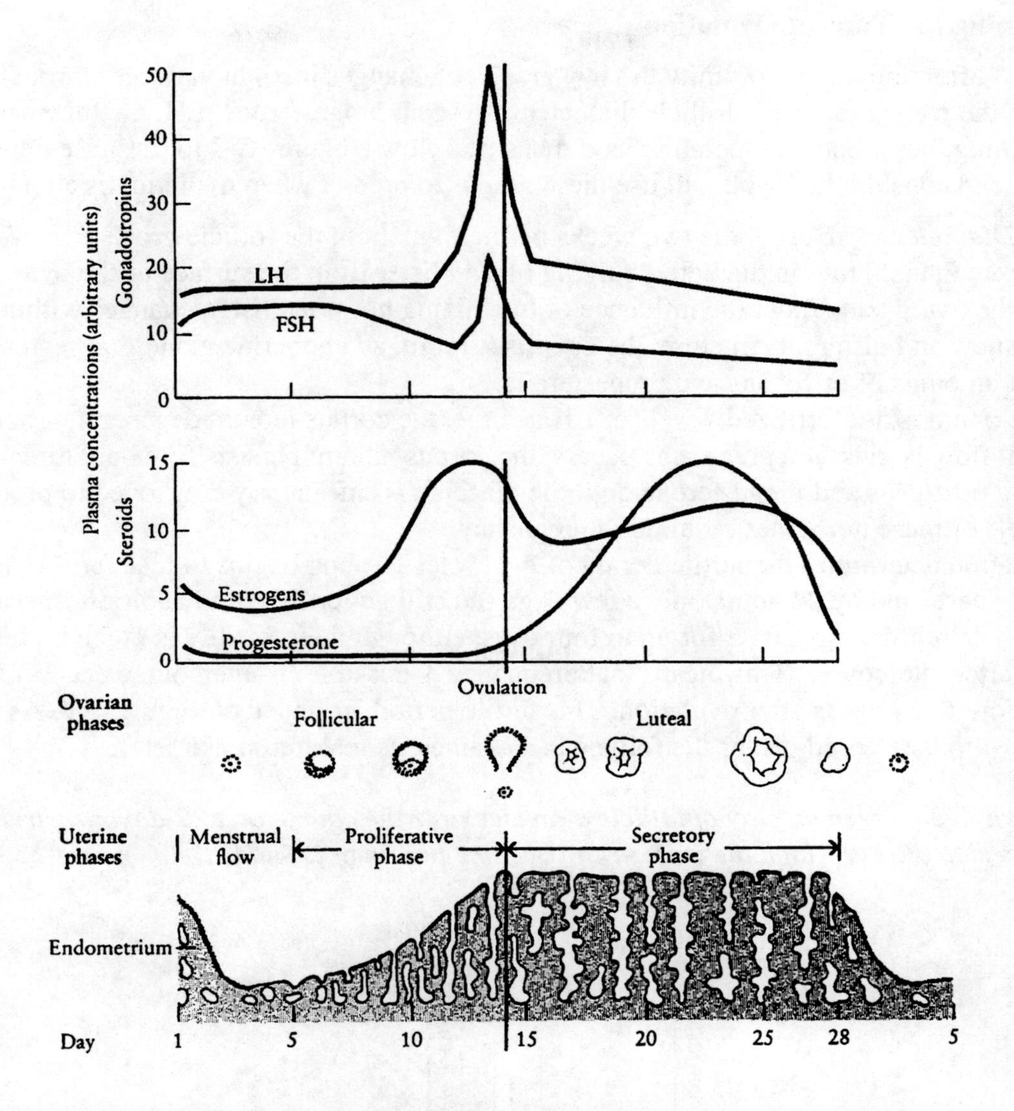

Figure C-2 *The menstrual cycle involves changes in hormone levels, in structures at the surface of the ovary, and in the uterine lining. The cycle begins with the first day of menstrual flow, the shedding of the endometrium that lines the uterine wall. The increase of FSH and LH at the beginning of the cycle promotes the growth of the ovarian follicle and its secretion of estrogens. Under the influence of estrogens, the endometrium regrows. The sudden rise in estrogens just before midcycle triggers a sharp increase in the release of LH from the pituitary, which stimulates the release of the oöcyte (ovulation). Following ovulation, LH and FSH levels drop. The follicle is converted to the corpus luteum, which secretes estrogens and progesterone. Progesterone further stimulates the endometrium, preparing it for implantation of the embryo. If pregnancy does not occur, the corpus luteum degenerates, the production of progesterone and estrogens falls, the endometrium begins to slough off, FSH and LH concentrations increase once more, and the cycle begins anew.*

Determining the Time of Ovulation

Cycle will offer you the opportunity to view graphs of changes in eight variables throughout the course of the menstrual cycle: follicle diameter, estrogen, progesterone, LH, the thickness of the uterine lining, basal body temperature, and menstrual flow (Figures C-3 to C-9). Review this material, and consider how you will use these graphs to predict when ovulation is about to occur.

Follicle Diameter During the two weeks before ovulation, the follicle grows from pinhead size to more than 20 mm in diameter, bulging like a blister from the surface of the ovary. Then, it releases the ovum and, under the influence of luteinizing hormone (LH), changes within a few days to a new and different structure, the **corpus luteum,** an endocrine gland that secretes abundant amounts of estrogens and progesterone.

If the ovum is not fertilized, levels of LH decline, the corpus luteum degenerates, and menstrual flow begins. If a pregnancy occurs, the corpus luteum persists for about three months, producing estrogens and progesterone until the placenta is sufficiently developed to produce the high levels of these hormones required in pregnancy.

Ovulation determines the fertile period of the cycle. The egg begins to lose fertility just hours after its release, and by 24 hours only a few eggs can still be fertilized. In addition, sperm can live in the female reproductive tract for up to four days, although their fertility is much higher in the first day after intercourse. This means that pregnancy is possible (if intercourse occurs) from 96 hours before to 24 hours after ovulation. This fertile period is shaded on Figure C-3. As you can see, the ovum is released just as the follicle has attained its maximum diameter.

Figure C-3 *Screen display of follicle diameter over the course of a 29-day menstrual cycle. The period during which the woman can become pregnant is shaded.*

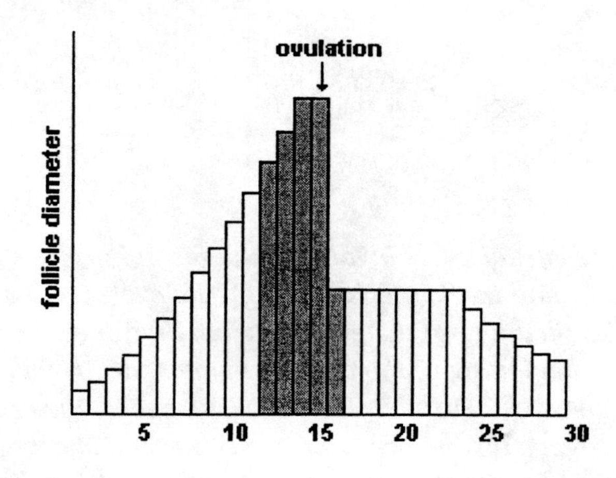

Estrogens Estrogen concentrations rise sharply through most of the follicular phase as the follicle grows, but then they drop sharply approximately two days before ovulation (Figure C-4). One explanation is that up until this point, several follicles may have been developing almost simultaneously, but now, the smaller follicles are suppressed by the one that will mature during this cycle.

Figure C-4 *Screen display of estrogen levels over the course of a 29-day menstrual cycle. The period during which the woman can become pregnant is shaded.*

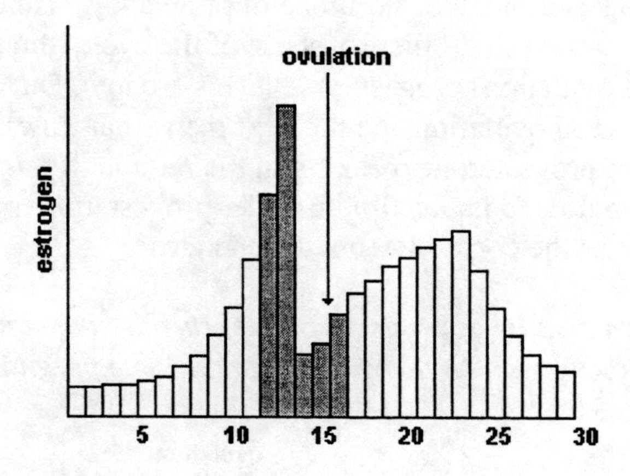

The estrogen level is very useful to you because it provides good advance warning of ovulation. After ovulation, estrogen concentrations rise as the corpus luteum develops and secretes this hormone. If fertilization of the ovum does not occur, the corpus luteum degenerates, and estrogen levels decline. Monitoring levels of estrogens can inform you of the timing of many of the important events of the menstrual cycle.

Figure C-5 *Screen display of luteinizing hormone levels over the course of a 29-day menstrual cycle. The period during which the woman can become pregnant is shaded.*

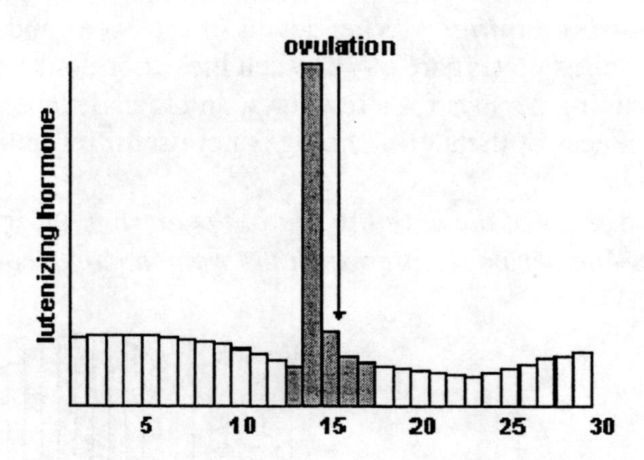

Luteinizing Hormone (LH) Luteinizing hormone, together with follicle-stimulating hormone, is involved in the maturation of the ovarian follicle. On its own, LH triggers ovulation and stimulates the production of estrogens and progesterone. Looking at the graph in Figure C-5, you may decide that the spectacular surge of luteinizing hormone preceding ovulation is one of the clearest indicators that ovulation is about to occur. This is true, but the LH surge occurs too close to ovulation (less than 24 hours before it) to be useful as an advance warning. If intercourse is avoided beginning at that point, residual sperm from intercourse that had occurred just previously would be likely to result in a pregnancy. LH remains nearly constant during the rest of the cycle, generally falling when estrogens and progesterone are high and rising again when they are low, illustrating the negative feedback relationship between LH and these hormones.

Progesterone Progesterone, the "hormone of pregnancy," is not very helpful for predicting the time of ovulation. During the follicular phase of the cycle (the interval between the beginning of menstrual flow and ovulation) progesterone levels are low. During the **luteal phase** of the cycle (the interval between ovulation and the next menstrual flow), the developing corpus luteum produces high levels of progesterone, peaking in the middle of this phase (Figure C-6). It is then that the uterine lining builds to its maximum under progesterone's influence. Progesterone concentrations decline as the corpus luteum degenerates.

Figure C-6 *Screen display of progesterone levels over the course of a 29-day menstrual cycle. The period during which the woman can become pregnant is shaded.*

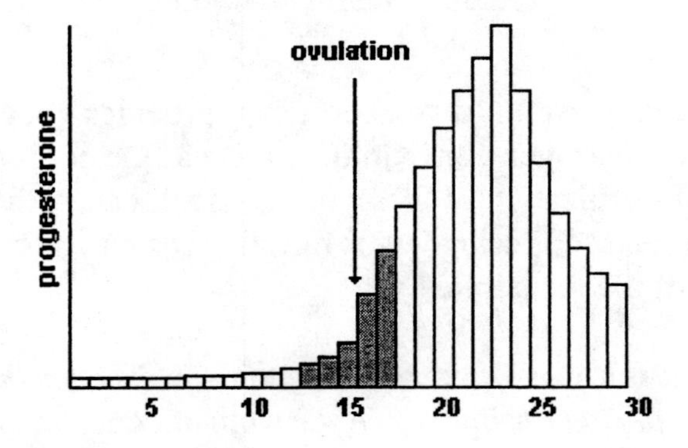

The Thickness of the Uterine Lining When levels of estrogens and progesterone are high, the uterine lining gradually builds up (Figure C-7). When these hormones decline as the corpus luteum deteriorates, the lining persists for a few days, and then degenerates and is lost as the menstrual flow. The thickness of the uterine lining is not useful for helping you play *Cycle*.

Figure C-7 *Screen display of the development of the uterine lining over the course of a 29-day menstrual cycle. The period during which the woman can become pregnant is shaded.*

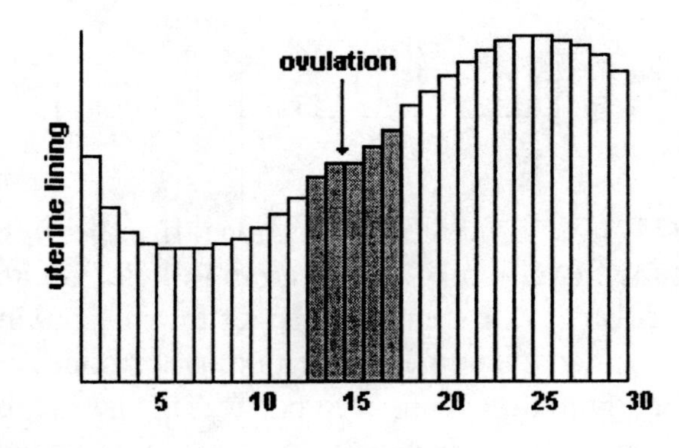

Menstrual Flow Menstrual flow, which occurs over about five days, gives no clue about the timing of ovulation (Figure C-8). Although the beginning of the flow signals the beginning of the hormonal and ovarian events of the new cycle, the flow itself is a product of the cycle that has just ended, not the one that is beginning. Because the corpus luteum has a lifespan of about 14 days, the time interval between ovulation and the next menstruation is much more constant in length than the time interval between menstruation and the next ovulation. Thus a woman can usually conclude that her last ovulation was 14 days before the onset of menstruation, but this gives no guidance about when she will ovulate next.

Figure C-8 *Screen display of the period of menstrual flow.*

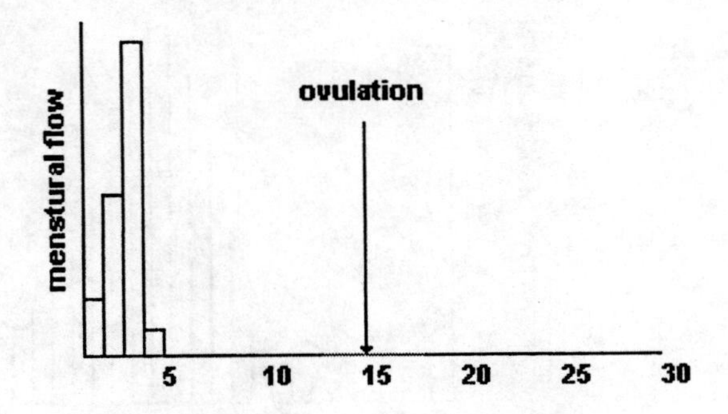

Basal Body Temperature The temperature of the body undergoes a daily cycle, with the lowest point usually occurring upon awakening in the morning. This "basal" body temperature is influenced by hormones and can provide a good indication of ovulation. Estrogens, which dominate the follicular phase, depress basal body temperature; progesterone, which increases during the luteal phase, raises basal body temperature. Thus basal body temperature undergoes a small (0.9°F, 0.5°C) rise within a few days following ovulation and remains elevated while progesterone is high (Figure C-9). Since this rise occurs mainly after ovulation, it can help to indicate when ovulation has occurred, but it gives little warning of when it will occur. Basal body temperature can be useful for playing *Cycle*, because once the rise has occurred, the user can be sure that ovulation is over and the egg is too old to be fertilized. However, the user is still "in the dark" before the temperature rise has occurred.

Figure C-9 *Screen display of variations in basal body temperature over the course of a 29-day menstrual cycle. The period during which the woman can become pregnant is shaded.*

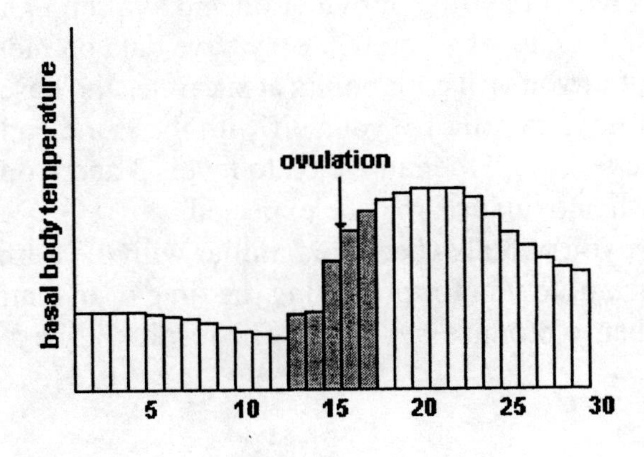

Cycle Variability While it may seem that all these data would make it easy to accurately predict ovulation, the problem is complicated by the fact that cycles vary in length. In 1968 a large study of cycle length collected data on 30,655 individual cycles. In Figure C-10, you can see that although 80% of the cycles fall between 23 and 32 days, cycles as short as 15 days and longer than 45 days were reported.

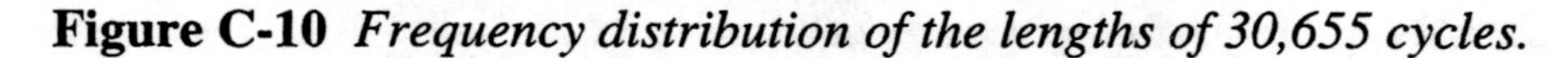

Figure C-10 *Frequency distribution of the lengths of 30,655 cycles.*

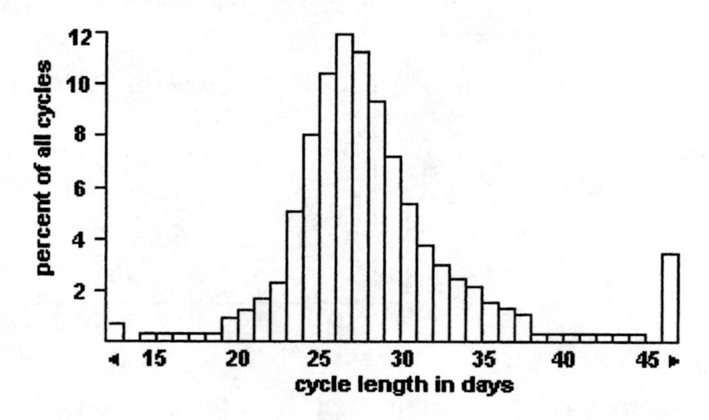

In *Cycle*, 50% of the ovulations occur on days 14, 15, and 16 after menstrual flow begins, but ovulation could occur at any time from 6 to 24 days after the start of the flow. In addition, 18% of all cycles never produce an egg because the follicle gets too old to ovulate, undergoes degeneration *(atresia),* and is said to be "atretic."

A high degree of cycle-length variability is usually not seen over short periods in individual women, but such a model would be realistic over several years in women with variable cycles. *Cycle* presents a sample of the range of variability.

Exercise A The *Cycle* Game

The challenge of *Cycle* is to balance uncertainty about the time of ovulation with the desire to accumulate as many points as possible. On level 1, cycles are always 29 days long, ovulation is always on day 15, you will have abundant useful data, and points are given at the "base" rate. Thus it is not hard to predict the time of ovulation and avoid pregnancy, but the rewards are not great. On the other hand, on level 4 you will only have data on menstrual flow (useless for predicting ovulation), but you will earn points at six times the base rate.

The key to success is to familiarize yourself with the game on level 1, gather data on cycle length variability on level 2, and then move on to levels 3 and 4 only when you have developed dependable rules on when ovulation may be expected.

As you play *Cycle*, you will also become familiar with the hormonal patterns of the menstrual cycle. These will be useful to you for predicting the time of ovulation, and you will learn how the hormones work together to produce ovulation. The result will be a better understanding of human reproduction.

Objectives

- Use hormonal and basal body temperature patterns to predict the time of ovulation.
- Derive a workable "rhythm method" that will allow successful *Cycle* play even if cycle length is varying and only data on menstrual flow and basal body temperature are available.
- Explain the significance of follicle diameter, estrogens, progesterone, LH, the thickness of the uterine lining, menstrual flow, and basal body temperature in determining the time of ovulation.

Procedure

1. Start the *Cycle* program (instructions for installing and running *Cycle* are given in the *Preparator's Guide* before Laboratory I (pages CD-ROM-1 to 4) and will be given to you by your instructor.
2. A brief summary of *Cycle's* instructions are found on the left side of the screen. If you need more extensive instructions on how the program operates, press the **Instructions** button, If you need a review of the menstrual cycle (most of which has been covered on pages 2-7 above, press the **Menstrual Cycle Review** button.
3. Press the **Begin** button. When the **Difficulty Level** dialog appears, choose level 1.
4. The most important information is on the right of the screen: the state of the egg and your current prediction about whether or not pregnancy can occur. As soon as the simulation starts, press the **Pause** button to pause the simulation.
5. Note that the pregnancy prediction when *Cycle* starts is **Yes** (in the program screen on the line, **Can pregnancy occur?**). Since it is very early in the cycle and menstruation is occurring, pregnancy cannot, in fact, occur and you are losing an opportunity to gain points. However, ignore this for now.
6. Press the **Resume** button to start the simulation again. Before the tenth day of the cycle, select the **Follicle Diameter** tab to examine the graph of follicle diameter and the **Estrogen** tab to examine the graph of estrogen levels. Note that the bright cursor is following the expected path for an "ideal" 29-day cycle. Using the follicle diameter display, trace the growth of the follicle toward the size at which ovulation will occur.
7. When ovulation occurs, "**State of the Egg**" will change from **Developing Follicle** to **Fertile Ovum**. This means that the egg could be fertilized if sperm are present. That egg's fertility will only last from 6-24 hours and then the "**State of the Egg**" will be reported as **Infertile Ovum**.
8. At the end of the cycle, a message will appear informing you that the cycle has ended. You may now review the cycle by selecting the **Follicle Diameter** tab to see the history of follicle diameter, the **Estrogen** tab to view estrogens, the **Progesterone** tab to view progesterone, and so on.
9. When you have finished reviewing, press **Begin** to bring back the level of difficulty screen. You are ready to begin the next cycle. For the next cycle, you will try to accumulate points without "getting pregnant." Select level 1 again, and click on the follicle diameter tab to view a graph of follicle development.
10. Early in this new cycle, pregnancy cannot occur. Press the **Change Prediction** button. This will change the pregnancy prediction to **No**, and points will begin to acculalate.

11. Watch the follicle diameter graph. Well before the follicle reaches its peak size, anc certainly before day 10, press the **Change Prediction** button. This will change the prediction back to **No**, and points will no longer be added to your score.
12. Ovulation will occur. Once "**State of the Egg**" changes to **Infertile Ovum**, press the **Change Prediction** button again. Points will be added to your score again.
13. Increase your proficiency by using level 2 for the next cycle. View either the graph of follicle diameter or estrogen, and change your pregnancy prediction from **No** to **Yes** when the follicle is approaching the size when ovulation will occur or when estrogen is just about to peak, which may occur earlier or later than on level 1.
14. As you work on level 2, collect data on the dates of ovulation. This will help you develop a safe "rhythm method" for later in the exercise.
15. When you can play successfully on level 2, move on to level 3. Here you will be able to view only menstrual flow (fairly useless) or basal body temperature. Remember that when basal body temperature rises markedly and stays high, ovulation is past.
16. Finally, when you feel that you have collected sufficient data on the lengths of the cycles, you may wish to try "flying blind" on level 4. Your instructor may hold a five-minute *Cycle* competition between lab teams, and playing (safely) on level 4 is the best way to win. However, be careful: a pregnancy on level 4 costs more than 2800 points!
17. If pregnancy occurs, your score will start decreasing immediately. There is a 33% chance that the zygote will die in the first 12 days after fertilization. If the zygote survives, you will be shown the developmental history of the fetus. After the pregnancy is over, review changes in the levels of estrogen, progesterone, and LH. You can compare differences between the hormonal events of the pregnant and the nonpregnant cycle.

    a. *Why is a woman fertile for only a brief period in the middle of her menstrual cycle?*

    ____

    ____

    ____

    b. *Which variable in* Cycle *(follicle diameter, progesterone, basal body temperature, and so on) is the most useful in helping you to predict the time of ovulation? Why?*

    ____

    ____

    c. *The egg remains fertile for about 24 hours after release; sperm can live in the female reproductive tract for four days. How does this affect the length of the fertile period?*

    ____

    ____

    d. *Why are observations of basal body temperature of limited use as a contraceptive aid?*

    ____

    e. *Couples who wish to start a pregnancy sometimes use basal body temperature to determine when to have intercourse. Why would this measure be useful as an aid to conception?*______________________________

    ____

    f. *Why is it much easier to estimate accurately when a woman's last ovulation occurred than to predict when the next ovulation will occur?*______________________________

    ____

Exercise B The Effect of Birth Control Pills

As was demonstrated in Exercise A, trying to avoid pregnancy by simply avoiding intercourse during a woman's fertile period is a method fraught with uncertainty, especially when the only information available is menstrual flow. Combination birth control pills, which contain both an estrogen and progesterone, are a more reliable method of preventing conception. These hormones prevent pregnancy by two mechanisms. First, their negative feedback relationship with follicle-stimulating hormone and luteinizing hormone suppresses secretion of those hormones, and thus prevents follicle development and ovulation. Second, the presence of high levels of these hormones changes the character of mucus in the cervix so that sperm cannot penetrate it. This also occurs naturally during nonfertile periods of the menstrual cycle and during pregnancy to help protect the fetus from infection.

The following exercise will simulate the action of the combination "Pill" by allowing you to change the rates of hormone secretion in *Cycle.*

Objectives

- Determine the effect of high concentrations of an estrogen and progesterone on follicle development, follicle-stimulating hormone, and LH, and explain the results.

Procedure

1. Press the **Begin** button to go to the Difficulty Level dialog.
2. Select level 1.
3. As soon as the cycle starts, select the **Experiments** tab. This will cause the display of the **Hormone Production Rates** page on the left side of the screen.
4. The display shows all hormone production rates as multiples of the normal rate. The screen in Figure CB-11 on the next page shows hormones being produced at the normal base rate.
5. Change the estrogen rate to 20 (20 times the normal secretion rate of estrogen). Then press the **Tab** key to move to the Progesterone field and change the secretion rate of progesterone to 50 times normal. Then press the **OK** button to return to the simulation.
6. Once back in simulation, check the graphs of estrogen and progesterone and verify that these hormones are at very high concentrations.
7. Next check the graphs of follicle diameter and LH concentration for differences from the normal, nonpregnant cycle.
8. Birth control pills are normally only taken for 21 days out of 28, so on day 21 select the **Experiments** tab again and change the secretion rates of estrogen and progesterone back to normal (rate = 1).
9. Note the changes in follicle diameter, and LH, estrogen, and progesterone concentrations once the pills are removed.
   a. *What is the effect of high estrogen and progesterone concentrations on follicle development?* ______________________________

Figure CB-11 *The experiment screen, showing how the user has changed estrogen to 20 times the normal secretion rate and progesterone to 50 times the normal secretion rate.*

HORMONE PRODUCTION RATES

Enter the new hormone production rate value (between 0.01 and 10000) in the corresponding field. Then press 'OK' to resume the game. The current production rates are shown in the corresponding fields.

(1 = normal production rate.)

Estrogen	20.00
Progesterone	50.00
FSH	1.00
LH	1.00

OK

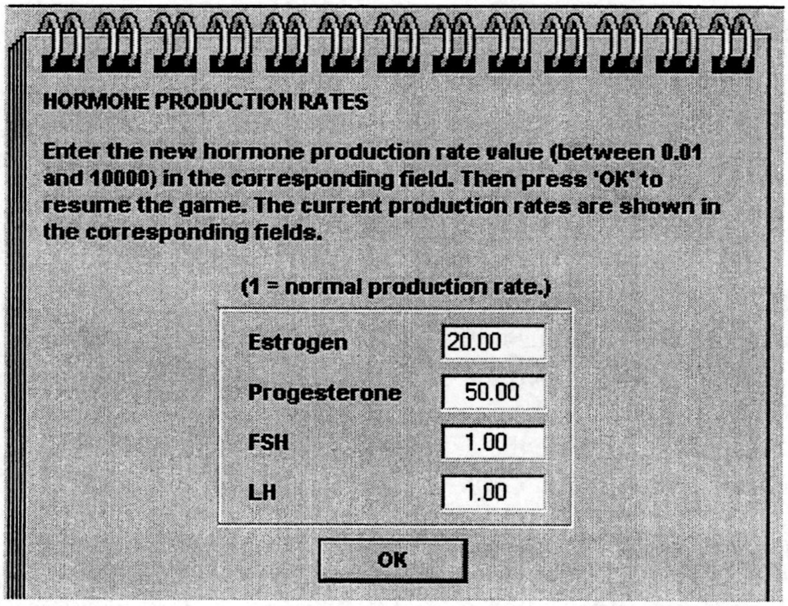

b. *Why does this effect occur?*__________________________________

c. *High concentrations of an estrogen and progesterone prevent pregnancy. Can low concentrations do the same? Repeat the steps above for estrogen and progesterone concentrations at any secretion rate less than 0.25 times normal. What happens? Why?*

Name________________________ **_Cycle_ Worksheet**

Section________________________

To answer the questions below, read pages 2-7 of the *Cycle* student exercises. Complete this worksheet and bring it to the class period in which the *Cycle* simulation will be used.

1. What are the two main functions of the menstrual cycle?

2. Why is a woman able to conceive for only a brief period during her menstrual cycle? In a 28-day cycle, when would this period occur?

3. What changes in estrogen levels occur during the course of the menstrual cycle?

4. Why is menstrual flow of limited use in predicting the fertile period, but is fairly accurate in determining when the *last* ovulation occurred?

5. Distinguish between the follicular phase and the luteal phase of the menstural cycle. Which is more variable in length, and why?

6. If you could predict the exact time of ovulation, for what time period before and after this would intercourse have to be avoided in order to prevent a pregnancy?

BioBytes *Dueling Alleles*

Instructional Objectives

Dueling Alleles illustrates the importance of population size as a factor governing the fluctuations of allelic frequencies. This simulation also explores how heterozygote advantage and recessiveness can maintain a deleterious allele in a population.

Dueling Alleles' Model of Changes in Allelic Frequency

The *Dueling Alleles* simulation uses several simplifying rules that may be unusual in real populations. For example, the frequency of both *B* and of *b* always start at 0.5. Furthermore, the population size is constant at the user-chosen number of mating pairs. In each generation, two parents breed and are replaced by two offspring (always one male and one female). The simulation randomly selects an individual in the population and randomly chooses a mate for it. Depending on the genotypes of these two parents (either *BB, Bb,* or *bb),* the program assigns genotypes to their two offspring, places these offspring in the next generation's mating pool, and removes the parents from the current generation's mating pool. Thus as computations for a generation proceed, the pool of unmated individuals shrinks until there are only two unmated organisms left. Once these two "find each other," the generation's results are displayed and selection of individuals from the new mating pool for the next generation begins.

If the student indicates that the death probabilities of all genotypes are zero, then every mating pair produces two offspring, thus replacing itself. But if the student assigns some probability of death for a genotype, then when an offspring of that genotype is generated, the program randomly determines whether it lives or dies. If it dies, random mating starts again until the dead offspring is replaced.

Thus, *Dueling Alleles* simulates an extremely orderly, constant-size, isolated population. If the student assigns no death probabilities to the different genotypes, every mating pair replaces itself and dies. The sex ratio is always 1:1. The environment is constant. Natural selection operates exclusively through differential death rates rather than through differences in both birth rates and death rates. The only chance an organism has to make a larger-than-average contribution to the next population is if an offspring of another mating pair dies and it is chosen by the random number lottery to be one parent of the replacement organism. Finally, there is no migration.

Although *Dueling Alleles* omits many of the important features of real populations, its simplifications allow the student to isolate the effects of the two processes on which the program concentrates: genetic drift and differential selection. When the student specifies no selection, then all gene frequency changes are the result of drift. Because the program does not include many possibly irrelevant complications, it is simple to operate. Finally, the program actually simulates individual organisms that are "searching for" and mating with other organisms. When the mating occurs, gametes and zygotes are randomly produced. Thus *Dueling Alleles* produces gene frequency changes in a manner very close to the way in which these changes occur in real populations.

Using *Dueling Alleles* in the Classroom

Before students use *Dueling Alleles,* have them read the introductory material to Exercise A, B, and C on pages 2-3, 4-5, and 8 in the "Manual_d" file on the CD-ROM (and also on the hard drive of the user's computer if installation was done correctly). Use the program after your usual lecture and discussion on population genetics and the Hardy-Weinberg principle.

If you have a sufficient number of computers available to allow one computer per group of 2-4 students, tell students to turn to page 3 of the "Manual_d" file on the CD-ROM and start following the directions for Exercise A. If only one or a few computers are available, an instructor demonstraton can be very effective. A large-screen monitor or TV set can help. Simply follow the directions in the exercises. Ask the students to predict the effects of changes in population size, changes in death probabilities for various alleles, and so on. Explaining the shapes of graphs and differences among them can result in good discussions.

Of course, it is possible to use one or two computers to provide enrichment material for small groups. Students may also work independently.

Using the Simulation

The program will ask the student:

1. How many mating pairs (1-500) the population should contain. When entering this, do not press the Enter key or the simulation will start immediately. Use the Tab key to move to the next field.
2. The probability that genotypes *BB, Bb,* and *bb* will die before reproducing.
3. If the student desires "continuous run." The other alternative is computing one generation at a time and then stopping until the student click a button to continue to the next generation. In most cases, "continuous run" is best.

A sample of the data to be entered is shown in Figure 1.

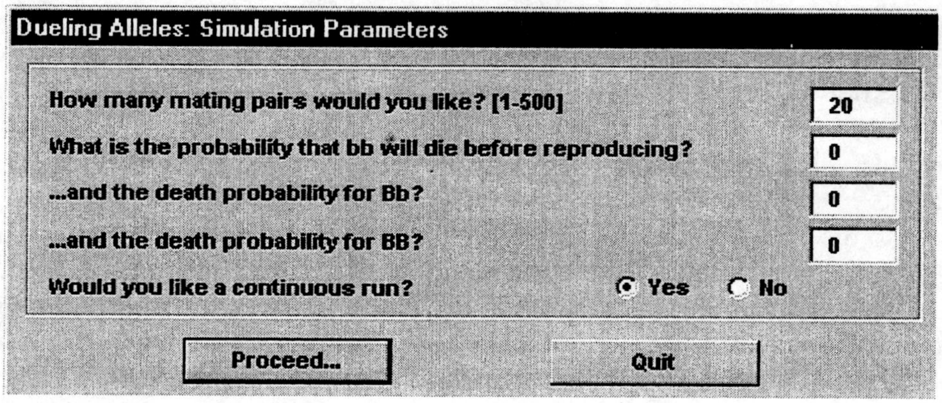

Dueling Alleles: Simulation Parameters

How many mating pairs would you like? [1-500] 20

What is the probability that bb will die before reproducing? 0

...and the death probability for Bb? 0

...and the death probability for BB? 0

Would you like a continuous run? Yes No

Proceed... Quit

Figure 1 *Data entry screen*

The frequency of the *b* allele will be graphed until the allele either reaches a frequency of 0.0 (goes extinct) or reaches a frequency of 1.0 (becomes fixed). The fluctuations in allelic frequency will be much more severe with smaller population sizes. Drift-drive fluctuations will be very noticeable with all population sized below 100 mating pairs. Because of random mating and random generation of gametes, each *Dueling Alleles* simulation is different, even from other simulations done on the same computer under identical conditions.

Typical Results

Figure 2 shows a portion of the screen as it would appear if the simulation was paused in the middle of a run with 20 mating pairs. Probability of death was 0 for all genotypes, so all fluctuations were the result of drift. The graph shows the frequency of the *b* allele, and the numbers below the graph show the frequency of both the *b* and the *B* allele (the 0.325 and 0.675 on the left), plus the numbers of each genotype (the 5, 16, and 19 on the right).

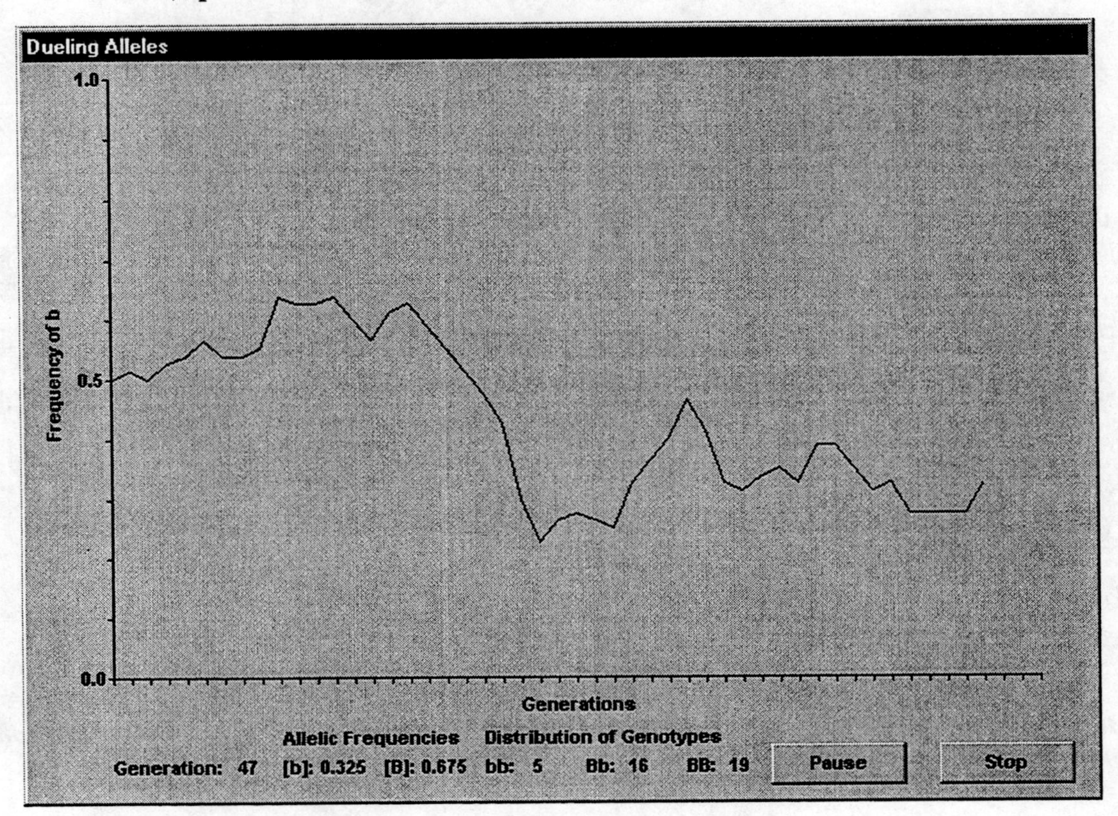

Figure 2 Duelling Alleles *after 47 generations.*

Figure 3 shows the results of a simulation in which there was selection pressure against the *bb* and *Bb* genotypes. The *b* allele went extinct in 28 generations, but note that the frequency occasionally *increased* due to drift. Fluctuations of *genotypic* frequencies are even more dramatic (and certainly more visible in nature) than fluctuations in allelic frequencies. The numbers of organisms of the *bb* and *Bb* genotypes for the simulation in Figure 3 are shown in Figure 4. Note that *bb* disappeared during several generations, but then reappeared because it was generated from *Bb* × *Bb* matings. However, once both *Bb* and *bb* had disappeared, the *b* allele was extinct.

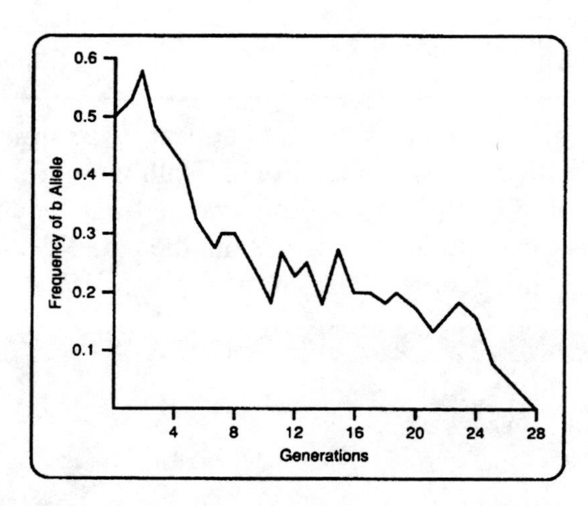

Figure 3 *The frequency of the b allele as it declines to extinction because of reduced viability of both the* bb *and* Bb *genotypes.*

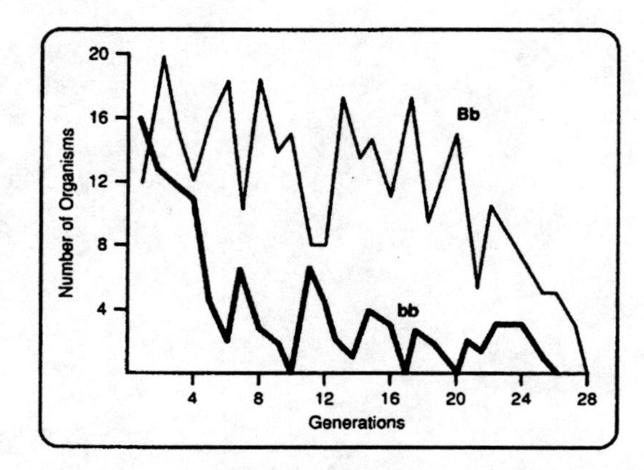

Figure 4 *The number of individuals with bb and Bb genotypes (out of a total population of 40) in the simulation shown in Figure 2.*

Instructor Discussion Guide for the Student Exercises

Student exercises include investigations on the effect of population size on the persistence of an allele, the effect of recessiveness on persistence of a deleterious allele, and a demonstration of the maintenance of deleterious alleles in a population by heterozygote advantage. The background material for each of these exercises is included in the introduction to each student exercise. The purpose of this discussion is to describe the important points of the results.

EXERCISE A Genetic Drift and the Extinction of Alleles

The smaller the population, the more violent the fluctuations in allelic frequencies, and the shorter the persistence of an allele. This exercise asks the students to test the time to extinction of either the *B* or the *b* allele with no selection and either 2, 4, 20, or 50 mating pairs in the population. Results, based on 25 simulations for each population size, are shown in Table 1.

Table 1 Mean, minimum, and maximum number of generations during which both the *B* and *b* alleles persisted

Mating Pairs	Mean	Minimum	Maximum
2	13.2	3	40
4	28.1	5	88
20	137.2	28	444
50	429.6	63	1084

Since there was no selection against either allele in these simulations, extinction of an allele was caused solely by drift

EXERCISE B Selection and the Effect of Recessiveness on Persistence of a Deleterious Allele

If a deleterious allele (*b*) is recessive, there is selection pressure against it only when it is homozygous (*bb*). It will decline in frequency, and as the fraction of *bb* individuals in the population decreases, selection pressure against the allele will almost disappear because *bb* individuals will become very rare. If having a *b* allele poses no disadvantage to the heterozygotes, the allele can be carried there indefinitely without any selection pressure against it. The rarer the recessive allele becomes, the more likely it is that heterozygotes will mate with homozygous dominant individuals, resulting in either homozygous dominant or heterozygous offspring. This phenomenon, the "recessive refuge" for deleterious alleles, allows a harmful but recessive allele to perpetuate itself in a population.

On the other hand, if an allele has selection pressure against it even in the heterozygous state (i.e., it is not entirely recessive), then it is no longer as protected by the recessive refuge and will disappear more rapidly.

The student exercise models this situation by assigning death probabilities of 1.0, 0.5, and 0.2 to the *bb* genotype. Then, where the allele is not entirely recessive, the *Bb* genotype has death rates of 0.5, 0.25, or 0.1 (half the *bb* death rate). Where the *b* allele is entirely recessive, the *Bb* genotype has a death probability of 0.0. A typical set of results is shown in Table 2.

Table 2 Persistence time in generations of the deleterious *b* allele. Simulations used 50 mating pairs. When *b* is completely recessive, *Bb* has a 0 death rate. When *b* is not completely recessive, *Bb* has a death rate half that of *bb*. *BB* always has a 0 death rate.

bb Death Rate	*b* Completely Recessive	*b* Not Completely Recessive
1.0	18.7	5.6
0.5	40.5	14.0
0.2	74.7	41.3

EXERCISE C Selection and Heterozygote Advantage

In this exercise, the student is asked to determine the persistence of both alleles using "balanced heterozygote advantage " (death rates of 0.5 for *BB*, 0.5 for *bb*, and 0.1 for *Bb*). Although there are many deaths occurring under this regime, *Bb* has the only selective advantage. This will maintain both the *b* and the *B* alleles in the population, and the only cause of the allele extinction will be drift. But when the selective advantage shifts to *BB* (death rates 0.5 for *bb*, 0.1 for *Bb*, and 0.0 for *BB*), the persistence of *b* is about 20 generations. Although this sounds like a relatively brief period, in the case of humans, 20 generations would be about 400 years.

The instructor should be aware that although the student exercise cites sickle cell anemia as an example of heterozygote advantage, the death rates used are not based on historical sickle cell anemia data. These death rates were chosen because they would maintain both alleles for long periods while balanced heterozygote advantage prevailed, and then would lead to rapid extinction of the *b* allele once the heterozygote no longer enjoyed an advantage.

Heterozygote advantage may also be unbalanced, giving the strongest selective advantage to the heterozygote but more of an advantage to one of the homozygotes than to the other. This case may be modeled in *Dueling Alleles* by assigning death probabilities of 0.5 for *bb*, 0.1 for *Bb*, and 0.2 for *BB*. This regime markedly prolongs the persistence of the b allele over the case where the death rates are 0.5 for *bb*, 0.2 for *Bb*, and 0.1 for *BB*. Once again, when the heterozygote has an advantage, both alleles are maintained in the population.

Dueling Alleles: A Simulation of Genetic Drift and Selection

Overview

We can track the fortunes of a population by noting how the number of individuals rises and falls. However, if we are more interested in the evolution of the population, it would be more productive to keep track of its genes, and how their numbers rise and fall. We track the abundance of an allele by a measure called the allelic frequency.

Say we are studying the *B/b* locus, and this locus has two alleles, *B* and *b*. The frequency of *B* is defined as the fraction of all the alleles at the *B/b* locus in a population that are *B*. For example, if the whole population has the *Bb* genotype, half the alleles at the *B/b* locus are *B*, and the frequency of *B* is 0.5. If 90% of the population is *bb* and 10% is *Bb*, then we compute the frequency of *B* by reasoning that none of the *bb* alleles is *B* and only half of the Bb alleles are *B*, so the frequency of *B*, written as [B], is given by

$$[B] = (0)(0.9) + (0.5)(0.1) = 0.05.$$

An allelic frequency can never get higher than 1.0 (at which point the allele is said to be "fixed") and cannot get lower than 0 (at which point the allele is said to be "extinct").

Evolution is a change in allelic frequencies. This change might be adaptive, as when natural selection gradually eliminates a harmful allele, or it might be neutral or even harmful, as when random fluctuations eliminate an allele, reducing the ability of the population to adapt to future changes in the environment.

Dueling Alleles will allow you to simulate both kinds of evolution. Exercise A will allow you to observe genetic drift, which often causes random loss of alleles. Exercise B allows you to experiment with the effect of selection on allelic frequencies.

The Role of Genetic Drift and Selection

The Hardy-Weinberg principle states that five conditions must be met for allelic frequencies to remain constant: (1) no mutation, (2) random mating, (3) large population, (4) no migration, (5) no selection. Under these conditions, evolution will not occur.

Why won't evolution occur under these conditions? ______________________________

____

These conditions are so restrictive that they are almost never met in nature, even for short periods. The consequence of this is that allelic frequencies are constantly changing. In this exercise you will use a microcomputer simulation called *Dueling Alleles* to explore other ways in which population size and selection cause changes in gene frequencies.

Exercise A GENETIC DRIFT

Genetic drift is the change in allelic frequencies that results from the random outcome of matings. An analogy would be a coin toss. If 100 people each tossed coins at the same time, we would be very surprised if exactly 50% of the coins turned up "heads" and 50% "tails" on toss after toss. The same chance deviations from predicted frequencies occur for combinations of alleles. For example, even if the frequency of both *B* and *b is* 50% in a population, and *BB, Bb,* and *bb* zygotes are all equally viable, in every "round" of mating, the percentage of zygotes with *BB, Bb,* and *bb* genotypes will undergo chance fluctuations from their calculated frequencies of 25%, 50%, and 25%.

Drift is exaggerated in small populations such as founder populations. Continuing the coin toss analogy, if we only have two people tossing coins, the frequency of heads will fluctuate from 0% to 50% to 100%. Likewise, small populations experience larger chance fluctuations in allelic frequencies.

One last coin toss analogy will confirm the importance of genetic drift. Say that if all coins in the toss turn up "heads," then all future coins will be made into "heads" on both sides, and "tails" will never be seen in the "population" again. If there are 100 coins in the toss, there is little chance that all 100 coins will land "heads" up at the same time. But if there are only two, then it is obvious that within a very few tosses both coins will come up "heads," and "tails" will vanish forever.

This is similar to the dynamics of genes in a population. If the population is small, there is a good chance that random fluctuations will, over many generations, result in the loss of an allele. We say that the allele has become extinct, and that the alternative allele has been "fixed" in the population. Unless migration or mutation brings the extinct allele back into the population, it has disappeared forever. Thus, genetic drift can cause drastic and permanent changes in allelic frequencies even if the population has no selection pressures, migration, or mutation.

In this exercise, you will examine the effect of population size on the fluctuation and persistence of an allele. You will study the frequency of the *b* allele at a locus where either the *b* or *B* allele could occur.

Objectives

- ❑ Define genetic drift and explain why it occurs.
- ❑ Define extinction and fixation of alleles in a population.
- ❑ Explain why drift produces more severe changes in allelic frequencies in small populations.
- ❑ Determine the relationship between population size and the persistence of an allele in the absence of selection.

Procedure

1. Start the *Dueling Alleles* program (instructions for installing and running *Dueling Alleles* are located just prior to Laboratory I).
2. Press the **Continue...** button to move past the *Dueling Alleles* title screen to the "Welcome" dialog. You might want to read the introductory text in this dialog to familiarize yourself with the operation of the program. Then press the **Proceed...** button to move to a dialog that will ask you a series of questions about the simulation you wish to perform. Answer these questions as follows:
    a. Use two mating pairs.
    b. Since this part of the exercise does not investigate selection, answer that the probability of dying before reproduction is 0 for *BB, Bb,* and *bb.*
    c. Indicate that you wish to use continuous run.
3. The simulation will begin. You can pause the run with the **Pause** button, and stop it entirely with the **Stop** button. The frequency of the *b* allele will always start at 0.5. Also, *Dueling Alleles* always maintains a constant population with the number of mating pairs you requested. Each mating pair replaces itself with its offspring and then dies. In this

experiment, the *b* allele will either become extinct or its frequency will become "fixed" at 1.0. Observe how many generations it takes for either event to occur, then record the number on the appropriate line in Table DA-1. Also, make notes about the fluctuation in frequency of the *b* allele.

4. Press the **Yes** button to request another simulation.
5. Indicate that you wish to use four mating pairs, and that all genotypes will have a 0 probability of dying before reproduction, and that you wish to use continuous run.
6. Record the results for four mating pairs in Table DA-1.
7. Repeat steps 4 through 6 for 20 mating pairs.
8. Repeat steps 4 through 6 for 50 mating pairs.
9. For each number of mating pairs, your instructor will average the class results for the number of generations before one of the alleles becomes extinct. Record these results in the last row of Table DA-1.

Table DA-1. Number of Generations to Extinction or Fixation of an Allele (No Selection)

	2 mating pairs	4 mating pairs	20 mating pairs	50 mating pairs
b extinct				
b fixed				
Mean time to extinction				

b. Did both b *and* B *alleles persist longer in the population with a small number of mating pairs or the one with a larger number of mating pairs?* ______________________
*Why?*____
____
____

c. Was the fluctuation of *allele* frequency greater with a small number of mating pairs or a larger number?__________________Why?______________________
____
____

<u>Exercise B</u> SELECTION

Evolution is a change in gene frequencies. This change can be brought about by mutation, immigration, emigration, drift (as illustrated in the previous exercise), or selection. Selection against a genotype occurs when its fitness (success at survival and reproduction) is not as great as the fitness of another genotype. Fitness is determined by the whole genotype, not a particular allele. Thus, if an African antelope has superior genes for every characteristic except watchfulness for lions, its genotype will have a reduced fitness if lions are very active in its habitat. All its "superior" genes will not be passed on to the next generation because they are bound up in one "package" with the nearly lethal "predator watchfulness" characteristic.

Genotypes that have selection against them will slowly decrease in frequency until the deleterious alleles responsible disappear from the gene pool of the population. This might raise the question why any deleterious alleles are left after millions of years of natural selection. Several factors tend to slow or stop the allele extinction process:

1. Elimination of deleterious alleles is slow, perhaps requiring hundreds or even thousands of generations, *except* when the selective disadvantage of the genotype is severe.
2. Mutation or immigration might introduce new copies of the deleterious alleles as fast as they are eliminated by selection.
3. The "disadvantaged" genotype may be selected against only when it is abundant, and may even enjoy a selective advantage when it is rare.
4. Recessive deleterious alleles may not experience any selective disadvantage when they are masked by the dominant allele.
5. Heterozygotes may experience a selective advantage over either homozygote, and thus both alleles will persist.

In this exercise you will use *Dueling Alleles* to explore recessiveness and heterozygote advantage as factors that tend to maintain deleterious alleles in a population.

The "Recessive Refuge" for Deleterious Alleles

As pointed out above, selection operates against the whole genotype. Assume that a locus can carry either a *b* allele or a *B* allele, and that *bb is* lethal. It might be expected that the *b* allele would rapidly be eliminated from the population. But if the *b* allele is recessive to the *B* allele, then the *b* allele may persist for long periods in the population by "hiding out" in the *Bb* heterozygote, which suffers no selective disadvantage. This heterozygote "refuge" will also become more and more effective as the *b* allele becomes rare. This occurs because the *b* allele will experience selective disadvantage only when a *bb* offspring occurs, and this will only happen when two heterozygotes mate. If the heterozygotes become rare, then almost all matings of heterozygotes will be with *BB* organisms.

The exercise below will illustrate this principle. First you will determine the persistence time of the deleterious *b* allele when the *B* allele is only incompletely dominant over it *(bb* and *Bb* experience selective disadvantage). Then you will do the same series of simulations with the *B* allele completely dominant (only *bb is* at a disadvantage).

Objectives

- ❑ Explain why recessive deleterious alleles tend to persist longer in populations than dominant deleterious alleles.
- ❑ Use the *Dueling Alleles* simulation to determine the degree to which recessiveness contributes to an allele's persistence in a population.

Procedure

1. If you have already used the *Dueling Alleles* program for the genetic drift simulation, indicate that you wish to perform another simulation. If you have not used the program before, start the *Dueling Alleles* program (instructions for installing and running *Dueling Alleles* are found elsewhere). Press the **Continue...** button to move past the *Dueling Alleles* title screen to the "Welcome" dialog. You might want to read the introductory text in this dialog to familiarize

yourself with the operation of the program. Then press the **Proceed...** button to move to a dialog that will ask you a series of questions about the simulation you wish to perform.

2. Answer the questions as follows:
   a. Use 50 mating pairs. This population is large enough so that drift will not be an overwhelming factor.
   b. Indicate that you wish the probability of death of *bb* to be 1.0, of *Bb* to be 0.5, and of *BB* to be 0. Thus both *bb* and *Bb* suffer from the effects of the *b* allele *(B* cannot completely mask *b).*
   c. Indicate that you wish to use continuous run.
3. The simulation will begin. The frequency of the *b* allele will be shown. You can pause the run with the **Pause** button, and stop it entirely with the **Stop** button.
4. The frequency of the *b* allele will always start at 0.5. Note the course of the frequency of the *b* allele and decide whether it is declining most rapidly when it is still abundant or when it is rare. Then record the number of generations to extinction of the *b* allele in the half of Table DB-2 headed "*B* does not mask *b*," under the column headed "Your results." Your instructor will give you the class averages when the exercise has been completed. Next, indicate that you wish to perform another simulation.

Table DB-2 The Effect of Recessiveness on the Persistence (in Generations) of a Deleterious Allele

	Generations to extinction of *b* allele			
	B* does not mask *b		***B* masks *b***	
Death Prob. for bb	**Your Results**	**Class Average**	**Your Results**	**Class Average**
1.0				
0.5				
0.2				

5. Repeat steps 2 through 4 with the following additional death probabilities: 0.50 for *bb,* 0.25 for *Bb,* and 0 for *BB*; 0.20 for *bb*, 0.10 for *Bb*, and 0 for *BB*. Write the times to extinction of the *b* allele in the left half of the table, under "Your results."
6. Repeat steps 2 through 4, but this time with a death probability of 0 for both *BB* and *Bb*, and the following death probabilities for *bb*: 1.0, 0.5, and 0.2. Write the times to extinction of the *b* allele in the half of the table headed "*B* masks *b*," under "Your results."
7. Your instructor will compute the class averages for the times to extinction of the *b* allele. Write these times in the columns headed "Class average." Answer the following questions:
   a. *Why do recessive deleterious alleles persist longer in a population than alleles that cannot be masked by other alleles?*____

   ____

   ____

   ____

   b. Did you see this effect in *your* data *(b* persisting longer when it could be masked by *B) ?*

   ____

c. *What is the effect on both recessive and nonrecessive deleterious alleles when the selection pressure against them decreases?* ______________________________

____

____

____

d. Does the frequency of the *b allele* seem to decline most rapidly when the *b* allele is rare or when it is abundant? ______________________________Why does this occur?______

____

____

____

e. *If recessiveness will help a deleterious allele persist longer in a population, will recessiveness speed or hinder the spread of a beneficial allele?* ______________

f. *Why?*______________________________

____

____

Heterozygote Advantage

Another way in which deleterious alleles are preserved is if heterozygotes have a selective advantage over either of the homozygotes (assuming that only two alleles are possible at a locus). In the extreme case in which both homozygotes (say *bb* and *BB)* die before reproducing and only the heterozygote survives, both the *B* and *b* alleles will persist in the heterozygotes and both *bb* and *BB* individuals will continue to be produced from *Bb* × *Bb* matings.

A classic example of heterozygote advantage is the persistence of sickle-cell anemia in equatorial Africa. Sickle-cell anemia is a severe blood disease present in *ss* individuals. *Ss* individuals have some mild sickle-cell symptoms, but they are also resistant to malaria. Finally, normal *(SS)* individuals are not anemic at all but may suffer disability or death if they become infected with malaria. In areas of Africa where malaria is prevalent, heterozygotes seem to have a selective advantage of 26% over normal individuals, which seems to preserve the sickle-cell allele in the gene pool. However, in areas where malaria is not common, *Ss* individuals have no advantage and the sickle-cell allele is eliminated. It is estimated that 300 to 350 years ago, 22% of the slaves in the American South carried the sickle-cell allele. But by the early 1950s, the frequency of the allele in the black American population had fallen to 9%.

In this exercise, you will demonstrate the preservation of a deleterious allele by heterozygote advantage. You will also observe the disappearance of a deleterious allele when the heterozygous condition is no longer advantageous.

Objectives

- ❑ Explain heterozygote advantage and how it may maintain alleles in a population.
- ❑ Use the *Dueling Alleles* simulation to determine the degree to which heterozygote advantage contributes to the persistence of an allele in a population.

Procedure

1. Indicate that you wish to perform another simulation.
2. Answer the questions as follows:
   a. Use 50 mating pairs.

b. Indicate that you wish the probability of death of *bb* to be 0.5, of *Bb* to be 0.1, and of *BB* to be 0.5. Thus both *bb* and *BB* suffer equal selective disadvantage and *Bb* has an advantage.

c. Indicate that you wish to use continuous run.

3. The frequency of the *b* allele will always start at 0.5. Note the course of the frequency of the *b* allele for 100 generations (over 2,500 years for humans). Record the results here as extinction or fixation of the *b* allele (at how many generations?), or as a declining or increasing trend, or as mere fluctuation without a long-term trend.______________________

____

4. Press the **Stop** button to exit from the simulation. Now set up the simulation for the conditions in which heterozygote advantage is removed. Indicate that you wish to use 50 mating pairs, and set the death probabilities at 0.5 for *bb,* 0.1 for *Bb,* and 0.0 for *BB* (in the case of sickle-cell anemia, the *BB* individuals have normal blood, so this change simulates moving them to a malaria-free environment).

5. Indicate that you wish to use continuous run, and observe simulation. Record the results here as extinction or fixation of the *b* allele (at how many generations?), or as a declining or increasing trend, or as mere fluctuation without a long-term trend.______________________

____

a. *What is heterozygote advantage, and how can it maintain two or more alleles in a population?*____

____

____

b. *Why would you expect that a deleterious allele that persists because of heterozygote advantage would disappear once the heterozygote advantage is removed?*_____________

____

____

c. *In this simulation, you used balanced heterozygote advantage (both* BB *and* bb *were at the same selective disadvantage relative to* Bb). *Do you think that unbalanced heterozygote advantage (say, with the death probabilities for* BB, Bb, *and* bb *set at 0.2, 0.1, and 0.5, respectively) could also maintain both the* B *and* b *alleles? Write your prediction here:*____

____

Use the program to test your prediction.

Notes:

Name___________________________ ***Dueling Alleles* Worksheet**

Section___________________________

To answer the questions below, read the introduction, page 1, and the introductions to Exercises A and B (pages 1, 3, 4, and 6). Complete this worksheet and bring it to the class period in which the *Dueling Alleles* simulation will be used.

1. What is an allelic frequency?

2. What is genetic drift?

3. Why is genetic drift more pronounced in small populations?

4. Why can genetic drift have a severe impact on the genetic diversity of a population?

5. Why would a recessive deleterious allele persist longer in a population than would be possible for a dominant deleterious allele?

6. Explain how heterozygote advantage can maintain a deleterious allele in a population.

BioBytes *Seedling*

Using the BioBytes Simulations to Extend the Laboratory Experience

The BioBytes simulations are designed to carry the laboratory experience further than would be possible under normal laboratory conditions. Complex or impractical procedures can be performed by pressing a key. Data that could take hours or even months to collect can be obtained in seconds. In a study of plant growth, for example, there is no practical way for students to measure a plant's photosynthetic output, nor to make radical changes in a plant's environment and then to adjust the plant's growth strategies to cope with these changes.

All the BioBytes simulations have been designed to provide many opportunities for decision-making and data interpretation. To assure that students have enough knowledge to use the simulation productively, you may want to assign parts of the written exercises as background reading. The introductions to the four *Seedling* exercises are found on pages 3-4, 9, 12, and 15 in the file "Manual_s" on the CD-ROM (and also on the computer's hard drive following installation). In addition, you may want to assign the *Seedling* worksheet on page 2 to be completed before the lab in which *Seedling* will be used.

In addition, because *Seedling* will frequently challenge students to extend the reach of their understanding, results are best when the instructor takes an active role—asking questions and providing further explanation—as students use the programs. This guide provides background material and pedagogical hints for enriching classroom discussions.

A Note About Simulations

The purpose of *Seedling* is to demonstrate the relationship between several of the more important variables of plant growth. The program leaves out many details. Despite the fact the simulations are not exact copies of reality, the skills gained in observing, organizing, and interpreting the data provided by the program can be directly transferred to hands-on laboratory investigation.

An Overview of *Seedling*

Seedling is a simulation of plant competition and plant physiology. The program models the growth of a group of seedlings, and can be used both as an experimental data generator (Exercise A, B, and C) and as a game (Exercise D).

The *Seedling* Experiments

Exercise A. Students grow plants in a growth chamber to determine the interplant distance that maximizes productivity per unit area *(not* productivity per plant). These experiments may be conducted in either a moist (high humidity and soil water) or a dry environment, and give different results depending on plant size and water supply.

Exercise B. Determination of the effects of light intensity and temperature on net photosynthesis. Students determine the combinations of light and temperature to use. Light intensity can vary from 0% to 200% of full sunlight, and temperature can vary from 1° C to 45° C.

Exercise C. Similar to Exercise B, but students determine the effects of temperature and relative humidity on transpiration.

The *Seedling* Game

Exercise D motivates students to learn which factors are most critical to plant success by engaging them in a simulated life-and-death struggle with their classmates. The game puts each team in control of a seedling that is growing in an area crowded with other seedlings controlled by other teams. The players may elect to grow in a growth chamber, or take their chances in a variable outdoor environment in either arid western Texas, coastal South Carolina, or tropical Puerto Rico. Players determine the success of their seedling by allocating each day's photosynthate either to growth of leaves, growth of stem height, growth of stem diameter, or growth of roots. Good allocations can produce a plant that can shade its competitors, steal their water, and withstand drought and cropping by hungry cows. Bad allocations produce frail plants that starve, die of dehydration, and blow over.

Some students show little interest in plants, and may not even regard them as truly living things. After the *Seedling* game puts them in control of a struggling plant and allows them to feel its stresses, their view of the natural world will be broadened. As one student told the author after her seedling died a lingering death, "From now on I'll look at trees with new respect."

Instructional Objectives

The purpose of the *Seedling* exercises is not only to teach plant physiology, but also to teach the principles of problem-solving and orderly, efficient, effective experimentation. These exercises provide step-by-step instructions, but require students to choose the values for the variables they wish to test. For example, in Exercise B students must decide which combinations of light and temperature conditions will yield the most useful information about how changes in these variables affect net photosynthesis.

In the experimental mode, the program can be used to generate results with a realistic amount of variability between replicate simulations run under the same conditions. Thus, students will gain experience in graphing and discussing variable data and will also learn about the value of factorial experimental design.

Seedling's Model of Plant Growth

Seedling simplifies, and, in some cases, distorts, botanical reality. For example, *Seedling* plants are assemblages of regular geometric shapes. Stems are uniform cylinders. Roots are cylinders 0.2 mm in diameter and fill the soil at a density of 100 mm/cm^3 of soil. Mature leaves are discs 0.3 mm thick and 28 mm in diameter and always grow in pairs. New leaves are started at the top of the plant, but they cannot begin until the newest leaf below the tip is mature. Thus, the moment a *Seedling* leaf reaches its full size, a new leaf is automatically started at the top of the plant. In the *Seedling* game, the student may drop leaves, even immature ones. If an immature leaf pair is dropped, a new leaf pair begins immediately at the tip of the plant.

The only nutrient in *Seedling* is water. Carbon dioxide and all soil nutrients are assumed to be present in excess.

Also, real temperate zone plants have a yearly cycle of activities that begins with vegetative growth in the spring. But at certain times in the growing season the plant "shifts gears" and devotes a larger and larger share of its photosynthate to flower and fruit production. When this occurs, even if conditions for growth remain favorable, vegetative growth dwindles. *Seedling* plants have no flower or fruit production and no life cycle, so they grow at the maximum rate until they die from lack of water' frost, or other causes. But since the program is modeling small, young plants, this is not a grave defect.

Although *Seedling* reports results on a per-plant basis, the program models a whole population of plants that are shading one another, protecting each other from wind, etc. In the experiments (Exercises A, B, and C), there is only one population, but in the game (Exercise D), there may be up to four different populations depending on the number of players.

Photosynthesis

Seedling models a temperate zone, sun-adapted, C_3, herbaceous crop plant such as a soybean. "C_3" here means that when the plant fixes CO_2, the first stable molecule into which CO_2 is incorporated has three carbons. C_4 plants fix CO_2 first into a four-carbon molecule. The C_3 - C_4 distinction has far-reaching implications because C_4 plants can reduce or entirely avoid a process called photorespiration, which wastes from 30% - 50% of a C_3 plant's photosynthate when temperatures are high and CO_2 concentrations are low. Thus, C_4 plants have a higher rate of photosynthesis in warmer regions of the globe. In contrast, the photosynthetic optimum of the C_3 *Seedling* plants occurs at 25°C.

The photosynthetic rate of a *Seedling* plant is determined by leaf area, leaf temperature, the amount of light falling on the leaves, the degree of water stress, and leaf age.

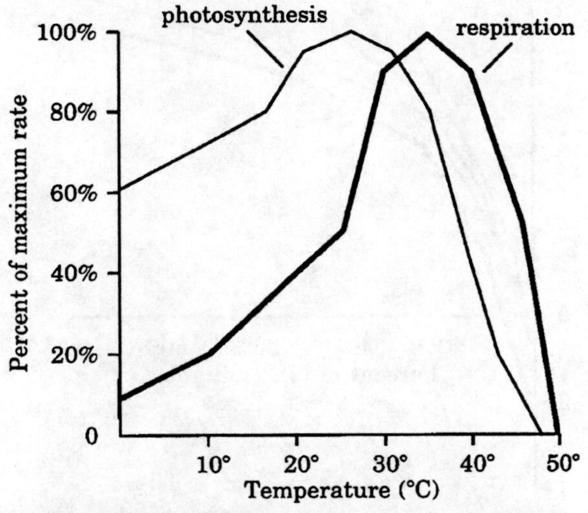

Figure 1 *The response of* Seedling *photosynthesis and respiration to temperature. The curves show relative rates only; in* Seedling, *the maximum photosynthetic rate per unit of leaf area is 8 times as high as the maximum respiratory rate.*

Photosynthesis and Temperature

Both respiration and photosynthesis of *Seedling* plants have an optimal temperature, but the responses of the two rates are somewhat different (Figure l).

First, as pointed out above, the optimum temperature for photosynthesis is 25°C, but the temperature for maximum respiration is 35°. In C_3 plants, the main reason for this low temperature optimum for photosynthesis is the increasing importance of photorespiration at high temperatures. Because C_4 plants can minimize photorespiration, they may have photosynthesis temperature optima of 30°C - 35°C. One C_4 shrub in Death Valley has an optimum temperature for photosynthesis of 47° C!

Second, photosynthesis maintains near-maximum values for a wider temperature range than respiration does. This effect is also caused by photorespiration. As warmer temperatures speed up the Calvin-Benson cycle, they also speed up photorespiration by the same amount. C_4 plants have a much more marked increase in photosynthesis in response to rising temperature than the C_3 plant depicted in Figure 1 because C_4 plants can control photorespiration.

Photosynthesis and Light

The photosynthetic response will "saturate" at high light intensities because the process of carbon fixation is being limited not by the products of the light-dependent reactions (chiefly NADPH and ATP) but by the ability of the enzymes of the Calvin-Benson Cycle to use NADPH and of ATP to reduce CO_2 to sugar. At cooler temperatures, these enzymes operate even more slowly, thus the maximum rate of photosynthesis is lower at lower temperatures. At high temperatures, respiration increases while photosynthesis decreases, so net photosynthesis declines (see the line for 35° C in Figure 2). For this reason, the worst environment for *Seedling* plants is high temperature combined with low light.

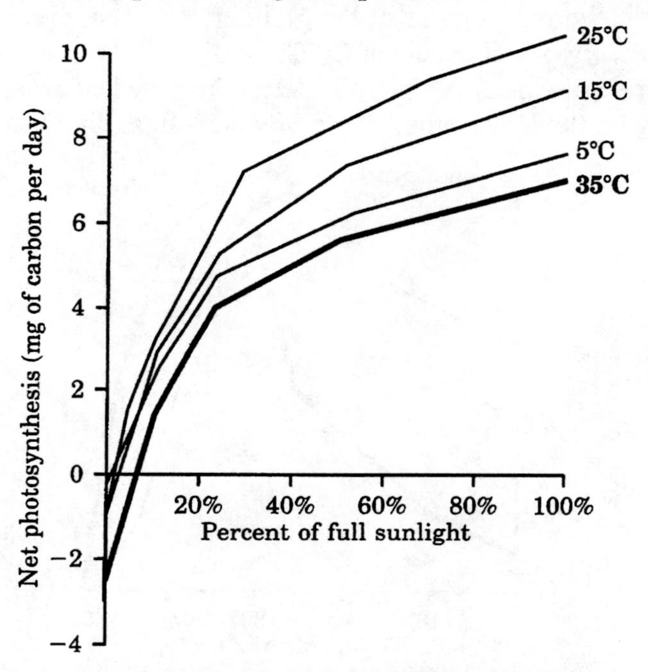

Figure 2 *Daily net photosynthesis (carbon fixed by photosynthesis minus carbon used up in respiration) for plants with one leaf pair under various conditions of growth-chamber sunlight and leaf temperature. In these simulations relative humidity was 70% and dawn soil water was 150 mg/cc of soil. Plants were growing 4 cm apart (mild competition).*

In plant physiology, it is customary to measure photosynthetic rates in micromoles of CO_2 assimilated per square meter of leaf per second. At 25° C, the seedling in Figure 2 would have a maximum photosynthetic rate of 27 μmoles $CO_2/m^2/sec$. For comparison, most C_3 crop plants have maximum rates of from 10 – 20 μmoles $CO_2/ m^2/sec$, and C_4 crop plants have rates from 20 - 40 μmoles $CO_2/m^2/sec$.

One of the most important influences on a leaf's light environment is shading by other leaves. *Seedling* leaves transmit only 15% of the light falling on them (a realistic value), and the program calculates the shading effect of every leaf on every other leaf below it. Shading can be quantified by the leaf-area index (LAI), the ratio of leaf area to the underlying soil area. A leaf-area index exceeding 1 indicates that leaf overlap is occurring. Row crops typically develop LAI's of 8 and forests reach 12; in forests, understory plants develop many reduced-light adaptations. When the largest plants in *Seedling* grow under "fierce" competitive conditions in the subtropical environment (Charleston, South Carolina), each plant supports 7328 mm^2 of leaf area over only 626 mm^2 of soil surface, an LAI of almost 12. On the other hand, if the same plants were growing under "mild" competitive conditions in the semiarid environment (El Paso, Texas), the LAI would be only 0.18.

Photosynthesis and Water Stress

Another influence on photosynthesis is water stress. Leaves absorb carbon dioxide from the atmosphere through stomata, and they also lose water (by transpiration) through these same openings. If a plant cannot obtain enough water from the soil to balance transpiration losses, it closes its stomata. This conserves water, but also cuts off its carbon dioxide supply and reduces photosynthesis. Thus when a *Seedling* plant dies from low soil water, the actual cause of death may be starvation rather than dehydration.

The water budget of a *Seedling* plant depends on its water gains (which depend on the root system and the soil water) and its water losses (which depend on humidity, temperature, and stomatal opening). *Seedling* uses formulas calculated from a well-established relationship known as Raoult's Law to predict the rate of loss of water from the leaf. This water movement is driven by the difference between mesophyll relative humidity (100%) and that of the air. For example, water loss is 15 times greater on a day with 20% relative humidity than on a day with 90% relative humidity. Temperature has a large effect too. The amount of water air can hold increases with temperature, and so if air is warmed without increasing its water content, its relative humidity will drop (for example, if saturated air at 20° C is warmed to 21° C without picking up more water, its relative humidity drops from 100% to 94%). In *Seedling*, water loss increases 1.85 times for every 10° C rise in temperature.

The extent to which stomata hinder this water loss (called "stomatal resistance") is a large topic, and influences photosynthesis because of the plant's inability to take up carbon dioxide if the stomata are closed. In real plants, the guard cells around stomata open and close in response to leaf water and carbon dioxide concentration, light, temperature, humidity, and even wind, allowing fine control of water loss. However, in *Seedling*, water loss is the only controlling factor. If the plant can make up this loss by uptake from the soil, the stomata stay wide open all day and photosynthesis is undiminished by water stress. However, if, for example, only 25% the required water can be taken up, the stomata close for 75% of the day and photosynthesis is only 25% of normal.

Photosynthesis in real plants is a source-sink phenomenon. If the concentration of photosynthate in the plant drops because a large amount is being consumed (the plant contains a large photosynthate "sink"), the plant automatically increases its rate of carbon fixation. On the other hand, carbon fixation automatically declines if photosynthate is in ample supply. Thus it is sometimes observed that if a large photosynthate sink such as a growing fruit is removed from a plant, the rate of photosynthesis will decline because the plant can satisfy its reduced needs with lower photosynthetic activity. *Seedling* plants are not this sophisticated; they constantly photosynthesize at the maximum rate the environment will allow.

Photosynthesis and Leaf Age

Leaf age is the final factor influencing photosynthesis. In real plants, leaves reach their peak photosynthetic activity just before they have fully expanded, and then begin to senesce. For example, the first leaf of the common bean, a C_3 plant, reaches its peak photosynthetic activity 5 days after it begins to unfold, attains mature size at 8 days (when it has already lost 10% of its photosynthetic ability), and by an age of 15 days is photosynthesizing at only 25% of its peak rate. Senescence can also be slow. For example, leaves of C_4 plants retain their "youthful vigor" for long periods, and conifer leaves last for years. *Seedling* leaves follow the C_3 pattern and lose half of their remaining photosynthetic activity every 8 days after maturity. A somewhat unrealistic feature of the game is that if a player stops allocating photosynthate to a leaf before it has attained full size, the leaf continues photosynthesizing at the peak rate indefinitely.

Leaf senescence is not necessarily a negative development for the plant. Older leaves tend to be shaded, so it does not matter if they lose photosynthetic potential. Also, leaves are stocked with nutrients when they are expanding, but after they have matured the nutrients may be needed by growing tissues elsewhere, especially by other expanding leaves. This shifting of nutrients is very important for the success of young leaves, but may contribute to senescence of the older leaves.

Respiration

There are two kinds of plant respiration—growth respiration and maintenance respiration. In *Seedling*, growth respiration, the consumption of photosynthate casused by the energy-requiring synthetic reactions of growth, is modeled by reducing photosynthesis by 25%, a realistic amount, and adding this amount of photosynthate to respiration.

Maintenance respiration oxidizes a temperature-dependent fraction of the carbon in a plant part each day, irrespective of growth. For new leaves at 25° C, *Seedling* maintenance respiration is 2.5% of the carbon in the leaf per day. This compares well with a figure of 2%/ day at 25°C which is commonly assumed for maintenance respiration in crops. For comparison, the respiration required to maintain a unit area of leaf for one day in *Seedling* is only 7.5% of the growth respiration required to construct it.

Respiration in real plants depends on temperature (Figure 1), but also on the age and growth status of the plant part. The most actively respiring tissues in plants are expanding leaves and growing tissues like apical meristems and cambium, and these tissues have the highest respiration rates in real plants. Respiration is also greater when the stock of photosynthate is high. For example, it is a common observation that respiration is higher at sunset (after a day of active photosynthesis) than at sunrise (when carbohydrate reserves are lowest). The respiratory rate of *Seedling* leaves decreases as they age, but other complexities are not taken into account.

Advice on Using *Seedling* in the Classroom

Preparing Students to Use *Seedling*

As mentioned above, you may want to assign parts of the written student exercises as background reading before *Seedling* is used. The introductions to the four *Seedling* exercises are found on pages 3-4, 9, 12, and 15 in the file "Manual-s" on the CD-ROM (and also on the computer's hard drive following installation). In addition, you may want to assign the *Seedling* worksheet on page 2.

Seedling should be used after your usual lecture and discussion on plant structure and function. The topics that will help most with *Seedling are*

a) the simple idea (sometimes not understood by students) that plants photosynthesize but they also respire. This will lead to the most important concept for *Seedling:* net photosynthesis = photosynthesis – respiration. If net photosynthesis is positive, the plant is growing; if it is negative, the plant is starving.
b) the fact that cumulative net photosynthesis over a period of several days shows a plant's overall success at keeping photosynthesis above respiration;
c) what transpiration is and the factors (high temperature, low humidity, large leaf area) that increase it;
d) the fact that lack of water inhibits photosynthesis by closing the plant's stomata and shutting off its CO_2 supply;
e) the factors that might cause water stress, such as dry soil, an inadequate root system, or simply such hot, dry weather that the plant cannot take up water fast enough to replace transpiration losses.

The definitions of *Seedling* terms on page 1 of the student exercises may help with these discussions.

Assigning Students to Computers

The best computer arrangements are different for Exercises A, B and C (the experiments) than they are for Exercise D (the game). For the experiments, each group of three students should be at one computer, and each student should have a copy of the student exercises (pages 3-19 in "Manual_s" on the CD-ROM). If only one computer is available, the best approach would be to rotate groups of three students through the computer while the rest of the class does related work. An instructor demonstration of an experiment is also possible, but since many students will probably be unable to see the screen, attention may wane. The instructor should circulate among the groups to provide advice and explanations.

The best format for the *Seedling* game is to divide the class into groups of six, assign each group a computer, and allow teams of two students to control each of the three seedlings on each computer. (The game becomes more stimulating when two team members debate the best strategy for their plant.) The instructor should circulate through the room to provide advice. If there is only one computer, probably the best strategy is to divide the class into three teams, allow the members to confer, and have a different member come forward at each turn to make changes and announce developments to the rest of the group. Net photosynthesis is the most important development, and when it becomes negative, the cause should be determined and corrective action taken.

Another one-computer strategy is to allow four to six students to use the game while the rest of the class does related work.

Doing the *Seedling* Experiments

All the *Seedling* exercises are accessed by clicking on the Exercise A, Exercise B, Exercise C, or Exercise D tabs on the bottom left of the first *Seedling* screen after the title screen (Figure 3).

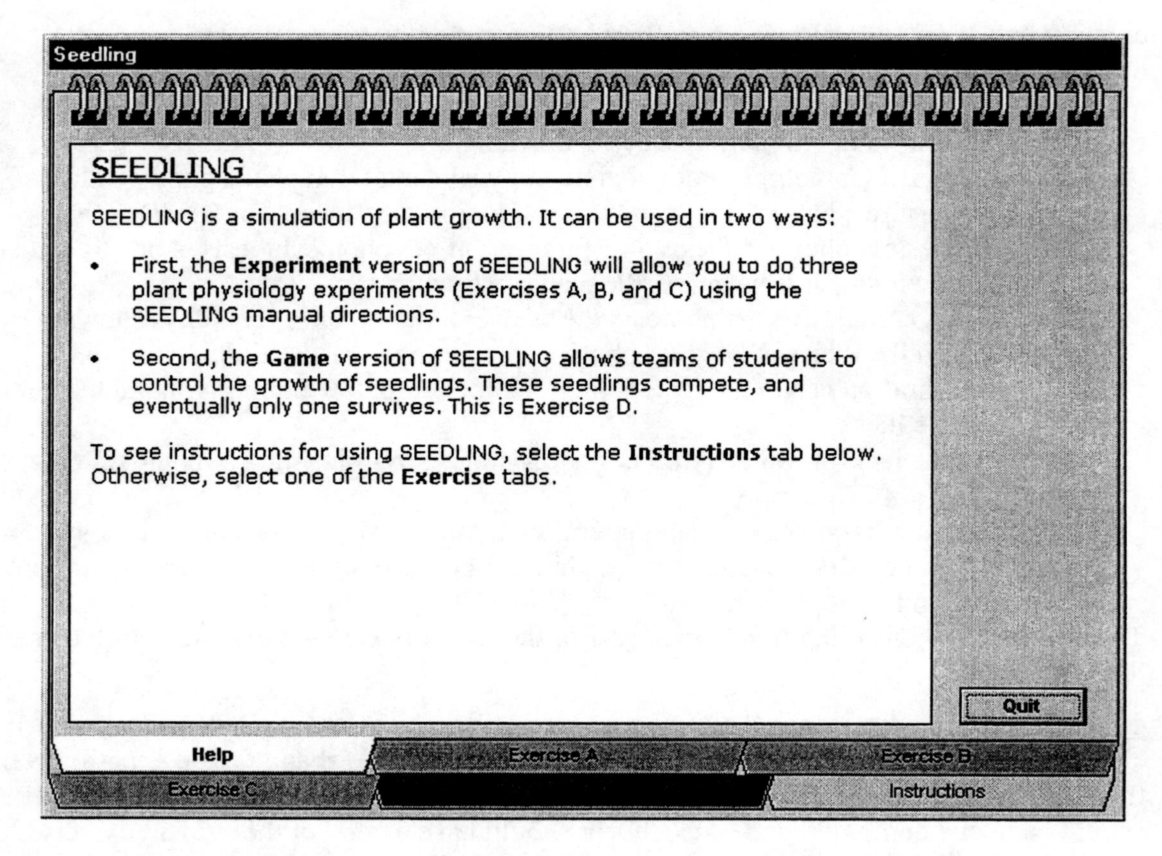

Figure 3 *The* Seedling *screen that follows the title screen. Clicking on tabs A, B, C, or D sends the student to those exercises.*

The Experiments (Exercises A, B, and C)

In all the experiments, the student must collect enough data to describe some phenomenon (such as the variation of net photosynthesis with light and temperature). What values of the independent variable to use is the student's choice. In all cases, the student should concentrate observations in the areas where the response is changing most rapidly.

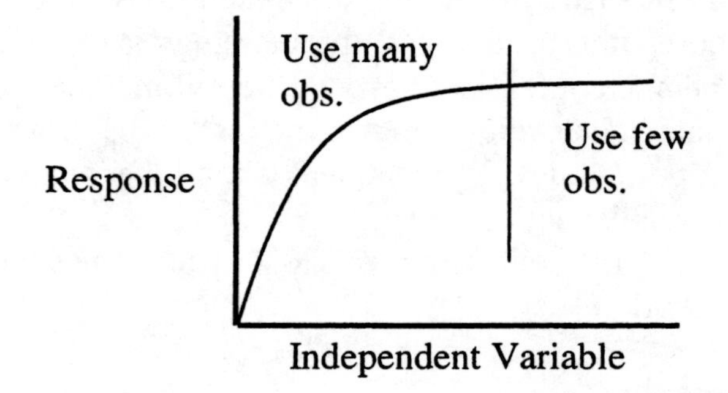

Figure 4 *A generalized response curve. Many data points should be taken in the rapidly changing region on the left rather than the stable region on the right.*

In Exercises B and C, students must also work with a factorial experiment in which every combination of a series of values is tested. In Table 1, for example, every empty block represents a group of plants that must be grown at the given light intensity and temperature. In Table 1, the light intensity response of the plant at a particular temperature is given by a row of the table; the temperature response at a particular light intensity is given by a column of the table.

Table 1 Net Photosynthesis for a Series of Light Intensities and Temperatures

		Light intensities				
		0%	5%	10%	30%	100%
Temperatures (° C)	5					
	15					
	25					
	35					
	40					

Lastly, in Exercises B and C, replicate simulations under the same conditions give slightly different results.

Discussion Guide for Exercise A

Exercise A asks students to determine the distance between plants that optimizes photosynthesis per square meter. When the student requests Exercise A, the program will ask the size of the plants desired and whether growth will take place in a moist or dry growth chamber environment. Then the student will see the screen in Figure 5.

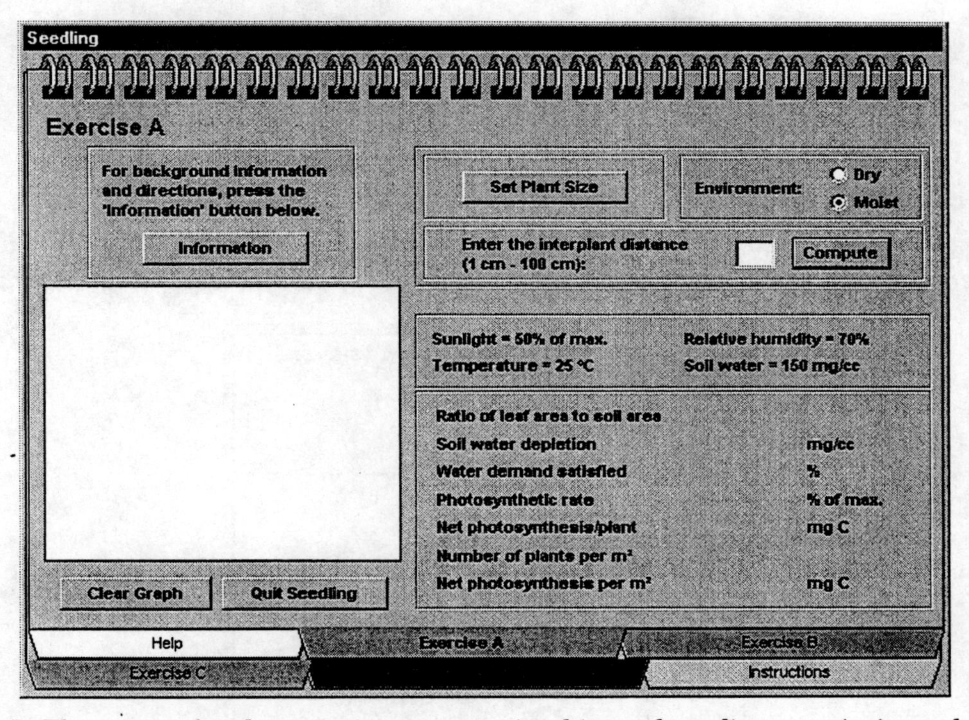

Figure 5 *The screen for the experiment on optimal interplant distance. As interplant distances are entered, a graph of photosynthesis/unit area vs. interplant distance appears in the blank space on the left, and detailed data on the last interplant distance used is displayed on the lower right.*

When the distance between plants is so small that plants are crowded, the data table on the lower right would show data like this:

Ratio of leaf area to soil area	12.00
Soil water depletion	50.08 mg/cc
Water demand satisfied	100 %
Photosynthetic rate	0 % of max.
Net photosynthesis/plant	-1.20 mg C
Number of plants per m²	10000
Net photosynthesis per m²	-11992 mg C

Figure 6 *Exercise A data table for small plants 1 cm apart in the moist environment. Note that the plants are so crowded that net photosynthesis per plant is negative.*

On the other hand, if distance between plants is great, individual plants do well, but the number of plants per square meter is so low that net photosynthesis per square meter is low:

Ratio of leaf area to soil area	0.12
Soil water depletion	0.50 mg/cc
Water demand satisfied	100 %
Photosynthetic rate	85 % of max.
Net photosynthesis/plant	10.56 mg C
Number of plants per m²	100
Net photosynthesis per m²	1056 mg C

Figure 7 *Exercise A data table for small plants 10 cm apart in the moist environment. While the plants are photosynthesizing at 85% of the theoretical maximum and water depletion is low, there are only 100 plants per square meter so net photosynthesis per unit area is low.*

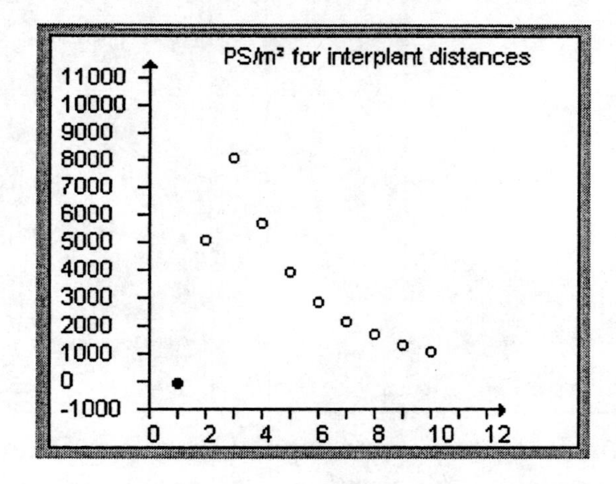

Figure 8 *Photosynthesis per square meter plotted against interplant distance in cm for the smallest plant size. The point for 1 cm was really much more negative, but it is plotted at –100 mg C/m² in order to emphasize the rest of the graph.*

These opposing tendencies combine to create an intermediate interplant distance at which productivity per unit area is highest. After interplant distances for 1-10 cm are simulated, the graph in the lower left of the Exercise A screen is shown in Figure 8 on the previous page.

The data above are for the moist environment (70% relative humidity and 150 mg/cc soil water). In the arid environment, 20% relative humidity causes a water demand 4.5 times as great per day as in the moist climate, and the soil supplies only 20% as much water. Because their stomata are closed most of the day in an effort to conserve water, they photosynthesize less than the moist environment plants and their optimal interplant distance is greater.

Seedling's prediction that production will achieve a maximum value at some intermediate interplant distance is not seen in nature for production of plant dry matter, but it is seen for plant parts such as seeds and fruits. In experiments where seeds are sown at increasing densities, some seeds always germinate before others, and then these early plants have an advantage that lasts throughout the experiment. At high densities, they may even kill the late-germinating plants. Thus as the population density of plants gets greater and greater, plant survival and the weight of individual plants declines, but some plants always survive and the production per square meter maintains a constant value as interplant distances get very small. *Seedling* does not duplicate this effect because all *Seedling* plants are identical—there are no superior plants to gain an early advantage and survive even after all the plants around them have died.

Despite the fact that in the field, production of plant dry matter per unit of area does not have a specific optimal interplant distance, production of seeds and fruits does follow the pattern in Figure 8. This occurs because if a plant is very stunted by crowding, it may not be able to flower at all. On the other hand, if plants are very far apart, there will be so few of them per square meter that fruit production per unit area will be low. Therefore there is an intermediate interplant distance that maximizes production of fruit (for example) per unit area.

Discussion Guide for Exercise B

Exercise B asks the student to describe the response of net photosynthesis to both light intensity and temperature. After deciding on the size of the plants, the student will be asked to fill in the text fields along the top and bottom of an empty table. Use the tab key to move between the fields:

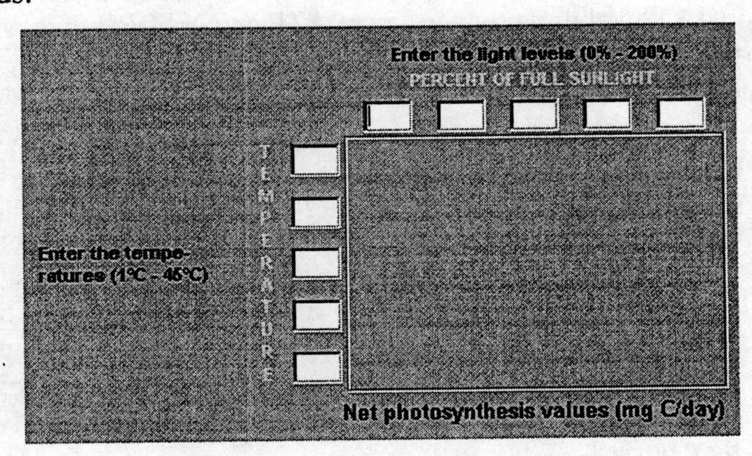

Figure 9 *Data entry for Exercise B. Light and temperature values to be used are up to the student. The lighter text is TEMPERATURE and PERCENT OF FULL SUNLIGHT.*

After all fields along the top and side are filled, the student presses the "Compute" button and the following table appears:

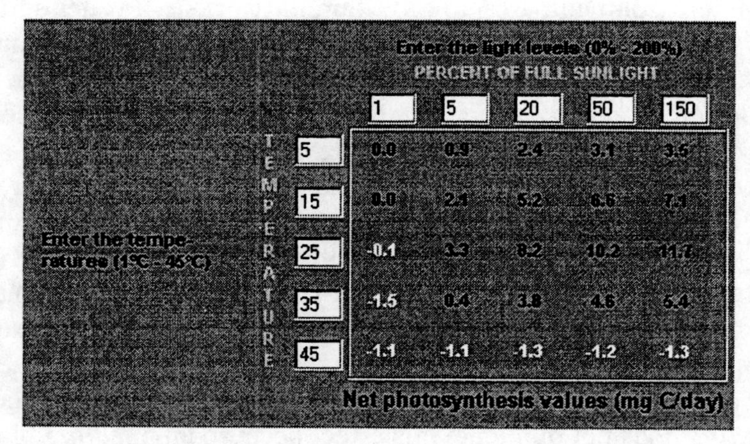

TEMPERATURE	1	5	20	50	150
5	0.0	0.9	2.4	3.1	3.5
15	0.0	2.1	5.2	6.6	7.1
25	-0.1	3.3	8.2	10.2	11.7
35	-1.5	0.4	3.8	4.6	5.4
45	-1.1	-1.1	-1.3	-1.2	-1.3

Figure 10 *The net photosynthesis values produced by the combinations of light and temperature along the margins of the table. Note the negative net photosynthesis at high temperatures and low light intensities. The lighter text is TEMPERATURE (vertical axis) and PERCENT OF FULL SUNLIGHT (horizontal axis).*

The results in Figure 10 can be explained by referring to Figures 1 and 2. The rates of photosynthesis and respiration increase slowly as the temperature increases because of the kinetic effect of heating on chemical reactions (heat provides activation energy and increases the number of collisions between molecules). However, once an optimum is passed, both photosynthesis and respiration decline rapidly. In *Seedling*, 25°C is the optimum temperature for photosynthesis and 35°C is the optimum for respiration. This low temperature optimum for photosynthesis is seen in many kinds of plants and is probably due to photorespiration, which increases greatly under hot, high-oxygen conditions.

Figure 2 shows the response of photosynthesis to both light and temperature. When light is intense, the enzymatic light-independent reactions limit the rate of photosynthesis. Increasing the light intensity to a level that produces more ATP and NADPH than can be used by the Calvin-Benson cycle results in no increase in photosynthesis, and we say that photosynthesis is "light-saturated." However, when light is dim, the small amounts of ATP and NADPH generated can easily be processed by the Calvin-Benson cycle, and it is the light-dependent reactions that are the limiting factor. At very low light intensities, a doubling of intensity can lead to a doubling of photosynthetic rate; we say that photosynthesis is "light-limited."

The compensation point is temperature-dependent because at temperatures above 25° C, the rate of photosynthesis declines while the rate of respiration increases. Thus a plant in dim light may be able to survive when temperatures are low but starve when temperatures rise. The student exercise requires students to select light and temperature values that encompass the full range of plant activity. Be sure they collect enough points to recognize the temperature optimum at 25° C (also obvious in Figure 10) and the negative part of the photosynthesis-light curve.

Discussion Guide for Exercise C

Exercise C asks the student to describe the response of transpiration to both relative humidity and temperature. The user interface is very similar to that for Exercise B. A typical set of results for the largest plants is as follows:

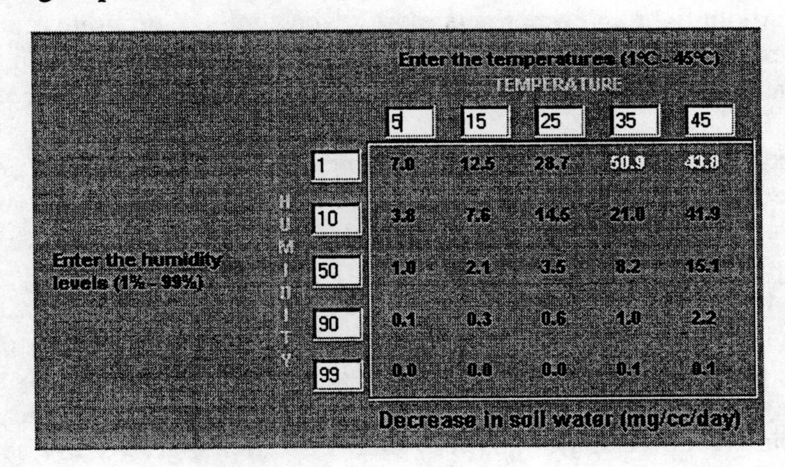

Figure 11 *Results for Exercise C. White numbers in the upper left indicate failure of the plant to satisfy its water demand because the root system could not take up water rapidly enough. The lighter text in this figure reads HUMIDITY (vertical axis) and TEMPERATURE (horizontal axis).*

Transpiration consumes a tremendous amount of water. Typically, a plant uses hundreds of grams of water for each gram of biomass it synthesizes. A leaf in an agricultural field can transpire its own weight in water in just an hour; an acre of corn can lose 400,000 gallons of water (equivalent to 16 inches of rainfall) in a growing season. If humans-required as much water per unit of mass as a young corn plant, we'd all need to drink 10 to 15 gallons per day.

Transpiration cannot be eliminated because the plant must keep its stomata open to obtain CO_2. However, desert plants have developed ways of reducing transpiration. Their leaves may have thick cuticles or be very small or even absent or may roll up to lessen contact with dry air. The stomata are sunken in pits and are closed most of the time. Many desert plants open their stomata at night, when transpiration is lowest and store CO_2 in organic acids until dawn, when photosynthesis begins. The price paid for all these adaptations is a low rate of photosynthesis; and where water is abundant, desert plants cannot compete with moist zone plants.

In *Seedling*, water demand increases 1.85 times for each 10° C rise in leaf temperature and is about 22 times greater on a day with 10% relative humidity than on a day with 90% relative humidity, even if the temperature is the same on both days. These results are shown in Figure 11. Note that on days with the hottest temperatures and lowest humidities, the plant could not meet its water demand and closed its stomata for part of the day. The result was a sharp reduction in photosynthetic rate.

Discussion Guide for Exercise D

As explained on pages 15-16 of the student manual in the file "Manual_s" on the CD-ROM, in the *Seedling* game, several populations of seedlings growing in a small area are simulated on one computer. Each of the seedlings is controlled by a "team" of one or more students.

The team directs the growth of its seedling by specifying how the seedling's photosynthate will be allocated to the growth of roots, leaves, stem thickness and stem height. By favoring one of these areas over another, and by dropping existing leaves, a team can produce a plant that is either competitively superior or seriously maladapted. As the seedlings grow, competition for both light and water gets more severe, and the program reports the stresses the plants are experiencing from shading, lack of water, and so forth. Then, one by one, the maladapted plants die. Finally, only one plant survives, and the program congratulates the winning team.

Unlike the experimental mode, there is no "experimental variability" in *Seedling*'s game mode. Two seedlings with identical allocations to roots, stems, and leaves will grow into identical plants. Therefore students can be sure that their seedling's fate is entirely in their hands.

Seedling can he successfully played at many levels. Although students with little or no background can play, the best results occur when the game is preceded by class discussion and a thorough reading of the laboratory materials. Encourage the students to think seriously about the strategies their plants should employ and then allow them to enjoy themselves.

Because of the great diversity of possible game condition—growth in a semiarid, a subtropical or a tropical setting; three different plant sizes, and any desired distance between plants—students will not quickly exhaust *Seedling*'s strategy challenges, especially if the level of competition is set at "severe."

The Game—Preliminary Choices

Number of plants. There can be up to three team-controlled plants, plus a computer-controlled plant. This computer plant is only a moderately good player.

Climate. The game always starts on March 16, but the student may choose to grow in one of three geographic locations: El Paso, in the semidesert of western Texas; Charleston, in the humid coastal lowlands of South Carolina; or near tropical San Juan, Puerto Rico. These sites offer sharply contrasting weather and selective pressures for plant growth in the March-April-May period, as shown in Table 2.

Table 2 Weather conditions in El Paso, Charleston, and San Juan

	El Paso	Charleston	San Juan
Mean temperature, March-May (° C)	12.5-22.3	13.6-22.3	24.6-26.2
Mean daytime relative humidity (%)	21	68	70
Cloudless days in March, April, and May	50	28	17
Rainy Days (out of 92) in March, April, and May	6	26	42
mm of rainfall in March, April, and May	24	292	304
Mean mm of rain per rain event	4.0	11.2	7.2
Initial soil water (mg/cc)	40	100	150

These conditions could be summarized as follows: E1 Paso's climate is seasonal, sunny, and dry. Low humidity and brisk winds are the major sources of stress. Although the temperature regime is similar in Charleston, plants there have both a much higher water supply and a lower water demand (due to high humidity). Heavy rains may alternate with dry spells, however. San Juan has constant warm temperatures, high humidity, and frequent light rains. The major sources of stress in San Juan are low sunlight due to rainstorms and cloudy days, and depression of photosynthesis due to tropical temperatures too hot for the temperate-zone crop plant modeled in *Seedling*.

The studentss may also choose to "grow" in a growth chamber with controlled temperature, light, humidity, and dawn soil water content. Growth chamber conditions can be made favorable or harsh for the game, but the main purpose of the growth chamber is to allow simulated experiments.

Plant Size. Next, the studentss must decide the starting size of their plants. They have a choice of young seedlings starting their first leaf pair, bigger plants starting their second leaf pair, or plants just about to finish their third leaf pair. Plant measurements are listed in Table 3. Note that these plants are quite small. The smallest is about half an inch high; the largest is less than three inches high. The larger plants give a more complex (and interesting) game because players may drop old, nonproductive leaves.

Table 3 *Seedling* Plant Sizes

	Small	Medium	Large
Height (mm)	14	32	75
Stem diameter (mm)	1.2	1.5	2.0
Number of leaf pairs	1	2	3
Leaf area (mm^2)	1200	3114	7328

Severity of Competition. This is probably the most important choice and is determined by the distance between plants. Each plant is given a circular plot of soil with a radius equal to half the interplant distance, so the amount of water available is proportional to the square of the interplant distance. Thus in El Paso, the large plants have access to twenty-five times as much water when competition is mild (interplant distance, 20 cm) as when it is fierce (interplant distance, 4 cm).

Except in El Paso, light is more limiting to plants than water is. In dense stands, leaves overlap and shade each other severely. For example, under "fierce" competitive conditions for the largest plants in Charleston, any point on the ground will have over 12 layers of leaves between it and the sky. The interplant distances are set so that plants under conditions of fierce competition can barely break even with a positive net photosynthesis, even on sunny days. A succession of rainy days will cause them to starve. On the other hand, light is not a problem under mild competition.

The student also has the option of entering user-chosen interplant distances, which is useful when *Seedling* is used as a data generator rather than as a game.

The Game—Allocation and Data Display

After the initial choices are made, the student sees the game screen shown on the following page in Figure 12.

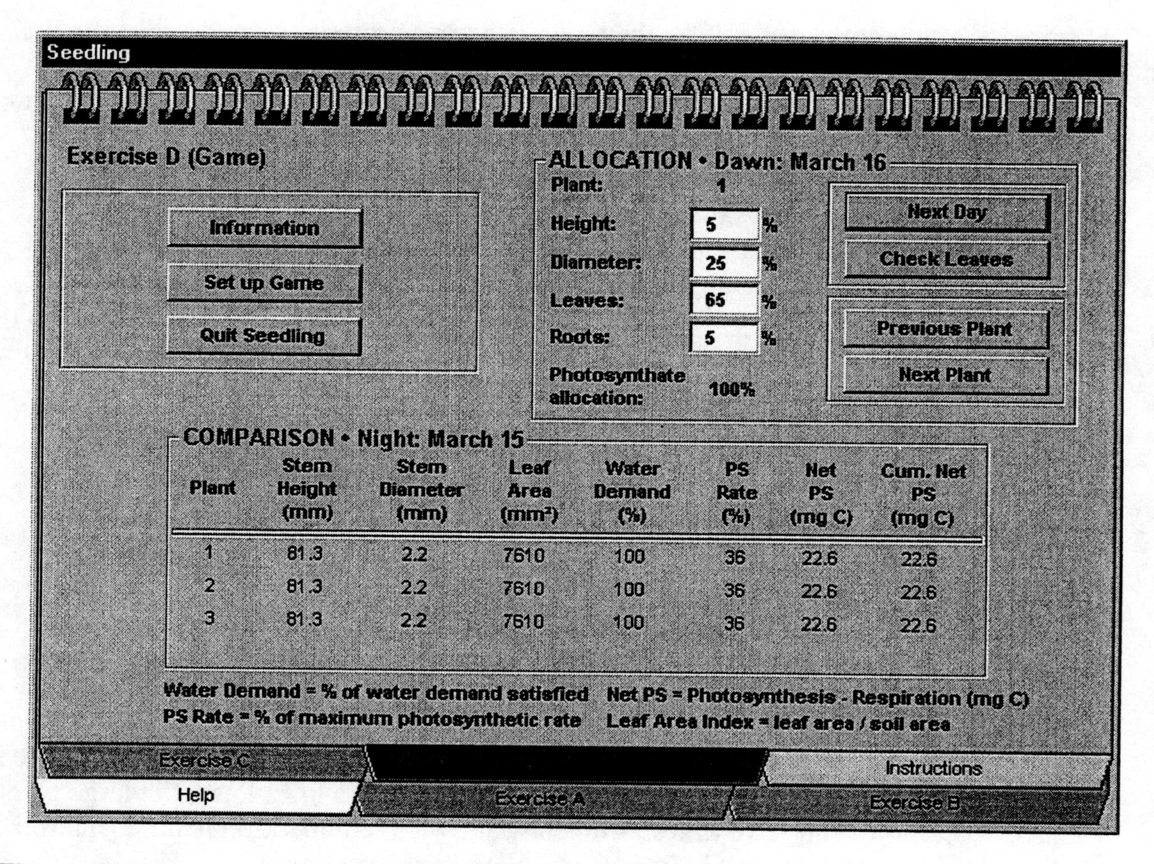

Figure 12 *The game screen. The allocation portion is in the upper left and the plant comparison chart is along the bottom.*

The default allocation of photosynthate is 25% to stem height, stem thickness, leaf growth, and roots. This is not a particularly good allocation and Figure 13 shows an "aggressive" allocation strategy:

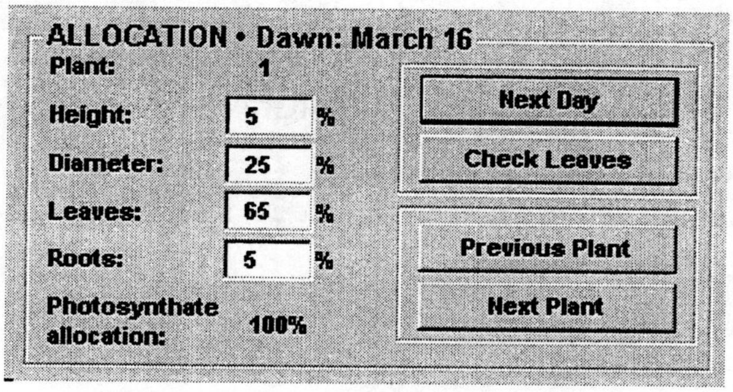

Figure 13 *The portion of the game screen used to make allocations.*

An important feature of the allocation screen is the "Check Leaves" button. If the student presses this, a dialog box like the following will be seen:

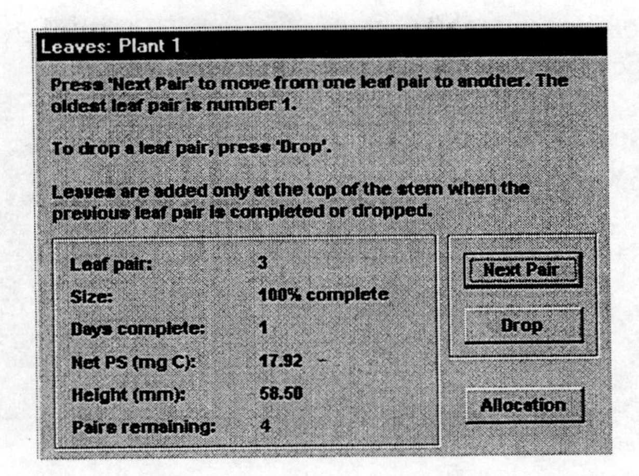

Figure 14 *The leaf-check screen shows whether all leaves still have a positive net photosynthesis. The student moves from one leaf pair to another by pressing the "Next Pair" button. Leaf pairs many also be dropped.* ***Always drop the lowest, oldest leaf pair.*** *Don't drop a leaf pair unless it shows a negative net photosynthesis on a sunny day. Don't drop all your leaf pairs.*

When all students are satisfied with their allocations, someone presses the "Next Day" button and the group gets a weather report. The most important part of this is the rainfall and soil water. Soil water is all important because dehydration and starvation are the major cause of death in *Seedling.* Soil water can reach a maximum of 150 mm/cc, but can decline to 0 mg/cc in just a few days under conditions of severe competition. A plant graphic is shown below in Figure 15.

Figure 15 *The plant graphic screen showing that plants 1 and 3 have dropped more leaf pairs than plant 2.*

The most important feature to consider on the Next Day's screen is the comparison chart along the bottom:

COMPARISON • Night: March 27

Plant	Stem Height (mm)	Stem Diameter (mm)	Leaf Area (mm^2)	Water Demand (%)	PS Rate (%)	Net PS (mg C)	Cum. Net PS (mg C)
1	85.2	2.5	2219	100	22	3.2	84.7
2	98.9	2.0	8045	100	3	-2.3	46.3
3	84.2	2.4	2651	100	18	3.0	82.8

Figure 16 *The comparison chart. Plant 2, with much more leaf area than plants 1 and 3, is suffering a low rate of photosynthesis, and had a negative net photosynthesis on this day.*

"Water Demand" shows the percent of its water demand that the plant was able to satisfy that day. "PS Rate (%) shows the percent of the theoretically maximum photosynthesis (given its leaf area) that each plant realized. Because of a cloudy day, all photosynthetic efficiencies in Figure 16 are small.

Net photosynthesis in mg of carbon fixed is the most important reading on the screen because it determines whether a plant is growing or starving. Although net photosynthesis may become negative on a very overcast or rainy day, if it is low (less than 2 mg C fixed per day) or negative on a sunny day, something is wrong. Either the plant is not getting enough water, or it is shaded, or it is carrying old, inactive leaves that should be dropped. Chronic negative net photosynthesis rapidly leads to death from starvation.

Cumulative net photosynthesis (the last line in the table) is the sum of all net photosynthesis values since the beginning of the game, and is useful for comparing the success of the different plants.

In Figure 16, Plant 2 was controlled by a team that refused to drop old, senescent leaves. The very low percent photosynthetic efficiency compared to the other plants, and its chronically negative net photosynthesis should have warned its team that something was radically wrong. In a few days it starved.

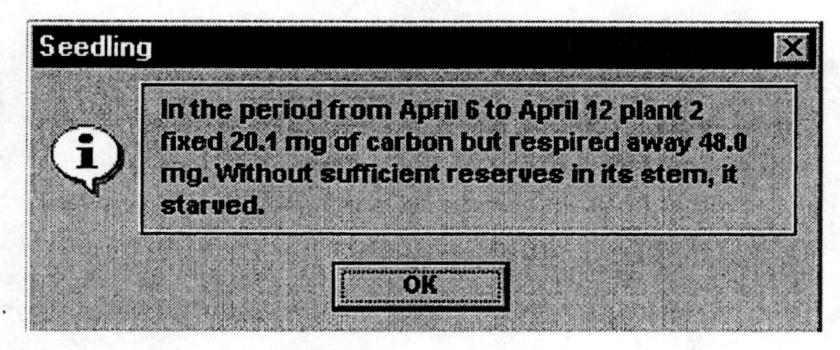

Figure 17 *A death message for Plant 2.*

The Game—Causes of Death

A plant's career can end in *Seedling* due to starvation, dehydration, blowing over, or even grazing or trampling by cows.

Starvation. When a plant is consuming more photosynthate during respiration than it produces during photosynthesis, net photosynthesis is negative and the plant compensates by consuming some of the food reserves stored in its stem. This use of reserves causes stem diameter to decline. When the stem shrinks 90% of the maximum size it has ever attained, a message indicates that the plant is starving. When the diameter declines to 80% of its maximum size, death from starvation results. The most common cause of starvation is shading, especially when aggravated by cloudy days. Wilting also leads to starvation by stomatal closure. A larger stem diameter makes plants resistant to starvation. Starvation is the most common cause of death in *Seedling*.

Dehydration. A healthy plant has a water content of 77%. If water stress occurs, the stomata will close, but water will still be lost at 10% of the maximum rate due to stomatal leakage, loss through the cuticle, and so forth. If even this 10% loss cannot be made up, the water content will drop below 77% and a "wilting" message will appear. If the water content dips to 50%, the plant will die. A large stem diameter, an adequate root system, and careful pruning of old leaves are the best defenses against dehydration. However, even the most provident plant will die if soil water drops to 0 mg/cc of soil and remains there for several days.

Blowing Over. The weather includes many days with moderate winds and a few days with high winds. The force that wind exerts on the stem is greatest when a plant has many large leaves high on its stem. The ability of a plant to resist blowing over is proportional to the stem's cross-sectional area. So growing low to the ground, elimination of unnecessary leaves, and a large stem will reduce the danger of blowing over. Blowing over is a rather uncommon cause of death.

Grazing and Trampling. In *Seedling*, cows are attracted by plants with tall stems and high leaves. Even if a cow arrives, there is only a 25% chance that any damage will occur. If damage does occur, it will either be cropping of the upper portions of the plants or trampling. Trampling is instant death and is totally random. Grazing usually results in subsequent starvation because all the newest, most productive leaves are removed. Low plants have the smallest grazing losses and are less likely to attract the cows in the first place.

The Game—Strategy

In each of the causes of death above, short plants with thick stems and few leaves are less likely to die. A small number of leaves reduces water and wind resistance and does not attract cows. Careful pruning of old leaves also reduces self-shading and the chances of starvation. Being low to the ground avoids wind damage and keeps the cows at a distance. Having a thick stem increases the plant's resistance to blowing over and serves as both a food reserve and a water reserve to resist starvation and drought.

These considerations are the key to the "patient" strategy for winning at *Seedling*: Pour resources into stem thickness and roots, don't add leaves to keep existing leaves from senescing and wait for opponents to die. This strategy can be effective, but if all players adopt it, the game will be slow-moving and dull because no plants will die. The "aggressive" strategy of growing tall and adding leaves as quickly as possible is more spectacular. Opponents are shaded, soil water is quickly depleted, and the game comes to a rapid conclusion—not always in favor of the aggressive plants. Encourage students to try an aggressive strategy. An active game with a brisk rate of plant death is not only more exciting but is also more instructive than a slow, dull game.

Seedling: A Computer Simulation of Plant Growth and Plant Competition

Overview

Seedling is a computer simulation of plant growth and plant physiology. The program can be used both as an experimental data generator and as a competitive game.

In this laboratory, you may use both functions of *Seedling*. First you will perform one or more of three experiments (Exercise A, B or C, as assigned by your instructor). These exercises explore the responses of plants to their environment. Exercise A demonstrates the effects of increased competition among plants by simulating the effects of growing the plants closer and closer together. Exercise B determines the responses of plants to different temperature and light conditions. Exercise C examines the influence of temperature and humidity on water use by plants.

If time permits, you can try your skill at the *Seedling* game (Exercise D), which allows you to apply knowledge gained from the first three exercises to "managing" the growth of a plant. By accurately evaluating your plant's requirements and environmental conditions, you can develop a strategy that will allow your plant to thrive and outcompete the plants of your classmates "growing" with your plant on the same computer.

Environmental Influences on Plant Growth

Environmental conditions limit the success of all organisms. For example, land plants must obtain light, heat, water, and nutrients from their environment. Light is necessary for photosynthesis, by which plants reduce atmospheric carbon dioxide to carbohydrates (photosynthate). Photosynthate furnishes energy and materials for the plant's metabolism. An optimum temperature range is important for a variety of reasons. Temperatures that are too cold may cause frost damage or slow down the plant's enzymatic reactions to the extent that the plant cannot grow. Temperatures that are too warm increase water demand, slow photosynthesis, speed up respiration, and may even denature enzymes.

a. *What happens when an enzyme is denatured?*____________________________________

____

b. *Why would the denaturation of enzymes interfere with plant metabolism?*____________

____

The availability of water often limits land plants. Water is required for the chemical reactions of photosynthesis, but most of the water taken up by plants is lost through transpiration (evaporation through stomata when they are open to extract CO_2 from the air). A plant must take up hundreds of grams of water in order to produce 1 gram of dry matter, mainly because of transpiration losses. If soil water becomes scarce, the plant closes its stomata to restrict water losses from the leaves, and this stops photosynthesis because the plant can no longer obtain CO_2. So plants under water stress might not only wilt, but also starve. Water also plays a role in the expansion of growing cells by the growth of the central vacuole. Whereas animals grow mainly by cell division, plants grow mainly by cell enlargement. It has been calculated that if a giant sequoia grew only by cell division rather than by cell division followed by cell enlargement, it would only attain a height of 2 m.

Finally, plants obtain many nutrients from the soil. However, plants cannot take up such necessary ions as nitrogen, sulfur, and phosphorus unless these are first dissolved in soil water.

c. *What are some of the important macromolecules requiring nitrogen, sulfur, and phosphorus in all living organisms?*____

In this laboratory you will use a computer program called *Seedling* to investigate the effect of several important environmental variables (light, temperature, soil water, humidity, and plant crowding) on plant growth.

Exercise A: **Spacing Plants for Optimal Yield**

The distance between plants greatly affects their **areal yield** (their final biomass per square meter of field surface). If plants are too close, intense competition for light and water causes them to grow poorly or die, and areal yield from a field of these crowded plants will be low. If the plants are too far apart, individual plants will grow well, but there will be so few plants per square meter that yield will be low. There is an intermediate distance between plants at which yield per square meter is maximized (Figure 1).

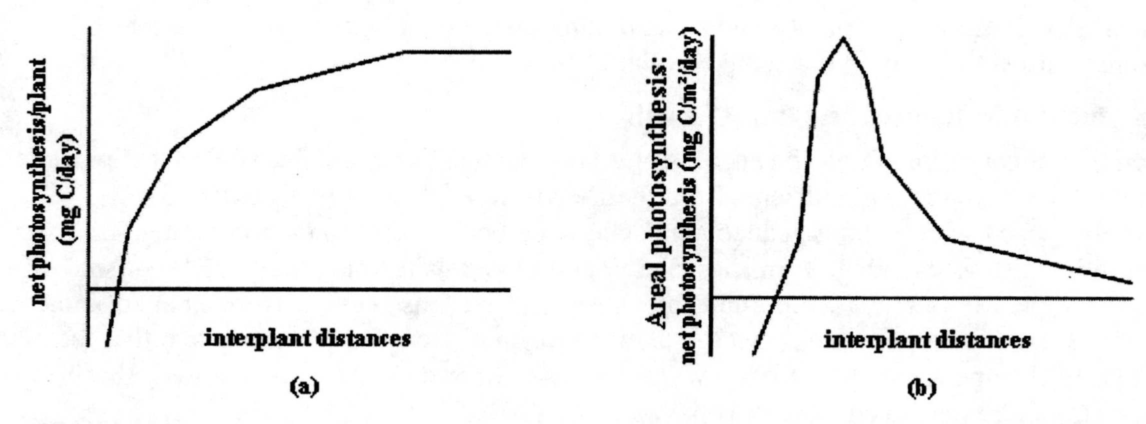

Figure 1 *Sample graphs (without specific intervals marked) for (a) net photosynthesis per plant for various interplant distances and (b) areal photosynthesis for various interplant distances.*

Yield (or biomass) increases as a result of photosynthesis—the storage of the sun's energy in carbon compounds used to build plant tissues—and respiration oxidizes and thus consumes the carbon compounds produced by photosynthesis. Net photosynthesis is a measure of biomass growth, expressed in milligrams of carbon (mg C) added per time interval such as a day. **Net** photosynthesis is calculated by taking the total mass of carbon compounds produced by photosynthesis and then subtracting the total mass of carbon oxidized in respiration.

Net photosynthesis is affected by several factors that you can test with the *Seedling* simulation. For example, under dry conditions, plants will be forced to close their stomata, they won't be able to take up CO_2, and photosynthesis will decline. Also, a small interplant distance affects larger plants more negatively than smaller ones because larger plants need more room. Finally, areal yields will be depressed either by the poor photosynthesis of individual plants, or a large distance between plants that reduces the number of plants per square meter.

Objectives

- ❑ Determine and plot the relationship between interplant distance and the daily net photosynthesis of individual plants for both moist and dry conditions.
- ❑ Determine and plot the relationship between interplant distance and yield per square meter of field surface (areal yield) for both moist and dry conditions. Use these data to determine optimum interplant distance for both environments.

PART 1—Moist Conditions

Procedure

1. Start the *Seedling* program (instructions for installing and running *Seedling* are found elsewhere). Select the **Exercise A** tab.
2. Select a plant size. You may choose to use either small, medium, or large plants. Record the size you choose:_______________
3. Indicate that you will be growing your plant in moist conditions. Your plants will be placed in a growth chamber with a temperature of 25°C, sunlight at 50% of maximum, relative humidity at 70%, and soil water at 150 mg/cc.
4. You must now select at least five different interplant distances at which to simulate plant growth. Your objective will be to find the interplant distance that maximizes *net* photosynthesis per unit area per day, as illustrated in Figure 1. You may select your own distances, but distances between 1 and 15 cm will probably work best. A distance of 1 cm is a good place to start, so type 1 as your first interplant distance and press the **Compute** button. Record "1 cm" at the top of column 1 in Table 1.
5. A table (similar to a single column in Table 1) will be displayed. This table lists several important indicators of plant success.
   - **Ratio to leaf area to soil area (leaf area index)** A measure of the number of square centimeters (cm^2) of leaf area shading each square centimeter of soil area. Values greater than 1 indicate leaf overlap. *Seedling* simulates a field of plants even if only one team is playing, so a team's own plants may be shading one another.
     a. *What are some of the causes and possible consequences of a leaf area index that is much larger than 1?* ____
     ____
     ____
   - **Soil water depletion** The decrease of soil water (in milligrams per cubic centimeter of soil: mg/cc) that has occurred due to the uptake of water by plants.
   - **Percent of water demand satisfied** The percentage of the plant's water needs that can be met by the uptake of water from the soil.
   - **Percent of the maximum net photosynthetic rate** This rate would be 100% if all leaves were new, if there was no lack of water, and if there was no shading.
   - **Net photosynthesis per plant** A measurement equal to photosynthesis (carbon fixed into carbon compounds) minus respiration (carbon used up by oxidation). If the result is positive, the plant has **photosynthate,** carbon products produced by photosynthesis, to devote to growth after satisfying its respiratory costs. If it is negative, the plant is starving and not growing.

- **Number of plants per square meter (plants/m^2)** A measurement computed from the distance between plants.
- **Net photosynthesis per square meter (m^2)** Net photosynthesis per plant multiplied by the number of plants per square meter.

Table 1 Plant Performance under Moist Conditions

	Interplant distances					
	______	______	______	______	______	______
Ratio of leaf area to soil area						
Soil water depletion						
Percent of water demand satisfied						
Percent of maximum net photosynthesis rate						
Net photosynthesis (mg C/plant/day)						
Number of plants/m^2						
Net photosynthesis/m^2						

6. Record the information on the screen in the first column of Table 1.
7. Follow steps 4 through 6 for at least four other interplant distances. Record each distance at the top of the columns in Table 1, and enter the information given on the screen. Your objective is to obtain enough points to plot curves similar to those shown in Figure 1.
8. Plot the individual and per square meter net photosynthesis curves on the axes in Figure 2. Note that you must place numbers on the axes. Label each of these curves "moist conditions."

The interplant distance that produces the greatest rate of areal photosynthesis (mg C per m^2 per day), and thus the largest areal yield, is influenced by several factors. Larger plants, for example, require larger interplant distances. Also, harsh conditions such as temperature extremes, low light intensities, and lack of water depress the rate of photosynthesis and increase the space required between plants.

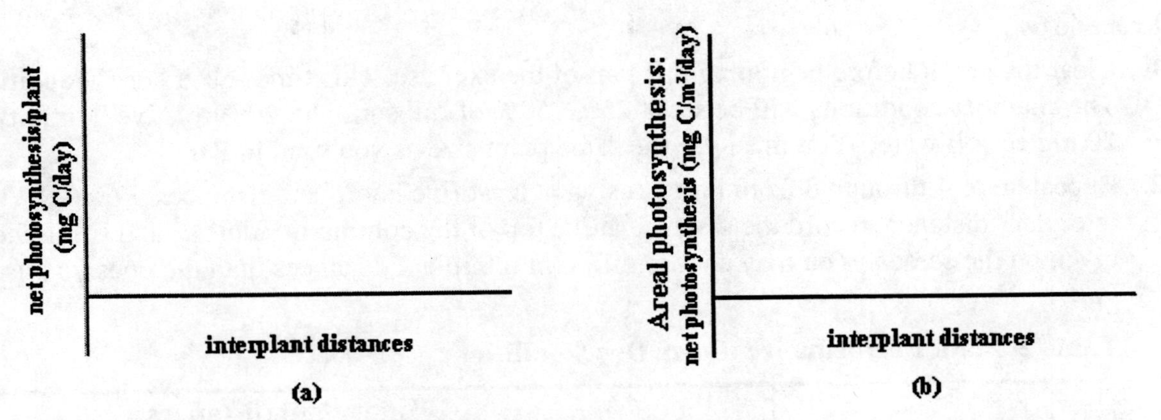

Figure 2 *Your data for (a) net photosynthesis per plant for various interplant distances and (b) areal photosynthesis for various interplant distances.*

a. *Why do plants starve and die at very close interplant distances?* ________________

b. *Why do the leaf area index and soil water depletion change with increasing in interplant distance?* ________________

c. *What is the leaf area index at which plants are first able to realize positive net photosynthesis?* ________________

d. *How does net photosynthesis for individual plants change as interplant distance increases?* ________________

e. *Does individual plant photosynthesis reach a maximum value?* ________ *Why?* ________________

f. *Which interplant distance gives the greatest net photosynthesis per square meter?* ____________ *Why does areal net photosynthesis decrease once this optimum interplant distance is exceeded?* ________________

PART 2—Dry Conditions

Repeat Exercise A, Part 1, using a cool but dry environment rather than the cool, moist environment used in Part 1.

Procedure

1. Clear the graph before beginning this part of the exercise. This time select "dry" conditions. The chamber conditions will be set at 25°C, 50% of full sunlight, 20% relative humidity, and 20 mg/cc soil water. You must use the same plant size as you used in Part 1.
2. Repeat steps 4 through 8 from Part 1, using at least five interplant distances. For each interplant distance, record the distance at the top of the column in Table 2, and fill in the data given on the screen. You may choose different interplant distances than the ones you used in Part 1.

Table 2 Plant Performance under Dry Conditions

	Interplant distances					
	______	______	______	______	______	______
Ratio of leaf area to soil area						
Soil water depletion						
Percent of water demand satisfied						
Percent of maximum net photosynthesis rate						
Net photosynthesis (mg C/plant)						
Number of plants/m^2						
Net photosynthesis/m^2						

3. Plot the curves for the dry environment on the same axes (page 5) on which you plotted the curves for the moist environment. Label the curves "dry conditions."
   a. *The low relative humidity under these dry conditions increases transpiration by a factor of 4.5 over the moist conditions, while water supply in the soil is only 13% of that available in moist conditions. What evidence do you have that these conditions are stressing the plants?* ______________________________

   ____

   ____

   b. *Plants normally reduce water loss on dry days by closing their stomata. What effect would this have on photosynthesis? Why?* ______________________________

   ____

   c. *How is the individual plant net photosynthesis curve for the dry environment different from the one for the moist environment?* ______________________________

   ____

   ____

   d. *Suppose plants could be sprayed with hormones that would force them to keep their stomata open despite dry conditions. Would this result in higher rates of photosynthesis*

*initially?*____________ *Over the long term?*____________
*Explain.*____

e. *How is the areal photosynthesis (net photosynthesis per square meter) curve for the dry environment different from the one for the moist environment?* ____________

Why do these differences occur? ________________________

Exercise B — Temperature, Sunlight and Net Photosynthesis

Light and temperature are two of the most important factors determining the rate of photosynthesis. Light energy is captured by chlorophyll and is used to split water and generate NADPH in the light-dependent reactions. The role of temperature is not so obvious, but environmental heat is a source of energy too. Heat causes the molecular motion necessary to the enzymatic pathway of the light-independent reactions, in which NADPH is oxidized and CO_2 is reduced to carbohydrate.

The Effect of Light on the Rate of Photosynthesis

When light is dim, a doubling of light intensity can lead to a doubling of photosynthetic rate. Thus we say the photosynthesis is "light-limited." At high light intensity, the enzymes of the light-independent reactions cannot process NADPH any faster, so raising the light intensity results in no change in photosynthetic rate because light is no longer a limiting factor. Under these conditions, photosynthesis is said to be "light-saturated."

The Effect of Temperature

The effect of temperature on net photosynthesis is complicated by the fact that net photosynthesis is measured as the difference between two processes: photosynthesis and respiration. When the temperature is low, rates of both photosynthesis and respiration increase with increasing temperature because heat increases the speed of chemical reactions. However, the photosynthetic rate peaks at about 25°C; above this temperature a process called photorespiration begins to consume photosynthate as fast as it is created. Meanwhile the cellular respiration rate continues to increase up to about 35°C, and then declines because the enzymes become inactivated by heat. Thus relatively cool temperatures favor photosynthesis over respiration. In this exercise you will explore the full response of the *Seedling* plants' net photosynthesis to variations in both light and temperature. You will employ a common technique known as a "factorial experiment" in which all possible combinations of several light and temperature levels are systematically tested. Table 3 will help you to organize your experiments.

Objectives

- ❑ Plot the response of *Seedling* net photosynthesis to light intensity at each of several temperatures, and explain the results.
- ❑ Use known information to plan further investigation.

Procedure

1. Start the *Seedling* program (instructions for installing and running *Seedling* are found elsewhere). Select the **Exercise B** tab.
2. Select a plant size. You may choose to use either small, medium, or large plants. Record the size you choose:_______________
3. You must now choose a range of temperatures and light intensities that you think will encompass the whole range of plant responses to these two variables. A good range of light intensities might extend from 1% to 100% of full sunlight, and temperature should extend from 5°C to about 40°C. Choose temperatures and light intensities with the following two considerations in mind:
   - Your choices should cover only the range of plant response. For example, since plants can photosynthesize from about 0°C to about 40°C, using 100°C, 200°C, and 300°C as three of your data points would not be useful—there is no photosynthesis at those temperatures.
   - You should concentrate your observations in "interesting areas." For example, if plants respond to light as illustrated in Figure 3, then you can learn the most about this response by concentrating observations in the "interesting area" where photosynthesis is changing rapidly rather than in the less interesting region where photosynthesis is constant.

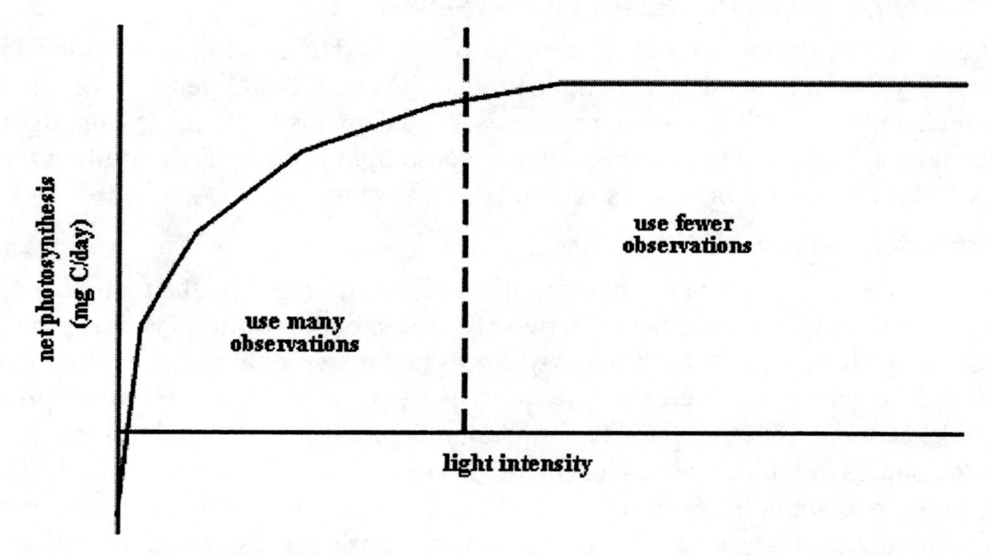

Figure 3 *Concentration of data points in areas in which the response is changing rapidly.*

4. After choosing the temperatures and light intensities you wish to use, enter them in the fields across and down in the middle of the screen. Press the **Tab** key to move from field to field. As you enter your temperature and light intensity values, write them at the top and along the side of Table 3 in the spaces provided.
5. Once you have entered the last temperature value, press the **Compute** button; the program will begin computations that will simulate plant growth at every temperature/light combination in Table 3 (under conditions of 70% relative humidity, 150 mg/cc soil water, and wide spacing between plants). As each combination is completed, the daily net photosynthesis (photosynthesis-respiration) for those conditions, measured as milligrams of

carbon (mg C), will be displayed below the light intensity values used and to the right of the temperature values used. Copy these data into Table 3 as they appear. Net photosynthesis will be positive whenever the plant is growing (producing more carbohydrates than it is consuming) and negative whenever it is starving (consuming more carbohydrates than are being produced by photosynthesis).

Table 3 Net Photosynthesis at Various Temperatures and Light Intensities

Temperatures	Light intensities (percent of full sunlight)				
	______	______	______	______	______

6. When you have obtained the 25 data points, graph your results on Figure 4, below. You may choose to do this in one of two ways: you may graph net photosynthesis (Y-axis) against light intensity (X-axis) with a different curve for each temperature, or you may graph net photosynthesis (Y-axis) against temperature (X-axis) with a different curve for each light intensity. In either case, there will be five curves on one graph. Label each of these curves clearly (for example, "15°C" or "50% sunlight").

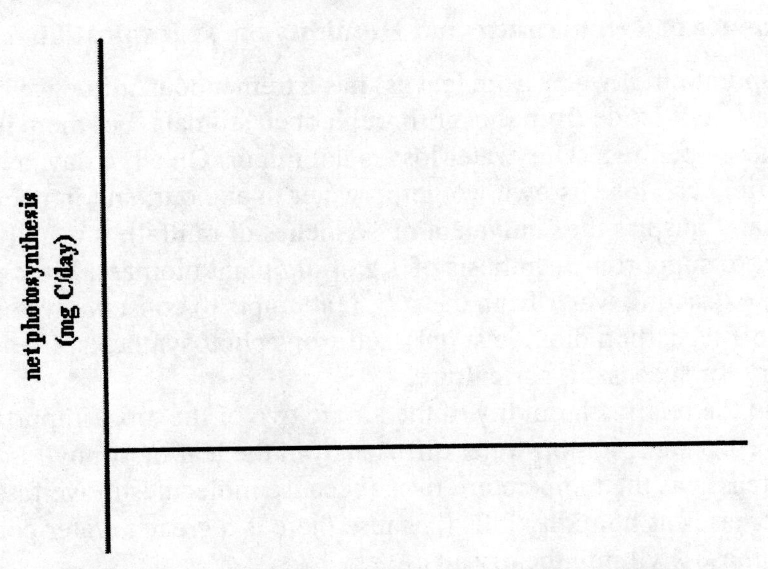

Figure 4 *Your data on response of net photosynthesis to light intensity and temperature*

a. *Why does the rate of photosynthesis increase with light intensity?* ______________________

____

b. Why does increasing the light intensity have only a small effect on the rate of net photosynthesis at low temperatures but a much greater effect at high temperatures?

____

____

c. Why does daily net photosynthesis decrease when temperatures become high?

____

____

____

d. The light intensity at which the rate of photosynthesis just balances the rate of respiration, and net photosynthesis is zero, is known as the compensation point. Use your data to describe how the compensation point varies with temperature.

____

____

____

e. Explain <u>why</u> the compensation point varies with temperature.

____

____

____

f. In nature, hot temperatures often occur in conjunction with dry conditions. Plants reduce water loss under dry conditions by closing their stomata. Why is the combination of heat and dry conditions (low humidity) especially damaging to plant growth?

____

____

____

<u>Exercise C</u> **Influence of Temperature and Humidity on Transpiration**

Transpiration (the evaporation of water from leaves) has a tremendous influence on plant growth. Plants obtain carbon dioxide from the air through open stomata, but there is unavoidable loss of water through these openings. This water loss is not minor. On a hot day, a leaf on a crop plant in an agricultural field can lose its own weight in water in an hour, and in one growing season, a field of corn can transpire the equivalent of 16 inches of rainfall. It usually takes about 500g of transpired water to support the synthesis of 1 g of dry plant biomass. If the plant cannot make up these losses by extracting water from the soil, it attempts to conserve water by closing its stomata, which cuts off its carbon dioxide supply and stops photosynthesis. Clearly, a good water supply is necessary for successful agriculture.

Leaf temperature and the relative humidity of the air are two of the most important influences on transpiration rate. During transpiration water diffuses from the leaf mesophyll into the air. The rate of transpiration increases as the temperature rises (because molecules move faster in warmer temperatures), and it increases as humidity falls (because there is a greater water potential gradient from the moist mesophyll into the dry air).

In this exercise you will explore the relation between transpiration rate, leaf temperature, and relative humidity. You will use a technique called a "factorial experiment" to systematically test every possible combination of several levels of humidity and temperature.

Objective

- ❑ Determine the relationship between transpiration rate, humidity, and temperature, and explain the results.

Procedure

1. Start the *Seedling* program (instructions for installing and running *Seedling* are found elsewhere). Select the **Exercise C** tab.
2. Select a plant size. You may choose either small, medium, or large size plants. Record your chosen size:_______________
3. You must now choose a range of temperatures and humidities that you think will encompass the whole range of plant response to these two variables. A good range of temperatures should extend from 5°C to about 40°C, and humidities should range from 5% to 95% (but not 100%). Choose the temperatures and humidities with the following two considerations in mind:
    - Your choices should cover only the range of plant response. For example, since plants can transpire from about 0°C to about 40°C, using 100°C, 200°C, and 300°C as three of your data points would not be useful--transpiration does not occur at those temperatures.
    - Your choices should concentrate observations in "interesting areas." For example, if you were studying photosynthesis and how plants respond to light as illustrated in Figure 3, page 8, then you could learn the most about the phenomenon by concentrating your observations in the "interesting area" where photosynthesis is changing rapidly rather than in the "less interesting" region where photosynthesis is constant.
4. After choosing the temperatures and light intensities you wish to use, enter them in the fields across and down in the middle of the screen. Press the **Tab** key to move from field to field.
5. Once you have entered the last humidity value, press the **Compute** button; the program will begin computations that will simulate plant growth at every temperature/humidity combination in Table 4 (computed with 50% of full sunlight, 150 mg/cc soil water, and wide spacing between plants). As each combination is completed, soil water depletion resulting from the process of transpiration will be displayed below the temperature value used and to the right of the humidity value used. Copy these data into Table 4 as they appear. A data point that appears in white it indicates that the plant was unable to satisfy its water demand under those conditions.

Table 4 Soil Water Depletion at Various Temperatures and Humidities

Relative humidity	Leaf temperature (°C)				

6. When you have obtained the 25 data points, graph your results below, in Figure 5. You may choose to do this in one of two ways: you may graph soil water depletion (Y-axis) against humidity (X-axis) with a different curve for each temperature, or you may graph soil water depletion (Y-axis) against temperature (X-axis) with a different curve for each humidity. In either case, there will be five curves on one graph. Label each of these curves clearly (for example, "15°C" or "50% humidity"). Label the axes with appropriate units.

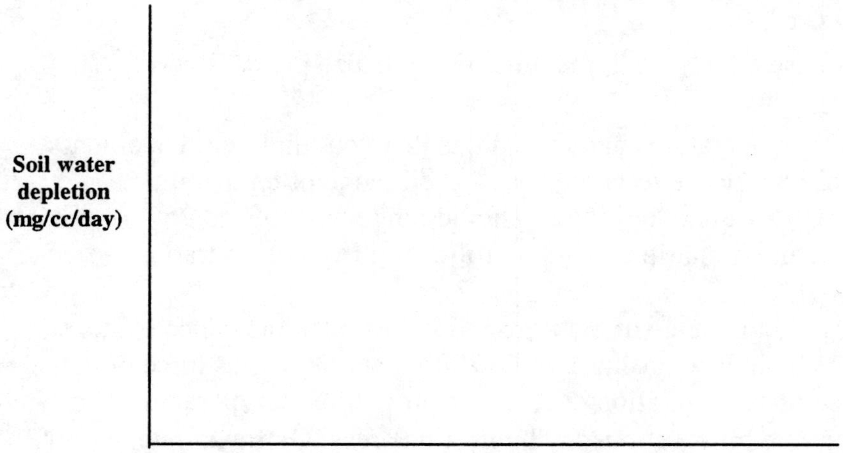

Soil water depletion (mg/cc/day)

Figure 5 *Your data on response of soil water depletion to temperature and humidity.*

a. *Why does transpiration rate increase as the temperature rises?*____________________

b. *Why does transpiration rate increase as the humidity falls?* ____________________

c. *On days when the plants experience unsatisfied water demand, why is the soil water depletion lower than might be expected, given the temperature and humidity? How does the plant restrict water loss on these hot, dry days?* ____________________

d. *You provided the plant with 150 mg/cc soil water in all of these experiments. How would the curves you drew be different if you had supplied less soil water (for example, 25 mg/cc)?*____________________

e. *Transpiration requires tremendous amounts of water. Why can't plant breeders develop plants that do not need to transpire?* ____________________

f. *Describe some adaptations that help desert plants, such as cacti, cope with low humidity, high temperatures, and low soil moisture.*____________________

Exercise D The *Seedling* Game

If you did Exercise A, B. or C, you had the opportunity to explore some of the ways a plant responds to its environment. In this exercise, you will test your understanding of plant responses by playing a competitive game with your classmates. Each student or team of students will control a seedling growing close to other seedlings. The players control the growth of their plant by specifying how its photosynthate (carbon products produced by photosynthesis) will be allocated to increase root mass, stem height, stem thickness, and number or size of leaves. By favoring some areas of the plant's anatomy over others and by dropping existing leaves, a team can produce a plant that is competitively superior (or one that is severely maladapted). As the seedlings grow, competition for light and water becomes more intense, and the less well-adapted plants will begin to die, succumbing to dehydration or starvation, or simply blowing over. One random cause of death is grazing cows that may either crop the leaves of the plant or trample it. Finally, only one plant remains and the winning team is congratulated.

The central theme of the *Seedling* game is the allocation of resources. In living organisms, these "decisions" are the result of the interactions among the organism's genetically programmed range of responses, environmental conditions, and the physiological state of the organism. In real plants, allocation varies with the season and depends on the age and lifestyle of the plant. For example, forest trees invest more than half of their photosynthate in the production of a permanent supporting structure, the woody stem and roots, and may delay the production of flowers or cones for ten or twenty years after germination. In contrast, a weedy annual plant devotes a minimum of photosynthate to supporting structure and a maximum to rapid leaf production, and produces flowers and seeds by the end of its first (and only) growing season.

Most plants favor vegetative growth early in the growing cycle and growth of reproductive parts later in the cycle. For example, trees invest large amounts of stored photosynthate in producing leaves as rapidly as possible early in the cycle. Tender new leaves are more vulnerable both to herbivores and insects, so it is important for leaves to mature and toughen up quickly. Flowers and seeds, which depend on the supply of photosynthate provided by leaves, are produced later. Sometimes trees even curtail fruit or cone production for several years, and then produce a heavy crop of fruit. This is thought to help control the population of fruit-eating animals which might increase rapidly if there were a dependable crop of fruit every year. But many weeds with short lifespans invest in seed production as rapidly as possible after germination.

Seedling plants are modeled after a generalized, herbaceous, temperate zone annual crop plant such as a soybean. Thus, its allocation strategy will have more in common with that of weeds than with that of trees.

These allocation strategies, evolved through millions of years of natural selection, are usually very effective. We rarely see plants "mistakenly" germinate on a warm day in winter, flower when their pollinators are out of season, starve because they grew a gigantic root mass but few leaves, or fall over because their stem was too thin for their height.

In *Seedling*, you will see how well your knowledge and strategic skill can substitute for a plant's genetic programming. Your problem will be simpler than that of a real plant, because you will need to consider only vegetative growth (*Seedling* plants do not reproduce). Also, although cows appear occasionally in the program, herbivory is not a serious problem and you cannot do anything to minimize it anyway. Your objective will be solely to maximize your plant's vegetative growth in the first days of the growing season.

The area of the plant's anatomy to which photosynthate is allocated (such as roots or stem height) is called a "sink," and each sink in *Seedling* has advantages and disadvantages as shown in Table 5.

Your task in the *Seedling* game will be to use allocation of the plant's photosynthate and an ability to drop nonproductive leaves to maximize your photosynthetic rate and hasten the death of your rivals.

Table 5 Advantages and Disadvantages of Photosynthate Allocations

Increasing allocation to:	Advantages	Disadvantages
Leaves	Leaf growth provides photosynthate for the plant, which, in this simulation, can then be allocated as desired. Leaves shade the leaves of competitors, thus depriving them of some of the sunlight needed for photosynthesis.	A plant's top leaves shade its own lower leaves. Transpiration occurs through leaves, so a leafy plant demands large amounts of water. Shaded or old leaves do not photosynthesize efficiently, but still respire at a high rate, thus consuming photosynthate rather than generating it. Leaves create wind resistance and may cause a plant to blow over. Leaves may attract predators (in this simulation, cows).
Stem Height	High leaves can shade the leaves of competitors.	Instability in high winds. Increases the amount of respiring tissue the plant must support.
Stem Thickness	Thicker stems provide resistance to blowing over. Stems provide a reserve of food and water. Thus during very cloudy or dry periods, plants with thicker stems have a better chance of surviving. They also have a better chance of recovery following damage, such as severe grazing.	Thicker stems respire more than thin ones, even though their respiratory rate per unit weight is lower. The thicker the stem, the higher the cost, in photosynthate, to add stem height.
Roots	Extensive roots allow the plant to obtain water in dry conditions and to deplete the water supply of competitors.	Add respiring tissue that the plant must support, but the respiratory rate of roots is low compared to that of leaves.

Objectives

- ❑ Explain the concept of the allocation of photosynthate, how allocation varies in different plants, and how allocation changes over the life of a plant.
- ❑ Give the advantages and disadvantages of allocating photosynthate to leaves, to stem height, or to stem thickness and to roots.
- ❑ Discuss how differing environments (for example, moist, low-light conditions versus dry and sunny conditions) affect allocation strategy.
- ❑ Manage the growth of a seedling so that it grows rapidly, is resistant to starvation and drought, can recover from grazing, and outlives its competitors.

Procedure

1. Start the *Seedling* program (instructions for installing and running *Seedling* are found elsewhere). Select the **Exercise D (Game)** tab. If you want to review the instructions for the game, select the **Instructions** tab, then press the **Game** button.)
2. Now, decide the following:
   a. the number of players, and whether you want to include a plant controlled by the computer;
   b. growth location: either Texas, South Carolina, Puerto Rico, or in a growth chamber with controlled conditions;
   c. initial plant size (one, two, or three leaf pairs);
   d. whether you desire mild, moderate, or severe competition between plants. The more severe the competition, the smaller the distance between the plants.

   Some points to keep in mind as you make these decisions:
   - The computer plant is only a moderately good player. It tends to rigidly stay with a strategy even after it has become counterproductive. Including a computer plant is often a good idea.
   - If you intend to use severe competition, you will be troubled in Texas by soil water depletion, in Charleston by cloudy periods that lead to starvation, and in Puerto Rico by cloudiness and excessively hot temperatures. "Default" conditions in the growth chamber will be ideal for plant growth, but the game will be less interesting there.
   - Using the largest plants results in the most complex and interesting game because dropping old, shaded leaf pairs no longer as active in photosynthate production is an important component of success. Using the smallest plants results in the simplest game.
   - Degree of competition is a very important decision. Almost any strategy will allow plant survival under "mild" competition because the plants are so far apart. But the game might be boring. Water depletion and shading will rapidly reach the crisis point under "severe" competition, and the game will be exciting--and probably short.
3. After you make these decisions, the main screen will be displayed. The given starting allocations are 25% to each photosynthate sink—<u>not</u> a good choice. Consider that a small allocation to height will probably be sufficient, that in Charleston and San Juan water availability (and the need for roots) will probably not be a problem, and that leaves are the key to photosynthetic output.
4. At the bottom of the screen is the "nighttime comparison chart," and reports what happened to each plant the previous day. The most important datum on the chart is "Net PS," or net

photosynthesis. This is the difference between photosynthesis and respiration: a negative reading indicates that the plant is starving and has no photosynthate to allocate to growth. Rapid comparison of the net photosynthesis **(Net PS)** figures shows the relative success of the plants that day, and **Cum. Net PS** shows the cumulative net photosynthesis of each plant so far. Watch stem diameter carefully. It shrinks during starvation, and if it declines to 80% of the largest size it had attained, the plant will die.

5. You have the opportunity to change your allocations. It would be good policy to check your leaf pairs at this point and drop any that show negative net photosynthesis on a reasonably sunny day. You may check your leaves by pressing the **Check Leaves** button from the main screen.
6. After the allocation screen, note the soil water and leaf area index figures on the "weather report" screen. Soil water depletion is rapid for small interplant distances, and must be watched carefully. If the ratio of leaf area to soil area is over 1.0, leaves are overlapping and shading each other.
7. Continue in this fa`shion until seedlings begin to die. Death in *Seedling* is usually due either to starvation or to dehydration. If either of these two stresses occurs, dropping old leaf pairs (and not revising allocations) is the best remedy. Remember that if net photosynthesis is negative, the plant has nothing to allocate. On the other hand, dropping a leaf pair can reduce water demand, shading, and respiration.

   a. *What is resource allocation?* ________________________________

   ____

   b. *Why does allocation strategy change throughout the life of a plant?* ____________

   ____

   ____

   ____

   c. *Do animals also use a changing resource allocation strategy?*________ *If so give an example.*________________________________

   ____

   ____

   d. *Many desert plants have large, succulent stems, large root systems, and tiny leaves. Because of their small leaves, they have very low photosynthetic rates and growth rates. Why does this strategy serve them well in their dry environment? Why wouldn't maximal leaf size and growth rate be selected for?* ________________________________

   ____

   ____

   ____

   e. *What kind of* Seedling *allocation strategy would be successful in the semiarid Texas environment?* ________________________________

   ____

   ____

   *In the tropical Puerto Rican environment?*________________________________

   ____

   ____

   ____

f. *Did any seedling win in the game you played? ________ If so, why did its opponents die?*

____

____

Glossary of Terms Used in *Seedling*

Allocation The assignment of photosynthate to the growth of roots, stem height, stem thickness, or leaves.

Areal photosynthesis Photosynthesis per unit area (in *Seedling*, mg C/m^2/day).

Fixed carbon Carbon which has been incorporated into sugar by photosynthesis.

Leaf-area index (LAI) The ratio of total leaf area to underlying soil area. When leaf-area index exceeds 1, leaves must be shading one another. High LAIs mean both severe shading and serious water loss through transpiration.

Net photosynthesis Carbon fixed by photosynthesis minus carbon consumed by respiration. Net photosynthesis is positive when photosynthetic production exceeds the requirements of respiration and negative when respiration outstrips photosynthetic productivity. Net photosynthesis must be positive for growth to occur. Negative net photosynthesis (due to shading, water stress, or old leaves) will cause death by starvation.

Photosynthate The sugars made by photosynthesis.

Photosynthetic rate In *Seedling*, the rate of carbon fixation expressed as a percentage of the maximum rate that would occur under optimum conditions of full sunlight, optimum temperature (25°C), abundant water, and all leaves new and unshaded.

Relative humidity The amount of water in the air expressed as a percentage of the maximum amount of water that air of this temperature could hold. Low relative humidities greatly increase water loss through transpiration.

Respiration The consumption of sugars and other organic molecules in the process of metabolism. Every cell in a plant requires photosynthate merely to stay alive. If respiration requires more photosynthate than photosynthesis can produce, net photosynthesis is negative and the plant must use its food reserves. In *Seedling*, food reserves are stored in the stem. When the food reserves are exhausted, the plant dies of starvation.

Stomata The microscopic openings on the underside of a leaf through which the leaf absorbs CO_2 from the air and loses water. During water stress, these openings close and photosynthesis stops.

Transpiration Evaporation of water through the stomata of leaves.

Water demand The percentage of its water needs that a plant is currently satisfying by water uptake from the soil. A water demand reading of less than 100% indicates water stress.

Name ______________________________ ***Seedling* Worksheet**

Section ______________________________

To answer the questions below, read the *Glossary of Terms Used in* Seedling, on the previous page; the introduction to Exercise A, page 2; the introduction to Exercise B, page 7; and the introduction to Exercise C, page 10. Complete this worksheet and bring it to the class period in which the *Seedling* simulation will be used.

1. What is net photosynthesis? What is indicated by positive values of net photosynthesis? By negative values?

2. Why do high temperatures (over about 30°C) decrease net photosynthesis? Why do low temperatures (below about 15°C) *also* decrease net photosynthesis?

3. Why does lack of soil water reduce a plant's net photosynthesis? Why is water loss worsened on days with low relative humidity?

4. What is leaf area index (LAI), and why do high LAI values indicate intense competition for light and water?

5. What is areal yield? In a field with widely spaced plants, why is areal yield low even though the yield of individual plants is high?

6. Say that the interplant distance that produces the highest areal yield is 5 cm with bright illumination and moist soil. Would the optimal interplant distance be higher or lower if:
   a. The plant size were made smaller?

   b. The illumination became dimmer?

   c. The soil became dry?